Subcellular Biochemistry

Volume 88

Series editor
J. Robin Harris
Institute of Zoology, University of Mainz, Mainz, Germany

The book series SUBCELLULAR BIOCHEMISTRY is a renowned and well recognized forum for disseminating advances of emerging topics in Cell Biology and related subjects. All volumes are edited by established scientists and the individual chapters are written by experts on the relevant topic. The individual chapters of each volume are fully citable and indexed in Medline/Pubmed to ensure maximum visibility of the work.

Series Editor
J. Robin Harris, University of Mainz, Mainz, Germany

International Advisory Editorial Board
T. Balla, National Institutes of Health, NICHD, Bethesda, USA
Tapas K. Kundu, JNCASR, Bangalore, India
A. Holzenburg, The University of Texas Rio Grande Valley, Harlingen, USA
S. Rottem, The Hebrew University, Jerusalem, Israel
X. Wang, Jiangnan University, Wuxi, China

More information about this series at http://www.springer.com/series/6515

J. Robin Harris • David Bhella

Editors

Virus Protein and Nucleoprotein Complexes

 Springer

Editors
J. Robin Harris
Institute of Zoology
University of Mainz
Mainz, Germany

David Bhella
MRC-University of Glasgow Centre
for Virus Research
Glasgow, United Kingdom

ISSN 0306-0225
Subcellular Biochemistry
ISBN 978-981-10-8455-3   ISBN 978-981-10-8456-0   (eBook)
https://doi.org/10.1007/978-981-10-8456-0

Library of Congress Control Number: 2018937669

Printed on acid-free paper

This Springer imprint is published by the registered company Springer Nature Singapore Pte Ltd. part of Springer Nature.
The registered company address is: 152 Beach Road, #21-01/04 Gateway East, Singapore 189721, Singapore

Preface

This volume of the Springer Subcellular Biochemistry Series, entitled *Virus Protein and Nucleoprotein Complexes*, follows on from the previous volumes dealing with soluble *Macromolecular Protein Complexes* (Volume 83) and *Membrane Protein Complexes* (Volume 87). Unlike Volume 68 of the series *Structure and Physics of Viruses*, emphasis in the present volume is upon the components of viruses rather than their intact structure. As in the previous volumes on *Complexes*, a diverse collection of interesting chapter topics is included, within which there is both structural and functional emphasis. All topics are under current investigation by active scientists. The 17 chapters included in the book cover many but not all of the relevant virus-related topics; as is inevitable with multi-author books some chapters have been lost during the preparation period, thus leaving open the possibility of a future related volume.

The book commences with a broad overview that places emphasis on virus assembly, followed by the capsid proteins of spherical viruses, helical virus nucleocapsids and viral envelope fusion. Then this book includes 16 more specific chapters dealing with the following: the ribonucleoprotein complex of negative-sense single-stranded RNA viruses; the viral RNA-dependent RNA polymerases; the filovirus filament proteins; influenza virus RNP; the nucleoproteins of single-stranded RNA viruses; Zika virus envelope protein and antibody complexes; the retrovirus capsid core; nucleoprotein intermediates in HIV-1 DNA integration; the oligomerization of retrovirus integrases; the human respiratory syncytial virus M2–1 protein; the filamentous bacteriophage proteins and their assembly; protein-RNA interactions in the single-stranded RNA bacteriophages; the bacteriophage head-to-tail interface; protein-protein interactions involving viroporins; protein complexes and virus-like particle technology; and finally the role of flaviviral proteins in the induction of innate immunity.

The diversity of information given in these chapters provides an in-depth insight into the present-day molecular virology, from a structural and functional perspective. The data from high-resolution cryo-electron microscopy and X-ray crystallography can be seen to provide a powerful contribution to structural virology. Whilst some important chapter topics were lost due to authors backing out, it is thought that the content of the book remains strong and it is hoped it will be of interest to many. Indeed, the theme of macromolecular complexes, introduced in Volume 83 of the series, is greatly extended within the present volume.

Mainz, Germany   J. Robin Harris
Glasgow, UK   David Bhella
December, 2017

Contents

Chapter 1
Virus Proteins and Nucleoproteins: An Overview

David Bhella

Introduction

Viruses are by far the most abundant of life forms on the planet Earth. Studies of viral diversity in the marine environment have led to estimates that there are around 10^{31} virus particles (*virions*) in the biosphere (Paez-Espino et al. 2016). Recent metagenomic studies of the viromes of diverse ecological niches indicate similarly staggering levels of genetic diversity (Breitbart and Rohwer 2005).

To set out to write an authoritative text on the proteins of *all viruses*, a subject matter so inconceivably vast might therefore be viewed as foolhardy to say the least. Detailed knowledge of the structure and function of viral proteins is however comparatively modest. Our understanding of viral replication strategies is largely derived from detailed studies of a small number of viruses: those that infect humans, economically important animal and plant species as well as bacteriophages, the study of which has been historically more tractable. These have informed our understanding of many critical viral functions nonetheless, as well as providing surprising insights into viral evolution.

Viral replication strategies are intimately linked with the biology of the host. They are characterised by a finely balanced interplay between viral and host factors, regulating both viral expression and host responses to infection. All viruses, regardless of how their genomes are encoded, must achieve certain fundamental functions to perpetuate themselves. This chapter will therefore describe proteins that enable some of these functions, the structural components of viruses. Subsequent chapters will focus on specific aspects of virus biology in greater detail, to reveal how viruses have evolved exquisite mechanisms to ensure their continued survival and success.

D. Bhella (✉)
MRC-University of Glasgow Centre for Virus Research, Scotland, UK
e-mail: David.Bhella@glasgow.ac.uk

© Springer Nature Singapore Pte Ltd. 2018
J. R. Harris, D. Bhella (eds.), *Virus Protein and Nucleoprotein Complexes*,
Subcellular Biochemistry 88, https://doi.org/10.1007/978-981-10-8456-0_1

Capsid Proteins: More than Just a Pretty Coat

The viral infectious cycle can be divided into two phases, intra- and extracellular. Within the cell, viruses are genetic elements, sometimes naked, sometimes enclosed in a protein coat. Replicative functions such as genome replication, transcription and translation, and regulation of host functions and virion morphogenesis can occur in distinct regions of the cell; thus, the virus is completely integrated with the host. In the extracellular environment, the virus exists as a more tangible entity, the virion. Often described as being analogous to a plant seed or bacterial spore, the virion is a metabolically inert means of transmitting the genetic element to a new host. To achieve this, viruses have evolved robust structures that protect their genomes from the hazards posed by the extracellular environment – protein coats termed 'capsids'. The capsid together with the packaged viral genome makes up the nucleocapsid. Many viruses enclose their nucleocapsid within a host-derived membrane that is studded with glycoproteins to mediate attachment and entry to the host cell.

It is tempting to think of viral capsid proteins as primarily protective in their function; however this would be a gross oversimplification. These proteins are dynamic molecules, adapted to fulfil a variety of critical functions including mediating self-assembly; intracellular trafficking; egress; evasion of host intrinsic, innate and adaptive immunity; attachment; and entry and genome release. Viral capsid proteins must maintain a multitude of specific interactions with host factors, often in the face of sustained attack by host immunity. They must be capable of assembling at the correct time and place to specifically enclose nascent viral genomes. Once assembled the capsids must resist environmental insult, delivering their precious cargo to precisely the right cellular compartment of a new host, where the capsid must then either release the genome or admit the necessary factors to permit initiation of gene expression. Owing to the constraints of genetic economy, these functions are accomplished by a limited number of gene products that come together to form a highly symmetrical assembly.

Viruses may be broadly classified as assembling either spherical or filamentous capsids: spherical capsids having the point-group symmetry of the icosahedron, a twenty-sided platonic solid, and filamentous capsids having helical symmetry. Variations on these themes are abundant, such as prolate phage heads or fullerene-like cone-shaped retroviral capsids; however the underlying principles of capsid assembly are well conserved across viral species.

Spherical Virus Structure

The symmetrical nature of spherical viruses was postulated by Crick and Watson in 1956 (Crick and Watson 1956), based on limited experimental data from both transmission electron microscopy and X-ray diffraction studies (Caspar 1956a). These revealed both the spherical shape of virions under investigation and the presence of cubic symmetry. To produce a closed shell of sufficient capacity to enclose a viral genome and from a limited number of gene products requires a capsid to assemble with the symmetry of one of the platonic solids (tetrahedron, cube, octahedron, dodecahedron or icosahedron). The presence of fivefold, threefold and twofold rotational symmetry in X-ray diffraction analyses of virus particles pointed to the likelihood that spherical viruses assemble with the symmetry of dodecahedra/icosahedra. Crick and Watson proposed spherical virus particles as being 'polyhedral or perhaps with bumps on, like a rather symmetrical mulberry', having fivefold, threefold and twofold rotational symmetry and comprising 60 asymmetric units.

The theory of icosahedral symmetry in the assembly of spherical viruses was refined by Caspar and Klug (Caspar and Klug 1962), to account for the growing number of viruses that were found to enclose very large genomes in spherical capsids comprising many more than 60 capsid proteins. Larger capsids are assembled by incorporating multiples of 60 capsid proteins. To achieve this, larger asymmetric units comprising multiple protomers are generated by the incorporation of local sixfold symmetry in a process known as triangulation. The spatial relationship of every capsid protein to every other one is consequently no longer strictly defined by icosahedral symmetry. Rather they are said to be 'quasi-equivalent' having subtle variations in their local bonding environments. Icosahedral capsids are described as having a triangulation number (T). Larger icosadeltahedra are assembled from T×60 capsid proteins – meaning that the asymmetric unit comprises T quasi-equivalent protomers. Allowed T-numbers are given by $T = Pf^2$, where $P = h^2 + hk + k^2$ (for all pairs of integers h and k, having no common factor) and f is any integer. Figure 1.1 shows the first four icosahedral assemblies in the T-number series; a $T = 4$ capsid, for example, has four quasi-equivalent positions, denoted A, B, C and D (Fig. 1.1c). The formation of 'morphological units' or capsomeres is brought about by clustering of capsid proteins into either hexamers/pentamers, trimers or dimers.

Small RNA Containing Spherical Viruses

The earliest spherical virus structures to be solved at atomic resolution by X-ray crystallography were plant viruses such as tomato bushy stunt virus (TBSV) (Harrison et al. 1978). TBSV is a positive-sense RNA-containing virus that assembles $T = 3$ icosahedral capsids, i.e. comprising 180 copies of the major capsid protein arranged in 3 quasi-equivalent bonding environments, designated A, B and C (Fig. 1.2a, b). The capsid protein has a molecular mass of 40 kDa and comes together to form dimeric capsomeres, giving rise to pronounced spikes on the capsid surface. AB dimers are arranged about the fivefold icosahedral symmetry axes, while CC spikes are at the twofold symmetry axes. The protein is divided into three domains, the N-terminal RNA-binding domain, the S (shell) domain that makes up the contiguous capsid floor and the P (protruding) domain that gives rise to the spikes (Fig. 1.2c). The TBSV capsid protein S-domain adopts a fold known as the β-jelly roll, a wedge-shaped eight-stranded antiparallel beta barrel (Fig. 1.2d).

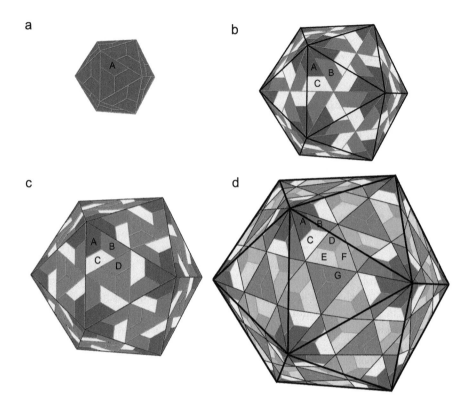

Fig. 1.1 Triangulation of icosadeltahedra
Spherical capsids assemble from 60 asymmetric units, arranged with icosahedral symmetry. To enable the creation of larger capsid shells, each asymmetric unit may comprise more than one capsid protein, arranged with quasi-equivalent packing. The permitted numbers of subunits in each asymmetric unit are given by the T-number series. Here icosadeltahedra are represented with T-numbers of 1 (**a**), 3 (**b**), 4 (**c**) and 7 (**d**). Each trapezium represents a single capsid protein coloured according to its quasi-equivalent position. Figure reproduced from (Bakker and Bhella 2013)

The β-jelly roll has been identified in a wide variety of viral capsid proteins, including the first animal virus capsids to be solved – the picornaviruses poliovirus and human rhinovirus 14 (HRV14) (Rossmann et al. 1985; Hogle et al. 1985). Picornaviruses diverge from the strict rules of quasi-equivalence by encoding four capsid proteins, three of which occupy the A, B and C positions of the T = 3 capsid.

$$\longrightarrow$$

Fig. 1.2 (continued) coloured in rainbow representation (blue, N-terminus; red, C-terminus), while the other is coloured grey. This representation highlights the β-jelly roll topology of the S-domain. The S-domain β-jelly roll topology is also shown diagrammatically in Figure (**d**). A nomenclature has been adopted for this fold in which the major β-strands are labelled A–I, such that the two β-sheets are defined as comprising strands BIDG and CHEF. Many other T = 3 RNA containing viruses use the β-jelly roll fold in their S-domains, such as feline calicivirus (**e–f**). Figures produced with PDB 2TBV (Harrison et al. 1978) and PDB 3M8L (Ossiboff et al. 2010)

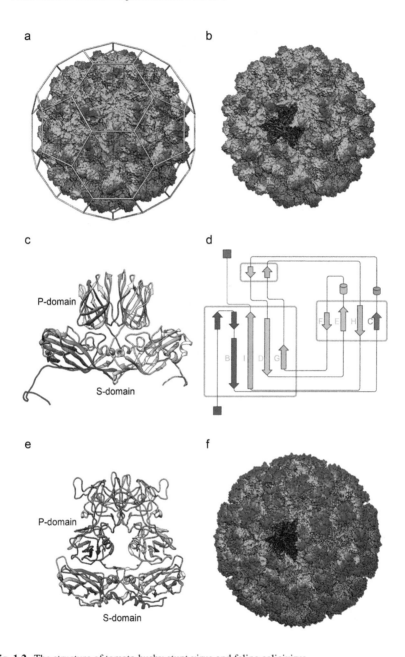

Fig. 1.2 The structure of tomato bushy stunt virus and feline calicivirus

TBSV assembles a T = 3 capsid (**a**) shows a solvent excluding surface representation of the capsid, overlaid with a 'soccer-ball' representation of the T = 3 symmetry that indicates the position of fivefold and local sixfold symmetry axes. In figure (**b**) one asymmetric unit is shown as a ribbon diagram, with quasi-equivalent positions coloured as A, purple; B, pink; and C, magenta. (**c**) A ribbon diagram of the TBSV CC dimer is viewed parallel to the capsid surface, one monomer is

Nonetheless the fold of each of the major capsid proteins bears a striking resemblance to that of TBSV. T = 3 positive-sense RNA-containing viruses that employ the β-jelly roll topology include the above-mentioned *Tombusviridae* and *Picornavirales* as well as *Nodaviridae*, which infect predominantly fish and insects, and *Caliciviridae* which include human norovirus the cause of winter-vomiting disease (Fig. 1.2e, f).

Many capsid proteins of small RNA-containing viruses can be induced to assemble into virus-like particles following recombinant expression or in vitro disassembly. It is thought that capsid assembly follows a local rule-based pathway, e.g. in the case of TBSV assembly, dimers adding to a growing capsid shell would adopt the appropriate conformation (AB or CC) at the point of association (Berger et al. 1994). Assembly of such VLPS is however characterised by the presence of many defective particles. There is growing evidence that genomic RNA plays a critical role in directing assembly of these small virions through specific interactions between packaging sequences throughout the genome and the capsid proteins (Shakeel et al. 2017).

It is interesting to note that viruses that infect hosts having adaptive immunity have evolved more complex capsid proteins, characterised by the presence of hypervariable regions and ornate surface loops (compare Figs. 1.2c, e). This is most likely a consequence of sustained immune attack driving high mutation rates, particularly in regions immediately surrounding receptor-binding sites. An important function of capsid proteins in non-enveloped viruses is mediating attachment and entry; conserved receptor-binding sites are therefore highly vulnerable to neutralisation by virus-specific antibodies.

Structural Insights into the Evolution of Large DNA-Containing Viruses (Adenovirus-PRD1 Lineage)

Use of the β-jelly roll fold to assemble spherical virus capsids is not limited to the small RNA-containing viruses. The major capsid protein of the large DNA-containing virus, adenovirus, also has been found to incorporate this topology (Rux et al. 2003). These large (~90 nm diameter) virions assemble from multiple structural proteins as T = 25 icosahedral capsids (Fig. 1.3a, b). The major capsid protein, hexon, is found at local sixfold symmetry positions, where it forms *trimeric* capsomeres (Fig. 1.3c–e). Assembling with sixfold quasi-symmetry is accomplished by a domain duplication at the base of the capsomere. Two β-jelly roll motifs are found in a vertical orientation, i.e. the barrel axis is oriented normal to the shell of the virion, as opposed to the parallel orientation of the barrel in the structure of TBSV (compare Fig. 1.3f with Fig. 1.2c).

Inferring evolutionary relationships between groups of viruses is challenging, owing to the rapid mutation rates observed in viral genomes. It has been suggested however that in the complete absence of sequence similarity, fold conservation in viral proteins may be used to discern common heritage for divergent viral taxa. This emerged from the discovery of common capsid architectures in viruses that infect

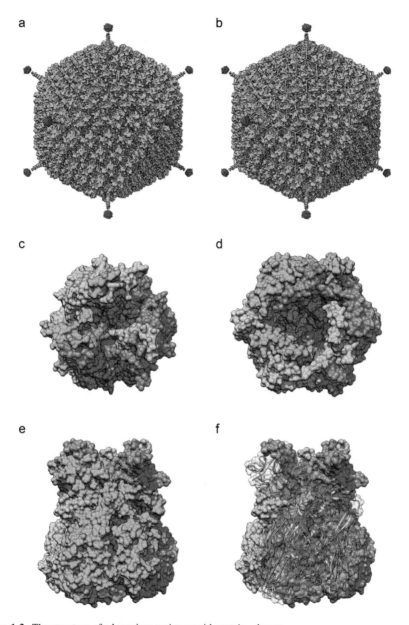

Fig. 1.3 The structure of adenovirus major capsid protein – hexon
Adenovirus, a large DNA-containing virus assembles a T = 25 icosahedral capsid comprising several structural proteins. The building blocks of the virion are the major capsid protein – hexon and penton – which forms the pentameric vertices of the capsid (**a**). A soccer-ball representation of the T = 25 symmetry (**b**) shows that at each local sixfold symmetry axis, there is a hexon capsomere. However, despite its name and location, this capsomere is a trimer. Threefold symmetry is most evident when the hexon structure is viewed from capsid exterior (*top* view – **c**). At the base of the capsomere, the structure appears sixfold symmetric (*bottom* view – **d**), owing to the presence of two β-jelly roll motifs shown in a side view of the hexon trimer and coloured orange (**e–f**).
Figure produced with (**a–b**) EMD-8471 (Yu et al. 2017b) and (**c–f**) PDB 1P30 (Rux et al. 2003)

hosts from different domains of life – prokaryotes, archaea and eukaryotes. Investigation of the capsid structure of bacteriophage PRD1 identified the same double β-jelly roll motif previously described for adenovirus, leading to the suggestion that they shared a common ancestor (Benson et al. 1999). The double β-jelly roll topology has since been found to be a feature of a wide range of viruses that infect divergent groups of hosts. Interestingly, the virophage *sputnik* was recently added to the growing list of viruses in the adenovirus-PRD1 lineage (Zhang et al. 2012). Although most viruses of this group had been found to have double-stranded DNA genomes, the lineage's most recent addition is a single-stranded DNA-containing virus FLiP (*Flavobacterium* infecting, lipid-containing phage), suggesting an evolutionary link between these two viral groups (Laanto et al. 2017).

H97 Lineage

A second group of DNA-containing viruses has been proposed to represent a distinct viral lineage based on observed conservation of structure and function. The order *Caudovirales* are tailed bacteriophages and may be the most ubiquitous group of viruses on the planet. They assemble large spherical or prolate capsids from pentameric/hexameric capsomeres that are packed according to the principles of icosahedral symmetry. The known structures of major capsid proteins of viruses in this order are described as HK97-like, owing to the presence of a conserved protein fold (Helgstrand et al. 2003) (Fig. 1.4). Interestingly this fold has recently been identified in the floor of the capsid in the herpesvirus human cytomegalovirus (Yu et al. 2017a), lending weight to previous suggestions that *Herpesvirales* and *Caudovirales* share a common ancestor (Baker et al. 2005). In addition to the common capsid protein topology, the viral replication strategies of these two groups of viruses share some surprising similarities.

Both orders of viruses package large double-stranded DNA genomes into preassembled procapsids – the immature precursor to the capsid. Genome packaging is at a very high-density in these viruses and requires a robust capsid to prevent premature release (Bhella et al. 2000). This is accomplished by an assembly pathway in which the pentameric and hexameric capsomeres assemble in a manner that has been likened to chain-mail armour and are sometimes covalently linked. Procapsid assembly is nucleated by formation of a portal, a molecular motor that pumps the viral genome into the assembled procapsid and occupies a unique fivefold vertex of the icosahedral assembly. Condensation of capsid proteins is often directed by a scaffold to ensure shells of the appropriate dimensions are formed; this may be a separate gene product or part of the major capsid protein. The scaffold proteins are subsequently proteolytically cleaved and removed from the capsid interior during capsid maturation. Upon completion of procapsid assembly, the viral genome is translocated through the portal structure, an ATP-dependent process that triggers capsid maturation. In the tailed bacteriophages, the portal is part of a larger connector structure at the interface between the genome-containing head and the tail assembly,

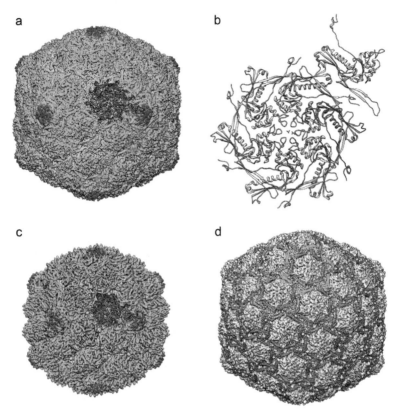

Fig. 1.4 The structure and maturation of the HK97 phage capsid
The bacteriophage HK97 assembles a T = 7 icosahedral capsid. Phage heads may be produced by recombinant expression of the major capsid proteins gp5 and the protease gp4. This system has allowed the efficient structural characterisation of the capsid assembly pathway. The mature phage head is assembled from pentamer/hexamer morphological units, making a capsid that is approximately 65 nm in diameter, shown in a radial colour scheme (**a**). The asymmetric unit comprises seven copies of gp5 (**b**). Assembly of the mature capsid involves a complex processing pathway. This starts with the formation of the ~55 nm diameter procapsid (**c**) and culminates in the formation of covalent linkages between capsid proteins giving rise to a chain-mail structure. Individual links in the armour are coloured to highlight this feature (**d**). Figure produced with (**a, b** and **d**) PDB 1OHG and (**c**) PDB 3QPR (Helgstrand et al. 2003; Huang et al. 2011)

which mediates attachment to the host cell (Chaban et al. 2015). This feature is described in more detail in Chap. 14. A tail-like structure has also been described for the herpesviruses, suggesting that an as-yet unknown set of structural proteins assemble around the portal (ul6). These are postulated to ensure that the viral genome is retained within the capsid and mediate attachment to nucleopore complexes in a newly infected cell, leading to genome release into the nucleus (Schmid et al. 2012).

Protein RNA Interactions in Helical Nucleocapsids

The earliest viral structures to be analysed by X-ray diffraction methods included the helical coat protein of tobacco mosaic virus (Franklin 1955; Caspar 1956b; Namba and Stubbs 1986).

Helical viral nucleocapsids include assemblies, like TMV, that completely enclose the viral genome, protecting it in the extracellular environment. Helical nucleocapsids are also found in enveloped viruses such as the *Mononegavirales*. These helical assemblies do not necessarily protect the viral genome but serve as a scaffold to present the genome to the viral RNA-dependent RNA polymerase both for transcription and replication. In both cases the capsid proteins and viral RNA are intimately associated and follow complementary helical symmetry.

Tobacco Mosaic Virus

Tobacco mosaic virus (TMV) like TBSV is a positive-sense RNA-containing plant virus that was extensively analysed in the early days of structural virology (Franklin 1955; Caspar 1956b; Namba and Stubbs 1986). Moreover, the virus's highly ordered helical structure has led to it having been widely used as a calibration standard and test specimen for cryo-electron microscopy (Sachse et al. 2007; Clare and Orlova 2010; Fromm et al. 2015). The right-handed helical virion is ~300 nm in length and 18 nm in diameter. It is assembled from an estimated 2130 capsid proteins. The helix comprises 16.33 subunits per turn and has a pitch of 23 angstroms (the distance from one turn to the next). The TMV capsid protein is approximately 18 kDa, comprising 158 amino acids arranged with predominantly α-helical secondary structure in a four-helix bundle motif (Fig. 1.5). The α-helices that comprise the four-helix bundle are named according to their orientations relative to the virion helix axis. They are described as radially oriented or slewed, e.g. left radial (LR), left slewed (LS), right radial (RR) and right slewed (RS). The RNA is safely ensconced between successive turns of the helix such that it is completely enclosed and interacts with capsid proteins both above and below. Each capsid protein binds

$\longrightarrow$

Fig. 1.5 (continued) A ribbon diagram shows the structure of the TMV capsid protein with three bases of RNA shown in ball and stick representation. The four major helices are labelled (**a**). Top (**b**), tilted (**c**) and side (**d**) views of the helical virion with four capsid proteins shown as ribbon diagrams. Rendering the capsid proteins as transparent surfaces reveals the path of the genomic RNA through the helical assembly in top (**e**) and tilted (**f**) views. Three copies of the capsid protein are shown as ribbons to show how the RNA is sandwiched between successive helical turns; a single subunit (*blue*) is shown in the first turn, while two subunits (*pink* and *purple*) are shown in the second. (**g**) A wall-eyed stereo pair showing a radial view of the RNA looking from the virion's helix axis. This shows how pairs of bases stack between the LR α-helices of successive protomers (*purple* and *pink*), while the phosphate groups lay against the upper-most subunit (*blue*). Figure produced with PDB 4UDV (Fromm et al. 2015)

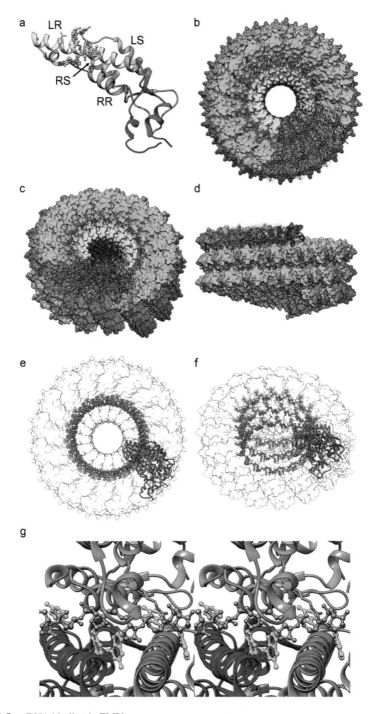

Fig. 1.5 – RNA binding in TMV

to three RNA nucleotides, and the RNA appears draped over the LR α-helix of the lower-most protomer. The central base lays flat against the LR α-helix, while those bases on either side stack against their neighbours in a cleft between the LR α-helices of successive subunits (Fig. 1.5f). Arginine residues, located in the loop that connects the LR and RR α-helices of the upper-most capsid protein, neutralise the negative charge of the phosphate groups. Such an arrangement of protein and RNA ensures that the genome is well protected in the harsh extracellular environment. Many viruses encapsidate their genomes in a helical assembly that is not protective, however, but serves as a scaffold that regulates both RNA synthesis and virion morphogenesis.

Respiratory Syncytial Virus Nucleocapsid Assembly and Function

The non-segmented negative-sense RNA-containing *Mononegavirales* comprise the families *Bornaviridae*, *Filoviridae*, *Paramyxoviridae*, *Pneumoviridae* and *Rhabdoviridae*. These include many notable human and animal pathogens such as ebola virus, measles virus, rabies virus and respiratory syncytial virus. *Mononegavirales* are enveloped viruses that with the exception of the *Bornaviridae* replicate in the cytoplasm via a positive-sense anti-genome intermediate. Here the structure and function of *Mononegavirales* nucleocapsids are discussed using respiratory syncytial virus (RSV) as an example, a more detailed account is given in Chap. 2.

The RSV genome is encapsidated by the nucleocapsid protein (N) to form a left-handed helical ribonucleoprotein assembly (RNP or nucleocapsid NC – Fig. 1.6a) that serves as the template for the viral RNA dependent RNA-polymerase (RdRp – see Chap. 3). Naked viral genomic RNA is not efficiently transcribed or replicated in the absence of N. The RdRp comprises two viral proteins, the 'large' protein (L), which contains the enzymatic activities necessary for RNA synthesis, and a multifunctional phosphoprotein (P). The interaction between the NC and RdRp is mediated by the C-terminal domain of P, one of two distinct functions for this largely disordered viral gene product.

Fig. 1.6 (continued) The RSV NC forms a left-handed helix in which the RNA wraps around the external face of the assembly (**a**) in a cleft formed between the N-terminal domain (**b** – *blue*) and the C-terminal domain (*orange*). The structure is stabilised by domain insertions of both the N-terminal arm (purple) and C-terminal arm (*red*) into neighbouring protomers. Each N subunit binds to seven nucleotides. Three nucleotides have their bases facing towards the protein, while four face outwards (wall-eyed stereo view – **c**). A comparison of the N-RNA structure for RSV with the N⁰P structure of the closely related HMPV reveals how the N-terminal region of P (*yellow*) binds to N, preventing non-specific RNA encapsidation (**d–e** viewed perpendicular to the RNA binding cleft, **f–g** viewed parallel). The C-terminal arm of N is seen to pack tightly into the RNA-binding cleft. Figure produced with PDB 2WJ8 (Tawar et al. 2009) and 5FVD (Renner et al. 2016)

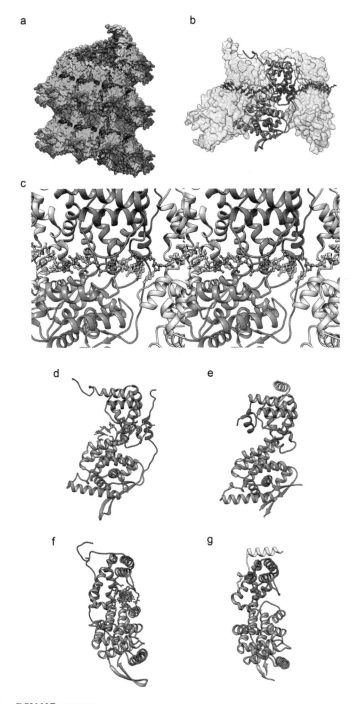

Fig. 1.6 – RSV NC structure

The second function of P is to serve as a chaperone of free N. Nucleocapsids are formed concomitantly with genome (and anti-genome) synthesis by the addition of N monomers to the nascent RNA strand. Unbound N (termed N^o) binds the N-terminal domain of P, forming the N^oP complex and preventing non-specific encapsidation of cellular RNA. It is thought, therefore, that P mediates the specific addition of N^o to nascent viral RNA through a second distinct interaction with the RdRp.

The RSV N protein comprises two globular domains – the N-terminal domain (NTD) and C-terminal domain (CTD) (Fig. 1.6b). Oligomerisation involves insertion of N and C terminal extensions (or 'arms') into the adjacent protomers, the N-arm comprising residues 1–28 and the C-arm 360–375. The viral genomic RNA wraps around the outside of the NC, embedded in the cleft formed between the NTD and CTD. Seven nucleotides bind to each N subunit with bases 2–4 facing into the cleft and bases 1 and 5–7 facing outwards from the NC (Fig. 1.6c) (Tawar et al. 2009).

Interestingly the crystal structure of N^o bound to the N-binding domain of P (P_{1-28}), for the closely related human metapneumovirus, reveals the mechanism by which P restricts non-specific RNA encapsidation. P_{1-28} binds to the C-terminal domain of N, the negatively charged C-terminal arm is seen to fold downwards into the positively charged RNA-binding cleft (Fig. 1.6d–f). The N^oP structure also shows the RNA-binding cleft to be more open, with a 10° rotation between the NTD and CTD. This suggests a mechanism for NC assembly in which the nascent RNA binds into the cleft, displacing the C-terminal arm and causing a rearrangement of the hinge region between the NTD and CTD. The N protein clamps around the RNA, while the freed C-terminal arm displaces P and is then able to bind the next incoming N (Renner et al. 2016).

Virus Attachment and Entry: Viral Envelope Fusion Proteins

The host cell plasma membrane represents a significant barrier to infection by an invading virus. A widely used strategy to overcome this barrier is for a virus to possess its own membrane. Enveloped viruses acquire membranes from their host, usually when virions bud into cellular compartments or directly from the plasma membrane. Viral entry is then accomplished by fusing the viral membrane with host membranes, a process mediated by viral-encoded fusion proteins that stud the viral envelope.

Three structural classes of viral fusion proteins have been identified that are able to accomplish this task. Class one is exemplified by the haemagglutinin (HA) protein of influenza A virus (IAV), class two by the E protein of dengue virus (Chap. 7) and class three by rabies virus G protein. The mechanisms of fusion in these three classes are however remarkably similar. The process of fusion involves the insertion of a hydrophobic domain, the fusion peptide, into the host membrane. Conformational rearrangements then bring the viral envelope and host membrane

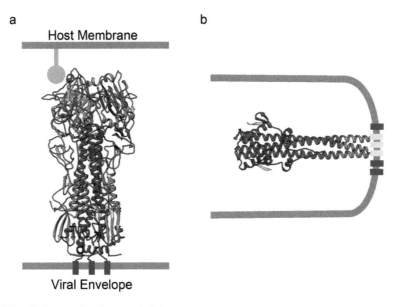

Fig. 1.7 – Influenza virus haemagglutinin
The IAV HA molecule assembles into trimers that stud the viral envelope (**a**). The protein is post-translationally cleaved into HA1 (*blue*) and HA2 (multicoloured). HA1 binds to sialic acid moieties on the host cell (*green*), triggering endocytosis. HA2 includes the fusion peptide (*yellow*) and undergoes extensive structural rearrangements at low pH to bring the viral envelope and host plasma membrane together and initiate fusion (**b**). Regions of HA2 are coloured magenta and pink in Figures **a** and **b** to highlight the changes in secondary structure that occur, leading to the formation of an extended coiled coil structure. Figure produced with PDB 1HMG (Weis et al. 1990) and 1HTM (Bullough et al. 1994)

together, leading to the formation of a fusion pore through which the virion contents may enter the cytoplasm.

Fusion proteins exist in a metastable 'pre-fusion' conformation in the mature virion prior to engaging the host. Under these conditions, the fusion-peptide is not exposed. Triggering of the protein is brought about by detection of the appropriate host-environment, such as engaging with a receptor molecule, and leads to the fusion protein adopting its fusogenic form – exposing the fusion peptide.

Perhaps the best characterised fusion protein is IAV HA (Fig. 1.7a) (Weis et al. 1990). This protein acts as mediator of both attachment and entry, binding to sialic acid moieties on the host cell surface leading to uptake of the virion by clathrin-mediated endocytosis. HA exists as trimers that coat the outer surface of the virion. Although it is expressed as a single open reading frame (HA0), this immature form is post-translationally cleaved; a priming step that yields two products in the ectodomain – HA1 and HA2. These two domains are covalently linked by disulphide bonds. HA1 mediates receptor binding, while the HA2 encodes the fusion activity, having the fusion peptide at its N-terminus and the transmembrane domain, that anchors the protein in the viral envelope, at the C-terminus. Fusion of viral and host membranes occurs upon acidification of the endosome, which triggers the transition of mature HA to a fusogenic state. The fusion peptide inserts into the endosomal membrane. The envelope proximal and distal domains of the HA2 trimer are now both embedded in membrane. The two ends are then brought together as the HA molecule undergoes a conformational rearrangement, forming a hairpin structure that brings the two membranes together (Fig. 1.7b) (Bullough et al. 1994).

Summary

Viral proteins have evolved to efficiently perform critical functions in the viral replication cycle. Viruses assemble robust containers that protect the viral genome from damage in the extracellular environment. These capsids must then deliver their contents efficiently to the target cell, binding specifically to host receptors that identify those cells to which the virus is well adapted. Capsid proteins must then mediate entry, traversing the considerable barrier of the host plasma membrane. Once inside, viral proteins work to overcome host defences, preventing intrinsic and innate immune responses from shutting the cell down or signalling to neighbouring cells that a virus is present. The viral genome is delivered safely to its site of replication and the processes of gene expression and viral replication get underway. The following chapters in this book will unveil the remarkable mechanisms by which viruses ensure their continued survival and success.

References

Baker ML, Baker ML, Jiang W et al (2005) Common ancestry of herpesviruses and tailed DNA bacteriophages. J Virol 79:14967–14970. https://doi.org/10.1128/JVI.79.23.14967-14970.2005
Bakker SE, Bhella D (2013) Pretty nasty - symmetry in virus architecture. Biochemist 35:14–19
Benson SD, Bamford JK, Bamford DH, Burnett RM (1999) Viral evolution revealed by bacteriophage PRD1 and human adenovirus coat protein structures. Cell 98:825–833
Berger B, Shor PW, Tucker-Kellogg L, King J (1994) Local rule-based theory of virus shell assembly. Proc Natl Acad Sci USA 91:7732–7736
Bhella D, Rixon FJ, Dargan DJ (2000) Cryomicroscopy of human cytomegalovirus virions reveals more densely packed genomic DNA than in herpes simplex virus type 1. J Mol Biol 295:155–161. https://doi.org/10.1006/jmbi.1999.3344
Breitbart M, Rohwer F (2005) Here a virus, there a virus, everywhere the same virus? Trends Microbiol 13:278–284. https://doi.org/10.1016/j.tim.2005.04.003
Bullough PA, Hughson FM, Skehel JJ, Wiley DC (1994) Structure of influenza haemagglutinin at the pH of membrane fusion. Nature 371:37–43. https://doi.org/10.1038/371037a0
Caspar DL (1956a) Structure of bushy stunt virus. Nature 177:475–476
Caspar DL (1956b) Structure of tobacco mosaic virus: radial density distribution in the tobacco mosaic virus particle. Nature 177:928
Caspar DLD, Klug A (1962) Physical principles in the construction of regular viruses. Cold Spring Harb Symp Quant Biol 27:1–24. https://doi.org/10.1101/SQB.1962.027.001.005

Chaban Y, Lurz R, Brasilès S et al (2015) Structural rearrangements in the phage head-to-tail interface during assembly and infection. Proc Natl Acad Sci 112:7009–7014. https://doi.org/10.1073/pnas.1504039112

Clare DK, Orlova EV (2010) 4.6Å Cryo-EM reconstruction of tobacco mosaic virus from images recorded at 300keV on a 4k×4k CCD camera. J Struct Biol 171:303–308. https://doi.org/10.1016/j.jsb.2010.06.011

Crick FH, Watson JD (1956) Structure of small viruses. Nature 177:473–475

Franklin RE (1955) Structure of tobacco mosaic virus. Nature 175:379–381

Fromm SA, Bharat TAM, Jakobi AJ et al (2015) Seeing tobacco mosaic virus through direct electron detectors. J Struct Biol 189:87–97. https://doi.org/10.1016/j.jsb.2014.12.002

Harrison SC, Olson AJ, Schutt CE et al (1978) Tomato bushy stunt virus at 2.9 a resolution. Nature 276:368–373

Helgstrand C, Wikoff WR, Duda RL et al (2003) The refined structure of a protein Catenane: the HK97 bacteriophage capsid at 3.44 Å resolution. J Mol Biol 334:885–899. https://doi.org/10.1016/j.jmb.2003.09.035

Hogle JM, Chow M, Filman DJ (1985) Three-dimensional structure of poliovirus at 2.9 a resolution. Science 229:1358–1365

Huang RK, Khayat R, Lee KK et al (2011) The Prohead-I structure of bacteriophage HK97: implications for scaffold-mediated control of particle assembly and maturation. J Mol Biol 408:541–554. https://doi.org/10.1016/j.jmb.2011.01.016

Laanto E, Mäntynen S, De Colibus L et al (2017) Virus found in a boreal lake links ssDNA and dsDNA viruses. Proc Natl Acad Sci 114:8378. https://doi.org/10.1073/pnas.1703834114

Namba K, Stubbs G (1986) Structure of tobacco mosaic virus at 3.6 a resolution: implications for assembly. Science 231:1401–1406

Ossiboff RJ, Zhou Y, Lightfoot PJ et al (2010) Conformational changes in the capsid of a Calicivirus upon interaction with its functional receptor. J Virol 84:5550–5564. https://doi.org/10.1128/JVI.02371-09

Paez-Espino D, Eloe-Fadrosh EA, Pavlopoulos GA et al (2016) Uncovering Earth's virome. Nature 536:425–430. https://doi.org/10.1038/nature19094

Renner M, Bertinelli M, Leyrat C et al (2016) Nucleocapsid assembly in pneumoviruses is regulated by conformational switching of the N protein. Elife 5:213. https://doi.org/10.7554/eLife.12627

Rossmann MG, Arnold E, Erickson JW et al (1985) Structure of a human common cold virus and functional relationship to other picornaviruses. Nature 317:145–153

Rux JJ, Kuser PR, Burnett RM (2003) Structural and phylogenetic analysis of adenovirus hexons by use of high-resolution x-ray crystallographic, molecular modeling, and sequence-based methods. J Virol 77:9553–9566. https://doi.org/10.1128/JVI.77.17.9553-9566.2003

Sachse C, Chen JZ, Coureux P-D et al (2007) High-resolution electron microscopy of helical specimens: a fresh look at tobacco mosaic virus. J Mol Biol 371:812–835. https://doi.org/10.1016/j.jmb.2007.05.088

Schmid MF, Hecksel CW, Rochat RH et al (2012) A tail-like assembly at the portal vertex in intact herpes simplex type-1 virions. PLoS Pathog 8:e1002961. https://doi.org/10.1371/journal.ppat.1002961

Shakeel S, Dykeman EC, White SJ et al (2017) Genomic RNA folding mediates assembly of human parechovirus. Nat Commun 8:5. https://doi.org/10.1038/s41467-016-0011-z

Tawar RG, Duquerroy S, Vonrhein C et al (2009) Crystal structure of a nucleocapsid-like nucleoprotein-RNA complex of respiratory syncytial virus. Science 326:1279–1283. https://doi.org/10.1126/science.1177634

Weis WI, Brünger AT, Skehel JJ, Wiley DC (1990) Refinement of the influenza virus hemagglutinin by simulated annealing. J Mol Biol 212:737–761. https://doi.org/10.1016/0022-2836(90)90234-D

Yu X, Jih J, Jiang J, Zhou ZH (2017a) Atomic structure of the human cytomegalovirus capsid with its securing tegument layer of pp150. Science 356:eaam6892. https://doi.org/10.1126/science.aam6892

Yu X, Veesler D, Campbell MG et al (2017b) Cryo-EM structure of human adenovirus D26 reveals the conservation of structural organization among human adenoviruses. Sci Adv 3:e1602670. https://doi.org/10.1126/sciadv.1602670

Zhang X, Zhang X, Sun S et al (2012) Structure of sputnik, a virophage, at 3.5-Å resolution. Proc Natl Acad Sci U S A 109:18431–18436. https://doi.org/10.1073/pnas.1211702109

Chapter 2
A Structural View of Negative-Sense RNA Virus Nucleocapsid Protein and Its Functions Beyond

Zhiyong Lou

Classification of Negative-Sense Single-Stranded RNA Viruses

The genome of negative-sense single-stranded RNA virus (NSRV) consists of one or several antisense RNA segment(s). According to the latest international virus taxonomy, NSRV are mainly divided into *Mononegavirales* order, *Bunyavirales* order, and *Arenaviridae* and *Orthomyxoviridae* families those are not assigned to an order, depending on the number of their genome segments (Davison 2017) (Table 2.1).

The order *Mononegavirales* is characterized by a single long non-segmented RNA genome and consists of four families: *Rhabdoviridae* family [e.g., rabies virus (RV) and vesicular stomatitis virus (VSV)]; *Paramyxoviridae*, with two subfamilies, *Paramyxovirinae* [e.g., measles virus (MeV) and Sendai virus (SeV)] and *Pneumovirinae* (e.g., respiratory syncytial virus, RSV); *Bornaviridae* (e.g., Borna disease virus, BDV); and *Filoviridae* [e.g., Marburg virus (MARV) and Ebola virus (EBOV)]. These viruses are also named as non-segmented NSRV (nsNSRV).

The order *Bunyavirales* is a newly classed viral order in the year of 2017 (Davison 2017). It is originated from the individual *Bunyaviridae* family, which consists five genera, including *Hantavirus* (e.g., Hantaan virus, HNTV), *Nairovirus* (e.g., Crimean-Congo hemorrhagic fever virus, CCHFV), *Orthobunyavirus* (e.g., Bunyamwera virus, BUNV), *Phlebovirus* (e.g., Rift Valley fever virus, RVFV), and *Tospovirus* (e.g., tomato spotted wilt virus, TSWV). All *Bunyaviridae* members have typical tripartite genomes (L, M, and S segments) (Plyusnin et al. 2010). In the latest taxonomy, *Bunyavirales* order is subdivided into nine families, including

Z. Lou (✉)
MOE Key Laboratory of Protein Science, School of Medicine, Tsinghua University, Beijing, China

Laboratory of Structural Biology, School of Medicine, Tsinghua University, Beijing, China
e-mail: louzy@mail.tsinghua.edu.cn

© Springer Nature Singapore Pte Ltd. 2018   19
J. R. Harris, D. Bhella (eds.), *Virus Protein and Nucleoprotein Complexes*,
Subcellular Biochemistry 88, https://doi.org/10.1007/978-981-10-8456-0_2

Table 2.1 Representative members in NSRV and the research progress of their RNP

Group	Order	Family	Virus	Progress in NP	Progress in RNP	Biological function
nsNSRV	*Mononegavirales*	*Bornaviridae*	BDV	Crystal structure		
		Filoviridae	MARV, EBOV	Crystal structure		
		Mymonaviridae	Sclerotimonavirus			
		Nyamiviridae	Nyamanini nyavirus			
		Paramyxoviridae	MeV, SeV	Crystal structure		
		Pneumovirinae	RSV	Crystal structure		
		Rhabdoviridae	RV, VSV	Crystal structure		
		Sunviridae	Sunshinevirus			
sNSRV	*Bunyavirales*	*Feraviridae*	Orthoferavirus			
		Fimoviridae	Emaravirus			
		Hantaviridae	HNTV, SNV	Crystal structure		Endonuclease
		Jonviridae	Orthojonvirus			
		Nairoviridae	CCHFV	Crystal structure		Endonuclease
		Peribunyaviridae	BUNV	Crystal structure		
		Phasmaviridae	Orthophasmavirus			
		Phenuiviridae	RVFV	Crystal structure		
		Tospoviridae	TSWV	Crystal structure		
		Arenaviridae	LAFV	Crystal structure		Exonuclease
		Orthomyxoviridae	Influenza virus	Crystal structure	Cryo-EM	

Feraviridae, *Fimoviridae*, *Hantaviridae*, *Jonviridae*, *Nairoviridae*, *Peribunyaviridae*, *Phasmaviridae*, *Phenuiviridae*, and *Tospoviridae* (Davison 2017).

The genomes of *Arenaviridae* (e.g., Lassa fever virus, LAFV) and *Orthomyxoviridae* (e.g., influenza virus) have two (Qi et al. 2011) and six to eight -ssRNA segments (Moeller et al. 2013). Moreover, three viral genera with – ssRNA genome are not assigned into a viral family. *Deltavirus* genus contains a circular negative-sense single-stranded RNA genome, and most of the genome has complementary strands; *Emaravirus* genus contains a four-segmented linear negative-sense single-stranded RNA genome; and *Tenuivirus* genus has four or five genome segments (Rizzetto 2009; Mielke-Ehret and Muhlbach 2012; Kormelink et al. 2011). All of them, together with bunyaviruses, are named as segmented NSRV (sNSRV).

The Ribonucleoprotein Complex of NSRA Viruses

The negative-sense RNA genome of NSRV is sensitive to environment and must be immediately transcribed into mRNA by the viral RNA-dependent RNA polymerase (RdRp) upon entry into host cells. To protect genomic RNA and ensure the progress of the entire virus life cycle, the genomic RNA is encapsidated by a virally encoded nucleocapsid protein (NP) and is further formed a stable ribonucleoprotein (RNP) complex together with RdRp (and/or with some accessory viral proteins) (Fig. 2.1). This RNP, instead of the naked RNA, is the only active template for virus replication and transcription (Kranzusch and Whelan 2012). Under this strategy, the genomic RNA is buried in the RNA-binding groove of viral NP, being protected against exogenous nucleases or the innate immune system in the host cell (Zhou et al. 2013).

Structural knowledge of viral RNPs was initiated by studying nsNSRVs (including BDV, RV, VSV, and RSV) since the year of 2003 and has been greatly enhanced by recent investigations of sNSRVs, including arenaviruses, bunyaviruses, and influenza virus. These achievements not only reveal how NP, genomic RNA, and RdRp form a highly ordered RNP but also indicate unexpected enzymatic functions

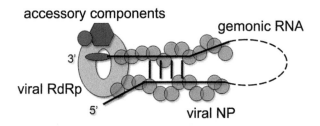

Fig. 2.1 A schematic diagram of NSRV RNP. Viral RdRp and NP are shown as orange ring and *blue* spheres. The genomic RNA is presented as a *black* line. Other accessory proteins participating in RNP formation are also indicated

of viral NPs. In particular, advances in the visualization of native or authentic RNP through electron microscopy (EM), combined with the structures of RNP components at atomic resolution, led to the understanding of the dynamic processes of RNP formation.

Structure of NPs Encoded by nsNSRVs

The structural study of nsNSRV-encoded NPs was initiated from BDV-encoded NP (Rudolph et al. 2003) followed by the NP-RNA complexes from VSV (Green et al. 2006), rabies virus (Albertini et al. 2006), and RSV (Tawar et al. 2009) (Fig. 2.1a). Up to date, except for the members in family *Mymonaviridae*, *Nyamiviridae*, and *Sunviridae*, the molecular details of other representative nsNSRV-encoded NPs have been clearly elucidated.

The framework of virus NP was first established with the structural study of BDV NP in the year of 2003 (Rudolph et al. 2003). In this work, the crystal structure of BDV NP revealed a canonical folding of virus NP family consisting N- and C-lobes to clamp a highly positive-charged groove for RNA binding flanked by N- and C-terminal extensions for inter protomer interactions (Fig. 2.2a). In such a canonical virus NP structural family, N- and C-lobes are mainly formed by a set of α-helices. N- and C-extensions are constituted by loop and/or α-helix. Although no RNA was found in the RNA-binding groove (Fig. 2.2b), highly positive-charged residues, e.g., Arg/Lys/His, from N- and C-lobes clamp a deep groove, suggesting an RNA-binding site that is approved by other viral NP-RNA complex structures.

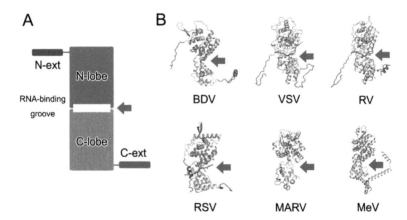

Fig. 2.2 Molecular assignment of a classical virus NP family. (a) A cartoon diagram of the folding of virus NP family. **(b)** Crystal structures of several representative nsNSRV-encoded NPs, including NP of BDV, VSV, RV, RSV, MARV (in complex with VP35 peptide), and MeV, are aligned and presented in the same orientation. N-lobe, C-lobe, N-extension, and C-extension are colored as *light-blue*, *green*, *purple*, and *red*, respectively. The RNA-binding grooves are indicated by red arrows

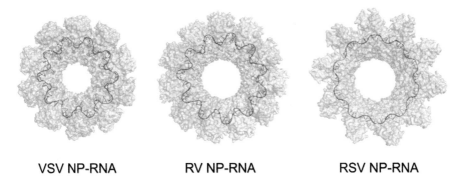

VSV NP-RNA   **RV NP-RNA**   **RSV NP-RNA**

Fig. 2.3 Ring-shaped NP-RNA complexes of nsNSRVs. Crystal structures of NP-RNA complexes from VSV, RV, and RSV (Albertini et al. 2006; Green et al. 2006; Tawar et al. 2009) are shown in the same orientation. The bound RNAs were shown as red lines

In the following years, the NP-RNA complexes from VSV, rabies virus, and RSV are reconstructed in expression cells, and their structures were determined (Albertini et al. 2006; Green et al. 2006; Tawar et al. 2009). All of these NP-RNA complexes form heterogeneous ring-shaped structures with 9 to 15 NP protomers per ring along with the bound RNA in the structure (Albertini et al. 2006; Green et al. 2006; Tawar et al. 2009) (Fig. 2.3). These three viral NPs show similarities in their molecular folding and are identical with the canonical virus NP folding defined by BDV NP. But significant difference appears in the position of bound RNA (Figs. 2.2b and 2.3). The RNA-binding grooves of VSV, rabies virus, and RSV NPs are located at the interface of N- and C-lobes, but the lateral NP contacts in VSV and rabies virus NP-RNA complex are positioned such that the curvature is opposite to that of RSV. This condition results in an inside-out nucleocapsid ring, with the RNA inside and the NP molecule-oriented outside-in in RSV NP-RNA complex (Tawar et al. 2009).

Filovirus is a key group of nsNSRV and contains numbers of pathogens to cause severe disease in humans and primates with high morbidity and mortality. Ebola virus (EBOV) and Marburg virus (MARV) are representative members of *Filoviridae* family. Although the structural studies on nsNSRV NP have been performed for a long time, the structural details of EBOV and MARV NP do not have breakthrough until very recently (Fig. 2.2b). The structure of the core domain of EBOV NP (NP$_{core}$) was first solved in complex with the fragment of VP35 (Leung et al. 2015; Kirchdoerfer et al. 2015). Filovirus NP$_{core}$ presents a canonical folding of virus NP family, though the C-lobe is much smaller with other reported viral NPs (Fig. 2.2b). A peptide derived from VP35 that binds the C-lobe of NP$_{core}$ with high affinity, thus inhibiting NP oligomerization and releasing RNA from NP-RNA complexes in vitro (Leung et al. 2015). The individual structure of EBOV NP$_{core}$ was subsequently solved and shows that the hydrophobic groove in the C-lobe for the interaction with VP35 is occupied by an α-helix of EBOV NP$_{core}$ itself, suggesting the mobility of EBOV NP element in the progress of RNP formation (Dong et al. 2015) (Fig. 2.4). The same observation was further validated in Lloviu virus (LLOV) and MARV, revealing a conserved mechanism of RNP formation within *Filoviridae* family.

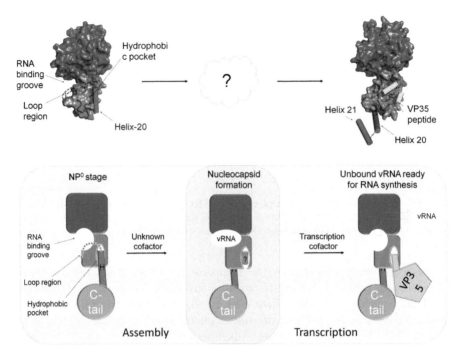

Fig. 2.4 Structural mobility of filovirus NP. The molecules of EBOV NP$_{core}$ alone and in complex with VP35 peptide are shown in the left and right parts of the upper panel. The N-lobe and C-lobe were colored as *blue* and *red*, respectively. Helices with distinct conformational shifts are shown as cylinders. VP35 peptide is shown as yellow cylinders. A model of NP conformation change is proposed in the bottom panel. This figure is cited from ref (Dong et al. 2015) with permission

Structure of NPs Encoded by Bunyavirus

Bunyavirus constitutes the largest RNA virus, as well as sNSRV, and family and is composed of over 350 members that cause severe infectious diseases throughout human, animal, and plants. Bunyaviruses are originally classed as *Bunyaviridae* family with *Hantavirus*, *Nairovirus*, *Orthobunyavirus*, *Phlebovirus*, and *Tospovirus* genus. Now, they are reorganized as *Bunyavirales* order with *Feraviridae*, *Fimoviridae*, *Hantaviridae*, *Jonviridae*, *Nairoviridae*, *Peribunyaviridae*, *Phasmaviridae*, *Phenuiviridae*, and *Tospoviridae* family (Davison 2017).

All bunyaviruses have trisegmented RNA genome to encode RdRp (L segment), the glycoproteins Gc and Gn (M segment), and NP (S segment). Unlike NPs of nsNSRVs, bunyavirus-encoded NPs have large differences, either in the primary sequence and structures or in their biological functions (Fig. 2.5). For example, CCHFV NP has the largest molecular weight of 52 kDa, and HNTV NP is 40 kDa, while the NPs from the other families are ranged from 26 kDa to 31 kDa (Li et al. 2013). Because of the diversity of bunyaviral NPs, they are interesting targets to study the structure-function relationship of viral NP.

Structure of Phlebovirus NP

Phlebovirus genus is classed in a new *Phenuiviridae* family of *Bunyavirales* order. The structural studies on phlebovirus-encoded NP initiate the research of virus NP among bunyaviruses. RVFV is a prototypic member of *Phlebovirus* genus and is the causative agent of Rift Valley fever. The structure of RVFV NP was first solved in monomeric form and presents an unusual compact structure that lacks a positively charged crevice to encapsidate RNA and N- or C-terminal extension for oligomerization (Raymond et al. 2010) (Fig. 2.5a). Later on, an hexameric structure of RVFV NP was reported (Ferron et al. 2011). Although the core regions of two structures are almost identical to have conserved N- and C-lobes as other viral NPs, significant structural variation was observed at the N-terminal extension (Fig. 2.5b). In the monomeric structure, N-extension interacts with the body region at the RNA-binding groove, thus preventing potential interacting with exogenous RNA (Raymond et al. 2010). But N-extension in the hexameric structure moves out and interacts with an adjacent protomer at the opposite side of RNA-binding groove to form a ring-shaped oligomer (Fig. 2.5c). In this oligomeric structure, the RNA-binding groove is therefore exposed to solvent. The distinct positions of the N-extension reflect the structural flexibility of phlebovirus NP during its RNP formation. Meanwhile, two studies have reported the crystal structure of NP from

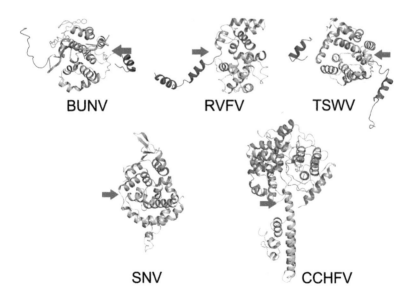

Fig. 2.5   Structures of bunyavirus-encoded NPs. Crystal structures of several representative nsNSRV-encoded NPs, including NP of BUNV, RVFV, TSWV, SNV (core domain), and CCHFV, are aligned and presented in the same orientation. N-lobe, C-lobe, N-extension, C-extension are colored as *light-blue*, *green*, *purple*, and *red*, respectively. The bound RNA in BUNV NP-RNA complex is shown as orange cartoons. The RNA-binding groove in each structure is indicated by red arrow

severe fever with thrombocytopenia syndrome virus (SFTSV) NP, which is another member of *Phlebovirus* genus that causes newly emerging infectious diseases in China and verified a conserved mechanism of NP multimerization within *Phlebovirus* genus (Jiao et al. 2013; Zhou et al. 2013). Subsequently, the NP-RNA complex structures of RVFV and Toscana viruses were determined and revealed a unique sequester mechanism for RNA encapsidation compared with nsNSRVs (Raymond et al. 2012). Moreover, though the recombinant NP-RNA complex showed different oligomeric state, the monomeric form is most likely to match the width of the native viral RNP. Notably, the structures of phlebovirus NPs do not have C-extension as other nsNSRV NPs and the interprotomer contacts are mainly employed by N-extension and core domain. This is a key difference exist in nsN-SRV and sNSRV NPs.

Structure of Nairovirus NP

Nairovirus-encoded NP has the largest molecular weight among *Bunyavirales* order. The structure of CCHFV NP was first solved in a monomeric form, which presents an unusual folding compared with other reported canonical viral NPs and reveals an unexpected nuclease activity (Guo et al. 2012) (Fig. 2.6a). A positively charged groove clamped by the head and stalk domains has been suggested to be responsible for RNA binding (Guo et al. 2012). Subsequently, the structure of NP from Baghdad-12 strain revealed a significant transposition of the stalk domain through a rotation of 180° and a translation of 40 Å, suggesting the structural flexibility to switch between alternative NP conformations during RNA binding or oligomerization (Carter et al. 2012) (Fig. 2.6b). In a subsequent work, the oligomeric CCHFV NP-RNA complexes were separated from expression cells and were reconstructed by cryo-EM (Wang et al. 2016) (Fig. 2.6c). This cryo-EM structure demonstrated that CCHFV NP-RNA presents a ring-shaped architecture in a head-to-stalk fashion. This structure also suggested a modified gating mechanism for viral genome encapsidation, in which both head and stalk domains participate in RNA binding. Again, as that has been observed in phlebovirus RNP, the monomer-sized NP-RNA complex should be the building block of the native CCHFV RNP (Wang et al. 2016) (Fig. 2.7).

Structure of Orthobunyavirus NP

Orthobunyavirus genus is the prototypic genus of bunyavirus, and it is re-classed in *Peribunyaviridae* family (Davison 2017). Orthobunyavirus encodes the smallest NP among the entire *Bunyavirales* order. Several groups analyzed NP or NP-RNA complex structures from BUNV (Li et al. 2013), Leanyer virus (LEAV) (Niu et al. 2013), La Crosse orthobunyavirus (LACV) (Reguera et al. 2013), and Schmallenberg virus

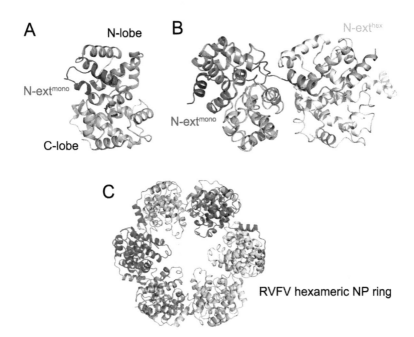

Fig. 2.6 Structure of phlebovirus NP. (a) Crystal structures of the monomeric RVFV NP. The N-arm, N-lobe, and C-lobe are colored red, *blue*, and green, respectively. (b) Comparison of RVFV NP structure in the monomeric and hexameric ring. The N- and C-lobes of one protomer are colored *blue* and *green*, whereas an adjacent protomer in the hexameric ring is shown as white cartoons. The N-arms of the NP monomer and one protomer in the hexameric form are colored red and gold, respectively. (c) Structure of the hexameric ring of RVFV NP (Raymond et al. 2012)

(SBV) (Dong et al. 2013) and reveal that orthobunyavirus NP presents a canonical virus NP folding, which contains N- and C-lobes to form RNA-binding groove and N- and C-terminal extensions for interprotomer interaction (Fig. 2.5a).

The structures of BUNV, LACV, and LEAV NP-RNA complexes revealed that both N- and C-terminal extensions contribute to NP oligomerization through a head-to-head mode (Li et al. 2013; Niu et al. 2013; Reguera et al. 2013). The structure of SBV NP was solved in a tetrameric form which is obtained from native conditions and a hexameric form in denaturation/refolding condition (Dong et al. 2013). In particular, the C-terminus of SBV NP is free in the native tetramer but is not involved in interprotomer interactions (Fig. 2.8a) (Dong et al. 2013). Moreover, the N-terminal extension in the tetrameric structure folds backward to reach the RNA-binding groove that is clamped by the N- and C-lobe while extending out in the hexameric form. This structural variation is similar to that has been observed in RVFV NP-RNA oligomer (Dong et al. 2013).

The high-order organization of orthobunyavirus NP-RNA has variations. In the study of BUNV NP, EM visualization supports that the monomer-sized NP-RNA complex is likely to be the building block of native RNP (Li et al. 2013) (Fig 2.8b, c). But a following work showed the helical architecture made of the tetrameric

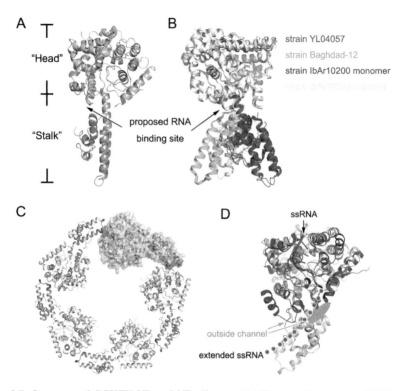

Fig. 2.7 Structure of CCHFV NP and NP oligomer. (**a**) The crystal structure of full-length CCHFV NP from strain YL04057 is shown as a cartoon diagram. The head and stalk domains are colored *green* and *blue*, respectively. (**b**) Comparison of the four different CCHFV NP structures. The head domains are colored white, whereas the stalk domains of CCHFV strain YL04057, strain Baghdad-12, strain IbAr10200 (monomeric form), and strain IbAr10200 (oligomeric form) NP are colored *red*, *green*, *blue*, and *yellow*, respectively. (**c**) The cryo-EM reconstruction of the pentameric CCHFV NP-RNA complex. (**d**) A modified gating mechanism for RNA encapsidation of CCHFV NP. Polypeptides of LASV NP and CCHFV NP are colored as *blue* and pale *green* and are aligned together. Structure elements in CCHFV NP, which are corresponding to the RNA-gating element in LASV NP, are colored as purple. The additional channel for RNA binding of CCHFV NP is colored as gold

NP-RNA building block of BUNV RNP (Ariza et al. 2013). Interestingly, the structure of LACV NP was additionally determined in a *P4₁* space group and suggested a filamentous structure along the crystallographic 4₁ axis.

Structure of Tospovirus NP

Tospoviridae genus constitutes the sole group of plant-infecting viruses in *Bunyavirales* order. TSWV is the prototypic member of *Tospovirus* genus and is one of the most devastating plant pathogens that cause severe diseases in numerous

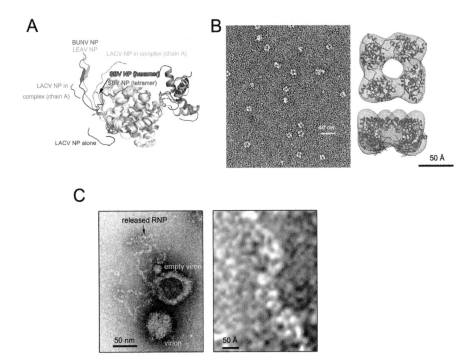

Fig. 2.8 Structure of orthobunyavirus NPs and its oligomerization for RNP formation. (a) Structural comparison of the reported structures of the orthobunyavirus NPs. The structures of SBV NP in tetrameric form (PDB code: 4IDU); SBV NP in hexameric form (PDB code: 4IDX); BUNV NP (PDB code: 4IJS), LEAV NP (PDB code: 4J1G), and LACV NP in monomeric form (PDB code: 4BGP); and chain A and chain B of LACV NP in tetrameric form (PDB code: 4BHH) are colored *blue, yellow, red, green, magenta, cyan*, and *orange*, respectively. **(b)** EM image of the negatively stained recombinant BUNV NP-RNA complex (*left*) and docking of the crystal structure of tetrameric complex into the EM density map (*right*). **(c)** Negatively stained EM image of the native BUNV RNP extracted from virions

agronomic and ornamental crops (Kormelink et al. 2011). According to the distinct hosts of tospovirus, our group studied the structure of tospovirus NP and anticipated to see new structural features of NP encoded by a plant-infecting nsNSRV. We solved the structure of the full-length TSWV NP and found three protomers in one asymmetric unit (ASU) forming a ring-shaped oligomer in a head-to-head interaction mode with a continuous RNA-binding groove in the inner side (under review) (Fig. 2.9). The body of TSWV NP possesses an N-lobe and a C-lobe to clamp an RNA-binding groove, and it is flanked by extended elements in both the N- and C-terminal parts for homotypic interactions (Fig. 2.9a). The RNA-binding groove comprises a set of positively charged residues and main chain nitrogen atoms from both N- and C-lobes. Moreover, the N- and C-extensions contacts with adjacent two NP protomers for oligomerization, as defined by nsNSRV NP (Fig. 2.9b). Most interestingly, the C-terminus of protomer A forms an additional robust contact with a protomer in another adjacent ASU at the interface of the N- and C-lobes on the

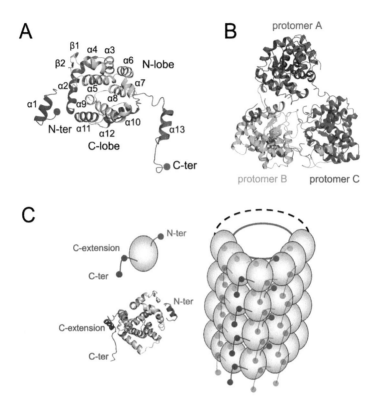

Fig. 2.9 Structure of TSWV NP. (a) Structure of TSWV NP monomer is shown a cartoon diagram with a rainbow coloring from the N- (*blue*) to the C-terminus (*red*). **(b)** Overall structure of the trimeric TSWV NP oligomer. Three protomers are shown in the cartoon diagram and are colored as *red, green*, and *blue*. **(c)** A model for TSWV RNP formation with a distinct homotypic interaction. The body part, N-extension, C-extension, and C-terminus of TSWV NP are shown as different diagrammatic schemes. The N-N interaction, C-C interaction, and the additional interaction at the C-terminus are shown as solid colors in the high-ordered model, but other sets are shown with transparency for clear presentation. Nucleic acids encapsidated by NPs are indicated by a red line

reverse side of the RNA-binding groove. This structural feature proposes a distinct mechanism for a homotypic NP interaction that forms high-ordered RNP in tospovirus (Fig. 2.9c). Whether this additional interprotomer contact is related with the unique function of a plant-infecting virus warrants further investigation.

Structure of NP Encoded by Arenavirus

LAFV is a prototypic member of the *Arenaviridae* family, which causes severe hemorrhagic fever in humans with high fatalities, and it is featured by two segmented RNA genome (Buchmeier MJ 2007; Martinez-Sobrido et al. 2007). The

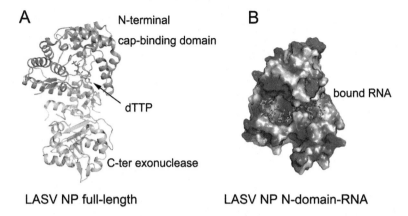

Fig. 2.10 Structural information of LAFV NP. (a) Crystal structure of full-length LAFV NP is shown as colored cartoon diagram. The N-terminal cap-binding domain and C-terminal exonuclease domain are colored as light-blue and green, respectively. The bound dTTP, which is considered to be a cap analog, is presented as a colored stick. (b) The crystal structure of the N-terminal domain of LAFV NP in complex with RNA. The polypeptide of the N-terminal domain of LAFV NP is covered with potential surface, while the bound RNA molecule is shown as colored sticks

structure of LAFV NP greatly extended the knowledge of virus NP family. Qi et al. first reported the structure of full-length LAFV NP, revealing two distinct N- (NTD) and C-terminal (CTD) domains, in contrast to other reported viral NPs (Qi et al. 2011) (Fig. 2.10a). A highly positively charged groove located at the interface of the NTD and CTD was predicted to be the genomic RNA-binding site based on the trimeric structure of LAFV NP, which is consistent with a small-angle X-ray scattering analysis(Brunotte et al. 2011). In a parallel study, the structure of LAFV NP N-terminal domain in complex with a single-stranded RNA was solved, revealing that RNA binds in a deep positively charged crevice located in N-terminal domain (Hastie et al. 2011b) (Fig. 2.10c). The C-terminal domain may not take a major responsibility for RNA encapsidation, but it moves slightly away from its position in the RNA-free trimer, allowing helices α5 and α6 to open away from the RNA-binding crevice in a gating mechanism (Hastie et al. 2011b).

Structure of NP Encoded by Orthomyxovirus

Influenza virus is the most well-studied member of *Orthomyxoviridae* family as an sNSRV. The first crystal structure of influenza A virus NP in a RNA-free form was solved by two independent groups (Ng et al. 2008; Ye et al. 2006). Influenza A virus NP presents a canonical folding of virus NP family, containing N- and C-lobes to clamp a deep positively charged RNA-binding groove, and tail portions to mediate oligomerization by interacting with the neighboring molecule (Ng et al. 2008; Ye et al. 2006; Chenavas et al. 2013) (Fig. 2.11). However, the high-ordered structures

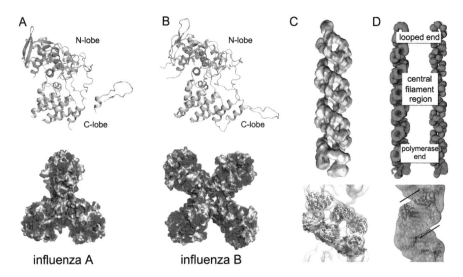

Fig. 2.11 Structure of influenza virus NP. Crystal structures of NP complex from influenza A virus (**a**) and influenza B virus (**b**) are shown in the same orientation in the upper panels. The N-lobe and C-lobe for each NP protomer are colored as *blue* and *green*, respectively. The homotrimer and homotetramer of influenza A virus and influenza B virus NP are covered with potential surface in the bottom panels. (**c**) Cryo-EM reconstruction of a native RNP by Arranz et al. (Arranz et al. 2013). The viral polymerase complex is located at the bottom end of the RNP and is shown in green and orange. The two opposite running NP-RNA strands are colored blue and pink, respectively. NP and RNA are fitted in the EM map in the lower panel. (**d**) Cryo-EM reconstruction of an authentic RNP by Moeller et al. (Moeller et al. 2013), The viral polymerase is highlighted in red, whereas the NP-RNA is colored dark green. The structure of NP protomer is fitted into the EM map in the lower panel. (**f**) and (**g**) are modified with permission (Arranz et al. 2013; Moeller et al. 2013), respectively

of the influenza virus NP are distinct. The first structure of influenza A virus NP was solved in a homotrimeric form (Ng et al. 2008; Ye et al. 2006), but a following research showed that influenza B virus NP forms a homotetrameric oligomer in crystal structure, though both NPs have high sequence and structural similarities (Fig. 2.11a, b). The oligomerization of influenza virus NP was further investigated and defined the residue R416 playing an essential role in influenza NP multimerization (Chenavas et al. 2013). In a R416A-induced monomeric structure, the folding of residues 386 to 401, the trimer exchange domain (residues 402 to 428), and residues 429 to 498 present a significant shift compared to that in the oligomer.

With the breakthrough of cryo-EM technics, progress of the study on an authentic or native influenza virus RNP has been made in recent years. A ring-shaped influenza virus RNP with PA-PB1-PB2 RdRp complex, nine NP molecules, and a short RNA was initially reconstructed in expression cells and studied by cryo-EM (Area et al. 2004; Coloma et al. 2009). Two recent works visualized the authentic influenza virus RNPs, which were extracted from native virions or reconstructed in expression cells at ~20 Å resolution (Arranz et al. 2013; Moeller et al. 2013) (Fig. 2.11c, d). The RNP model presented in these works indicated that the RNP adopts a double-helical structure with two antiparallel strands leading to and away from the polymerase, which is located at one end of the RNP. The double-helical stem region has a rise of 32.6 Å between two neighboring NPs with 4.9 NP molecules per turn (Moeller et al. 2013). However, Arranz et al. presented a left-handed helix, while Moeller et al. reported a right-handed helix. As a result, Arranz et al. proposed that the body domain of NP mainly facilitates the interprotomer interaction, whereas Moeller et al. indicated that the head domain of NP is likely to stabilize the RNP. Further study with recent developed high-resolution cryo-EM techniques would be useful to elucidate the clear mechanism of influenza virus RNP formation.

Unexpected Biological Function of Viral NPs

It is well known that virus NP functions as a structural protein in RNP formation and the entire virus life cycle. In this context, virus NP only encapsidates RNA and oligomerizes to form high-ordered RNP, but does not have other biological function. However, a few recent studies extended our knowledge.

The structure of LAFV NP firstly revealed that a virus NP has an enzymatic activity beyond a structure protein (Qi et al. 2011). The C-terminal domain of LAFV NP bears a molecular folding of 3'-5' exonuclease/exoribonuclease superfamily and was further demonstrated to have a metal-dependent exoribonuclease activity which is responsible for suppressing host immune responses (Hastie et al. 2011a; Qi et al. 2011; Jiang et al. 2013; Hastie et al. 2012). Although the enzymatic activity of LAFV NP is located in a separate domain, this is a first to find a biological function of viral NP.

Later on, an enzymatic activity of virus NP was reported in CCHFV NP. Guo et al. reported that CCHFV NP has intrinsic metal-dependent endonuclease activity on both single- and double-stranded DNA (Guo et al. 2012). Very interestingly, the core domain of CCHFV NP has very high structural similarity to the LAFV NP N-terminal domain, though they have no primary sequence homologs (Guo et al. 2012), indicating LAFV NP N-terminal domain may have a similar function. Furthermore, to our most interest, NP encoded by Sin Nombre virus (SNV), a member of *Hantaviridae* family in *Bunyavirales* order, was recently reported to have an endonuclease activity like CCHFV NP (Moncke-Buchner et al. 2016). However, the structure of SNV NP displays a conserved folding of virus NP family (Guo et al. 2015; Olal and Daumke 2016) but does not like CCHFV NP or LAFV NP. The key residues that play an essential role in SNV endonuclease activity are located in RNA-binding groove and organize a similar architecture with conserved endonuclease catalytic center (Moncke-Buchner et al. 2016). Actually, we also checked other NP encoded by hantaviruses and found the same activity (unpublished data). Because SNV NP has a canonical folding of virus NP family, it is interesting to verify whether other virus NP has a same endonuclease activity and further investigate its function in virus life cycle.

Antiviral Development Targeted at Virus NP and RNP Formation

NP is the most conserved viral protein throughout different genotypes of one virus species and the mechanism for RNP formation is also known to have the strict homology. Moreover, the correct RNP form and function is a key step for the replication, transcription, and assembly of NSRV. The antivirals targeting NP or the formation of virus RNP is thus conceivable to be an ideal goal for the development of small-molecule therapies against viral resistance to currently available drugs (Lou et al. 2014). However, this strategy does not have success until recently great progress has been made on influenza virus.

A small-molecule compound, nucleozin, was reported to trigger influenza virus NP aggregation, inhibit the nuclear accumulation of NP, and thus inhibit the replication of influenza virus at a nanomolar median effective concentration (EC_{50}) (Kao et al. 2010). Meanwhile, Gerritz et al. also presented influenza replication inhibitors to induce the formation of unnatural higher-order NP oligomer and inhibition influenza virus replication with an EC_{50} up to 60 nM (Gerritz et al. 2011). This is the first success to demonstrate that virus NP is a druggable target and develop antivirals targeting at NP. Notably, the structure of NP in complex with a representative compound of these inhibitors revealed that two inhibitors in an antiparallel orientation lock two adjacent NP protomers, which are consistent with the observation of cell biology (Gerritz et al. 2011). Additional progress has also been reported with the study of SFTSV NP. The structure of SFTSV NP in complex with suramin, an antiviral inhibitor, revealed that the blocker that binds at the RNA-binding cavity can attenuate SFTSV replication and indicated a new therapeutic antiviral approach to impact RNP formation (Wang et al. 2012).

Conclusion and Perspective

With the effort in the past decade, the understanding of molecular details of virus NP and the dynamic processes of RNP formation achieve great progresses. Although the structural and functional studies summarized here suggest an overall picture of NP and RNP in NSRV, many important questions warrant further investigations. First, the real architectures of native RNP that is packaged in virions or functions in host cells remain unclear. In particular, the atomic structure of a real virus RNP is also missing. It is necessary to further investigate the structure of a real RNP to elucidate the conformational shifts of NP and other components during the dynamic process of RNP formation. Second, the structure of CCHFV NP suggests a close relationship of CCHFV with arenavirus but is distant with other bunyaviruses. It may need further consideration to include this structural information in viral taxonomy. Finally, the success of antiviral development targeted at virus NP is very rare. Because virus NP is one of most conserved protein among different genotypes of one virus, it is possible to combine structural findings in the development of antiviral therapeutics that target RNP formation and function. We anticipate that future developments and discussions will continue to explore the structural aspects of viral RNPs and the antiviral reagents that impact RNP formation that could have clinical applications in the future.

References

Albertini AA, Wernimont AK, Muziol T, Ravelli RB, Clapier CR, Schoehn G, Weissenhorn W, Ruigrok RW (2006) Crystal structure of the rabies virus nucleoprotein-RNA complex. Science 313(5785):360–363

Area E, Martin-Benito J, Gastaminza P, Torreira E, Valpuesta JM, Carrascosa JL, Ortin J (2004) 3D structure of the influenza virus polymerase complex: localization of subunit domains. Proc Natl Acad Sci U S A 101(1):308–313. https://doi.org/10.1073/pnas.0307127101

Ariza A, Tanner SJ, Walter CT, Dent KC, Shepherd DA, Wu W, Matthews SV, Hiscox JA, Green TJ, Luo M, Elliott RM, Fooks AR, Ashcroft AE, Stonehouse NJ, Ranson NA, Barr JN, Edwards TA (2013) Nucleocapsid protein structures from orthobunyaviruses reveal insight into ribonucleoprotein architecture and RNA polymerization. Nucleic Acids Res 41(11):5912–5926

Arranz R, Coloma R, Chichon FJ, Conesa JJ, Carrascosa JL, Valpuesta JM, Ortin J, Martin-Benito J (2013) The structure of native influenza virion ribonucleoproteins. Science 338(6114):1634–1637

Brunotte L, Kerber R, Shang W, Hauer F, Hass M, Gabriel M, Lelke M, Busch C, Stark H, Svergun DI, Betzel C, Perbandt M, Gunther S (2011) Structure of the Lassa virus nucleoprotein revealed by X-ray crystallography, small-angle X-ray scattering, and electron microscopy. J Biol Chem 286(44):38748–38756

Buchmeier MJ dlTJ, Peters CJ (2007) Arenaviridae: The viruses and their replication. Fields Virology eds Knipe DM, Howley PM (Lippincott Williams & Wilkins, Philadelphia), 5th edn. pp 1791–1827

Carter SD, Surtees R, Walter CT, Ariza A, Bergeron E, Nichol ST, Hiscox JA, Edwards TA, Barr JN (2012) Structure, function, and evolution of the crimean-congo hemorrhagic fever virus nucleocapsid protein. J Virol 86(20):10914–10923. https://doi.org/JVI.01555-1210.1128/JVI.01555-12

Chenavas S, Estrozi LF, Slama-Schwok A, Delmas B, Di Primo C, Baudin F, Li X, Crepin T, Ruigrok RW (2013) Monomeric nucleoprotein of influenza a virus. PLoS Pathog 9(3):e1003275

Coloma R, Valpuesta JM, Arranz R, Carrascosa JL, Ortin J, Martin-Benito J (2009) The structure of a biologically active influenza virus ribonucleoprotein complex. PLoS Pathog 5(6):e1000491. https://doi.org/10.1371/journal.ppat.1000491

Davison AJ (2017) Journal of general virology - introduction to 'ICTV virus taxonomy profiles. J Gen Virol 98(1):1. https://doi.org/10.1099/jgv.0.000686

Dong H, Li P, Elliott RM, Dong C (2013) Structure of Schmallenberg orthobunyavirus nucleoprotein suggests a novel mechanism of genome encapsidation. J Virol 87(10):5593–5601

Dong S, Yang P, Li G, Liu B, Wang W, Liu X, Xia B, Yang C, Lou Z, Guo Y, Rao Z (2015) Insight into the Ebola virus nucleocapsid assembly mechanism: crystal structure of Ebola virus nucleoprotein core domain at 1.8 a resolution. Protein Cell 6(5):351–362. https://doi.org/10.1007/s13238-015-0163-3

Ferron F, Li Z, Danek EI, Luo D, Wong Y, Coutard B, Lantez V, Charrel R, Canard B, Walz T, Lescar J (2011) The hexamer structure of Rift Valley fever virus nucleoprotein suggests a mechanism for its assembly into ribonucleoprotein complexes. PLoS Pathog 7(5):e1002030

Gerritz SW, Cianci C, Kim S, Pearce BC, Deminie C, Discotto L, McAuliffe B, Minassian BF, Shi S, Zhu S, Zhai W, Pendri A, Li G, Poss MA, Edavettal S, McDonnell PA, Lewis HA, Maskos

K, Mortl M, Kiefersauer R, Steinbacher S, Baldwin ET, Metzler W, Bryson J, Healy MD, Philip T, Zoeckler M, Schartman R, Sinz M, Leyva-Grado VH, Hoffmann HH, Langley DR, Meanwell NA, Krystal M (2011) Inhibition of influenza virus replication via small molecules that induce the formation of higher-order nucleoprotein oligomers. Proc Natl Acad Sci U S A 108(37):15366–15371

Green TJ, Zhang X, Wertz GW, Luo M (2006) Structure of the vesicular stomatitis virus nucleoprotein-RNA complex. Science 313(5785):357–360

Guo Y, Wang W, Ji W, Deng M, Sun Y, Zhou H, Yang C, Deng F, Wang H, Hu Z, Lou Z, Rao Z (2012) Crimean-Congo hemorrhagic fever virus nucleoprotein reveals endonuclease activity in bunyaviruses. Proc Natl Acad Sci U S A 109(13):5046–5051

Guo Y, Wang W, Sun Y, Ma C, Wang X, Wang X, Liu P, Shen S, Li B, Lin J, Deng F, Wang H, Lou Z (2015) Crystal structure of the Core region of hantavirus Nucleocapsid protein reveals the mechanism for ribonucleoprotein complex formation. J Virol 90(2):1048–1061. https://doi. org/10.1128/jvi.02523-15

Hastie KM, Kimberlin CR, Zandonatti MA, MacRae IJ, Saphire EO (2011a) Structure of the Lassa virus nucleoprotein reveals a dsRNA-specific 3′ to 5′ exonuclease activity essential for immune suppression. Proc Natl Acad Sci U S A 108(6):2396–2401

Hastie KM, Liu T, Li S, King LB, Ngo N, Zandonatti MA, Woods VL, Jr., de la Torre JC, Saphire EO (2011b) Crystal structure of the Lassa virus nucleoprotein-RNA complex reveals a gating mechanism for RNA binding. Proc Natl Acad Sci U S A 108 (48):19365–19370

Hastie KM, King LB, Zandonatti MA, Saphire EO (2012) Structural basis for the dsRNA specific-ity of the Lassa virus NP exonuclease. PLoS One 7(8):e44211

Jiang X, Huang Q, Wang W, Dong H, Ly H, Liang Y, Dong C (2013) Structures of Arenaviral nucleoproteins with triphosphate dsRNA reveal a unique mechanism of immune suppression. J Biol Chem 288(23):16949–16959

Jiao L, Ouyang S, Liang M, Niu F, Shaw N, Wu W, Ding W, Jin C, Peng Y, Zhu Y, Zhang F, Wang T, Li C, Zuo X, Luan CH, Li D, Liu ZJ (2013) Structure of severe fever with thrombocytopenia syndrome virus Nucleocapsid protein in complex with Suramin reveals therapeutic potentials. J Virol 87(12):6829–6839

Kao RY, Yang D, Lau LS, Tsui WH, Hu L, Dai J, Chan MP, Chan CM, Wang P, Zheng BJ, Sun J, Huang JD, Madar J, Chen G, Chen H, Guan Y, Yuen KY (2010) Identification of influenza a nucleoprotein as an antiviral target. Nat Biotechnol 28(6):600–605

Kirchdoerfer RN, Abelson DM, Li S, Wood MR, Saphire EO (2015) Assembly of the Ebola virus nucleoprotein from a chaperoned VP35 complex. Cell Rep 12(1):140–149. https://doi. org/10.1016/j.celrep.2015.06.003

Kormelink R, Garcia ML, Goodin M, Sasaya T, Haenni AL (2011) Negative-strand RNA viruses: the plant-infecting counterparts. Virus Res 162(1–2):184–202. https://doi.org/10.1016/j. virusres.2011.09.028

Kranzusch PJ, Whelan SP (2012) Architecture and regulation of negative-strand viral enzymatic machinery. RNA Biol 9(7):941–948

Leung DW, Borek D, Luthra P, Binning JM, Anantpadma M, Liu G, Harvey IB, Su Z, Endlich-Frazier A, Pan J, Shabman RS, Chiu W, Davey RA, Otwinowski Z, Basler CF, Amarasinghe GK (2015) An intrinsically disordered peptide from Ebola virus VP35 controls viral RNA synthesis by modulating nucleoprotein-RNA interactions. Cell Rep 11(3):376–389. https://doi. org/10.1016/j.celrep.2015.03.034

Li B, Wang Q, Pan X, Fernandez de Castro I, Sun Y, Guo Y, Tao X, Risco C, Sui SF, Lou Z (2013) Bunyamwera virus possesses a distinct nucleocapsid protein to facilitate genome encapsida-tion. Proc Natl Acad Sci USA 110(22):9048–9053

Lou Z, Sun Y, Rao Z (2014) Current progress in antiviral strategies. Trends Pharmacol Sci 35(2):86–102. https://doi.org/10.1016/j.tips.2013.11.006

Martinez-Sobrido L, Giannakas P, Cubitt B, Garcia-Sastre A, de la Torre JC (2007) Differential inhi-bition of type I interferon induction by arenavirus nucleoproteins. J Virol 81(22):12696–12703

Mielke-Ehret N, Muhlbach HP (2012) Emaravirus: a novel genus of multipartite, negative strand RNA plant viruses. Virus 4(9):1515–1536

Moeller A, Kirchdoerfer RN, Potter CS, Carragher B, Wilson IA (2013) Organization of the influenza virus replication machinery. Science 338(6114):1631–1634

Moncke-Buchner E, Szczepek M, Bokelmann M, Heinemann P, Raftery MJ, Kruger DH, Reuter M (2016) Sin Nombre hantavirus nucleocapsid protein exhibits a metal-dependent DNA-specific endonucleolytic activity. Virology 496:67–76. https://doi.org/10.1016/j.virol.2016.05.009

Ng AK, Zhang H, Tan K, Li Z, Liu JH, Chan PK, Li SM, Chan WY, Au SW, Joachimiak A, Walz T, Wang JH, Shaw PC (2008) Structure of the influenza virus a H5N1 nucleoprotein: implications for RNA binding, oligomerization, and vaccine design. FASEB J 22(10):3638–3647

Niu F, Shaw N, Wang YE, Jiao L, Ding W, Li X, Zhu P, Upur H, Ouyang S, Cheng G, Liu ZJ (2013) Structure of the Leanyer orthobunyavirus nucleoprotein-RNA complex reveals unique architecture for RNA encapsidation. Proc Natl Acad Sci U S A 110(22):9054–9059

Olal D, Daumke O (2016) Structure of the hantavirus nucleoprotein provides insights into the mechanism of RNA Encapsidation. Cell Rep 14(9):2092–2099. https://doi.org/10.1016/j.celrep.2016.02.005

Plyusnin A, Beaty BJ, Elliott RM, Goldbach R, Kormelink R, Lundkvist Å, Schmaljohn CS, Tesh RB (2010) Bunyaviridae. In: King AMQ, Lefkowitz EJ, Adams MJ, Carstens EB (eds) Virus taxonomy: classification and nomenclature of viruses. Ninth Report of the International Committee on Taxonomyof Viruses. Elsevier, San Diego,

Qi X, Lan S, Wang W, Schelde LM, Dong H, Wallat GD, Ly H, Liang Y, Dong C (2011) Cap binding and immune evasion revealed by Lassa nucleoprotein structure. Nature 468(7325):779–783

Raymond DD, Piper ME, Gerrard SR, Smith JL (2010) Structure of the Rift Valley fever virus nucleocapsid protein reveals another architecture for RNA encapsidation. Proc Natl Acad Sci USA 107(26):11769–11774

Raymond DD, Piper ME, Gerrard SR, Skiniotis G, Smith JL (2012) Phleboviruses encapsidate their genomes by sequestering RNA bases. Proc Natl Acad Sci USA 109(47):19208–19213

Reguera J, Malet H, Weber F, Cusack S (2013) Structural basis for encapsidation of genomic RNA by La Crosse Orthobunyavirus nucleoprotein. Proc Natl Acad Sci USA 110(18):7246–7251

Rizzetto M (2009) Hepatitis D: thirty years after. J Hepatol 50(5):1043–1050

Rudolph MG, Kraus I, Dickmanns A, Eickmann M, Garten W, Ficner R (2003) Crystal structure of the borna disease virus nucleoprotein. Structure 11(10):1219–1226

Tawar RG, Duquerroy S, Vonrhein C, Varela PF, Damier-Piolle L, Castagne N, MacLellan K, Bedouelle H, Bricogne G, Bhella D, Eleouet JF, Rey FA (2009) Crystal structure of a nucleocapsid-like nucleoprotein-RNA complex of respiratory syncytial virus. Science 326(5957):1279–1283

Wang Y, Dutta S, Karlberg H, Devignot S, Weber F, Hao Q, Tan YJ, Mirazimi A, Kotaka M (2012) Structure of Crimean-Congo hemorrhagic fever virus nucleoprotein: superhelical homo-oligomers and the role of caspase-3 cleavage. J Virol 86(22):12294–12303

Wang X, Li B, Guo Y, Shen S, Zhao L, Zhang P, Sun Y, Sui SF, Deng F, Lou Z (2016) Molecular basis for the formation of ribonucleoprotein complex of Crimean-Congo hemorrhagic fever virus. J Struct Biol 196(3):455–465. https://doi.org/10.1016/j.jsb.2016.09.013

Ye Q, Krug RM, Tao YJ (2006) The mechanism by which influenza a virus nucleoprotein forms oligomers and binds RNA. Nature 444(7122):1078–1082

Zhou H, Sun Y, Liu M, Wang Y, Liu C, Wang W, Liu X, Li L, Deng F, Guo Y, Lou Z (2013) The nucleoprotein of severe fever with thrombocytopenia syndrome virus processes an oligomeric ring to facilitate RNA encapsidation. Protein Cell 4(6):445–455

Chapter 3
Viral RNA-Dependent RNA Polymerases: A Structural Overview

Diego Ferrero, Cristina Ferrer-Orta, and Núria Verdaguer

Introduction

RNA virus infections are the main cause of epidemic diseases in humans and animals. The humanitarian disaster caused by the Ebola virus in 2014 or the last Zika virus outbreak in 2015 is only one of the examples in a long series of unexpected or recurring RNA virus outbreaks. One way to limit the impact of RNA viruses is to prevent their replication, and an exhaustive knowledge of the replication mechanisms used by these pathogens is therefore essential. Viruses with RNA genomes are divided into four groups: single positive-strand ((+)ssRNA), single negative-strand ((−)ssRNA), double-strand RNA (dsRNA) viruses, and retroviruses. These viruses use specific strategies to replicate and transcribe their genetic material, but with the exception of retroviruses, all of them have a common element: the RNA-dependent RNA polymerase (RdRP).

Currently, the high-resolution structures of RdRPs and RdRP-RNA catalytic complexes have been determined for a large number of RNA virus groups, including the *Picornaviridae*, *Caliciviridae*, *Flaviviridae*, and *Permutatetraviridae* families of (+)ssRNA viruses; the *Cystoviridae*, *Reoviridae*, *Picobirnaviridae*, and *Birnaviridae* families of dsRNA viruses; and the *Orthomyxoviridae* and *Bunyaviridae* (segmented) and *Rhabdoviridae* (non-segmented) (−) ssRNA virus families (Table 3.1). These studies contributed enormously to our comprehension about the mechanisms of action of these enzymes. RdRPs belong to the superfamily of template-directed nucleic acid polymerases (TdPPs), including DNA-dependent DNA polymerases, DNA-dependent RNA polymerases, and reverse transcriptases. All these enzymes share similar architecture and a conserved two-metal ion mechanism of phosphodiester bond formation (Brautigam and Steitz 1998). The shape of

D. Ferrero · C. Ferrer-Orta · N. Verdaguer (✉)
Structural Biology Unit, Institut de Biologia Molecular de Barcelona (IBMB-CSIC),
Barcelona, Spain
e-mail: dfecri@ibmb.csic.es; cfocri@ibmb.csic.es; nvmcri@ibmb.csic.es

© Springer Nature Singapore Pte Ltd. 2018

J. R. Harris, D. Bhella (eds.), *Virus Protein and Nucleoprotein Complexes*,
Subcellular Biochemistry 88, https://doi.org/10.1007/978-981-10-8456-0_3

Table 3.1 High-resolution structures of RdRPs and RdRP catalytic complexes solved by X-ray crytallography available in the PDB to date

+RNA	*Picornaviridae*		
	PV 3D^{pol} partial structure	Hansen et al. 1997	1RDR
	PV 3D^{pol}	Thompson and Peersen 2004	1RA6, 1RA7, 1RAJ, 1TQL
	FMDV 3D^{pol} isolated and a template-primer complex	Ferrer-Orta et al. 2004	1U09, 1 WNE
	HRV1B, HRV14, and HRV16 3D^{pol}	Love et al. 2004	1XR5, 1XR6, 1XR7
	HRV16 3D^{pol}	Appleby et al. 2005	1TP7
	Initiation complex of FMDV	Ferrer-Orta et al. 2006b	2D7S, 2F8E
	Elongation complexes of FMDV	Ferrer-Orta et al. 2007	2E9R, 2E9T, 2EC0, 2E9Z
	PV 3D^{pol} with NTPs	Thompson et al. 2007	2ILY, 2ILZ, 2IM0, 2IM1, 2IM2, 2IM3
	PV 3CD protein	Marcotte et al. 2007	2IJD, 2IJF
	CVB3 3D^{pol}	Campagnola et al. 2008	3DDK
	CVB3 3D^{pol} with VPg	Gruez et al. 2008	3CDU, 3CDW
	EV-71 3D^{pol} with NTP and analog	Wu et al. 2010	3N6L, 3N6M, 3N6N
	Elongation complexes of PV, HRV, and CVB3	Gong et al. 2013	4K4S, 4K4T, 4K4U, 4K4V, 4K4W, 4K4Y, 4K4X, 4K4Z, 4K50
	EV-71 3D^{pol} with VPg	Chen et al. 2013	4IKA
	EMCV 3D^{pol}	Vives-Adrian et al. 2014	4NYZ, 4NZ0
	Elongation complex of EV-71	Shu and Gong 2016	5F8G, 5F8H, 5F8I, 5F8J, 5F8L, 5F8M, 5F8N
	EV-68 3D^{pol}	Wang et al. 2017	5XE0
	Caliciviridae		
	RHDV RdRP	Ng et al. 2002	1KHV, 1KHW
	NV RdRP	Ng et al. 2004	1SH0, 1SH2, 1SH3
	NV elongation complex	Zamyatkin et al. 2008	3BSO, 3BSN
	Flaviviridae		
	HCV NS5B	Lesburg et al. 1999	1C2P
		Bressanelli et al. 1999	1CSJ
		Ago et al. 1999	1QUV
	HCV NS5B with NTPs	Bressanelli et al. 2002	1GX6, 1GX5

(continued)

Table 3.1 (continued)

	BVDV NS5	Choi et al. 2004	1S48, 1S49, 1S4F
	WNV NS5 RdRP domain	Malet et al. 2007	2HCN, 2HCS, 2HFZ
	DenV NS5 RdRP domain with NTPs	Yap et al. 2007	2J7U, 2JUW
	HCV NS5B with primer-template RNA	Mosley et al. 2012	4E76, 4E78, 4E7A
	JEV NS5 full length	Lu and Gong 2013	4K6M
	JEV RdRP with and without NTPs	Surana et al. 2014	4MTP, 4HDG, 4HDH
	HCV initiation complex	Appleby et al. 2015	4WT9, 4WTA, 4WTC, 4WTD, 4WTE, 4WTF, 4WTG, 4WTI, 4WTJ, 4WTK, 4WTL, 4WTM
	DenV NS5 full length with cap O-RNA and SAH	Zhao et al. 2015a	5DTO
		Zhao et al. 2015b	4V0Q, 4V0R
	DenV NS5 full length dimer with SAH	Klema et al. 2016	5CCV
	ZIKV NS5 RdRP domain	Godoy et al. 2017	5U04
	ZIKV NS5 full length	Upadhyay et al. 2017	5TFR
	Permutatetraviridae		
	TaV RdRP with and without NTPs	Ferrero et al. 2015	5CX6, 5CYR, 4XHA, 4XHI
ds RNA	*Reoviridae*		
	λ3 Initiation complex and RdRP with cap analog	Tao et al. 2002	1MWH, 1MUK, 1N1H, 1N35, 1N38
	Rotavirus SA11 VP1 with and without RNA and NTPs	Lu et al. 2008	2R7O, 2R7Q, 2R7R, 2R7S, 2R7T, 2R7U, 2R7V, 2R7W, 2R7X
	CPV capsid and polymerase complex	Li et al. 2017	5H0R, 5H0S
	Birnaviridae		
	IBDV VP1-VP3 C-terminus peptide complex	Garriga et al. 2007	2PUS, 2R70, 2QJ1, 2R72
	IBDV VP1	Pan et al. 2007	2PGG
	IPNV VP1	Graham et al. 2011	2YI8, 2YI9, 2YIA, 2YIB
	IPNV VP1-VP3 peptide complex	Bahar et al. 2013	3ZED
	Picobirnaviridae		
	hPBV RdRP	Collier et al. 2016	5I61, 5I62

(continued)

Table 3.1 (continued)

	Cystoviridae		
	φ6 Initiation complex	Butcher et al. 2001	1HHS, 1HHT, 1HI0, 1HI1, 1HI8
		Salgado et al. 2004	1UVL, 1UVN, 1UVI, 1UVJ, 1UVK, 1UVM
	φ12 P2 protein	Ren et al. 2013	4GZK, 4IEG
-RNA	*Bunyaviridae*		
	LACV L protein with and without viral RNA	Gerlach et al. 2015	5AMR, 5AMQ
	Orthomyxoviridae		
	Influenza A and influenza B virus RdRP	Reich et al. 2014	4WSA, 4WRT
	Influenza C virus RdRP	Hengrung et al. 2015	5D98, 5D9A
	Rhabdoviridae		
	VSV L protein	Liang et al. 2015	5A22

these molecules, which imaginatively looks like a cupped right hand with "fingers," "palm," and "thumb" subdomains, provides the correct geometrical arrangement of substrate molecules and metal ions at the active site for catalysis (Fig. 3.1). In addition, RdRPs show unique extensive interactions between the fingers and thumb subdomains, which completely encircle the active site of the enzyme and contribute to the formation of a well-defined channel, where the template binds.

The RdRPs from picornavirus and calicivirus are the smallest polymerases known, and their highly conserved architecture is found at the core of other larger viral RdRPs of known structure (Fig. 3.1). In the larger RdRPs, the fingers, palm, and thumb subdomains are surrounded by elaborate additional elements derived from long N- and/or C-terminal extensions, which in most cases have other enzymatic activities, required for replication and/or transcription of the viral RNA (Fig. 3.1). This chapter summarizes the structural and biochemical studies of different viral RdRPs reported during the past years (Table 3.1). A particular emphasis will be placed on the structure-function relationships for different RdRPs that cover representatives of major virus families, displaying different replication mechanisms. The functional roles of polymerase self-interaction and interactions of these enzymes with other viral and/or host proteins will be also discussed.

Replication and Transcription Strategies

The genomic RNA of (+)ssRNA viruses acts as mRNA that is directly translated by the cellular ribosomes to produce the viral proteins during the very first steps of infection. Then the RdRP copies the (+)ssRNA strand into a complementary (−) ssRNA strand forming a dsRNA replication intermediate. The (−)ssRNA chain

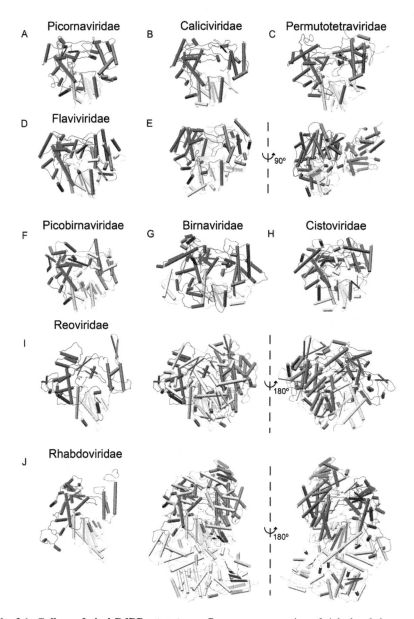

Fig. 3.1 Gallery of viral RdRPs structures. Cartoon representation of right handed arrangements of viral polymerases belonging to different families, highlighting fingers (*red*), palm (*yellow*), and thumb (*green*) subdomains. N-terminal and C-terminal extensions are shown, when present, in slate blue and gray, respectively. (+)ssRNA RdRPs are shown in panels A–E. (A) FMDV (pdb, 2E9R), (**b**) HNV (pdb, 5TSN), (**c**) TaV (pdb, 4XHI), (**d**) HCV (pdb, 5PZL), and (**e**) front and lateral views of full-length NS5 from DENV3 (pdb, 4V0Q), with the N-terminal MTase domain shown in slate *blue*. The RdRPs of distinct dsRNA virus families are shown in panels F–I. (**f**) Human picobirnavirus (hPBV; pdb, 5I61), (G) IBDV (pdb, 2PUS), (H) φ6 (pdb, 4B02), (I) front view of the RdRP core domain of the (−) ssRNA virus VSV (pdb, 5A22). Right panels show the front and back views of the full-length L protein with the accompanying extensions at the N- and C-terminus. The latter containing the capping, connector, MTase, and C-terminal domains

serves then as a template to produce a large excess of (+)RNA which acts as genomic RNA (Ball 2007). Among the various (+)ssRNA virus families, a different set of strategies has been selected for the genome replication and transcription. The common objective is that the genome itself should be stable and recognized by the cellular machinery as mRNA. These strategies include (i) the incorporation of a m7Gppp cap-1 structure on the 5′-end of the genomic RNAs (e.g., flaviviruses but not hepaciviruses and pestiviruses). RNA capping in flaviviruses occurs in a series of reactions involving the multifunctional NS5 protein which comprises a N-terminal S-adenosyl-L-methionine (SAM)-dependent methyltransferase (MTase) domain and a RdRP domain at the C-terminus (Fig. 3.1e). The MTase moiety of NS5 is responsible for the formation of a cap-1 structure through N7 and 2'O-methylation events (Davidson 2009). (ii) The incorporation of internal ribosome entry sites (IRES) at the 5′ untranslated region of the genomic RNA, as occurred in picornavirus, calicivirus, and most probably in permutatetravirus (Jackson et al. 1990; Zeddam et al. 2010). RNA protection in these viruses is achieved through a covalently attached protein at the 5′-end of the genome that in the case of permutotetraviruses is the same replicase molecule which acts as VPg. (iii) These viruses also incorporate poly-A tail loops (picornavirus and calicivirus) or untranslated RNA sequences (flavivirus, permutotetravirus) at the 3′-end of their genomes.

A common feature of dsRNA viruses is that they retain several copies of their RdRP within the icosahedral particle throughout the infectious cycle (McDonald et al. 2009; Estes and Kapikian 2007; Schiff et al. 2007). The vast majority of dsRNA viruses, including the *Reoviridae*, *Totiviridae*, and *Cystoviridae* families, contain a second pseudo-T = 1 core shell that anchors the RdRP and associated enzymes, serving as a platform for RNA synthesis. This core also protects the dsRNA genome by preventing the activation of the RNA-induced antiviral host response. Among this group, the polymerases of the *Reoviridae* family are better characterized. During viral transcription, the RdRPs of reoviruses use the minusstrand of dsRNA as template for the synthesis of multiple copies of plus-strand RNA. Following their packaging into the core particle, the RdRPs initiate a single round of (−)ssRNA synthesis on each (+)ssRNA, creating progeny virions that contain a complete set of dsRNA genome segments (reviewed in McDonald et al. 2009). Of the dsRNA virus group that do not possess T = 1 cores, the most studied examples belong to the *Birnaviridae* family. In these viruses, the bisegmented dsRNA genome, covalently attached to the VP1 polymerase (acting as VPg), is bound to the protein VP3, which acts as a nucleocapsid protein, mimicking that which occurs in many (−)ssRNA viruses. Two or more copies of the bipartite genome are encapsidated along with isolated copies of the VP1 RdRP into a singleshelled *T* = 13 capsid (Luque et al. 2009a; b). The VPg-linked genome of birnaviruses also appears to be reminiscent of the *Picornaviridae* and *Caliciviridae* families of (+)ssRNA viruses.

(−)ssRNA viruses can be broadly categorized as segmented and non-segmented. Orthomyxoviruses such as influenza A contain six to eight RNA genomic segments, bunyaviruses such as hantavirus contain three, and arenaviruses such as Lassa virus contain two segmented RNAs. Non-segmented viruses (*Mononegavirales*) comprise some of the most lethal human and animal pathogens, including Ebola virus and rabies virus from the *Filoviridae* and *Rhabdoviridae* families, respectively. Our current structural understanding of RNA synthesis in NSVs comes principally from the recently solved structures of the trimeric polymerase complex from influenza viruses A, B, and C polymerases (Pflug et al. 2014; Reich et al. 2014; Hengrung et al. 2015), the monomeric L protein from La Crosse orthobunyavirus (LACV) (Gerlach et al. 2015), and the L protein from the rhabdovirus, a vesicular stomatitis virus (VSV) which is a representative from the *Mononegavirales* group (Liang et al. 2015). All these structures, which appeared after many years of intense research in laboratories worldwide, revealed that the architecture of the RdRP core remains essentially invariant and closely resembles the corresponding enzymatic regions of dsRNA virus polymerases (Fig. 3.1). Moreover, a nucleocapsid (N) protein coat covers the genomic RNA, and the viral polymerases use this N-RNA complex as template, rather than uncoated RNA (Liang et al. 2015; Das and Arnold 2015; te Velthuis and Fodor 2016). During RNA synthesis, a few subunits of N dissociate from the template RNA for access to the catalytic site of RdRP and then reassociate as the process continues (Albertini et al. 2006; Green et al. 2006). Besides replicating the viral genome, polymerases of (−)ssRNA viruses also transcribe the positive-sense viral mRNAs using the same N-RNA template. To initiate translation from viral mRNAs, addition of a type-1 cap at the 5′-end and the polyadenylation of a 3′-terminal tail are mandatory. Non-segmented and segmented (−)ssRNA viruses differ significantly in how this is achieved. All segmented (−)ssRNA viruses employ a "cap-snatching" mechanism, involving host mRNA binding and cleavage to create a short-capped primer (Reich et al. 2014; Plotch et al. 1981). By contrast, the L proteins of nonsegmented viruses use the enzymatic activity of their GDP polyribonucleotidyl transferase (PRNTase) domain to catalyze an unusual sequence of capping reactions (Li et al. 2008; Ogino and Banerjee 2007). The structure and function of (−)ssRNA virus RdRPs and, in particular, that of influenza virus are extensively reviewed in Chap. 5.

RNA-Dependent RNA Polymerase Core

The closed "right-hand" design of RdRPs encircles seven structural motifs (A to G; Fig. 3.2a), containing highly conserved amino acids that are essential for polymerase function (Bruenn 2003); Ferrer-Orta et al. 2006a). Fingers, palm, and thumb subdomains of RdRPs collaborate with each other, supporting the binding of RNA and NTPs. In fact, the structures of ternary RdRP-RNA-NTP catalytic complexes revealed the presence of three well-defined channels in the polymerase structures of all (+)ssRNA and also in a number of dsRNA viruses (Ferrer-Orta et al. 2006a). These channels serve as the entry paths for template (template channel) and for nucleoside triphosphates (NTP channel) and as the exit path for the dsRNA product (dsRNA channel) (Fig. 3.2b, c). The N-and C-terminal extensions that surround the RdRP cores of reovirus RdRPs, as well as that of (−)ssRNA viruses (Fig. 3.1i, j),

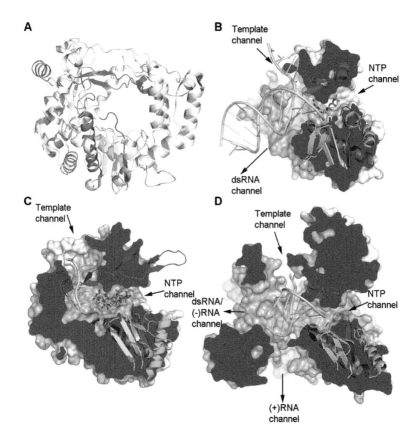

Fig. 3.2 Conserved structural elements in the RNA virus polymerase. Overall structure of a viral RdRP. (**a**) Ribbon representation of a typical RdRP PV 3D^{pol} (pdb: 3OL6). The seven conserved motifs are indicated in different colors: motif **a**, *red*; motif **b**, *dark green*; motif **c**, *yellow*; motif **d**, *cyan*; motif **e**, wheat; motif **f**, *blue*; and motif **g**, *magenta*. (**b–d**) Lateral views of surface representations of three representative enzymes (gray semitransparent). The surfaces have been cut to expose the channels that are the entry and exit sites of the different substrates and reaction products. The structural elements that support motifs A to G are also shown as ribbons, colored as in panel A, (**b**) PV elongation complex (pdb, 3OLB), (**c**) the bacteriophage φ6 initiation complex (pdb,1HI0), and (**d**) reovirus λ3 RdRP-RNA complex (pdb,1 N35)

create globular, cage-like structures in which the buried active sites are connected to the exterior through four well-defined channels (Fig. 3.2d) (Tao et al. 2002; McDonald et al. 2009; Reguera et al. 2016; see also Sect. "Additional Domains Linked to Polymerase Proteins").

The NTP and template entry channels meet at the catalytic site located in palm subdomain (Figs. 3.2b–d). The palm architecture, composed of a three-stranded antiparallel β-sheet core, flanked by three α helices, is the most highly conserved feature not only in RdRPs but also among all known TdPPs (Gorbalenya et al. 2002; O'Reilly and Kao 1998). It contains four structural motifs, arranged in sequential order A-B-C-D-E from amino to carboxyl terminus (Fig. 3.3a). Motif A is located

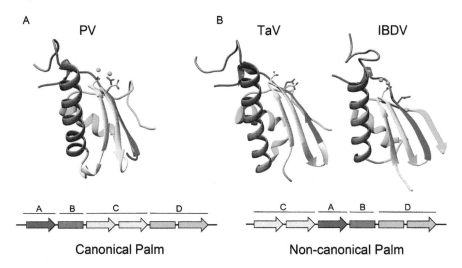

Fig. 3.3 Architecture palm subdomains. Cartoon of secondary structure elements forming the palm subdomain of viral RdRPs highlighting the motifs **a–d**, colored as in Fig. 3.2. (**a**) The canonical palm of the (+)ssRNA virus PV RdRP. The Mg^{+2} ions interacting with the catalytic active site residues are shown in green. (**b**) The noncanonical architecture of the palm subdomains in the Thosea asigna virus (TaV) (*left panel*) and in IBDV (*right panel*) RdRPs

at the end of a β-strand in the central core and has the consensus sequence $DX_{4-5}D$, while motif C is at the top of a β-hairpin, containing the triplet sequence XDD (Figs. 3.3c and 3.4a). In TdPPs other than RdRPs, only the amino terminal aspartic acid residues in motifs A (D_A) and C (D_C) are conserved. These acidic residues bind divalent ions and are crucial for catalysis (Ferrer-Orta and Verdaguer 2009; te Velthuis 2014). In RdRPs, the second aspartate of motif A is involved in the selection of NTPs over 2′ deoxyribonucleotide triphosphates (2′d-NTPs) by hydrogen bonding to the 2′ and 3′ hydroxyl groups of the incoming nucleotide (Gohara et al. 2004; Fig. 3.4b). Motif B forms an α-helix that packs against one strand of the β-sheet core. A conserved hydrophilic residue, asparagine in (+)ssRNA or serine in dsRNA viruses, also participates in NTP selection forming a critical hydrogen bond with the 2′-OH of the incoming NTP (Fig. 3.4b). Motif D is comprised by an α-helix and a short loop that bends back around to form the fourth strand of the β-sheet core (Fig. 3.4a, b). The signature of this motif is a basic residue (lysine or histidine) that appears to be protonated during the polymerization reaction and is involved in the regulation of the catalytic efficiency and fidelity (Castro et al. 2009; Yang et al. 2012; Verdaguer and Ferrer-Orta 2012). Motif E forms a tight loop which lies at the junction between the palm and thumb subdomains. The turn of this loop projects into the active site cavity, contributing to the positioning of the 3′ end of the RNA primer strand for attack on the α-phosphate of the NTP during phosphoryl transfer (Figs. 3.2a–d). The structure of HIV RT catalytic complex showed that residues immediately following the E motif acted as a pivot point for the thumb subdomain movement upon template-primer binding (Huang et al. 1998). However, the closed

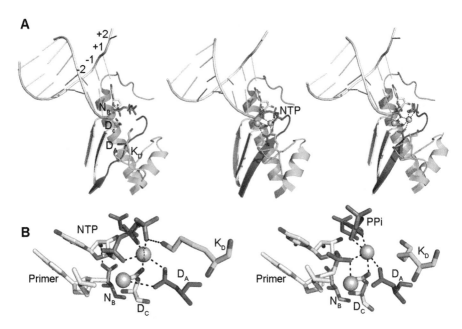

Fig. 3.4 The replication elongation process. Sequential structures illustrating the movement of the different residues within the palm domain in a RdRP-RNA-NTP open ternary complex (*left*), a RdRP-RNA-NTP closed ternary complex (*middle*), and RdRP-RNA-PPi closed ternary complex (*right*). The different structures correspond to the 3D^{pol}-RNA-CTP open complex from PV (pdb, 3OLB), the RdRP-RNA-CTP complex from Norwalk virus (pdb, 3BSO), and 3D–RNA-CTP closed complex from poliovirus (pdb, 3OL7). (**b**) The active site interactions in the closed (*left*) and open (*right*) elongation complexes

right hand of RdRPs prevents such large conformational changes. In fact, the sequential structures of different enterovirus catalytic complexes (Gong and Peersen 2010; Gong et al. 2013; Shu and Gong 2016) showed that a subtle movement of the strand containing the catalytic motif A relative to the strands harboring motif C is essential to catalysis and dependent upon correct NTP binding (Fig. 3.4a; see also Sect. "Replication Elongation and Regulation" for details).

Exceptions to the canonical A-B-C-D organization have been reported in members of the *Birnaviridae* and *Permutotetraviridae* families of dsRNA and (+)ssRNA viruses, respectively. In these enzymes, motif C is located upstream of motif A forming a noncanonical C-A-B-D arrangement with a unique connectivity of the major structural elements of the active site (Gorbalenya et al. 2002; Zeddam et al. 2010). However, the crystal structures of permuted RdRPs from two members of the *Birnaviridae* family, infectious bursal disease virus (IBDV) and infectious pancreatic necrosis virus (IPNV), and that of the permutotetravirus *Thosea asigna* virus (TaV) revealed that in spite of their permuted connectivity, the overall architecture of their catalytic sites is identical to those of canonical RdRPs (Garriga et al. 2007; Pan et al. 2007; Graham et al. 2011; Ferrero et al. 2015) (Figs. 3.3a, b).

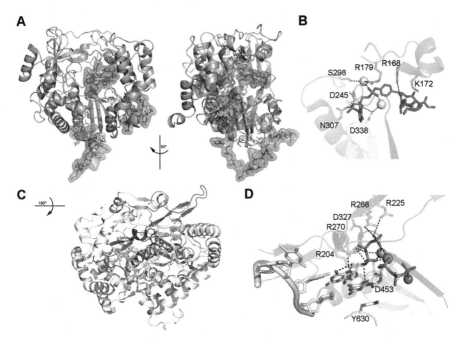

Fig. 3.5 Replication initiation complex. (a) Comparison of identified VPg binding sites in picornavirus 3D^{pol}. The structure of the FMDV 3D^{pol} in complex with an uridylylated VPg (pdb: 2F8E) has been used as the 3D^{pol} model, represented as a gray cartoon. The bound VPgs from FMDV, CVB3 (pdb, 3CDW), and EV71 (pdb, 4IKA), are shown as sticks and semitransparent surfaces in slate *blue*, *dark green*, and *orange*, respectively. (b) Details of the interactions seen in the active site of the FMDV 3D^{pol} during uridylylation (pdb: 2F8E). The N-terminal residues of VPg and the UMP molecule covalently linked to the VPg Y3 side chain are shown in atom type sticks, the divalent cations as white spheres, and the 3D^{pol} side chains involved in the uridylylation reaction represented in sticks and explicitly labeled. The structural conserved motifs **a**, **b**, **c**, and **f** are colored as in Fig. 3.2. (c) Top-down view of bacteriophage φ6 RdRP initiation complex (pdb: 1HI0) with a template RNA shown in pale *cyan* and the two GTPs initiating nucleotides depicted in atom type sticks. The seven motifs **a** to **e** are also shown, colored as in Fig. 3.2. (d) Close-up of the active site residues involved in interactions in the de novo initiation complex. Residues are represented in sticks and colored as in panel **c**

The fingers subdomain, located at the RdRP N-terminus, contains two conserved motifs: motif G, with a consensus sequence T/SX$_{1-2}$GP (X symbolize any residue), is located in a loop outlining the template channel entry, and motif F, defined as R-X$_{1-2}$-I/L. The conserved R, together with other partially conserved basic residues form the roof of the NTP channel (Figs. 3.2b–d).

The architecture of the thumb domain, located at the C-terminus of the RdRP core, is poorly conserved (Fig. 3.1). Picornavirus and calicivirus RdRPs have small thumbs, allowing the formation of a large central cleft in the front side of the molecule, which facilitates the accommodation of the newly synthesized dsRNA during the elongation process (Figs. 3.1a, b and 3.2b). This cavity can also accommodate the protein primer during the initiation of RNA replication (Fig. 3.5a)

(Ferrer-Orta et al. 2006b). In contrast, the RdRPs of flaviviruses and permutotetra-viruses, as well as, those of all dsRNA and (−)ssRNA virus families known have significantly larger thumb subdomains (Figs. 3.1c–j). These large thumbs also contain elements that protrude into the template channel that serve as the priming platforms that stabilize the de novo initiating complexes (Butcher et al. 2001; Tao et al. 2002; Appleby et al. 2015) (see also Sect. "Initiation of RNA Synthesis").

Initiation of RNA Synthesis

Correct initiation of RNA synthesis is crucial to maintain the integrity of the viral genome. Although diverse RNA viruses use several replication scenarios, there are only two principally different mechanisms by which RNA polymerases can initiate RNA synthesis: primer dependent initiation or de novo initiation (van Dijk et al. 2004; Kao et al. 2001). RNA viruses can use either one or sometimes both mechanisms.

Primer-Dependent Initiation

The picornavirus 3D^{pol} is a good representative of RdRPs using a primer-dependent replication initiation mechanism. In this mechanism, an oligonucleotide or a viral encoded protein is the provider of the 3′-hydroxyl group for the addition of the next nucleotide. Picornavirus VPg (or 3B) is a short polypeptide ~20 amino acids long, covalently attached to the 5′-end of the viral genome that not only protects the RNA genome from degradation but also serves as a primer for the synthesis of both negative and positive RNA strands (Paul et al. 1998, 2003; Pathak et al. 2008). Hence, picornavirus RNA synthesis is initiated by the successive attachment of two UMP molecules to the hydroxyl group of a strictly conserved tyrosine residue of VPg. Then, the VPg-pUpU molecule primes the synthesis of the complementary strand (reviewed in Paul and Wimmer 2015). The uridylylation reaction is also catalyzed by 3D^{pol}, employing the carboxylic groups present at the active site (D$_A$ and D$_C$) and using as a template an AA-containing RNA from the 3′-poly(A) tail or from the *cis-acting replication element* (*Cre*) located in different regions of the viral genome. However, in caliciviruses it has been described that the nucleotidylylation of VPg can also occur in absence of templates (Goodfellow 2011).

VPg binding and uridylylation in picornaviruses have been extensively characterized by biochemical and structural studies in different members of the family (PV, HRV16, FMDV, CVB3, and EV71). These studies indicated the presence of three different VPg binding sites on 3D^{pol} (Lyle et al. 2002; Appleby et al. 2005; Ferrer-Orta et al. 2006a; Gruez et al. 2008; Chen et al. 2013; Ferrer-Orta et al. 2006b; Fig. 3.5a). Remarkably, while most picornaviruses express only a single VPg protein, FMDV possesses three similar copies of VPg (VPg1, VPg2, and VPg3), and all of them were found linked to the viral genome (Forss and Schaller 1982; King et al. 1980).

The structures of two complexes between FMDV 3D^{pol} and VPg1, showing both the uridylylated and a non-uridylylated forms of VPg, revealed that the primer protein accessed the active site through the large RNA binding cleft, occupying the binding site of the template-primer RNAs in the elongation complexes. Conserved residues in the fingers, palm, and thumb domains of the polymerase contacted the VPg peptide, contributing to the stabilization of the complex in its binding cavity. Mutational analyses of the 3D^{pol} and VPg interacting residues together with functional studies of these mutants showed drastic effects in uridylylation (Ferrer-Orta et al. 2006b). The structures show that the VPg N-terminal portion is located close to the NTP channel, approaching the hydroxyl group of the Y3 side chain to the 3D^{pol} catalytic residues D_A and D_C (Fig. 3.5b); then the uridylylation reaction appears to follow a similar mechanism to that described for the nucleotidyl transfer during RNA elongation (Steitz 1998). This "front-loading" model for VPg binding, compatible with a mechanism of VPg uridylylation in *cis*, was supported by the crystal structures of HRV16 and of the PV 3CD precursor (Appleby et al. 2005; Marcotte et al. 2007).

A second VPg binding site was found in the structure of CVB3, where the C-terminal half of VPg was located at the base of the thumb subdomain of 3D^{pol} in an orientation that did not allow uridylylation in *cis* (Fig. 3.5a). Based on these data, authors proposed that VPg bound at this site could be uridylylated in *trans* by another 3D^{pol} molecule or alternatively it could play a structural role, stabilizing the uridylylation complex (Gruez et al. 2008).

The third VPg binding site was shown at the bottom of the palm domain of the EV71 3D^{pol} where the VPg residues displayed an extended V-shape conformation, extending from the front side of the catalytic center to the back side of the enzyme (Chen et al. 2013; Fig. 3.5a). In this complex, Y3 is buried at the base of the palm of 3D^{pol}, and a conformational rearrangement would be necessary in order to expose this residue for uridylylation by other polymerase molecule.

Taking into account the important sequence homology between the picornaviral VPg sequences and the high similarities existing in the 3Dpol structures, it seems reasonable to assume that all picornaviral VPgs would bind to the same place of 3D^{pol}, for uridylylation. However, the structural data show wide variability of interactions between VPg and 3D^{pol}. Looking at the structures, it is tempting to speculate that the three different VPg binding sites observed might reflect distinct binding positions of VPg to both 3D^{pol} and its precursor 3CD at different stages of the virus replication initiation process. Further efforts directed toward the structural characterization of higher-order complexes, involving the polymerase 3D^{pol} and different proteins or protein precursors, VPg, 3AB, 3C^{pro}, 3CD, together with RNA templates, would be necessary to facilitate the understanding of the molecular events underlying the initiation of RNA synthesis in this group of viruses.

De Novo Initiation

In the de novo synthesis, one initiation nucleotide provides the 3'-hydroxyl, serving as a primer for the addition of the next nucleotide to form the first phosphodiester bond of the product strand. Structural comparisons showed that RdRPs of viruses using de novo initiation (e.g., flaviviruses, reoviruses, and all known (−)ssRNA virus families) share distinctive characteristic ensuring competent replication initiation that includes larger thumb subdomains containing structural elements that fill most of the active site cavity (Figs. 3.1d–i and 3.2c, d), providing a platform for the initiating nucleotides, and also serving as a barrier that prevents chain elongation (Fig. 3.5b) (Butcher et al. 2001; reviewed in Lescar and Canard 2009; Ferrer-Orta et al. 2006a). Therefore a large conformational change, moving the initiation platform away from the active site, would be required to allow the transition from the initiation to the elongation states of RNA synthesis (Butcher et al. 2001; Mosley et al. 2012; Appleby et al. 2015).

Replication Elongation and Regulation

RNA elongation is based on sequential nucleotidyl transfer reactions, involving the nucleophilic attack of the NTP α-phosphate by the 3'-OH of the primer strand. The process leads to phosphodiester bond formation and release of the pyrophosphate side product. According to the two-metal-ion mechanism of catalysis (Steitz 1998), metal B binds the triphosphate moiety of the NTP, whereas metal A would reduce the 3'-OH primer pKa allowing its deprotonation. The acceptor for this proton has been shown to be a water molecule (Nakamura et al. 2012).

The plethora of replication elongation complex structures currently available has provided great insights into the conformational rearrangements associated with the different steps occurring during the elongation process of RNA synthesis (Fig. 3.4). The most complete pictures have been provided by the sequential structures of catalytic complexes determined for different members of the *Picornaviridae* and *Caliciviridae* families of (+)ssRNA viruses (Ferrer-Orta et al. 2004, 2007; Gong and Peersen 2010; Gong et al. 2013; Shu and Gong 2016; Ng et al. 2004; Zamyatkin et al. 2008; revised in Peersen 2017 and Deval et al. 2017). Based on these data, it has been proposed that full catalytic cycle that takes place repeatedly during processive elongation can be divided into six major states: S1, template/primer binding; S2, NTP binding; S3, active site closure; S4, catalysis; S5 opening of the active site; and S6 translocation and pyrophosphate release (Gong and Peersen 2010). Some of these steps were trapped in crystal structures and are summarized in Fig. 3.4a.

Initially the RdRP binds the RNA template-primer, with the T + 1 templating base locked in a position above the active site completely stacked on the upstream duplex. In this state, the polymerase remains essentially in the same conformations than in the unbound form. This conformation is characterized by an open active site conformation that is described by a partial formation of the central three-stranded

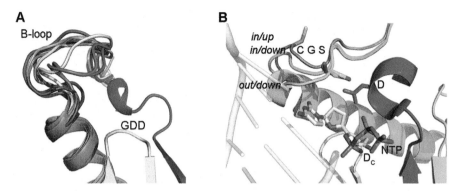

Fig. 3.6 Conformational flexibility of the B-loop. (**a**) Superposition of the diverse conformations found for the B-loop in different RdRP structures. The structural conserved motifs A, B, and C are represented as red, green, and yellow ribbons, respectively. The B-loop is shown in different colors from red (*up* conformation) to *blue* (*down* conformation). The RdRPs used in the superimposition follow NV, Mg^{2+} bound form (pdb, 1SH3, chain A), (PV unbound form (pdb, 1RA6), FMDV, RNAcomplex (pdb, 1WNE), PV C290V (pdb, 4NLP) IBDV VP1 + VP3Cterm. (pdb, 2R70), NV Mg^{2+} bound form (pdb, 1SH3, chain B), PV C290F (pdb, 4NLQ) IBDV VP1 unbound form (pdb, 2PUS), and FMDV K18E (pdb, 4WYL). (**b**) Superimposition of the three alternative conformations described for the PV $3D^{pol}$: *in/up* (dark green; pdb, 3OLB), *in/down* (green; pdb, 4NLO), and *out/down* (*light green*; pdb, 4NLQ). The RNA template primer is represented as a semitransparent cartoon, and the incoming NTP molecule is shown as sticks

β-sheet of the palm subdomain, containing motifs A and C. The initial binding of incoming NTP occurs also in this polymerase conformation. This NTP enters via the NTP channel located at the back side of the enzyme (Figs. 3.2 and 3.4a). This channel contains a number of basic residues, organized to interact with the nucleotide triphosphates. NTP selection takes place via specific interactions established between the ribose 2′ and 3′ hydroxyl groups of the incoming NTP and three conserved residues of motifs B (S and N) and a second conserved D, located at the N-terminus of motif A (Fig. 3.4b). These interactions would stabilize the subtle restructuring of the palm domain, resulting in the formation of a functional closed active site. An incorrect nucleotide can bind, but its ribose hydroxyls will not be correctly positioned for active site closure; in consequence, the incorporation efficiency will be reduced. After catalysis, the active site opens again by reversing the movement of motif A in the palm, and this movement is followed by a distinct translocation step to complete the catalytic cycle and position the next templating base in the active site (Peersen 2017). Structural comparisons between (+)ssRNA, −ssRNA, and dsRNA RdRP palm domains show that dsRNA and (−)ssRNA palms are completely structured in the RdRP unbound form, suggesting that this palm-based active site closure mechanism is exclusive of (+)ssRNA viruses.

Besides its function in the selection of incoming nucleotides, motif B and, in particular, the flexible loop located at the N-terminus of the α-helix (the B-loop; Fig. 3.6) also play a central role in template binding. This loop is able to adopt different conformations when it binds to different templates and incoming nucleo-tides, being one the most flexible elements of the active site of RdRPs in picornavi-ruses, as well as in other viral families (reviewed in Garriga et al. 2013). Structural comparisons evidenced large movements of the B-loop, ranging from a conforma-tion in which the loop is packed against the finger domain leaving the active site cavity fully accessible for template entry to a configuration where the loop pro-trudes toward the catalytic cavity and clashes with the template RNA (Fig. 3.6a). The key residue of this flexible region is a strictly conserved glycine, which acts as a hinge for the movement. Furthermore, the pattern of interactions established between the B-loop and the RNA phosphodiester backbone of the upstream duplex, between the −1 and −2 nucleotides, suggested the involvement of the B-loop in translocation (Ferrer-Orta et al. 2007; Gong and Peersen 2010). In fact, an extensive structural and functional work recently performed in PV evidenced the direct involvement of the conserved B-loop sequence S-G-C in polymerase translocation (Sholders and Peersen 2014). These studies identified two different movements for the B-loop: *in/out*, when the loop is packed against the fingers (*in*) or when the loop is protruding into the active site (*out* and *up/down* (*up* when residue C is buried into a hydrophobic pocket or *down* when C is exposed to the solvent). On the basis of these movements, the B-loop can adopt three conformations: *in/up*, *in/down*, *out/down* (Fig. 3.6b). Initially, an unbound structure shows an *in/up* conformation of the B-loop, allowing the rNTP entry. When the polymerase binds the incoming NTP, the B-loop changes to an *in/down* conformation; in this position the S residue flips down toward the active site establishing interactions with the conserved active site D_A, contributing to the nucleotide selection. After nucleotide incorporation, a move-ment of the B-loop to *out/down* would facilitate translocation among the RNA and prevent backtracking after translocation (Sholders and Peersen 2014)(Fig. 3.4a).

Finally, growing amounts of data indicate that the conformational changes in motif D play a crucial role in determining both efficiency and fidelity of nucleotide addition. Biochemical studies of nucleotidyl transfer reactions carried out by single subunit DNA polymerases, RTs or RdRPs, show that two protons (not just one as was expected; see above) are transferred during the reaction; this second proton derives from a basic residue of the polymerase and is transferred to the pyrophos-phate leaving the group. Pyrophosphate protonation is not essential for the polym-erization reaction but contributes from 50- to 2000-fold to the rate of nucleotide addition (Castro et al. 2009). Mutagenic and kinetic analyses of nucleotide incorpo-ration in different polymerases indicate that the proton donor is a conserved K resi-due within motif D of RdRPs and NMR; studies revealed that the reorganization of motif D in order to facilitate the second proton transfer is dependent to the correct formation of ternary RdRP-RNA-NTP complex (Castro et al. 2009; Shen et al. 2012; Yang et al. 2012). The protonation state of this conserved K is critical to achieve the closed conformation of the active site, and the ability of the motif D to adopt the catalytically competent conformation seems also affected by binding of an incorrect nucleotide. Altogether, these studies relate the efficiency and fidelity of nucleotide addition with the conformation of motif D.

Additional Domains Linked to Polymerase and Coordination of their Functions

Extra N- and C-terminal decorations, folded as independent domains, often surround the RdRP cores (Fig. 3.1e, j). The polymerases of pestivirus and flavivirus, within the *Flaviviridae* family, are illustrative examples. The structure of bovine viral diarrhea virus (BVDV) showed an N-terminal domain (~130 residues) that is located over the thumb and interacts with the fingertip region through a β-hairpin motif either from the same polypeptide chain or from the neighboring protein (Choi et al. 2004, 2006) (Fig. 3.7). The function of this N-terminal domain is not known, but the deletion of the first 106 residues causes 90% loss of activity, indicating its involvement in the RdRP reaction (Lai et al. 1999). The observed domainswapping and its position near the template channel entry suggest a possible role in template translocation (Choi and Rossmann 2009). Flaviviruses encode the 5'-RNA methyltransferase (MTase) and the RdRP domains in a single polypeptide chain (Fig. 3.1e), indicating that RNA synthesis and 5'-capping would be coupled in flaviviral genome replication. Hence understanding whether the two domains cooperate to regulate the RNA replication activity is a fundamental question. Interactions between the two domains have been demonstrated for dengue virus (DENV) and west Nile virus (WNV) by mutational and in vitro experiments (Potisopon et al. 2014; Lim et al. 2013). Stimulatory effects by the MTase domain on the overall steady-state RdRP activity have also been demonstrated for the DENV NS5 protein. This stimulation occurred both during initiation and elongation (Potisopon et al. 2014; Lim et al. 2013; revised in Selisko et al. 2014). Small-angle X-ray scattering (SAXS) data showed that NS5 was mainly extended in solution (Bussetta and Choi 2012; Saw et al. 2015; Subramanian Manimekalai et al. 2016). However the crystal structures of full-length NS5, determined for three different flaviviruses, DENV, Japanese encephalitis virus (JEV), and and Zika virus (ZIKV), showed an overall compact conformation of the enzyme (Lu and Gong 2013; Zao et al. 2015b; Klema et al. 2016; Upadhyay et al. 2017). Structural comparisons also showed that whereas in the three proteins, the RdRP domain uses almost the same backside surface to contact the MTase, the DENV MTase appears rotated about 110° relative to the JEV or ZIKV arrangements, forming a slightly larger interdomain interface with the RdRP (Fig. 3.8). Interestingly, the MTase residues involved in interdomain interactions in JEV and ZIKV monomers (Lu and Gong 2013; Upadhyay et al. 2017) are present at the monomer-monomer interface in the DENV dimeric structure shown by Klema et al. 2016 (Fig. 3.7). One fascinating feature of this structure is the possibility of NS5 autoregulation through its dimeric assembly (see also Sect. "RdRP Self-Interactions").

As mentioned in Sect. "The RdRP Core," the structures of the RdRPs from two dsRNA viruses, members of the *Reoviridae* family, revealed that the polymerase has a closed cage-like structure consisting of a central RdRP core that is sandwiched in between large N-terminal and C-terminal domains (Fig. 3.1i). The N-terminal domains surround the fingers and thumb subdomains, effectively closing the enzymes.

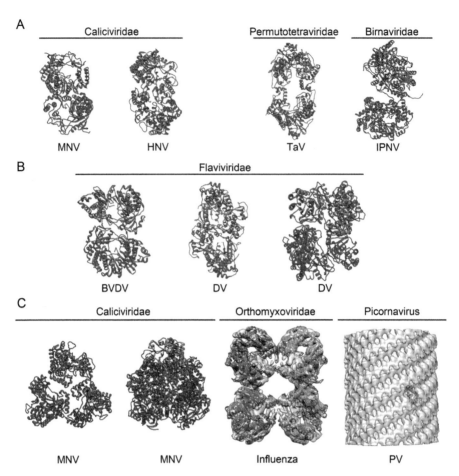

Fig. 3.7 Structures of RdRP assemblies in different viral families. (**a**) Dimeric arrangement of RdRPs in two members of the *Caliciviridae* family: murine norovirus, MNV (pdb, 3QID) and human norovirus; HNV (pdb, 2B43) and in two representatives of the *Permutotetraviridae* TaV (pdb, 4XHI) and *Birnaviridae*: infectious pancreatic necrosis virus; and IPNV (pdb, 2YIB). (**b**) Dimers found in the RdRP structure of bovine viral diarrhea virus, BVDV (pdb, 1S4F); in the dengue virus, DV RdRP domain (pdb, 4C11); and in the full-length NS5 structure of DV (pdb, 5CCV), members of the *Flaviviridae* family. (C) Oligomeric assemblies seen in representatives of virus families: the trimeric and hexameric RdRP structures of the calicivirus MNV (pdb: 3QID). The tetrameric structure of the influenza virus polymerase (*Orthomixoviridae*; pdb, 3J9B, EMD, 6203). Tubular arrangement of the PV 3D^{pol} (EMD: 2270)

The C-terminal domains have a bracelet-like shape, resembling the sliding clamps of DNA polymerases that encircle templates and contribute to the efficiency of nucleotide addition (Reviewed in Mac Donald et al. 2009). The catalytic center is connected to the exterior by four channels (Fig. 3.2d); the template and nucleotide entry channels are located at equivalent positions of other RdRPs, but the reovirus polymerases possess two other channels serving as RNA exit paths (Fig. 3.2d).

A   B

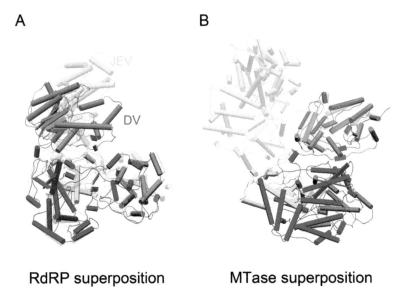

RdRP superposition   MTase superposition

Fig. 3.8 Additional domains bound to RdRPs. The NS5 protein of flaviviruses has a distinctive organization with a capping-related MTase naturally fused to the RdRP through a flexible linker. Comparisons with the NS5 structures from JEV (pdb, 4K6M) and DENV3 (pdb, 4V0Q) show that in both cases the MTase domain is located above the fingers subdomain at the backside of the RdRP. However, structural superimpositions show that both the relative orientation and intermolecular interactions between the MTase and RdRP domains in these two viruses differ significantly. (**a**) Superimposition of the RdRP domains and (**b**) superimposition of the MTase domains. Structural comparison with the ZIKV NS5 structure, recently solved (pdb: 5TFR), shows that ZIKV and JEV NS5 proteins are essentially identical (Upadhyay et al. 2017)

Only one RNA exit tunnel is believed to operate during replication; this tunnel extends through the bracelet domain and represents the pathway for release of newly made dsRNA to the core interior. During transcription, both RNA exit tunnels of the polymerase are expected to function. One is used for the release of the minus-strand template RNA from the polymerase and is the same tunnel used for the release of the dsRNA product during replication (dsRNA/(−)RNA exit channel). The other is used for the release of newly synthesized (+)ssRNAs and represents a conduit that directs nascent transcripts out of the core (reviewed in Gridley and Patton 2014). Another unique feature of reovirus polymerases is the presence of a shallow cleft near the template entry channel that represents an RNA cap-binding site (Tao et al. 2002; Lu et al. 2008). This site may help to recruit capped (+)ssRNA templates during replication and/or provide an anchoring point for the capped 5′-end of the (+)ssRNA of viral genome segments during multiple rounds of transcription (McDonald et al. 2009).

Additional domains linked to the RdRP are also found in the polymerases of (−) ssRNA viruses, e.g., the bunyavirus polymerase (L protein) consists of a N-terminal endonuclease domain joined to a C-terminal RdRP domain (Gerlach et al. 2015). A more complex example is found in the VSV polymerase (Liang et al. 2015).

The 3.8 Å cryo-EM structure of this enzyme shows that the central RdRP domain is surrounded by three globular domains, possessing complementary enzymatic activities that are required for viral mRNA 5' cap synthesis: a 2'-O methyltransferase (MTase), a guanine-N7-MTase, and a polyribonucleotidyl transferase (PRNTase) (Liang et al. 2015). In contrast influenza virus encodes an RdRP in which its multiple activities are separated out into three individual polypeptides, called PA, PB1, and PB2 (Reich et al. 2014). Detailed descriptions of the structure and interactions of influenza virus polymerases as well as other polymerases of negative-stranded RNA viruses are given in Chap. 5.

RdRP Self-Interactions

Protein oligomerization, characterized by reversible protein-protein associations mediated by electrostatic and hydrophobic interactions, hydrogen bonds, or by covalent stabilization by disulfide bonds, is a frequent property observed in proteins of all biological systems. A number of functional advantages (e.g., gain in stability or additional functions) underlay the selection of the oligomeric states of proteins during virus evolution. In the case of viral proteins, these advantages allow the construction of capsids, as well as large replication machineries, with high diversity and complex architectures, employing minimum genome occupancy. Moreover, the diversity of complexes that can be built from a single viral protein allows adaptation to diverse sets of functions through its ability to generate distinct oligomeric states, as well as tuning and regulation of existing functions. The rapid evolution of viruses has allowed them to develop many tricks to adapt and evolve proteins in the face of rapidly changing selection pressures. RdRPs are not exempt from this behavior; they can form homo-oligomers via RdRP self-interactions or hetero-oligomers, including copies of different proteins. The general tendency is to form symmetrical arrangements of globular or filamentous helicoidal shapes (Fig. 3.7).

The presence of RdRP-RdRP interactions to form dimers and also higher-order oligomers and the role of these assemblies in modulation of the enzyme activity have been extensively reported for several families of human and animal (+)ssRNA viruses, including *Picornaviridae*: PV and FMDV (Hobson et al. 2001; Spagnolo et al. 2010; Bentham et al. 2012); *Caliciviridae*, *Feline calicivirus* (FCV) and human and murine noroviruses (HNV, MNV) (Kaiser et al. 2006; Ng KK et al. 2004; Högbom et al. 2009); *Flaviviridae*, hepatitis C virus (HCV) (Wang et al. 2002; Chinnaswamy et al. 2010; Clemente-Casares et al. 2011) and DENV (Klema et al. 2016); and *Permutotetraviridae*, TaV (Ferrero et al. 2015). In addition, RdRP oligomerization was also found in replicases of (−)ssRNA viral families like *Rhabdoviridae*, VSV (Rahmeh et al. 2010); *Orthomyxoviridae*, influenza A virus (Chang et al. 2015); *Paramyxoviridae*, measles virus (MeV), Sendai virus (SV), and parainfluenza virus-3, (PIV-3) (Çevik et al. 2004, 2007; Smallwood and Moyer 2004); and in dsRNA viruses of the *Birnaviridae* family, infectious pancreatic necrosis virus (IPNV) (Bahar et al. 2013). Most of these RdRP oligomers were functionally and structurally studied in vitro. However, intracellular accumulations of oligomeric RdRPs were also observed in vivo or during viral infections in different RNA viruses (e.g., PV (Hobson et al. 2001), SV (Smallwood et al. 2002), Rift Valley fever virus (Zamoto-Niikura et al. 2009), and HNV (Högbom et al. 2009)), indicating the relevance of these quaternary structures in viral replication.

RdRPs structures determined by X-ray crystallography and by the emergent high-resolution cryo-electron microscopy (cryo-EM) are providing fine details about the interactions established in the contact interfaces, revealing the mechanisms and functions associated with this reversible process. According to the structural information available in the PDB and EMDB, RdRPs self-interaction is mostly represented by dimer formation, sometimes reaching the formation of higher-order oligomers. RdRP-RdRP interactions can be clustered in three main groups: (i) lateral interactions, predominantly involving contacts between fingers and thumb subdomain; (ii) interactions mediated by N-terminal tails, using either small secondary structural elements or full independent domains; and (iii) the assembly of these structures to forming higher-order oligomers (Fig. 3.7). Contacts between these domains in oligomer formation may cause small conformational changes that are transferred to the active site as an allosteric regulation or could even modify the accessibility of the substrate channels.

The first example of functional RdRP oligomerization was described in the crystal structure of PV 3D^{pol} (Hansen et al. 1997). The work described two extensive interfaces of interaction, driving the formation of higher-order filamentous structures that explained previous biochemical studies (Pata et al. 1995). Interface I derives from extensive interactions between the thumb subdomain of one molecule and the back of the palm subdomain of the adjacent molecule, and interface II involves contacts of the N-terminal region of one molecule with the thumb and fingers subdomains of the second molecule. Mutational experiments provided further evidences of the functional relevance of interfaces I and II in efficient RNA binding and formation of the catalytic sites, respectively (Hobson et al. 2001). Initial cryo-EM data also demonstrated that the purified PV 3D^{pol} was able to organize two-dimensional lattices and tubular arrangements formed by polymerase fibers (Lyle et al. 2002), and recently, the structure of these assemblies has been characterized at the pseudo-atomic level (Fig. 3.7c) (Tellez et al. 2011; Wang et al. 2013). The planar lattices, forming a ribbon-like structure, consist of linear arrays of dimeric RdRPs supported by strong interactions through the interface I as defined in the crystal structure of 3D^{pol}. These data suggested that poliovirus polymerase could change conformation upon forming oligomers and, in the tubular arrays, becoming these arrays compatible with the RdRP activity (Wang et al. 2013).

The presence of dimers as well as of high-order oligomeric states has also been described in members of the *Caliciviridae* family (Fig. 3.7a and c). The first structure reported from the HNV RdRP NS7 (Ng et al. 2004) suggested the possibility of dimer formation by extensive contact surfaces, involving a hydrophobic patch in the top of thumb subdomain. The hypothesis was later demonstrated by the functional and structural characterization of the NS7 homodimers (Högbom et al. 2009). Further work performed in MNV reinforced the concept that RdRP-RdRP interactions in solution are dynamic and the protein is able to reach small populations of different oligomers, including dimers, trimers, and hexamers (Lee et al. 2011).

The *Flaviviridae* family provides many examples of RdRP oligomerization. Wang and colleagues showed by combining different techniques that the HCV polymerase NS5B acts as a functional oligomer (Wang et al. 2002). The crystal structure revealed two large interacting interfaces involving the thumb region of one molecule and the finger region of its neighbor in the dimeric state, in accordance with the biochemical data. Reinforcing this idea, Chinnaswamy and colleagues established a link between NS5B oligomerization and de novo replication initiation (Chinnaswamy et al. 2010). These studies also showed that primer extension was not affected by dimer formation. Additionally, Clemente-Casares and colleagues addressed this issue in HCV genotypes (1 to 5) confirming the cooperative effect of dimer formation in de novo synthesis. These authors further identified a number of acidic residues within an α-helix from the fingers subdomain and a positive patch of residues from a α-helix of the thumb as the regions involved in dimerization. The electrostatic nature of this interaction explains why this assembly is dependent on ionic strength (Clemente-Casares et al. 2011).

Evidence of dimer formation, affecting RdRP activity, was also shown in the pestivirus BVDV. The crystal structure of BVDV RdRP shows a dimeric arrangement mediated by interactions between the N-terminal domain (residues 78–133) of one molecule and the fingertip region and the thumb domain of the neighbor molecule (Choi et al. 2004). This motif partially occludes the template channel entrance, explaining why some N-terminal truncations cause important changes in polymerase activity (Lai et al. 1999).

The full-length NS5 protein of DENV is able to form dimers in crystals (Zhao et al. 2015b; Klema et al. 2016) or even higher-order oligomers (Klema et al. 2016). The structure of DENV NS5 solved by Klema and colleagues showed eight independent NS5 molecules in the crystal asymmetric unit, organized in four dimers. Two distinct types of interactions were observed: type I dimers, involving the periphery of the MTase domain (opposite to the active site cleft) in one NS5 molecule that contacts the base of the palm subdomain of the second NS5 (Fig. 3.7 c, left panel), were consistently found in two different crystal forms of DENV NS5 (Klema et al. 2016; Zhao et al. 2015a, b). In contrast type II interactions appeared to be the consequence of crystal packing. Site-directed mutagenesis of residues involved in type I and type II interfaces demonstrated that only mutations altering type I dimers drastically reduced viral titters (Klema et al. 2016). In light of these data, the authors suggested that type I dimers would facilitate the coordination between MTase and RdRP activities without requiring the large conformational changes inherent in the monomer model. Moreover, it has been shown that the isolated RdRP domain of DENV is also able to form compact and stable dimers (Fig. 3.7b, central panel) that show enhanced de novo polymerization activity and thermostability (Lim et al. 2013).

Stable functional dimers were also observed in the RdRP structure of TaV, a member of *Permutotetraviridae* family, whose polymerase activity closely resembles that of flavivirus (Ferrero et al. 2015) but posseses a permuted (ABC→CAB) architecture of the palm domain equivalent to that of birnavirus RdRPs (Fig. 3.3). The TaV RdRP dimers are maintained by mutual interactions between the active site cleft and the flexible N-terminal tail of the two molecules involved in interactions (Fig. 3.7a right panel). This end protruding in the central cavity of its neighbor lies in the template channel suggesting a mechanism of regulation of RdRP activity. In fact, in agreement with the RdRP structure, the biochemical data revealed that the deletion of the N-terminal arm (21 residues) abolished the protein dimerization and resulted in a drastic increment in RNA synthesis. In a similar manner, the RdRP of the birnavirus IPNV employs its N-terminal arm to mediate self-interactions (Fig. 3.7a, right panel) (Graham et al. 2011).

Recent high-resolution cryo-EM studies evidenced the presence of oligomers of the heterotrimeric polymerase of influenza virus (Chang et al. 2015) (Fig. 3.7c, right panel). The reengineered version of the multimeric polymerase, composed of full-length PA, PB1, and the 130 N-terminal residues of PB2, exists as a dimer in solution. However, this stable state can be altered in the presence of vRNA or other ssRNAs to spontaneously form a tetrameric RdRP complex. Interestingly, this is a rare case where the oligomeric state of the polymerase depends on substrates such as RNA. Indeed, the tetrameric complex could be reverted upon addition of ssRNA3' suggesting that oligomerization state of influenza RdRP is regulated by binding to a promoter region. Chang and colleagues also determined that an N-terminal region in PB2 (residues 86–130) was crucial for oligomerization. Mutagenesis of residues involved in interactions resulted in the loss of the polymerase activity in vitro and in vivo, supporting the critical role of polymerase oligomerization for influenza virus viability (Chang et al. 2015).

RdRP Interactions with Viral and/or Host Proteins Regulating the Activity

RNA viruses usually associate their RdRPs with other viral proteins or host proteins that act as scaffold to form the replicative complexes in specific intracellular locations such as membranous compartments (reviewed in den Boon et al. 2010 and Harak and Lohmann 2015) or inside capsids (Estrozi et al. 2013; Zhang et al. 2015; Ilca et al. 2015). This association could also complement the RdRP activity, for example, helicases, NTPases, or capping enzymes. In addition, the RdRP protein interactions involving either viral or host proteins would contribute to the fine-tuning of the polymerase activity, ensuring the proper levels of RNA production. Despite the high occurrence of these macromolecular interactions in the virus world, the structural information available for these complexes is still limited, probably due to the transient character of these interactions or to the nature of the RdRP partners that often are membrane proteins. However, the recent methodological advances in high-resolution cryo-EM, including the deployment of direct electron detectors

(Cheng 2015), have enabled investigators to significantly increase the number and resolution of the RdRP viral protein complexes determined.

Structures of RdRP viral protein complexes mainly come from dsRNA viruses. A common feature of many dsRNA viruses is that the polymerases are always capsid associated performing their function of replicating and transcribing the viral genome within this structure. The confinement of RdRP into the viral capsid prevents the activation of the host's RNA-induced defenses. Structures of RdRP-capsid protein complexes have been determined for several members of the *Reoviridae* family, reovirus λ3 (Zhang et al. 2003), rotavirus (Estrozi et al. 2013), cytoplasmic polyhedrosis virus, (CPV) (Zhang et al. 2015; Liu and Cheng 2015), and for a representative of the *Cystoviridae* family, the bacteriophage φ6 (Ilca et al. 2015).

The structures of the transcribing and non-transcribing particles of CPV have been been determined by cryo-EM methods (Liu and Cheng 2015). The single-shelled CPV is one of the simplest dsRNA viruses that pack its segmented dsRNA genome and the transcriptional enzyme complex (TEC) within the capsid. TEC consists of two extensively interacting subunits: an RdRP and an NTPase, VP4. Liu and colleagues showed that the CPV interior has organized dsRNA densities associated with the RdRP that are anchored to the inner surface of the capsid, in a position slightly off center to the fivefold axis (Fig. 3.9a, b). Other densities were observed in close contact with the RdRP, NTP, and template entry channels that were assigned to VP4. The de novo VP4 tracing showed that the structure of this protein is closely related to that of orthoreovirus μ2 protein. Important differences were observed in the interactions between the packed dsRNA with the RdRP bracelet domain in the non-transcribing and transcribing particles. In the non-transcriptional particles, the conformation of the RdRP bracelet domain blocked the template and the transcript exit channels that become open in the transcriptionally active particles. These changes are required for the switch to transcription mode. Additionally, Zhang and colleagues determined the cryo-EM structures of TEC and the dsRNA organization inside quiescent and actively transcribing particles at 5.1 and 4 Å, respectively (Zhang et al. 2015). The resolution achieved was sufficient to define that VP4 was an NTPase and the absence of interaction with dsRNA suggested the lack of helicase activity. Comparisons also showed significant conformational changes between the TEC complexes in the two particles, including the above mentioned conformational changes of the RdRP bracelet domain. In addition, the N-terminal α-helix of the capsid protein interacts with TEC in both particles, inserted in the interface between the NTPase domain of VP4 and the RdRP fingers, directly affecting the TEC conformation. As the capsid protein also interacts with the bracelet domain of the RdRP, the conformational changes observed appear directly related to transcription regulation. (Zhang et al. 2015).

Rotavirus is a triple-layered icosahedral virus containing 11 segments of genomic dsRNA. The inner core is formed by the association of dimers of the structural protein VP2 in a T = 2 capsid. The cryo-EM structure of the entire rotavirus particle revealed that the RdRP (VP1) is anchored slightly offset to the fivefold axes formed by the VP2 core (Estrozi et al. 2013; Fig. 3.9c). Previous studies established that this VP1-VP2 interaction was essential for VP1 activity (Patton et al. 1997;

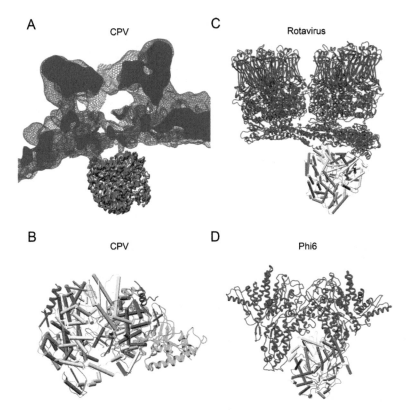

Fig. 3.9 RdRP interactions with other viral proteins. (**a**) Cross section of the cryo-EM map of cytoplasmic polyhedrosis virus (CPV) (EMD-6321) around one of the icosahedral fivefold axes, showing the capsid shell protein (*blue* mesh and in *solid blue* capped surface) bound to the RdRP (*red* solid surface). (**b**) High-resolution structure of ternary complex between the CPV RdRP (pipes and plank representation), the structural protein protein VP4 (*green*), and capsid shell protein (only the helices, directly contacting the RdRP are shown as ribbons in blue) (pdb: 3jb6). (**c**) The rotavirus RdRP VP1 (pipes and plank) bound to rotavirus capsid protein VP2 (blue ribbons). (**d**) P2, φ6 RdRP (pipes and plank representation) bound to capsid protein P1 from an in vitro assembled complex (pdb: 5fj7). In all RdRP structures, N- and C-terminal domains and fingers, palm, and thumb subdomains were highlighted in light *blue, gray, red, yellow*, and *green*, respectively

Tortorici et al. 2003). This dependence can be explained because the VP2-contacting region of VP1 includes residues located near the retracted priming loop that supports the de novo initiation RNA synthesis. Therefore, the VP1-VP2 contacts would induce rearrangements promoting the conformational shift in the priming loop to an extended form allowing RNA synthesis initiation.

Bacteriophage φ6 is another multilayer virus with the inner capsid formed by the P1 protein. The location of the φ6 RdRP, P2, into the P1 shell of in vitro assembled particles by cryo-EM methods, allowed the accurate fitting of the monomeric P2 bound to the icosahedral threefold axes, showing the exact orientation of P2 in the P1 core interior and P1-P2 interactions (Ilca et al. 2015; Fig. 3.9d).

Finally, although viral RdRP-host protein interactions have been extensively studied at the biochemical level (Le Breton et al. 2011; Mairiang et al. 2013; Mas et al. 2016), the structural information on RdRP-host protein complexes is extremely limited. The X-ray structure of the 18 C-terminal residues of DENV NS5 solved in complex with the nuclear importin-alpha (Tay et al. 2016) is one of the very few examples, making evident the absence of structural information in this field.

Conclusions and Perspectives

Currently, the atomic resolution structures of RdRPs and RdRP-RNA-NTP substrate complexes have been solved for several members of (+)ssRNA, (−)ssRNA, and dsRNA virus families, contributing decisively to our knowledge about the mechanisms of action of these enzymes. Picornaviruses and caliciviruses constitute the best known examples in (+) ssRNA viruses, with the structures of RdRP catalytic complexes available from major representatives of these families. These structures provided incredible insights into the interactions involved in template and primer binding, NTP selection and binding, catalysis, and chain translocation. Due to the high similarity between RdRPs, specially between viruses possessing (+) ssRNA genomes, these general mechanisms can be extrapolated to other members of the group.

Concerning flaviviruses, the structure of the full-length NS5 protein is known for three representatives of the group, providing useful data about the dynamic interplay of methyltransferase and polymerase domains. However, information regarding the interactions of these domains with RNA substrates or with their protein partners in the replication complex is still not available.

The spectacular advances in cryo-EM technology have only recently enabled the resolution of complicated structures of functional viral replicative complexes in dsRNA and (−)ssRNA viruses at atomic detail. The transcribing replication machines of various reoviruses and the high-resolution structure of the VSV L protein constitute some illustrative examples.

Future efforts will be directed toward the description of these multidomain and multiprotein assemblies in complex with their respective nucleic acid substrates and other viral regulatory proteins and host factors. This structural information will be essential to understand how these enzymes are able to bind their promoters and how the viral and host factors modulate the transcription and replication activities in these machineries. As viral polymerases are leading therapeutic targets for treating infections caused by RNA viruses, the new interactions determined would serve as promising targets for drug discovery.

Acknowledgments We acknowledge the funding from the Spanish Ministry of Economy Industry and Competitiveness (BIO2014-54588-P and Maria de Maeztu action MDM-2014-0435).

References

Ago H, Adachi T, Yoshida A, Yamamoto M et al (1999) Crystal structure of the RNA-dependent RNA polymerase of hepatitis C virus. Structure 7:1417–1426

Albertini AA, Wernimont AK, Muziol T, Ravelli RB, Clapier CRet al. (2006) Crystal structure of the rabies virus nucleoprotein-RNA complex. Science 313(5785):360–363

Appleby TC, Luecke H, Shim JH, Wu JZ, Cheney IW et al (2005) Crystal structure of complete rhinovirus RNA polymerase suggests front loading of protein primer. J Virol 79(1):277–288

Appleby TC, Perry JK, Murakami E, Barauskas O, Feng J et al (2015) Viral replication. Structural basis for RNA replication by the hepatitis C virus polymerase. Science 347(6223):771–775

Bahar MW, Sarin LP, Graham SC, Pang J, Bamford DH et al (2013) Structure of a VP1-VP3 complex suggests how birnaviruses package the VP1 polymerase. J Virol 87(6):3229–3236

Ball AL (2007) Virus replication strategies. In: Knipe DM, Howley PM (eds) Fields virology. Wolters Kluwer Health/ Lippincott Williams & Wilkins, Philadelphia, pp 119–140

Bentham M, Holmes K, Forrest S, Rowlands DJ, Stonehouse NJ (2012) Formation of higher-order foot-and-mouth disease virus 3D^{pol} complexes is dependent on elongation activity. J Virol 86(4):2371–2374

Brautigam CA, Steitz TA (1998) Structural and functional insights provided by crystal structures of DNA polymerases and their substrate complexes. Curr Opin Struct Biol 8(1):54–63

Bressanelli S, Tomei L, Roussel A, Incitti I, Vitale RL et al (1999) Crystal structure of the RNA-dependent RNA polymerase of hepatitis C virus. PNAS 96:13034–13099

Bressanelli S, Tomei L, Rey FA, De Francesco R (2002) Structural analysis of the hepatitis C virus RNA polymerase in complex with ribonucleotides. J Virol 76:3482–3492

Bruenn JA (2003) A structural and primary sequence comparison of the viral RNA-dependent RNA polymerases. Nucleic Acids Res 31:1821–1829

Bussetta C, Choi KH (2012) Dengue virus nonstructural protein 5 adopts multiple conformations in solution. Biochemistry 51:5921–5931

Butcher SJ, Grimes JM, Makeyev EV, Bamford DH, Stuart DI (2001) A mechanism for initiating RNA-dependent RNA polymerization. Nature 410(6825):235–240

Campagnola G, Weygandt M, Scoggin K, Peersen O (2008) Crystal structure of coxsackievirus B3 3Dpol highlights the functional importance of residue 5 in picornavirus polymerases. J Virol 82(19):9458–9464

Castro C, Smidansky ED, Arnold JJ, Maksimchuk KR, Moustafa I et al (2009) Nucleic acid polymerases use a general acid for nucleotidyl transfer. Nat Struct Mol Biol 16:212–218

Çevik B, Holmes DE, Vrotsos E, Feller JA, Smallwood S et al (2004) The phosphoprotein (P) and L binding sites reside in the N-terminus of the L subunit of the measles virus RNA polymerase. Virology 327:297–306

Çevik B, Smallwood S, Moyer SA (2007) Two N-terminal regions of the Sendai virus L RNA polymerase protein participate in oligomerization. Virology 363:189–197

Chang S, Sun D, Liang H, Wang J, Li J et al (2015) Cryo-EM structure of influenza virus RNA polymerase complex at 4.3 Å resolution. Mol Cell 57:925–935

Chen C, Wang Y, Shan C, Sun Y, Xu P et al (2013) Crystal structure of enterovirus 71 RNA-dependent RNA polymerase complexed with its protein primer VPg: implication for a trans mechanism of VPg uridylylation. J Virol 87(10):5755–5768

Cheng Y (2015) Single-particle Cryo-EM at crystallographic resolution. Cell 161:450–457

Chinnaswamy S, Murali A, Li P, Fujisaki K, Kao CC (2010) Regulation of de novo-initiated RNA synthesis in hepatitis C virus RNA-dependent RNA polymerase by intermolecular interactions. J Virol 84:5923–5935

Choi KH, Groarke JM, Young DC, Kuhn RJ, Smith JL et al (2004) The structure of the RNA-dependent RNA polymerase from bovine viral diarrhea virus establishes the role of GTP in de novo initiation. PNAS 101:4425–4430

Choi KH, Gallei A, Becher P, Rossmann MG (2006) The structure of bovine viral diarrhea virus RNA-dependent RNA polymerase and its amino-terminal domain. Structure 14:1107–1113

Choi KH, Rossmann MG (2009) RNA-dependent RNA polymerases from *Flaviviridae*. Curr Opin Struct Biol 19:746–751

Clemente-Casares P, López-Jiménez AJ, Bellón-Echeverría I, Encinar JA, Martínez-Alfaro E et al (2011) De novo polymerase activity and Oligomerization of hepatitis C virus RNA-dependent RNA-polymerases from genotypes 1 to 5. PLoS One 6(4):e18515

Collier AM, Lyytinen OL, Guo YR, Toh Y, Poranen MM et al (2016) Initiation of RNA polymerization and polymerase Encapsidation by a small dsRNA virus. PLoS Pathog 12(4):e1005523

Das K, Arnold E (2015) Negative-strand RNA virus L proteins: one machine, many activities. Cell 162(2):239–241

Davidson AD (2009) Chapter 2 new insights into Flavivirus nonstructural protein 5. In: Maramorosch K, Shatkin AJ, Purphy FA (eds) Advances in virus research, Vol 74. Elsevier Inc., pp 41–101

den Boon J, Diaz A, Alquist P (2010) Cytoplasmic vial replication complexes. Cell Host Microbe 8(1):77–85

Deval J, Jin Z, Chuang YC, Kao CC (2017) Structure(s), function(s), and inhibition of the RNA-dependent RNA polymerase of noroviruses. Virus Res 234:21–33

Estes MK, Kapikian AZ (2007) Rotaviruses. In: Knipe DM, Howley PM (eds) Fields virology pp. 1917–1974. Wolters Kluwer Health/ Lippincott Williams & Wilkins. Philadelphia

Estrozi LF, Settembre EC, Goret G, McClain B, Zhang X et al (2013) Location of the dsRNA-DependentPolymerase, VP1, in Rotavirus Particles. J MolBiol 425:124–132

Ferrer-Orta C, Arias A, Perez-Luque R, Escarmís C, Domingo E et al (2004) Structure of foot-and-mouth disease virus RNA-dependent RNA polymerase and its complex with a template-primer RNA. J Biol Chem 279(45):47212–47221

Ferrer-Orta C, Arias A, Escarmís C, Verdaguer N (2006a) A comparison of viral RNA-dependent RNA polymerases. Curr Opin Struct Biol 16:27–34

Ferrer-Orta C, Arias A, Agudo R, Pérez-Luque R, Escarmís C et al (2006b) The structure of a protein primer-polymerase complex in the initiation of genome replication. EMBO J 25(4):880–888

Ferrer-Orta C, Arias A, Pérez-Luque R, Escarmís C, Domingo E et al (2007) Sequential structures provide insights into the fidelity of RNA replication. Proc Natl Acad Sci U S A 104(22):9463–9468

Ferrer-Orta C, Verdaguer N (2009) Chapter 18 RNA virus polymerases. In: Cameron CE, Götte M, Raney KD (eds) Viral genome replication. Springer Inc, pp 383–401

Ferrero DS, Buxaderas M, Rodriguez JF, Verdaguer N (2015) The structure of the RNA-dependent RNA polymerase of a Permutotetravirus suggests a link between primer-dependent and primer-independent polymerases. PLoS Pathog 11(12):e1005265

Forss S, Schaller H (1982) A tandem repeat gene in a picornavirus. Nucleic Acids Res 10(20):6441–6450

Garriga D, Navarro A, Querol-Audi J, Abaitua F et al (2007) Activation mechanism of a noncanonical RNA-dependent RNA polymerase. Proc Natl Acad Sci U S A 104(51):20540–20545

Garriga D, Ferrer-Orta C, Querol-Audí J, Oliva B, Verdaguer N (2013) Role of motif B loop in allosteric regulation of RNA-dependent RNA polymerization activity. J Mol Biol 425(13):2279–2287

Gerlach P, Malet H, Cusack S, Reguera J (2015) Structural insights into Bunyavirus replication and its regulation by the vRNA promoter. Cell 161:1267–1279

Godoy AS, Lima GM, Oliveira KI, Torres NU, Maluf FV et al (2017) Crystal structure of Zika virus NS5 RNA-dependent RNA polymerase. Nat Commun 8:14764

Gohara DW, Arnold JJ, Cameron CE (2004) Poliovirus RNA-dependent RNA polymerase (3Dpol): kinetic, thermodynamic, and structural analysis of Ribonucleotide selection. Biochemistry 43(18):5149–5158

Gong P, Peersen OB (2010) Structural basis for active site closure by the poliovirus RNA-dependent RNA polymerase. Proc Natl Acad Sci U S A 107(52):22505–22510

Gong P, Kortus MG, Nix JC, Davis RE, Peersen OB (2013) Structures of coxsackievirus, rhinovirus, and poliovirus polymerase elongation complexes solved by engineering RNA mediated crystal contacts. PLoS One 8(5):e60272

Goodfellow I (2011) The genome-linked protein VPg of vertebrate viruses – a multifaceted protein. Curr Opin Virol 1(5):355–362

Gorbalenya AE, Pringle FM, Zeddam JL, Luke BT, Cameron CE et al (2002) The palm subdomain-based active site is internally permuted in viral RNA-dependent RNA polymerases of an ancient lineage. J Mol Biol 324:47–62

Graham SC, Sarin LP, Bahar MW, Myers RA, Stuart DI et al. (2011) The N-terminus of the RNA polymerase from infectious pancreatic necrosis virus is the determinant of genome attachment. PLoS Pathog 7:e1002085

Green TJ, Zhang X, Wertz GW, Luo M (2006) Structure of the vesicular stomatitis virus nucleoprotein-RNA complex. Science 313(5785):357–360

Gridley C, Patton J (2014) Regulation of rotavirus polymerase activity by inner capsid proteins. Curr Opin Virol 9:31–38

Gruez A, Selisko B, Roberts M, Bricogne G, Bussetta C et al (2008) The crystal structure of coxsackievirus B3 RNA-dependent RNA polymerase in complex with its protein primer VPg confirms the existence of a second VPg binding site on Picornaviridae polymerases. J Virol 82(19):9577–9590

Hansen JL, Long AM, Schultz SC (1997) Structure of the RNA-dependent RNA polymerase of poliovirus. Structure 5:1109–1122

Harak C, Lohmann V (2015) Ultrastructure of the replication sites of positive-strand RNA viruses. Virology 479:418–433

Hengrung N, El Omari K, Serna Martin I, Vreede FT, Cusack S et al (2015) Crystal structure of the RNA-dependent RNA polymerase from influenza C virus. Nature 527(7576):114–117

Hobson SD, Rosenblum ES, Richards OC, Richmond K, Kirkegaard K et al (2001) Oligomeric structures of poliovirus polymerase are important for function. EMBO J 20:1153–1163

Högbom M, Jager K, Robel I, Unge T, Rohayem J (2009) The active form of the norovirus RNA-dependent RNA polymerase is a homodimer with cooperative activity. J Gen Virol 90:281–291

Huang H, Chopra R, Verdine GL, Harrison SC (1998) Structure of a covalently trapped catalytic complex of HIV-1 reverse transcriptase: implications for drug resistance. Science 282:1669–1675

Ilca SL, Kotecha A, Sun X, Poranen MM, Stuart DI et al (2015) Localized reconstruction of subunits from electron cryomicroscopy images of macromolecular complexes. Nat Commun 6:8843

Jackson RJ, Howell MT, Kaminski A (1990) The novel mechanism of initiation of picornavirus RNA translation. Trends Biochem Sci 15:477–483

Kaiser WJ, Chaudhry Y, Sosnovtsev SV, Goodfellow IG (2006) Analysis of protein-protein interactions in the feline calicivirus replication complex. J Gen Virol 87:363–368

Kao CC, Singh P, Ecker DJ (2001) De novo initiation of viral RNA-dependent RNA synthesis. Virology 287(2):251–260

King AMQ, Sangar DV, Harris TJR, Brown F (1980) Heterogeneity of the genome-linked protein of foot-and-mouth disease virus. Virol 34:627–634

Klema VJ, Ye M, Hindupur A, Teramoto T, Gottipati K et al (2016) Dengue virus nonstructural protein 5 (NS5) assembles into a dimer with a unique Methyltransferase and polymerase Interface. PLoS Pathog 12(2):e1005451

Lai VC, Kao CC, Ferrari E, Park J, Uss AS et al (1999) Mutational analysis of bovine viral diarrhea virus RNA-dependent RNA polymerase. J Virol 73:10129–10136

Le Breton M, Meyniel-Schicklin L, Deloire A, Coutard B, Canard B et al (2011) Flavivirus NS3 and NS5 proteins interaction network: a high-throughput yeast two-hybrid screen. BMC Microbiol 11:234

Lee J, Alam I, Han KR, Cho S, Shin S et al (2011) Crystal structure of murine norovirus-1 RNA-dependent RNA polymerase. J General Virology 92:1607–1616

Lesburg CA, Cable MB, Ferrari E, Hong Z, Mannarino AF et al (1999) Crystal structure of the RNA-dependent RNA polymerase from hepatitis C virus reveals a fully encircled active site. Nat Struct Biol 6:937–943

Lescar J, Canard B (2009) RNA-dependent RNA polymerases from flaviviruses and Picornaviridae. Curr Opin Struct Biol 19(6):759–767

Li J, Rahmeh A, Morelli M, Whelan SP (2008) A conserved motif in region v of the large polymerase proteins of nonsegmented negative-sense RNA viruses that is essential for mRNA capping. J Virol 82:775–784

Li X, Zhou N, Chen W, Zhu B, Wang X et al (2017) Near-atomic resolution structure determination of a Cypovirus capsid and polymerase complex using Cryo-EM at 200kV. J Mol Biol 429(1):79–87

Liang B, Li Z, Jenni S, Rahmeh AA, Morin BM et al (2015) Structure of the L protein of vesicular stomatitis virus from electron Cryomicroscopy. Cell 162:314–327

Lim SP, Koh JHK, Seh CC, Liew CW, Davidson AD et al (2013) A crystal structure of the dengue virus non-structural protein 5 (NS5) polymerase delineates interdomain amino acid residues that enhance its thermostability and de novo initiation activities. J Biol Chem 288:31105–31114

Liu H, Cheng L (2015) Cryo-EM shows thepolymerasestructures and a nonspooledgenomewithin a dsRNA virus. Science 349(6254):1347–1350

Love RA, Maegley KA, Yu X, Ferre RA, Lingardo LK et al (2004) The crystal structure of the RNA-dependent RNA polymerase from human rhinovirus: a dual function target for common cold antiviral therapy. Structure 12(8):1533–1544

Lu G, Gong P (2013) Crystal structure of the full-length Japanese encephalitis virus NS5 reveals a conserved methyltransferase-polymerase interface. PLoS Pathog 9(8):e1003549

Lu X, Mc Donald SM, Tortorici MA, Tao YJ, Vasquez-Del Carpio R et al (2008) Mechanism for coordinated RNA packaging and genome replication by rotavirus polymerase VP1. Structure 16(11):1678–1688

Luque D, Rivas G, Alfonso C, Carrascosa JL, Rodriguez JF et al (2009a) Infectious bursal disease virus is an icosahedral polyploid dsRNA virus. PNAS 106(7):2148–2152

Luque D, Saugar I, Rejas MT, Carrascosa JL, Rodriguez JF et al (2009b) Infectious bursal disease virus: ribonucleoprotein complexes of a double-stranded RNA virus. J Mol Biol 386(3):891–901

Lyle JM, Clewell A, Richmond K, Richards OC, Hope DA et al (2002) Similar structural basis for membrane localization and protein priming by an RNA-dependent RNA polymerase. J Biol Chem 277(18):16324–16331

Mairiang D, Zhang H, Sodja A, Murali T, Suriyaphol P et al (2013) Identification of new protein interactions between dengue fever virus and its hosts, human and mosquito. PLoS One 8(1):e53535

Malet H, Egloff MP, Selisko B, Butcher RE, Wright PJ et al (2007) Crystal structure of the RNA polymerase domain of the West Nile virus non-structural protein 5. J Biol Chem 282(14):10678–10689

Marcotte LL, Wass AB, Gohara DW, Pathak HB, Arnold JJ et al (2007) Crystal structure of poliovirus 3CD protein: virally encoded protease and precursor to the RNA-dependent RNA polymerase. J Virol 81:3583–3596

Mas A, Clemente-Casares P, Ramirez E, Sabariegos R (2016) The HCV replicase interactome. American JVirol 5(1):8–14

McDonald SM, Tao YJ, Patton JT (2009) The ins and outs of four-tunneled RNA-dependent RNA polymerases. Curr Opin Struct Biol 19(6):775–782

Mosley RT, Edwards TE, Murakami E, Lam AM, Grice RL et al (2012) Structure of hepatitis C virus polymerase in complex with primer-template RNA. J Virol 86:6503–6511

Nakamura T, Zhao Y, Yamagata Y, Hua YJ, Yang W (2012) Watching DNA polymerase η make a phosphodiester bond. Nature 487:196–201

Ng KK, Cherney MM, Vazquez AL, Machin A, Alonso JM et al (2002) Crystal structures of active and inactive conformations of a caliciviral RNA-dependent RNA polymerase. J Biol Chem 277(2):1381–1387

Ng KK, Pendás-Franco N, Rojo J, Boga JA, Machín A et al (2004) Crystal structure of nor-walk virus polymerase reveals the carboxyl terminus in the active site cleft. J Biol Chem 279(16):16638–16645

O'Reilly EK, Kao CC (1998) Analysis of RNA-dependent RNA polymerase structure and function as guided by known polymerase structures and computer predictions of secondary structure. Virology 252:287–303

Ogino T, Banerjee AK (2007) Unconventional mechanism of mRNA capping by the RNA-dependent RNA polymerase of vesicular stomatitis virus. Mol Cell 25(1):85–97

Pan J, Vakharia VN, Tao YJ (2007) Structural of a birnavirus polymerase reveals a distinct active site topology. Proc Natl Acad Sci U S A 104:7385–7390

Pathak HB, Oh HS, Goodfellow IG, Arnold JJ, Cameron CE (2008) Picornavirus genome replication: roles of precursor proteins and rate-limiting steps in oriI-dependent VPg uridylylation. J Biol Chem 283:30677–30688

Pata JD, Schultz SC, Kirkegaard K (1995) Functional oligomerization of poliovirus RNA-dependent RNA polymerase. RNA 5:466–477

Patton JT, Jones MT, Kalbach AN, He YW, Xiaobo J (1997) Rotavirus RNA polymerase requires the core shell protein to synthesize the double-stranded RNA genome. J Virol 71:9618–9626

Paul AV, van Boom JH, Filippov D, Wimmer E (1998) Protein-primed RNA synthesis by purified poliovirus RNA polymerase. Nature 393:280–284

Paul AV, Peters J, Mugavero J, Yin J, van Boom JH et al (2003) Biochemical and genetic studies of the VPg uridylylation reaction catalyzed by the RNA polymerase of poliovirus. J Virol 77(2):891–904

Paul AV, Wimmer E (2015) Initiation of protein-primed picornavirus RNA synthesis. Virus Res 206:12–26

Peersen OB (2017) Picornaviral polymerase structure, function, and fidelity modulation. Virus Res 234:4–20

Pflug A, Guilligay D, Reich S, Cusack S (2014) Structure of influenza a polymerase bound to the viral RNA promoter. Nature 516:355–360

Plotch SJ, Bouloy M, Ulmanen I, Krug RM (1981) A unique cap(mGpppXm)-dependent influenza virion endonuclease cleaves capped RNAs to generate the primers that initiate viral RNA transcription. Cell 23(3):847–858

Potisopon S, Priet S, Collet A, Decroly E, Canard B et al (2014) The methytransferase domain of the dengue virus protein NS5 ensures efficient RNA synthesis initiation and elongation by the polymerase domain. Nucleic Acids Res 42(18):11642–11656

Rahmeh AA, Schenk AD, Danek EI, Kranzuch PJ, Liang B et al (2010) Molecular architecture of the vesicular stomatitis virus RNA polymerase. Proc Natl Acad Sci U S A 107(46):20075–20080

Reguera J, Gerlach P, Cusack S (2016) Towards a structural understanding of RNA synthesis by negative strand RNA viral polymerases. Curr Opin Struct Biol 36:75–84

Reich S, Guilligay D, Pflug A, Alexander MH, Berger I et al (2014) Structural insight into cap-snatching and RNA synthesis by influenza polymerase. Nature 516(7531):361–366

Ren Z, C Franklin M, Ghose R (2013) Structure of the RNA-directed RNA polymerase from the cystovirus φ12. Proteins, 81(8):1479–1484

Salgado PS, Makeyev EV, Butcher SJ, Bamford DH, Stuart DI et al (2004) The structural basis for RNA specificity and Ca2+ inhibition of an RNA-dependent RNA polymerase. Structure 12(2):307–316

Saw WG, Tria G, Gruber A, Subramanian Manimekalai MS, Zhao Y, Chandramohan A et al (2015) Structural insight and flexible features of NS5 proteins from all four serotypes of dengue virus in solution. Acta Cryst D71:2309–2327

Schiff LA, Nibert ML, Tyler KL (2007) Orthoreoviruses and their replication. In: Knipe DM, Howley PM (eds) Fields virology. Wolters Kluwer Health/ Lippincott Williams & Wilkins, Philadelphia, pp 1853–1916

Selisko B, Wang C, Harris E, Canard B (2014) Regulation of Flavivirus RNA synthesis and replication. Curr Opin Virol 0:74–83

Shen H, Sun H, Li G (2012) What is the role of motif D in the nucleotide incorporation catalyzed by the RNA-dependent RNA polymerase from poliovirus? PLoS Comput Biol 8(12):e1002851

Sholders AJ, Peersen OB (2014) Distinct conformations of a putative translocation element in poliovirus polymerase. J Mol Biol 426(7):1407–1419

Shu B, Gong P (2016) Structural basis of viral RNA-dependent RNA polymerase catalysis and translocation. PNAS 113(28):E4005–E4014

Smallwood S, Hovel T, Neubert WJ, Moyer SA (2002) Different substitutions at conserved amino acids in domains II and III in the Sendai L RNA polymerase protein inactivate viral RNA synthesis. Virology 304:135–145

Smallwood S, Moyer SA (2004) The L polymerase protein of parainfluenza virus 3 forms an oligomer and can interact with the heterologous Sendai virus L, P and C proteins. Virology 318:439–450

Spagnolo JF, Rossignol E, Bullitt E, Kirkegaard K (2010) Enzymatic and nonenzymatic functions of viral RNA-dependent RNA polymerases within oligomeric arrays. RNA 16:382–393

Steitz TA (1998) A mechanism for all polymerases. Nature 391(6664):231–232

Subramanian Manimekalai MS, Saw WG, Pan A, Gruber A, Gruber G (2016) Identification of the critical linker residues conferring differences in the compactness of NS5 from dengue virus serotype 4 and NS5 from dengue virus serotypes 1–3. Acta Cryst D72:795–807

Surana P, Satchidanandam V, Nair DT (2014) RNA-dependent RNA polymerase of Japanese encephalitis virus binds the initiator nucleotide GTP to form a mechanistically important pre-initiation state. Nucleic Acids Res 42(4):2758–2773

Tao L, Farsetta DL, Nibert ML, Harrison SC (2002) RNA Synthesis in a Cage—structural studies of reovirus polymerase λ3. Cell, 111:733–745

Tay MYF, Smith K, Ng IHW, Chan KWK, Zhao Y et al (2016) The C-terminal 18 amino acid region of dengue virus NS5 regulates its subcellular localization and contains a conserved arginine residue essential for infectious virus production. PLoS Pathog 12(9):e1005886

te Velthuis AJW (2014) Common and unique features of viral RNA-dependent polymerase. Cell Mol Life Sci 71:4403–4420

te Velthuis AJW, Fodor E (2016) Influenza virus RNA polymerase: insights into the mechanisms of viral RNA synthesis. Nat Rev Microbiol 14:479–493

Tellez AB, Wang J, Tanner EJ, Spagnolo JF, Kirkegaard K et al (2011) Interstitial contacts in an RNA-dependent RNA polymerase lattice. J Mol Biol 412:737–750

Thompson AA, Peersen OB (2004) Structural basis for proteolysis-dependent activation of the poliovirus RNA-dependent RNA polymerase. EMBO J 23:3462–3471

Thompson AA, Albertini RA, Peersen OB (2007) Stabilization of poliovirus polymerase by NTP binding and fingers-thumb interactions. J Mol Biol 366:1459–1474

Tortorici MA, Broering TJ, Nibert ML, Patton JT (2003) Template recognition and formation of initiation complexes by the replicase of a segmented double-stranded RNA virus. J Biol Chem 278:32673–32682

Upadhyay AK, Cyr M, Longenecker K, Tripathi R, Sun C, Kempf DL (2017) Crystal structure of full-length Zika virus NS5 protein reveals a conformation similar to Japanese encephalitis virus NS5. Acta Cryst F 73:116–122

van Dijk AA, Makeyev EV, Bamford DH (2004) Initiation of viral RNA-dependent RNA polymerization. J Gen Virol 85(5):1077–1093

Verdaguer N, Ferrer-Orta C (2012) Conformational changes in motif D of RdRPs as fidelity determinant. Structure 20(9):1448–1450

Vives-Adrian L, Lujan C, Oliva B, van der Linden L, Selisko B et al (2014) The crystal structure of a cardiovirus RNA-dependent RNA polymerase reveals an unusual conformation of the polymerase active site. J Virol 88(10):5595–5607

Wang C, Wang C, Li Q, Wang Z, Xie W (2017) Crystal structure and thermostability characterization of EV-D68-3Dpol. J Virol, JVI:00876–00817

Wang J, Lyle JM, Bullitt E (2013) Surface for catalysis by poliovirus RNA-dependent RNA polymerase. J Mol Biol 425:2529–2540

Wang QM, Hockman MA, Staschke K, Johnson RB, Case KA et al (2002) Oligomerization and cooperative RNA synthesis activity of hepatitis C virus RNA-dependent RNA polymerase. J Virol 76(8):3865–3872

Wu Y, Lou Z, Miao Y, Yu Y, Dong H et al (2010) Structures of EV71 RNA-dependent RNA polymerase in complex with substrate and analogue provide a drug target against the hand-foot-and-mouth disease pandemic in China. Protein Cell 1(5):491–500

Yang X, Smidansky ED, Maksimchuk KR, Lum D, Welch JL et al (2012) Motif D of viral RNA-dependent RNA polymerases determines efficiency and fidelity of nucleotide addition. Structure 20(9):1519–1527

Yap TL, Xu T, Chen YL, Malet H, Egloff MP et al (2007) Crystal structure of the dengue virus RNA-dependent RNA polymerase catalytic domain at 1.85-angstrom resolution. J Virol 81(9):4753–4765

Zamoto-Niikura A, Terasaki K, Ikegami T, Peters CJ, Makino S (2009) Rift valley fever virus L protein forms a biologically active oligomer. J Virol 83:12779–12789

Zamyatkin DF, Parra F, Alonso JM, Harki DA, Peterson BR et al (2008) Structural insights into mechanisms of catalysis and inhibition in Norwalk virus polymerase. J Biol Chem 283(12):7705–7712

Zeddam JL, Gordon KH, Lauber C, Alves CA, Luke BT et al (2010) Euprosterna elaeasa virus genome sequence and evolution of the Tetraviridae family: emergence of bipartite genomes and conservation of the VPg signal with the dsRNA Birnaviridae family. Virology 397:145–154

Zhang X, Walker SB, Chipman PR, Nibert ML et al (2003) Virus polymerase λ3 localized by cryo-electron microscopy of virions at a resolution of 7.6 Å. Nature StrucBiol 10(12):1011–1018

Zhang X, Ding K, Yu X, Chang W, Sun J et al (2015) In situ structures of the segmented genome and RNA polymerase complex inside a dsRNA virus. Nature 527:531–534

Zhao Y, Soh TS, Lim SP, Chung KY, Swaminathan K et al (2015a) Molecular basis for specific viral RNA recognition and 2'-O-ribose methylation by the dengue virus nonstructural protein 5 (NS5). Proc Natl Acad Sci U S A 112(48):14834–14839

Zhao Y, Soh TS, Zheng J, Chan KWK, Phoo WWet al. (2015b) A crystal structure of the dengue virus NS5 protein reveals a novel inter-domain interface essential for protein flexibility and virus replication. PLoS Pathog 11(3):e1004682

Chapter 4
Filovirus Filament Proteins

Daniel R. Beniac, Lindsey L. Lamboo, and Timothy F. Booth

Introduction

The filovirus family are so-called owing to the filamentous or thread like morphology of their virus particles, first observed in clinical specimens and in cell culture, by electron microscopy (Almeida et al. 1971). Filoviruses are agents of severe haemorrhagic fevers, can be highly contagious, and have the potential to cause serious outbreaks with high mortality (Sanchez et al. 2007). Filoviruses are members of the Mononegavirales, meaning that their genome consists of a single species of negative-sense RNA molecule. The Mononegavirales (also known as *non-segmented negative-sense viruses)* includes paramyxoviruses (such as measles virus) and rhabdoviruses (such as rabies virus). One feature shared by all of the *Mononegavirales* is that they have helical nucleocapsids, meaning that their RNA is wound into helical structures that include one or more virally encoded proteins. In filoviruses, as in other non-segmented negative-strand RNA viruses, the helical nucleocapsids also contain other viral proteins that achieve RNA transcription and replication. As well as the large polymerase protein (L), there are three other nucleocapsid-associated proteins in filoviruses that form the replicase complex. With five different proteins, the Filoviridae have the most complex helical nucleocapsid within the *Mononegavirales*.

D. R. Beniac
Viral Diseases Division, National Microbiology Laboratory, Public Health Agency of Canada, Winnipeg, MB, Canada
e-mail: Daniel.beniac@canada.ca

L. L. Lamboo · T. F. Booth (✉)
Viral Diseases Division, National Microbiology Laboratory, Public Health Agency of Canada, Winnipeg, MB, Canada

Department of Medical Microbiology and Infectious Diseases, Rady College of Medicine, University of Manitoba, Winnipeg, MB, Canada
e-mail: Lindsey.lamboo@canada.ca; tim.booth@canada.ca

© Springer Nature Singapore Pte Ltd. 2018
J. R. Harris, D. Bhella (eds.), *Virus Protein and Nucleoprotein Complexes*,
Subcellular Biochemistry 88, https://doi.org/10.1007/978-981-10-8456-0_4

73

As well as their structural functions, in forming and retaining helical viral nucleo-protein structures, these nucleocapsid-associated proteins possess other properties that promote viral replication, such as modulation of the host cell defence mechanisms that have been well documented elsewhere (e.g. Sanchez et al. 2007) and which we will not describe in detail here. It is of importance in understanding these structures to realize that negative-strand RNA virus genomes are usually not free within the cell or outside it (Reguera et al. 2016). These viruses, and their helical nucleoproteins, function as dynamic macromolecular complexes, which must be able to enter the cell through membrane fusion as well as to gain egress, driven by membrane budding. Their genomic RNA remains tightly bound with the nucleoprotein complexes, including the polymerase, during all stages of the infection cycle, except for brief transient periods where the RNA must detach for access to the template. In addition, the promoter and termination sequences require access to the polymerase complexes, and replication must be tied with the process of formation of progeny nucleocapsids. Thus, it is important to understand how these helical nucleocapsid structures are formed and how they function dynamically.

Systems for Studying Filovirus Proteins and Nucleoproteins

Systems for studying these structures and their formation and function include expression of different combinations of viral proteins in non-infectious, heterologous systems, to generate nucleocapsid-like and virus-like particles. In conjunction with the use of cryo-electron microscopy (cryo-EM), structural-functional relationships of large oligomeric viral structures, consisting of multiple proteins and nucleic acid, can be studied in a time-resolved manner. Such macromolecular structures may be refined by fitting higher resolution X-ray crystallography structures of the individual molecules or submolecular domains. The development of cryo-electron microscopy was recently recognized by the award of the 2017 Nobel prize in chemistry to Dubochet, Frank and Henderson. Much of the data presented here was obtained using their broad approaches, as described by Adrian et al. (1984) and also employing 3D image processing (Frank et al. 1996) with cryo-EM and negative-stain electron microscopy. Although there are many types of filamentous viruses present in plants (both RNA and DNA viruses), animal viruses with a filamentous virion morphology are relatively rare. Of these, the filoviruses are the only family where the entire virion can be said to be truly filamentous in morphology. By contrast, the paramyxoviruses have a helical filamentous nucleocapsid, but these are enveloped in roughly spherical envelopes. When the envelopes of paramyxoviruses are broken open, these filamentous helical nucleocapsids can be released into the supernatant and are readily visible in negative-stain electron microscopy. In the rhabdoviruses, the nucleocapsid is also helical, but its shape is relatively short and stubby, with a pronounced pointed end, leading to a bullet-shaped virion. In rhabdoviruses, the envelope is closely associated with the envelope matrix protein, such that the helical structure of the nucleocapsid is retained in the envelope protein layer containing the matrix protein and spike. Strong interactions between the matrix protein and the nucleocapsid appear to maintain this structure (Ge et al. 2010), such that lysis of rhabdovirus envelopes to release intact nucleocapsids appears to be not possible by biochemical means. The *Filoviridae* includes three genera: *Ebolavirus*, *Marburgvirus* and *Cuevavirus*. In most of this chapter, for simplicity, we will focus our discussion specifically on Ebola virus, since this species is the most studied structurally, noting that all viruses in this family share their basic molecular and morphological characteristics and have the same general genetic organization, coding and replication strategies.

Filovirus Proteins

The filamentous morphology of the Ebola virus is driven primarily by two structural components, the viral membrane protein VP40 and the flexible filamentous nucleo-capsid. The nucleocapsid (NC) is comprised of the 18.9 kb negative-sense RNA genome, the nucleocapsid protein (NP) and viral proteins L, VP35, VP30 and VP24 (Booth et al. 2013). In addition, there is a single species of surface glycoprotein spike (GP) which functions in cell attachment and membrane fusion for entry of the virion into the cell. The spike is trimeric and highly glycosylated and is a class 1 viral membrane fusion protein. The various structural features of Ebola virus are shown in Figs. 4.1 and 4.2.

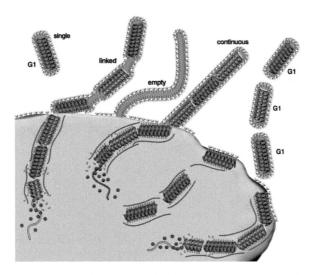

Fig. 4.1 Ebola virus structure and morphogenesis. Diagram showing the range of morphology and polyploidy present in Ebola virions. The model of budding illustrates the various filamentous Ebola particle structures that have been observed by electron microscopy. The colour-coding shows nucleocapsid helices in orange, red and yellow: VP40 as green ovals, nucleocapsid protein as purple spheres, VP24-VP35 bridges as blue ovals, microtubules in brown, and GP spikes in yellow. Partially reproduced from (Beniac et al. 2012)

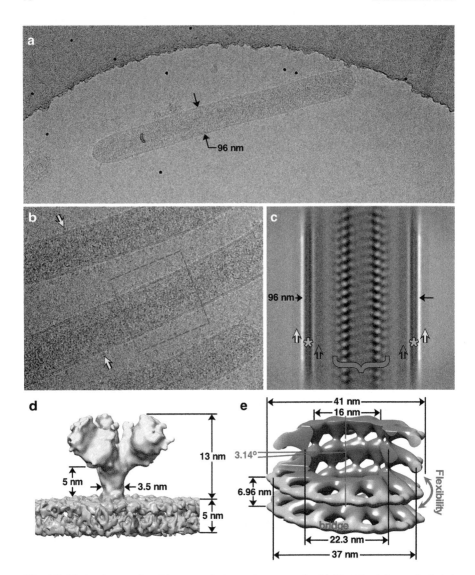

Fig. 4.2 Ebola virus analysed by cryo-electron microscopy. (**a, b**) Unstained frozen hydrated images. The red square in (**b**) shows the size of the images used to generate the image average in (**c**). Individual GP spikes are identified by yellow arrows. (**c**) Average from several hundred images of linear regions of Ebola virus filaments. The averaging procedure reduces the noise in the images, and conserved structural elements are revealed. The following regions are shown: yellow arrow, GP spike layer; blue asterisk, lipid envelope; red arrow, VP40 layer; and purple bracket, nucleocapsid. 3D reconstructions of the GP spike (**d**) and nucleocapsid (**e**). In (**d**) the GP spike is coloured yellow, and the adjacent viral envelope is coloured *blue*. In (**e**) the central region of the nucleocapsid is coloured purple, and the "bridge" structure on the exterior of the nucleocapsid is coloured blue. A section of the helical nucleocapsid is shown, with the front face of the top region cut away to reveal the interior

Filovirus Morphology

Filoviruses have several different morphologies that are broadly related, including the common filamentous forms including single, linear, continuous and linked virions (Fig. 4.1). In addition, there are also "comma" -shaped virions and sometimes torroidal particles (Fig. 4.3). Empty Ebola virions have also been observed, corresponding to filamentous enveloped particles, but without a nucleocapsid (Beniac et al. 2012). Moreover, Ebola viruses have a modular linear morphology that can result in very long filaments containing multiple genome copies (Fig. 4.1). In linked virions, alternating nucleocapsid-containing and empty regions result in an appearance reminiscent of a string of sausages, with each unit containing a single viral nucleocapsid. Length measurements of Ebola virus have revealed that approximately half of the viruses have a single genome, and the rest are polyploid, with the longest virus measured containing up to 22 copies of the genome (Beniac et al. 2012). This degree of polyploidy in an animal virus is unique. Although there are other animal viruses that are polyploid, for example, the *Birnaviridae* and *Paramyxoviridae*, these usually contain up to four copies of their genome (Hosaka

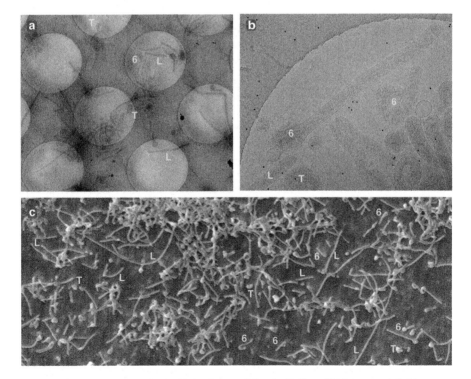

Fig. 4.3 Variable morphologies in Ebola virus particles. (**a, b**) Cryo-EM images. (**c**) Gold sputter-coated images of Ebola virus observed in a scanning electron microscope. Some of the characteristic comma (C)-, linear (L)-, and torroidal (T)- shaped viruses have been identified

et al. 1966; Luque et al. 2009), while in some paramyxoviruses, up to six copies of the viral genome have been shown to exist (Loney et al. 2009). An Ebola virus particle containing a single genome is on average 982 ± 79 nm long and 96 nm in diameter from each of the outer leaflets of the lipid bilayer of the viral envelope, with a prominent 41 nm diameter helical nucleocapsid running along the centre of the axis of the virus (Beniac et al. 2012). Adjacent to the nucleocapsid is a low-density region, which is then followed by a region of high density next to the lipid envelope which is attributed to the membrane-associated matrix protein VP40 (Bornholdt et al. 2013; Beniac et al. 2012). The other distinctive morphological features on the Ebola virus are the 13 nm glycoprotein spikes (GP spikes) on the surface of the virus (Beniac and Booth 2017). These GP spikes are responsible for attachment to the receptor on the host cell. The exterior of the virus is decorated with these trimeric GP spikes. Each monomer in the trimeric spike is composed of two disulphide-linked polypeptides GP1 and GP2. The receptor binding and mucin-like domains are within GP1. The GP spike is a class 1 fusion protein, has a receptor binding domain and a narrow stalk that attaches to the viral envelope and has three prominent lobes on the distal end of the spike, with each lobe being composed primarily of the highly glycosylated mucin-like domains (Brindley et al. 2007; Kuhn et al. 2006; Manicassamy et al. 2005; Mpanju et al. 2006; Beniac and Booth 2017). Although different entry pathways might be possible in various cell and tissue types, one apparently universal entry pathway that has been characterized for Ebola virus is via the endosome, where cathepsin cleavage of the GP spike receptor/fusion molecule reveals the Niemann–Pick C1 (NPC-1) binding site. In this pathway the human NPC-1 cholesterol transporter appears to be the receptor that the GP spike recognizes for attachment (Cote et al. 2011; Gong et al. 2016; Carette et al. 2011). This is then followed by NPC-1 binding and subsequent membrane fusion (Hood et al. 2010; Schornberg et al. 2006; Chandran et al. 2005; Dube et al. 2009). Unlike some other enveloped viruses, the GP spike is the only protein on the surface of the virus and is responsible for both viral attachment and membrane fusion.

In this chapter, we will focus on the viral proteins that compose the nucleocapsid: the NP protein, VP35, VP30, and VP24 as well as the VP40 the matrix protein and the outer GP spike. We discuss how these proteins interact with each other to form the filamentous and flexible virion. Although these viruses are somewhat variable in appearance, there is an underlying structural coordination that maintains a highly conserved 96 nm diameter distance between the outer leaflets of the lipid bilayer of the viral envelope which make up the cylindrical filament common to all regions of the virus that contain a nucleocapsid.

Models for Filovirus Morphogenesis

One of the striking features of Ebola virus is the extreme length of some of the filaments, which are present both in cell-cultured material and viruses derived from infected animals (Booth et al. 2013). However, some filovirus particles also exist in approximately round "doughnut-like" torroidal forms (Fig. 4.3), where the nucleo-capsid is bent into a rounded crescent shape, almost forming a circle as seen inter-nally within the envelope by cryo-EM. In addition, there is also what appears to be a type of virion that is intermediate between the filamentous and the torroidal forms, where only one end of the nucleocapsid is bent into a "check mark" or "comma" shape (Fig. 4.3: (Almeida et al. 1971; Geisbert and Jahrling 1995; Beniac et al. 2012)). In the case of the comma or torroidal viruses, the nucleocapsid is wound up around itself within the envelope of the spherical-shaped virus, demonstrating the flexibility of the nucleocapsid with a 360-degree bend. In addition to bent particles, there are also branched filaments which are frequently seen in cell culture superna-tant but are rarely seen in centrifuged samples. Careful observation indicates that the branched regions of tubular envelope are usually empty inside, without an inter-nal nucleocapsid. Thus, these are likely to be defective particles, and the presence of larger numbers of empty envelopes tubular structures could partially be an artefact of cell culture growth. Early studies of the Marburg virus identified the torroidal form to have diameters ranging from 300 to 400 nm (Almeida et al. 1971). To esti-mate the flexibility of the nucleocapsid required to achieve this, one can calculate this based on a torroidal virus using the average value of 350 nm diameter. Using the 96 nm filamentous Ebola virus as a guide to find the centre of the nucleocapsid from the edge of the lipid envelope of the virus would give one a distance of 48 nm from each side of the 350 nm torroidal virus. This is accounting for the space occupied by VP40 and the gap between VP40 and the 41 nm diameter nucleocapsid. If this value is subtracted from each side of the 350 nm torroid, one would have a nucleocapsid centred on a circle 254 nm in diameter. This diameter would have a circumference of 798 nm, and with a helical pitch of 6.96 for the Ebola nucleocapsid would give approximately 114 helical repeats in a 350 nm torroidal virus. Thus, each helical repeat needs to bend 3.14° to accommodate a complete 360° turn. This 3.14° bend can be easily accommodated by the structure of the Ebola virus nucleocapsid (Fig. 4.2e). Another interesting observation is that the estimated 798 nm circumference-length that the nucleocapsid would curve was calculated as just slightly shorter than the estimated 916 nm length of the Ebola virus nucleocapsid (Booth et al. 2015). Based on this observation, we can postulate that a 350 nm tor-roidal virus would most likely accommodate a single nucleocapsid that winds up upon itself by about 360°. This observation could in part explain why most of the torroidal virus particles are in the 300–400 nm size range (Almeida et al. 1971).

In some of the comma-shaped viruses, the nucleocapsid has been observed wrap-ping around either a vesicle or region of low density within the spherical globular end of the comma-shaped virus (Beniac et al. 2012). With all this observed struc-tural variability, there is also an extremely conserved structural motif present in linear regions of the virus that contain a nucleocapsid (Figs. 4.2 and 4.4). Single particle image analysis procedures (Frank et al. 1996) applied to thousands of cryo-EM images of straight linear regions have established that there is a conserved structure (Beniac et al. 2012). The results of this analysis have been modelled in Fig. 4.4. The image processing revealed several features, the first being the promi-nent helical nucleocapsid that runs along the central axis of the cylindrical linear

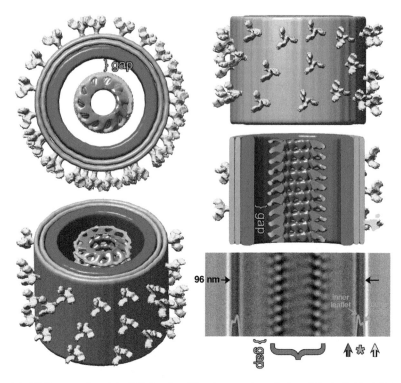

Fig. 4.4 3D Structural model of Ebola virus. Four images show fitting of the individual GP spike and nucleocapsid reconstructions. In addition, a 2D average from cryo-EM images of the Ebola virus is included for comparison. The following colour scheme is used: GP spike (*yellow* – reconstruction and arrow), viral envelope (*blue* – model and asterisk), VP40 (*red* – model and arrow), nucleocapsid (purple bracket and purple/light blue reconstruction)

virus. The nucleocapsid has a distinctive herringbone appearance with a diameter of 41 nm and a pitch of 6.96 nm. In addition to the nucleocapsid, the diameter of the cylindrical virus is fairly constant at 96 nm between the outer leaflets of the lipid bilayer of the viral envelope. This thickness is so constant that the two leaflets can be clearly resolved in both of the lipid bilayers (Fig. 4.4). If there were any degree of variation in the diameter of the virus, these features in the bilayer would appear blurred by the image averaging procedure, and the lipid bilayer would only appear as a single-smeared band of density, not two distinct layers as has been observed. Similarly if the nucleocapsid was not held in a central position within the interior of the virus, it would also have appeared blurred and in the averaging process, similar to other filamentous structures that are flexible or curved and therefore difficult to analyse structurally (Radermacher et al. 1987). In Ebola, the nucleocapsid can be seen clearly running along the central axis of the virus. Both the lipid bilayer and nucleocapsid were clearly preserved in the image averaging procedure (Beniac et al. 2012) indicating that the 3D structural organization of Ebola virus, especially in the nucleocapsid, seems to be well ordered in the majority of virus particles (Fig. 4.4).

In addition, the VP40 layer adjacent to the lipid bilayer was observed as a region of high density approximately 9 nm thick. This leaves a region of lower density between the VP40 layer and the nucleocapsid which is approximately 7–8 nm thick. Although both of these regions do not show clear averaged mass densities, the fact that the overall thickness and central positioning of the nucleocapsid are so highly conserved indicate that there must be some form of structural order in these regions. The VP40 layer has been postulated to be composed of hexamers of VP40 which would fit into this layer of high density (Bornholdt et al. 2013). In the low-desnisty gap between VP40 hexamers and the nucleocapsid, there must be contacts that maintain this fairly constant spacing of 7 nm, probably involving some form of quasi-equivalent contacts between the nucleocapsid and VP40 that keep the nucleocapsid centred but still allow bending of the viral particles (Booth et al. 2013). In the absence of the nucleocapsid, the diameter of the filament is variable and usually thinner in diameter than in regions containing a nucleocapsid (Figs. 4.5 and 4.6, 4.7). There must be specific contacts between the nucleocapsid that also facilitate envelopment and egress of mature particles containing the nucleocapsid. When virus-like particles are generated with the GP and VP40 proteins expressed together, one can see the influence of the VP40 matrix protein on the filamentous structure, in the absence of the nucleocapsid. These VP40 or VP40-GP structures are variable in length and diameter (Fig. 4.6), can be globular, and often appear as branched structures. The use of cryo-EM and cryo-electron tomography (cryo-ET) methods (Adrian et al. 1984; Dubochet et al. 1988) has the advantage over conventional electron microscopy of keeping the structure fully hydrated in its native state (Figs. 4.6 and 4.7), as compared to dehydrated filovirus specimens where the structure partially collapses during the drying process (Fig. 4.5). Virus-like particles show two distinctive appearances when imaged by cryo-EM (Fig. 4.6). The virus-like particle can be fairly straight, linear and with a constant diameter (Fig. 4.6a), or the particle can be straight but with a variable diameter along its length as observed by cryo-ET (Fig. 4.6b–e). These images show that the virus-like particle can have a highly variable diameter, in this case giving it an hourglass-like shape along its long axis, as compared to the virus particles, which have more regular, cylindrical shapes. The VP40 matrix protein can be seen adjacent to the inner leaf of the lipid bilayer in a linear pattern in these cryo-ET analyses. This distinctive single layer is similar to the blurred band of high-density adjacent to the inner side of the lipid bilayer in Ebola virus (Figs. 4.2 and 4.4) and the layer of proteins adjacent to the lipid bilayer seen in Ebola virus as observed by cryo-ET (Fig. 4.7). Tomography shows that the VP40 layer has a chequer board like lattice, when viewed end-on in Z-slices. The symmetry of the lattice may not be revealed by image averaging techniques, due to the flexibility of the membrane. There are also likely to be dislocations and imperfections in the layer of VP40 oligomers so that they have quasi-equivalent arrangement along the interior of the lipid bilayer (Fig. 4.2). It has been proposed that this lattice-like pattern may represent the arrangement of hexamers of VP40 along the lipid bilayer, since the approximately 5 nm spacing aligns with the density of the VP40 hexamer solved by X-ray crystallography (Bornholdt et al. 2013; Beniac et al. 2012). Exactly how VP40 interacts with the nucleocapsid is not yet known; however

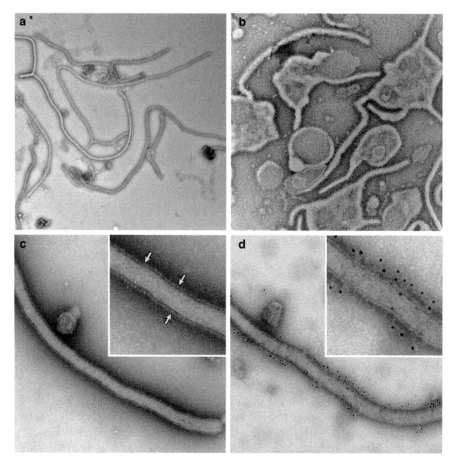

Fig. 4.5 Ebola virus-like particles. (**a, b**) Negative stained images show the variability in these structures. (**c**) High-magnification image of virus-like particles generated by expressing the matrix protein VP40 and the GP spike (yellow arrows). The presence of the Ebola GP spike was confirmed by immuno-gold electron microscopy (**d**)

it has been shown that both VP35 and the nucleocapsid protein can interact with VP40 (Johnson et al. 2006; Noda et al. 2007; Hoenen et al. 2005). In addition, VP40 is required for the envelopment of nucleocapsids during the assembly of the Ebola virus (Noda et al. 2006a). This clearly establishes a structural-functional relationship between the nucleocapsid and VP40. Expression of VP40, by itself, in the absence of the nucleocapsid, is capable of strongly driving the formation of budding of filamentous structures. However, virus-like particles are less well ordered and are smaller in diameter, as compared with Ebola virus particles.

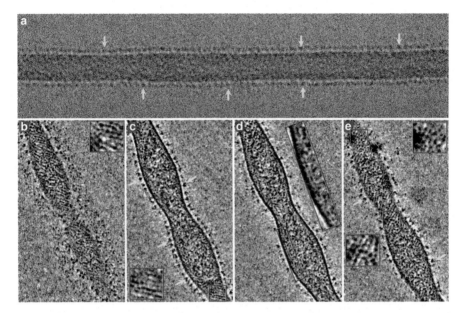

Fig. 4.6 Cryo-electron microscopy and tomography of Ebola virus-like particles. (**a**) A cryo-electron micrograph of a filamentous Ebola virus-like particle with individual GP spikes identified (*yellow* arrows). (**b–e**) A series of Z-slices through a cryo-electron tomogram of Ebola virus-like particles. GP spikes are identified by yellow arrows, the VP40 matrix proteins are identified by red arrows, and red boxes show the location of the magnified inset images

Polyploidy in Filoviruses

It was observed in many previous studies that Ebola virus particles had a wide variation in length. However, when large numbers of particles were measured carefully, excluding nonlinear virions, those consisting of empty lengths of envelope and any highly bent particles, it was found that the lengths of full nucleocapsid-containing virions fell into distinct size classes (Fig. 4.8: Beniac et al. 2012). The use of cryo-EM avoided shrinking artefacts and allowed clear delineation between full and empty filaments, since the nucleocapsid is easily identified inside virus particles. These analyses demonstrated conclusively that the length of virions tended to be a multiple of about 980 nm, indicating that multiple copies of the genome were enveloped in single long particles. About half of the virions consisted of two or more nucleocapsids (Fig. 4.8). Full virus particles containing a nucleocapsid are wider (96 nm in diameter) than empty filamentous viral envelopes, which are between 48 and 52 nm in diameter (Beniac et al. 2012). The linked viruses have clearly delineated regions, with a nucleocapsid, and then an empty linker region, followed by another nucleocapsid (Fig. 4.8d). The nucleocapsid runs the full length of the virus along the central axis of the viral filament and extends to the end of the virus, as shown by cryo-ET (Fig. 4.7c, d). It was also shown that the nucleocapsid consists of

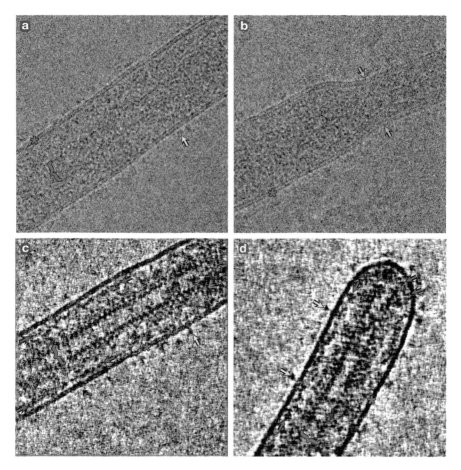

Fig. 4.7 Cryo-electron microscopy and tomography of Ebola virus particles. (**a**) A straight linear region of an Ebola virus. (**b**) Region where the nucleocapsid ends and the filament narrows to an empty tubular envelope. (c-d) Z-slices through cryo-electron tomograms of Ebola virus, showing a straight linear region (c) and the end of a viral filament (**d**). The GP spike (*yellow arrow*), viral envelope (*blue asterisk*), VP40 (*red arrow*), and nucleocapsid (*purple bracket*) are shown

two concentric helices (Figs. 4.2 and 4.9). The inner helix is at a diameter of approximately 22.3 nm, and the second is in the outer layer of the helix where the "bridge" structure is, at a diameter of approximately 37 nm. Since the genome is a single continuous length of RNA, only one of these two paths could facilitate the 18.9 kb RNA genome winding along the helix. If a model of RNA based on the winding of RNA in the respiratory syncytial virus (a paramyxovirus that also has a helical nucleocapsid) is used, one can estimate the putative length of an Ebola virus nucleocapsid containing one genome copy. If the Ebola genomic RNA is placed within the outer bridge, this would imply a nucleocapsid 561 nm long, assuming that the genome is continuously wound into the helix with no supercoiling. If the inner

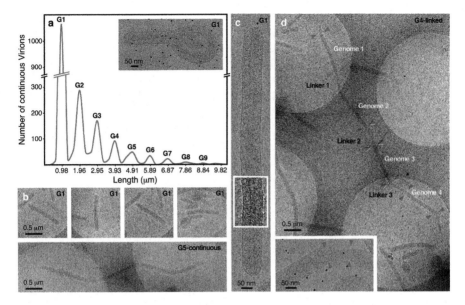

Fig. 4.8 Ebola virus polyploidy was estimated based on the length of continuous viruses which have no apparent gaps between adjacent nucleocapsids in a linear filamentous virus particle. (**a**) Measurement of the length of continuous viruses reveals a series of discrete size classes that are direct multiples of the length corresponding to one copy of the genome (G1). (**b-c**) Gallery of selected images of continuous Ebola viruses. (**d**) An example of a linked virus. In linked viruses there is a discreet thinning of the diameter of the virus after a nucleocapsid followed by a thin empty tube which then enlarges at the point where the next nucleocapsid begins. The linked virus in this image has four copies of the viral genome. Reproduced from Beniac et al. (2012)

diameter is used as the model, the nucleocapsid containing one genome copy would be approximately 916 nm long. This latter nucleocapsid length agrees well with the observed 982 nm length of the virus. If one considers that the virus is 96 nm wide, with a 41 nm diameter nucleocapsid, this would require 27.5 nm of additional length on each end to accommodate the low density gap, the VP40 layer and lipid envelope. If one adds this estimate for both ends (27.5 nm each) to the putative nucleocapsid length of 916 nm, one would have a total virus length of 971 nm which is very close to that observed, of 982 ± 79 nm, using cryo-EM (N = 1110) (Beniac et al. 2012; Booth et al. 2015). The closeness of the modelled nucleocapsid length, coupled with its similarity to the observed length of a single-genome copy Ebola virus, and the discreet length classes that are multiples of 982 nm provide strong evidence that the Ebola virus is polyploid. About 47% of the virions have multiple copies of the genome, with the longest measured virus having 22 estimated copies of the viral genome. Multiple genome-length virions, consistent with having more than one genome copy, have also been observed in infected human and animal specimens (Beniac et al. 2012) suggesting that polypoidy is a genuine feature of filoviruses, and not just an effect of cell culture growth.

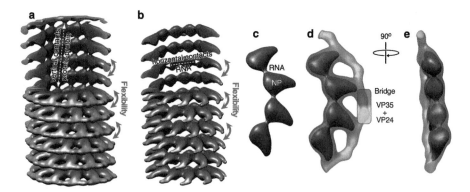

Fig. 4.9 3D structure of the Ebola virus nucleocapsid shown as a surface-shaded surface representation with radial colouring: the interior is purple and the exterior is blue. (**a, b**) The nucleocapsid structure shown at two density thresholds, both with the top section cut away to reveal the interior of the interior. Horizontal and vertical connectivity within the nucleocapsid is indicated. (**c**) A repeat of four nucleocapsid proteins is shown at a high-density cut off to reveal the putative location of the viral RNA that links the NP subunits. (**d, e**) The same repeat shown in (**c**) rendered at a lower density threshold to show the location of the VP24/VP35 bridge in relation to the NP protein and genomic RNA

The Filovirus Nucleocapsid: A Double-Layered Helix

The Ebola virus nucleocapsid is comprised of six principal components: the 18.9 kb negative-sense RNA genome, the nucleocapsid protein (NP), the polymerase protein L and viral proteins VP35, VP30 and VP24. The polymerase protein L is part of the RNA-dependent RNA polymerase complex. This polymerase complex is composed of VP35 and L, with L being the catalytic component of the complex (Ayub and Waheed 2016). The polymerase protein L is likely part of the nucleocapsid, but it exists in low copy numbers and therefore is unlikely to have a regular periodicity in step with the helical symmetry of the nucleocapsid. Thus it may not be easily revealed by image processing as a regular repeating component of the nucleocapsid (Booth et al. 2013). The NP protein, VP35 and VP30 have all been shown to bind RNA (Noda et al. 2010; Cardenas et al. 2006; John et al. 2007), whereas VP24 is involved in transcription and genome replication (Watanabe et al. 2007). It has been demonstrated that the nucleocapsid protein, VP35 and VP24 can generate nucleocapsid-like structures (Huang et al. 2002; Watanabe et al. 2006). In addition immunoprecipitation and immunofluorescence co-localization studies have demonstrated individual interactions between the NP protein and VP35, between the nucleocapsid protein and VP24 and between VP35 and VP24 (Banadyga et al. 2017). The relative stoichiometry of Ebola virus proteins have been characterized by gel electrophoresis (Elliott et al. 1985; Sanchez et al. 2007; Beniac et al. 2012; Booth et al. 2015). These studies show that the NP protein and VP35, VP30 and

VP24 are present in purified virions in equimolar ratios, with approximately 1400 copies per single-genome virus. It has been calculated using the average spacing of GP in the envelope as determined by cryo-ET that the matrix protein VP40 has approximately six times more copies then these proteins. By contrast, the polymerase protein L has only about 50 copies per virion. The equimolar ratios of the nucleocapsid protein, VP35, VP30 and VP24 suggest they are all components of a repeating, equivalent structure with helical symmetry in the nucleocapsid.

Investigations of the likely location of the RNA genome in the nucleocapsid, and nucleocapsid length modelling, have thus established that the most probable path for the genomic RNA is within the inner connecting ring of the nucleocapsid, at a diameter of 22.3 nm in the helix (Figs. 4.2 and 4.9). The structure of the nucleocapsid solved by cryo-EM at 19 Å resolution also indicates the likely location of the nucleocapsid proteins VP35 and VP24. Transfection studies, coupled with ultra-thin section electron microscopy, showed that NP, VP24 and VP35 are the minimal requirements to form a 50 nm diameter tubular structure, similar to that formed in Ebola virus-infected cells. Furthermore, when cells were transfected with only the NP protein, smaller tubular structures only 20–25 nm in diameter were formed that were nuclease sensitive (Watanabe et al. 2006; Huang et al. 2002; Noda et al. 2010; Noda et al. 2006a, b). This observation fits with the structure as determined by cryo-EM and agrees with the modelling of the RNA in a helical ring centred at 22.3 nm. In this case the self-assembled NP protein complexed with cellular RNA formed a 25 nm tube which corresponds to the inner part of the cryo-EM nucleocapsid structure (Fig. 4.9). The repeating mass on the inner helical layer of the nucleocapsid (which has a saw tooth-like pattern) has been identified as the location of the Ebola NP protein (Fig. 4.9c). This density is large enough to accommodate a partial structure of the nucleocapsid protein representing about 42% of the protein that was solved by X-ray crystallography (Dong et al. 2015). This protein has a somewhat elongated structure with a distinctive RNA binding groove between the N-lobe and C-lobe of this protein. In addition, the transfection expression data indicated that the production of VP24 and VP35 with NP causes the formation of a larger 50 nm diameter tubular filaments. This supports the interpretation of the cryo-EM structure described by Beniac et al. (2012), assigning the mass in the "bridge" structure to VP24 and VP35. This bridge forms the outer layer of the helix and maintains the same pitch and number of helical repeats. The bridge itself is thought to be composed of heterodimers of VP24 and VP35 that attach to radial protrusions of the NP protein units that are adjacent to each other in the helix (Figs. 4.2 and 4.9). The discovery of the bridge structure explains how the condensed structure of NP oligomerized with genomic RNA is able to be wind into a tight helix with a much shorter pitch than the coils formed with NP and cellular RNA alone (that produces oligomeric chains which are very loosely coiled). The addition of the outer layer imposes rigidity and a tighter helical pitch. This is clearly important for compaction of the genome so that it can be efficiently packaged into mature virions. The bridge undoubtedly also plays some role in the winding-up as well as unwinding or partial transient exposure of the genome template during replication. The details of how these functions are achieved need to be explored in the future, possibly using time-resolved cryo-EM studies.

The inclusion of VP30 with NP, VP24 and VP35 transfections had no apparent effect on the nucleocapsid-like structures generated. This suggests that although VP30 is part of the nucleocapsid, and likely binds to NP, it is not essential for the formation of the bridge structure on the exterior of the nucleocapsid and probably does not form part of the outer layer of the helix (Booth et al. 2015). This interpretation is consistent with other previous studies showing that VP30 is part of the nucleocapsid, but is not required for the formation of the 50 nm diameter nucleocapsid-like structures (Huang et al. 2002; Sanchez et al. 2007; Groseth et al. 2009). The co-location of all these proteins in the nucleocapsid is clearly important for their function during replication, something which has been often overlooked in earlier studies to explore the effects and interactions of these viral proteins with cellular proteins when expressed singly in isolation.

The final important feature of the nucleocapsid structure, as seen by cryo-EM, is that there are no vertical connections between the bridges of adjacent layers of the helical repeat. The only vertical connectivity between successive turns of the helix are between the stacked coils of the N protein, which form the inner helical layer (Fig. 4.8a). In this structure the nucleocapsid can easily pivot about the coils of the N-protein filaments, while at the same time the VP35-VP24 bridge structure appears to stabilize the nucleocapsid structure when it bends (Figs. 4.2 and 4.9: Booth et al. 2013). In many ways the Ebola virus nucleocapsid is similar to a hollow corrugated flexible pipe. This flexibility of the nucleocapsid is essential in order for the virus to be able to take on the curved linear, comma and torroidal shapes that have been observed (Fig. 4.3).

The nucleocapsid appears to be assembled in the cytoplasm of the cell, within apparent "virus factories", as observed in thin sections of virus infected cells or in transfected cells expressing the N protein or the N protein in conjunction with several other nucleoproteins (Watanabe et al. 2006). The VP40 protein plays an essential role in the transport of the nucleocapsids to the plasma membrane of the cell and in subsequent incorporation of the nucleocapsid into the virus (Noda et al. 2006a). A model for morphogenesis of Ebola virions has been developed, whereby nascently formed progeny nucleocapsids, once assembled within the cell, are transported to the cell surface where budding and envelopment is driven through specific interactions between one or more of the nucleocapsid proteins and VP40 (Fig. 4.1). This model also explains how multiple copies of the genome, arranged as repeating helical nucleocapsid cylinders, connected end-to end, might be incorporated into linked or continuous virus particles, containing both single or multiple copies of the virus genome, during envelopment at the cell surface during the budding and maturation process (Fig. 4.1).

Comparisons Between Filoviruses and Other Non-segmented Negative-Strand Viruses

As mentioned previously, all of the *Mononegavirales*, including rhabdoviruses, paramyxoviruses and filoviruses, have helical nucleocapsids. Both Ebola virus and vesicular stomatitis virus (VSV: a rhabdovirus) have a cylindrical-shaped virion with a helical nucleocapsid that runs along the central axis (Ge et al. 2010; Beniac et al. 2012). Also, they both have one species of GP spike on the exterior of the lipid envelope, with the internal matrix proteins adjacent to the interior of the envelope. Like all rhabdoviruses, VSV is bullet shaped so that each virion has a distinctive pointed and blunt end. The length of the virus is fixed, and the cylindrical shape is fairly rigid. Ebola virus differs in that both ends appear similar and are hemispherical; the virus exists in multiple lengths (Fig. 4.8), its filaments are extremely flexible and it has several distinctive morphological forms (Figs. 4.1 and 4.3). In VSV there is an ordered symmetrical relationship between the nucleocapsid proteins and the neighbouring matrix protein (M). The M protein has a unique hub region which connects to both neighbouring M and nucleocapsid (N) proteins. Thus in rhabdoviruses the M protein probably maintains the helical symmetry of the nucleocapsid. Ebola virus has a more complex, double-layered nucleocapsid, and while there are clearly specific contacts between the NC proteins and the matrix protein VP40, that promote virion morphogenesis and promote a stable filamentous structure with a uniform cylindrical diameter, the helical symmetry of the NC appears not to be repeated by VP40 in the envelope layer, implying a more complex structure possibly involving quasi-equivalent interactions. The double-layered helical nucleocapsid of filoviruses is unique, where VP24 and VP35 form a bridge structure in the outer layer of the nucleocapsid (Figs. 4.2 and 4.9). These additional components, plus the inherent flexibility of the Ebola virus compared to the rigid VSV, indicate at two related viruses have distinctively different methods of attachment between their matrix proteins and nucleocapsids. Thus rhabdoviruses have a short, stiff shape, with no polyploidy present, and an inflexible, rigid association between the nucleocapsid and M protein. Paramyxoviruses, like filoviruses, have an extremely flexible nucleocapsid, but instead of a filamentous envelope, these NCs are enveloped in large spherical enveloped virions, where polyploidy is possible and specific interactions between the NC and matrix proteins appear to be highly variable allowing random coiling of the NCs within the envelope. In addition, in paramyxoviruses, the helical nucleocapsid demonstrates some structural flexibility and the ability to have variation in both the pitch and twist per helical turn of the nucleocapsid, at least in some species (Bhella et al. 2004). By contrast we observed only one type of helical organization in Ebola virus nucleocapsids. Filoviruses thus represent an intermediate morphology between rhabdoviruses and paramyxoviruses, being filamentous and flexible, with a nucleocapsid that has a conserved helical structure that is centred within the tubular envelope. In filoviruses the NC lies on the central axis of the viral filament, adjacent to region of low density. This indicates relatively strong interactions with the matrix protein, but these contacts do not appear to impose any corresponding helical order on the matrix protein VP40 or the GP spike molecules in the envelope.

The nature of these contacts between the NC and envelope layers in filoviruses was investigated by cryo-EM and linear 2D image processing at radial depths along the length of virions (Figs. 4.4 and 4.10a). Images which encompassed half the diameter of the Ebola virus were used to generate class averages of the nucleocapsid and lipid bilayer of the viral envelope (Fig. 4.10a; left image: Beniac and Booth 2017). Inclusion of the NC layers lead to blurring of the GP spikes and VP40 layer. This data clearly shows that both the nucleocapsid and the viral envelope are the dominant repetitive features in these images. In contrast, mass density attributed to VP40 and GP appears as a smear. This "smearing" of density due to the VP40 and GP layers could be attributed to insufficient contrast or a low degree of order insufficient to be revealed by image processing or their having a regular repeating pattern that does not match (or is out of phase) with the regular repeating pattern of the nucleocapsid (Fig. 4.10a). Image processing was carried out using a series of masks designed to test whether repeating patterns were present in the virion (Fig. 4.10b–d, left column). Each mask contained the region over the viral envelope, as well as either the VP40 matrix protein layer (b), the nucleocapsid (c) or GP spike (d). For each of these three tests, the masked images were aligned and processed by multivariate statistical analysis and hierarchical ascendant classification (Frank et al. 1996; Beniac and Booth 2017). Once the parameters were determined for each masked image, they were also applied to unmasked images (Fig. 4.10b–d). The results show the orientation of the VP40 layer, the nucleocapsid layer and GP layer within the virion (Fig. 4.10).

The results in Fig. 4.10 failed to reveal any apparent stoichiometric structural co-ordination between VP40 or the GP spike layers, nor any ordered interactions between either of these layers and the nucleocapsid of the Ebola virion. This indicates that if there is any structural/stoichiometric relationship between the nucleocapsid or the GP spike and VP40, it was not sufficiently ordered to be revealed by linear analysis of the currently available data, or more likely, the NC and VP40 are arranged on different lattices where some mismatch in protein-protein interactions can be accommodated.

Fig. 4.10 (continued) structure (*green arrows*). However, in all the other cases, the adjacent regions of density did not show a periodic structure present (*red* X). In all cases a blur of density indicates that the protein is present but not sufficiently organized into a pattern that would put it in "phase" with the NC region included in the analysis. (**e**) Tomographic Z-section shows the location of these proteins when single particle analysis techniques are not used, for comparison. The colour-coding is nucleocapsid, purple/blue; VP40, red; and GP spike, yellow. Regions that do not correlate are labelled with a red X and regions that correlate, with a green arrow. Parts of this figure were reproduced from (Beniac and Booth 2017)

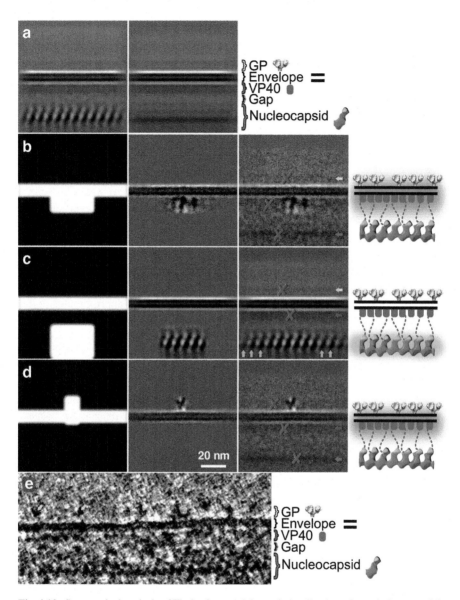

Fig. 4.10 Symmetrical analysis of Ebola virus particles and visualization of protein layers and the envelope. (**a**) Average of one side (half) of a section of a linear filamentous region of Ebola virus is shown on the left. On the right is a horizontally averaged image of the one on the left, to show how non-periodic densities can be blurred by averaging. (**b**–**d**) The effect of masking of specific regions of the virus on adjacent regions of the virus. The mask used is shown on the left, the masked image average that was used for single-particle analysis is presented in the centre column and on the right the corresponding unmasked image averages are shown. These images show the effect of including just the image regions corresponding to the lipid envelope plus VP 40 (**b**), the lipid envelope plus the nucleocapsid (**c**) and the lipid envelope plus the GP spike (**d**). When the nucleocapsid included (**c**), the adjacent regions of the nucleocapsid showed a repeating periodic

Conclusions

Analysis of the variety of viral structures and morphologies present in bona fide Ebola virus (Figs. 4.1, 4.9 and 4.10), and comparisons with that of virus-like particles (Fig. 4.6), has generated a consensus model for Ebola virus. This includes the structure of the nucleocapsid and suggests a general scheme for morphogenesis and maturation of the virion (Fig. 4.1). The model reconciles and visualizes how protein-protein and protein-nucleic acid interactions, which have also been observed by biochemical means, take place to generate oligomeric structures during virion formation. At the centre of the model is a highly organized nucleocapsid, which in Ebola virus has a pitch of 6.96 nm and 10.81 repeats per helical turn. The helix is double layered with the NP protein and RNA genome in the inner layer and has a distinctive bridge structure in the outer layer composed of VP24 and VP35. There are no vertical connections between bridges in successive vertical turns of the helix, which explains the observed flexibility of the NC and its ability to bend into tight curves, to form, for example, comma-shaped and torroidal forms of the virion. The VP40 protein is found in a layer which is adjacent to the inner leaf of the viral envelope. This VP40 layer is arranged in a lattice with a spacing of approximately 5 nm, but its exact symmetry is unclear. The matrix layer of filoviruses clearly allows flexibility of the envelope, and interactions between the NC and VP40 must also allow flexibility, unlike rhabdoviruses where a more rigid helical symmetry is maintained from the NC through the envelope to the GP spike layer. In addition, filoviruses have a low-density gap between the VP40 layer and the nucleocapsid, and the spacing of this gap is conserved, which keeps the NC in a centred position along the long axis of the Ebola virus filament. It is likely that this low-density gap represents a region of flexible contacts between VP40 and the nucleocapsid which maintain this space and also allow flexibility for the entire filovirus virion. The overall structural organization of filoviruses allows the possibility for a high degree of polyploidy, meaning that virus particles can have many copies of the viral genome, and up to 20 or more have been identified in a single virion. It is intriguing to speculate that, since this family includes a variety of agents that cause severe hemorrhagic fevers, polyploidy may partially account for the capability of filoviruses to replicate to very high titres and be easily transmitted between hosts. In addition, the filamentous morphology may facilitate invasion of tissues through penetration of endothelial membranes, and this may also play a role in widespread dissemination of filovirus infection in a wide variety of tissues and cells in the body. Future studies are needed to reveal how the functions of replication are achieved, presumably through structural changes in the nucleocapsid. The use of time-resolved cryo-EM analyses of in vitro-assembled subviral particles, as well as isolated structures generated during real infections, will provide valuable approaches.

References

Adrian M, Dubochet J, Lepault J et al (1984) Cryo-electron microscopy of viruses. Nature 308:32–36

Almeida JD, Waterson AP, Simpson DIH (1971) Morphology and Morphogenseis of the Marburg agent. In: Martini GA, Siegert R (eds) Marburg Virus Disease. Springer, Berlin, pp 84–97

Ayub G, Waheed Y (2016) Sequence analysis of the L protein of the Ebola 2014 outbreak: insight into conserved regions and mutations. Mol Med Rep 13:4821–4826

Banadyga L, Hoenen T, Ambroggio X et al (2017) Ebola virus VP24 interacts with NP to facilitate nucleocapsid assembly and genome packaging. Sci Rep 7:7698. https://doi.org/10.1038/s41598-017-08167-8

Beniac DR, Booth TF (2017) Structure of the Ebola virus glycoprotein spike within the virion envelope at 11 a resolution. Sci Rep 7. https://doi.org/10.1038/srep46374

Beniac DR, Melito PL, Devarennes SL et al (2012) The organisation of Ebola virus reveals a capacity for extensive, modular polyploidy. PLoS One 7:e29608. https://doi.org/10.1371/journal.pone.0029608

Bhella D, Ralph A, Yeo RP (2004) Conformational flexibility in recombinant measles virus nucleocapsids visualised by cryo-negative stain electron microscopy and real-space helical reconstruction. J Mol Biol 340:319–331

Booth TF, Beniac DR, Rabb MJ et al (2015) Filovirus structure and morphogenesis. In: Pattnaik AK, Whitt MA (eds) Biology and Pathogenssis of Rhabdo- and filoviruses. World Scientific Publishing Co. Pte. Ltd, Singapore, pp 427–451

Booth TF, Rabb MJ, Beniac DR (2013) How do filovirus filaments bend without breaking? Trends Microbiol 21:583–593

Bornholdt ZA, Noda T, Abelson DM et al (2013) Structural rearrangement of ebola virus VP40 begets multiple functions in the virus life cycle. Cell 154:763–774

Brindley MA, Hughes L, Ruiz A et al (2007) Ebola virus glycoprotein 1: identification of residues important for binding and postbinding events. J Virol 81:7702–7709

Cardenas WB, Loo YM, Gale M et al (2006) Ebola virus VP35 protein binds double-stranded RNA and inhibits alpha/beta interferon production induced by RIG-I signaling. J Virol 80:5168–5178

Carette JE, Raaben M, Wong AC et al (2011) Ebola virus entry requires the cholesterol transporter Niemann-pick C1. Nature 477:340–343

Chandran K, Sullivan NJ, Felbor U et al (2005) Endosomal proteolysis of the Ebola virus glycoprotein is necessary for infection. Science 308:1643–1645

Cote M, Misasi J, Ren T et al (2011) Small molecule inhibitors reveal Niemann-pick C1 is essential for Ebola virus infection. Nature 477:344–348

Dong S, Yang P, Li G et al (2015) Insight into the Ebola virus nucleocapsid assembly mechanism: crystal structure of Ebola virus nucleoprotein core domain at 1.8 a resolution. Protein Cell 6:351–362

Dube D, Brecher MB, Delos SE et al (2009) The primed ebolavirus glycoprotein (19-kilodalton GP1,2): sequence and residues critical for host cell binding. J Virol 83:2883–2891

Dubochet J, Adrian M, Chang JJ et al (1988) Cryo-electron microscopy of vitrified specimens. Q Rev Biophys 21:129–228

Elliott LH, Kiley MP, McCormick JB (1985) Descriptive analysis of Ebola virus proteins. Virology 147:169–176

Frank J, Radermacher M, Penczek P et al (1996) SPIDER and WEB: processing and visualization of images in 3D electron microscopy and related fields. J Struct Biol 116:190–199

Ge P, Tsao J, Schein S et al (2010) Cryo-EM model of the bullet-shaped vesicular stomatitis virus. Science 327:689–693

Geisbert TW, Jahrling PB (1995) Differentiation of filoviruses by electron microscopy. Virus Res 39:129–150

Gong X, Qian H, Zhou X et al (2016) Structural insights into the Niemann-pick C1 (NPC1)-mediated cholesterol transfer and Ebola infection. Cell 165:1467–1478

Groseth A, Charton JE, Sauerborn M et al (2009) The Ebola virus ribonucleoprotein complex: a novel VP30-L interaction identified. Virus Res 140:8–14

Hoenen T, Volchkov V, Kolesnikova L et al (2005) VP40 octamers are essential for Ebola virus replication. J Virol 79:1898–1905

Hood CL, Abraham J, Boyington JC et al (2010) Biochemical and structural characterization of cathepsin L-processed Ebola virus glycoprotein: implications for viral entry and immunogenicity. J Virol 84:2972–2982

Hosaka Y, Kitano H, Ikeguchi S (1966) Studies on the pleomorphism of HVJ virons. Virology 29:205–221

Huang Y, Xu L, Sun Y, Nabel GJ (2002) The assembly of Ebola virus nucleocapsid requires virion-associated proteins 35 and 24 and posttranslational modification of nucleoprotein. Mol Cell 10:307–316

John SP, Wang T, Steffen S et al (2007) Ebola virus VP30 is an RNA binding protein. J Virol 81:8967–8976

Johnson RF, McCarthy SE, Godlewski PJ et al (2006) Ebola virus VP35-VP40 interaction is sufficient for packaging 3E-5E minigenome RNA into virus-like particles. J Virol 80:5135–5144

Kuhn JH, Radoshitzky SR, Guth AC et al (2006) Conserved receptor-binding domains of Lake Victoria marburgvirus and Zaire ebolavirus bind a common receptor. J Biol Chem 281:15951–15958

Loney C, Mottet-Osman G, Roux L et al (2009) Paramyxovirus ultrastructure and genome packaging: cryo-electron tomography of sendai virus. J Virol 83:8191–8197

Luque D, Rivas G, Alfonso C et al (2009) Infectious bursal disease virus is an icosahedral polyploid dsRNA virus. Proc Natl Acad Sci U S A 106:2148–2152

Manicassamy B, Wang J, Jiang H et al (2005) Comprehensive analysis of ebola virus GP1 in viral entry. J Virol 79:4793–4805

Mpanju OM, Towner JS, Dover JE et al (2006) Identification of two amino acid residues on Ebola virus glycoprotein 1 critical for cell entry. Virus Res 121:205–214

Noda T, Ebihara H, Muramoto Y et al (2006a) Assembly and budding of ebolavirus. PLoS Pathog 2:e99. https://doi.org/10.1371/journal.ppat.0020099

Noda T, Hagiwara K, Sagara H et al (2010) Characterization of the Ebola virus nucleoprotein-RNA complex. J Gen Virol 91:1478–1483

Noda T, Sagara H, Yen A et al (2006b) Architecture of ribonucleoprotein complexes in influenza a virus particles. Nature 439:490–492

Noda T, Watanabe S, Sagara H et al (2007) Mapping of the VP40-binding regions of the nucleoprotein of Ebola virus. J Virol 81:3554–3562

Radermacher M, Wagenknecht T, Verschoor A et al (1987) Three-dimensional structure of the large ribosomal subunit from Escherichia Coli. EMBO J 6:1107–1114

Reguera J, Gerlach P, Cusack S (2016) Towards a structural understanding of RNA synthesis by negative starnd RNA viral polymerases. Curr Opin Struct Biol 36:75–84

Sanchez A, Geisbert TW, Feldmann H (2007) Filoviridae: Marburg and Ebola Viruses. In: Knipe DM, Howley PM (eds) Fields Virology. Lippincott Williams & Wilkins, Philadelphia, pp 1409–1448

Schornberg K, Matsuyama S, Kabsch K et al (2006) Role of endosomal cathepsins in entry mediated by the Ebola virus glycoprotein. J Virol 80:4174–4178

Watanabe S, Noda T, Halfmann P et al (2007) Ebola virus (EBOV) VP24 inhibits transcription and replication of the EBOV genome. J Infect Dis 196(Suppl 2):S284–S290

Watanabe S, Noda T, Kawaoka Y (2006) Functional mapping of the nucleoprotein of Ebola virus. J Virol 80:3743–3751

Chapter 5
Structure and Function of Influenza Virus Ribonucleoprotein

Chun-Yeung Lo, Yun-Sang Tang, and Pang-Chui Shaw

Introduction

The influenza virus is a highly contagious pathogen that causes annual epidemics and sporadic pandemics, leading to substantial human morbidity and mortality and imposing heavy burden on our healthcare system. Although vaccines and antivirals have been developed to counteract influenza virus, its rapid mutation rate results in the efficient generation of resistant strains. Reassortment of virus strains within reservoir species produces novel influenza virus that posesses serious threat to humankind.

Influenza belongs to the Orthomyxoviridae family. It is a negative-sense single-stranded RNA virus ((−)ssRNA virus). Its genome is divided into eight segments. Each negative-sense RNA segment is encapsidated into a ribonucleoprotein (RNP) molecule, consisting of a trimeric RNA-dependent RNA polymerase complex (including PA, PB1 and PB2) and numerous NP proteins. RNP plays a central role in the life cycle of influenza virus. In the virion, genetic information is stored in the form of RNP. In infected cells, RNP is required for the transcription and replication of viral genome. In the generation of viral progenies, newly synthesized viral RNA has to be assembled into RNP before being packaged into a new viral particle.

In this chapter, we summarize current understanding on the structure of RNP complex, as well as the structure of each subunit. In addition, we also incorporate the latest findings on how the viral transcription and replication are carried out. Besides, other assistive and modulatory functions of each subunit are discussed.

C.-Y. Lo · Y.-S. Tang · P.-C. Shaw (✉)
Centre for Protein Science and Crystallography, School of Life Sciences, The Chinese University of Hong Kong, Shatin, N.T, Hong Kong, China
e-mail: pcshaw@cuhk.edu.hk

© Springer Nature Singapore Pte Ltd. 2018
J. R. Harris, D. Bhella (eds.), *Virus Protein and Nucleoprotein Complexes*, Subcellular Biochemistry 88, https://doi.org/10.1007/978-981-10-8456-0_5

Structure of RNP and Its Subunits

RNP is composed of an RNA-dependent RNA polymerase (RdRP) with three subunits – PA, PB1, and PB2, nucleoprotein (NP), and vRNA. RNP has a rod-shaped structure. Its head contains the RdRP complex. The vRNA segment extends from the head and is encapsidated by a number of NP molecules (Reviewed in Eisfeld et al. 2014).

Several low-resolution 3D structures of recombinant influenza RNP and RdRP were determined with electron microscopy and image processing (Martín-Benito et al. 2001; Area et al. 2004; Torreira et al. 2007; Coloma et al. 2009; Resa-Infante et al. 2010). Various crystal structures of RNP subunits started to emerge around in 2007 (Ye et al. 2006; Ng et al. 2008; He et al. 2008; Obayashi et al. 2008; Guilligay et al. 2008; Sugiyama et al. 2009; Yuan et al. 2009; Dias et al. 2009). The structure of native RNP was also determined by cryo-EM to around 20 Å in 2012 (Moeller et al. 2012; Arranz et al. 2012).

In 2014, crystal structure of vRNA promoter-bound influenza RdRP complex was determined using bat-derived H17N10 influenza virus and influenza B virus (Pflug et al. 2014; Reich et al. 2014). Soon after, the structures of 5' cRNA-bound flu B RdRP and apo form of flu C RdRP were also solved (Hengrung et al. 2015; Thierry et al. 2016). Recently, the structure of an initiation state-resembling form of flu B RdRP was determined (Reich et al. 2017).

Structure of Influenza RNA-Dependent RNA Polymerase (RdRP) Complex

RdRP is a heterotrimeric complex with a U-shaped architecture (Fig. 5.1) (Pflug et al. 2014; Reich et al. 2014). PB1 is located at the center. Top region of the U contains two protruding arms formed by the PA-N endonuclease domain and the PB2 cap-binding domain. Both arms are in close proximity to the primer-entry/product-exit channel. The PA-C terminal domain and the promoter binding site are located at the bottom region of the U. The linker connecting PA-N and PA-C lies on the surface of PB1, forming extensive interaction with it. A large catalytic cavity is identified within PB1. There are four channels connecting the PB1 catalytic cavity from the exterior: the (1) template entry channel, (2) product exit channel, (3) NTP entry channel, and (4) template exit channel.

PB1 forms the central scaffold of RdRP. It interacts with PA and PB2 through its N-terminal and C-terminal region, respectively, echoing previous mapping studies and co-crystal structure determination of RdRP subunits (González et al. 1996; Ohtsu et al. 2002; He et al. 2008; Obayashi et al. 2008; Sugiyama et al. 2009). However, the inter-subunit interactions are much more extensive than previously thought.

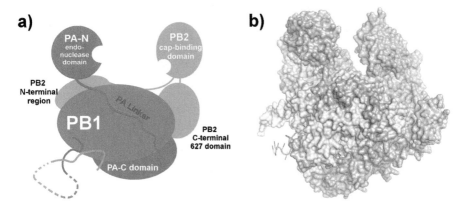

Fig. 5.1 Structure of influenza RdRP heterotrimer (PDB: 4WSB). (a) Schematic diagram showing the architecture of RdRP heterotrimer. RdRP subunits assemble a U-shaped architecture. Major domains of each subunit are shown in oval. (b) Surface representation of RdRP showing a compact structure of RdRP heterotrimer. PA, PB1, and PB2 are colored in pink, pale blue, and pale green, respectively. Extensive inter-subunit interactions can be identified in the structure. The open cleft at the top of the U-shape is the location for cap-snatching. A capped RNA is captured by the PB2 cap-binding domain and is directed to the PA-N endonuclease for cleavage at 10–13 nt downstream of the cap. As a result, a capped primer is generated for viral protein transcription

Various structures of RdRP reveal that the RdRP can undergo substantial conformational change (Resa-Infante et al. 2011; Pflug et al. 2014; Reich et al. 2014; Thierry et al. 2016). Two radically different configurations of PB2 could be observed (Thierry et al. 2016). The PA-N domain is also capable of in situ rotation (Reich et al. 2015; Hengrung et al. 2015; Thierry et al. 2016). This might be related to the activation of the cap-snatching mechanism or to the switch between replication and transcription mode of RdRP. In contrast, the structures of PB1, PA-C, and PA-linker remain relatively stable among various RdRP structures.

Structure of PA Subunit

PA is 716 aa long, and limited tryptic digestion reveals that it consists of two domains: (1) N-terminal (PA-N) domain and (2) C-terminal (PA-C) domain (Guu et al. 2008). PA-N contains the first 196 residues and is the smaller subunit (~25 kDa). The larger PA-C domain (~55 kDa) contains residues from 277 to 716. The two domains are connected by a long linker (residues 197–276).

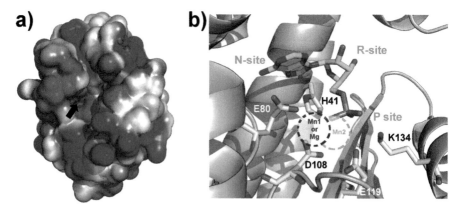

Fig. 5.2 Structure of PA-N domain (PDB: 2W69). (a) Surface representation of PA-N showing its electrostatic potential. The catalytic site is indicated by a black arrow; it is strongly negatively charged. (b) Detailed structure showing the active site with a bound AMP. Crucial catalytic residues, H41, E80, D108, E119, and K134 are shown in white. H41, D108, E119, and K134 are involved in positioning the divalent ions. The positions of divalent ions Mg^{2+} or Mn^{2+} are marked in the structure. The AMP molecule reveals the nucleotide-binding site (N-site), ribose binding site (R-site), and the phosphate-binding site (P-site). Drawn using PDB: 3HW5

PA-N Domain

PA-N forms a novel global fold. The PA-N structure has an α/β architecture, where five mixed β-strands form a twisted plane surrounded by seven α-helices. Structural alignment with type II restriction endonucleases reveals that PA-N contains a PD-(D/E)XK motif, a structural motif characteristic of many nucleases (Dias et al. 2009; Yuan et al. 2009).

A negatively charged cavity surrounded by helices α2–α5 houses one or two divalent cations (Fig. 5.2a). The structure by Yuan et al. includes one Mg^{2+}, while the structure by Dias et al. includes two Mn^{2+}. The divalent metal ions are coordinated by the same residues (H41, D108, E119, and K134) in both structures. The Mg^{2+} ion is coordinated by five ligands: E80, D108, and three water molecules that are stabilized by H41, E119, and the carbonyl oxygens of L106 and P107. For the Mn^{2+}-bound structure, one ion is coordinated by E80, D108, and two water molecules and the other ion by H41, D108, E119, and the carbonyl oxygen of I120. D108 acts as a bridge for both Mn^{2+} ions (Fig. 5.2b).

Co-crystal structures of PA-N with various nucleoside monophosphates (NMP) were determined by Zhao et al. (Zhao et al. 2009). NMPs were shown to bind at the catalytic center of PA-N, in close proximity with the divalent cations. These complexes represent the scenario nuclease cleavage reaction. E119, K134, K137, and Y130 form the phosphate-binding site (P site). H41 also contributes in stabilizing the phosphate moiety. Main chain atoms of A37, I38, C39, and the side chain of H41 form the ribose-binding site (R-site). A20, L38, L42, and E80 form the nucleoside-binding site (N site), which bind the nucleobase in a flexible manner (Fig. 5.2b).

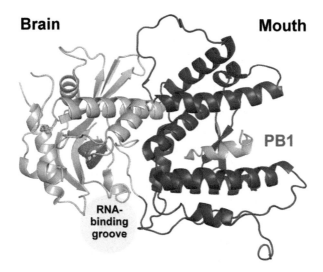

Fig. 5.3 Structure of PA-C (PDB: 2ZNL). Cartoon representation of PA-C showing the "brain" domain (colored in pink) and "mouth" domain (colored in dark red). The N-terminus of PB1 (colored in purple) inserts between the jaws of the mouth domain. A putative RNA-binding groove resides at the brain domain

The structure of residues 55–66 is highly variable. A truncated form of PA-N, where residues 52–64 were replaced by a single glycine, is able to fold correctly and can be crystallized. Besides, another study shows that the deletion of residues of 51–72 can retain normal endonuclease activity and proper folding (DuBois et al. 2012). The exact function of this loop is unclear.

PA-C Domain

PA-C forms a novel fold and can be subdivided into two parts, the "brain" domain and the "mouth" domain (Fig. 5.3). The "brain" domain contains a deep semicircular basic groove, hypothesized to be a putative RNA-binding groove (Obayashi et al. 2008; He et al. 2008). Highly conserved basic residues (K328, K539, R566, and K574) are found lining in the groove.

On the other hand, the "mouth" domain of PA-C interacts with PB1-N. The first 15 residues of PB1-N bind obliquely to a hydrophobic cavity of PA-C, located between the jaws (Obayashi et al. 2008; He et al. 2008). Mutagenic study shows that the PB1 LLFL motif (residues 7–10) is crucial for the binding of PB1-N to PA-C (Perez and Donis 2001). The apo form of PA-C without PB1-N also shows a highly similar structure (Moen et al. 2014).

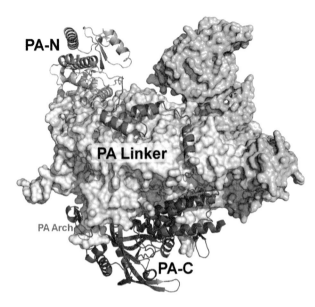

Fig. 5.4 Full length PA subunit in the context of RdRP (PDB: 4WSB). PA is shown in cartoon representation, while PB1 and PB2 are shown in surface diagram. PA-N (colored in pink) locates at the top of the U structure of RdRP, while PA-C (colored in dark red) locates at the bottom. The linker region (colored in bright red) lies on the surface of PB1. The previously uncharacterized PA arch (residue 366–397) locates at the bottom and wraps around a β-hairpin of PB1. The RdRP promoter binding site is located in close proximity to the PA arch

Full-Length PA

In the context of RdRP heterotrimer, the overall folds of PA-N and PA-C are highly similar to the structures determined from individual domain.

The previously uncharacterized flexible linker region (residues 197–276) was revealed to be lying across the PB1 surface and contributes to multiple inter-subunit interactions (Fig. 5.4).

New inter-subunit interactions were also identified. PA-N was shown to interact with both PB1 and PB2 through its helix α4. M92 and S93 form polar interactions with PB2, while E77 and T89 form polar interactions with PB1. On the other hand, the endonuclease flexible loop (residue 67–74) packs on PB1 helix α22, where K73 interacts with PB1 E731.

In PA-C domain, a formerly disordered loop (PA arch) was characterized in the full structure of PA (Fig. 5.5). The PA arch consists of residues 366–397 (residues of influenza A bat strain; corresponding to residues 371–402 of other human/avian strains). This loop forms an arch that allows a PB1 β-hairpin (residues 353–370) to insert. The PA arch is in close proximity to the 5'-vRNA hook. Residues 366–370 on the PA arch form a phosphate-binding loop, interacting with the backbone of 5'-vRNA promoter A10–A11.

Fig. 5.5 Structure of the PA arch. PA is shown in cartoon representation, while PB1 and PB2 are shown in surface diagram. The orange ribbon represents 5' vRNA. The hooked structure of 5' vRNA is held by both PB1 and PA-C. The PA arch forms a loop that encloses a PB1 hairpin (residue 353–370). On its own, the PA arch is disordered

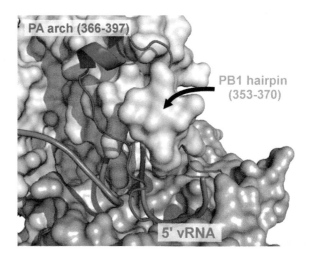

On the other hand, a previously disorganized region consisting of residues 544–553 (550-loop, corresponding to residues 549–558 of other human/avian strains) was shown to be a β-hairpin loop in the full-length PA structure. A recent co-crystal structure of bat influenza A full RdRP and serine-5-phosphorylated Pol II C-terminal domain (CTD) peptide shows that the 550-loop is in close proximity to the bound Pol II CTD peptide (Lukarska et al. 2017).

Furthermore, in various structures of full RdRP (flu A, flu B, and flu C), PA-N was shown to be capable of in situ rotation (Pflug et al. 2014; Reich et al. 2014; Hengrung et al. 2015; Thierry et al. 2016). However, the linker region remains at the same position. This suggested that the flexible loop region at the beginning of the linker region might play a role in the orientation of the PA-N domain. This might be related to the regulation between transcription and replication of the viral genome.

Structure of PB1 Subunit

Structure of PB1 has remained elusive for a long time due to its structural instability when expressed alone (Swale et al. 2016). Two co-crystal structures, PA-PB1 and PB1-PB2, have been determined and reveal the structure of the 15 N-terminal and 80 C-terminal residues of PB1 (Obayashi et al. 2008; He et al. 2008; Sugiyama et al. 2009).

The PB1 N-terminal (1–15) forms a 3_{10} helix that inserts into the PA-C hydrophobic groove. The 3_{10} helix contains a LLFL hydrophobic motif which is crucial for the interaction. Mutation of the first 12 residues could significantly disrupt PA-PB1 interaction (Perez and Donis 2001; Wunderlich et al. 2011).

The PB1 C-terminal (678–757) forms three α-helices that interacts with another three α-helices from the N-terminal region of PB2. Both polypeptides were unable to form tertiary structure on its own. PB1-PB2 interacts mainly through polar

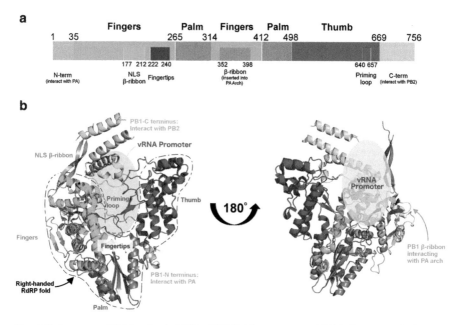

Fig. 5.6 Structure of PB1 subunit (PDB: 4WSB). The structure of PB1 shows a typical right-handed fold of RNA-dependent RNA polymerase, with both N-terminal and C-terminal extensions. The right-handed fold of PB1 (marked by dotted lines) consists of the fingers, palm, and thumb domain (colored in green, orange, and red, respectively). The N-terminus of PB1 (colored in gold) forms a short helix that interacts with PA-C, and the C terminus of PB1 (colored in light green) forms several helices that interact with PB2-N. A long, flexible β-ribbon carrying a bipartite NLS (colored in cyan) protrudes from the finger domain. It can interact with RanBP5 and is in close proximity with the viral promoter. A loop termed fingertips (colored in purple) is near the NTP entry channel. It is likely to be involved in NTP channeling and binding. Furthermore, a priming loop (colored in blue) is located in the PB1 cavity. It is crucial for de novo replication of viral genome. Lastly, the β-ribbon that inserts into PA arch (colored in dark yellow) is located on the exterior of the fingers domain

interaction, where PB1 K698 and D725 form salt bridges with PB2 (Sugiyama et al. 2009).

Full structure of PB1 was only determined in the context of the vRNA-bound full RdRP (Fig. 5.6) (Pflug et al. 2014; Reich et al. 2014). PB1 shows a typical right-handed RdRP fold, with the fingers, fingertips, palm, and thumb domains that are present in other RNA-dependent RNA polymerases (reviewed in te Velthuis 2014; Reguera et al. 2016).

A large cavity can be found within PB1, in conjunction with the PB2 N-terminal domain. This forms the active site and it is accessible through several channels. The NTP entry channel involves highly conserved PB1 basic residues located mainly in the palm domain and the fingertip region, in close proximity to PA-C. The template entry channel involves all three subunits and is located near the promoter binding region. The product exit site is at the opposite site template entry channel, where the PA endonuclease, PB2 lid domain, and PB2 cap-binding domain are present. The template exit channel is located near the PB2 lid domain, which was proposed to be involved in separating the template-product duplex. The four-tunnel structure resembles other viral RdRP (Reguera et al. 2016).

The active site contains residues D305 on motif A and D445/D446 on motif C that play a crucial role in coordinating divalent metal ions and promote catalysis. Mutation of these residues abolish polymerase activity of PB1 (Pflug et al. 2014; Biswas and Nayak 1994). A parallel β-loop protruding into the active site from the thumb domain forms a priming loop (residues 641–657). This loop resembles the HCV priming loop and is important for de novo RNA synthesis (Pflug et al. 2014; Appleby et al. 2015).

The full RdRP structure confirms the roles of PB1 N-terminus and C-terminus in the interaction with PA and PB2, respectively. However, other regions were also revealed to be involved in inter-subunit interactions: (1) the surface of the finger domain is wrapped by the PA-linker, (2) the C-terminal thumb domains interact with both PB2-N and PA-C, and (3) the palm-domain interacts with the C-terminal domain of PB2 (Pflug et al. 2014). In the cRNA promoter-bound full RdRP structure, other inter-subunit interactions involving PB1 are detected due to conformational change (Thierry et al. 2016).

A long, flexible β-ribbon (residues 177–212; bat strain residues) protrudes out from the finger domain of PB1 (Pflug et al. 2014). It contains the bipartite NLS that was previously proven to be involved in RanBP5 binding and nuclear import of the PA-PB1 heterodimer.

Another β-hairpin at the finger domain protrudes into an extended loop in PA called the PA arch (Pflug et al. 2014). Both regions are involved in 5′ promoter anchoring. PB1 was also found to bind 3′ and 5′ viral promoter with the help of PB2/PA and PA, respectively.

Structure of PB2 Subunit

PB2 subunit is 759 residues in size. The first one-third (residues 1–247) constitute a N-terminal domain, and the remaining two-third make up a C-terminal domain (PB2-C). Individual domain structures generally correlate well when aligned onto the holo complex. The structure displays a high degree of flexibility and adopts different conformations in order to sustain steps in replication and transcription (Fig. 5.7).

PB2 N-Terminal Domain (1–247)

Structure of first 37 residues of PB2-N was obtained at first as a complex structure with PB1-C. In fact, apo-PB2 binds like a clamp on the PB1 core. The first and major point of anchor is via PB2-N binding onto the C-terminus of PB1. The PB1-C-PB2-N structure by Sugiyama et al. (Sugiyama et al. 2009) largely correlated

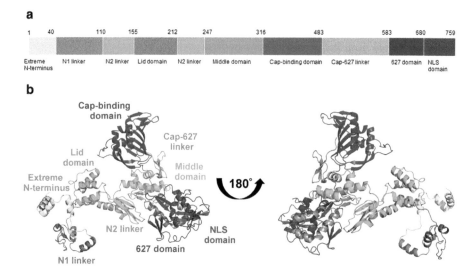

Fig. 5.7 Structure of PB2 subunit (PDB: 4WSB). The PB2 subunit consists of ten regions and adopts extended conformations when bound on PB1. PB2-N can be divided into an extreme N-domain and a "lid" domain, while PB2-C comprises a middle domain, the cap-binding domain, "627-domain" and a nuclear localization signal (NLS) domain. Cap-binding domain and 627-domain are linked up by a "cap-627 linke." Note that the NLS (738–755) is not visible in this structure and is assumed to be flexible

with later published heterotrimeric flu A, flu B, and flu C polymerase structures (Pflug et al. 2014; Reich et al. 2014; Hengrung et al. 2015). Briefly, the first helix (residues 1–22) on PB2 distorts at around residue 15 and bends to about 90°. As a result, the first two-thirds of this helix leans against a "helix-duo" at the extreme C-terminus of PB1 (residues 712–753), and the remaining one-third lies to cover this three-helix ensemble. This motif is stabilized mainly via salt bridges. The rest of PB2 molecule then folds onto the PB1 core via various linkages spanning the whole molecule.

Residues between 40–110, composed of three helices, three strands, and loops, make up the N1 domain which mainly serves the structural role. This area makes extensive contact with PB1.

Residues beyond 110 to 247 make up a N2 linker domain, where a subdomain termed lid domain (residues 155–212) resides in. The lid domain exists as four intertwining helices and covers the PB1 core including the template and primer channels, which will otherwise become unprotected and more exposed to solvent. The lid domain is also close to the putative template exit channel.

PB2 Cap-Binding Domain (320–483)

The cap-binding domain comes after a small middle domain which has four inter-twining helices spanning residues 251–316. This middle domain is neither in close contact with PB1 nor PA but is involved in a global domain reorientation. Hence it serves as a connection between the PB2-N and PB2-C.

The cap-binding domain consists of four helices folded against a beta sheet structure made up of five strands. Additional strands are found beyond the N- and C-termini of the beta sheet. The overall fold resembles a yacht made up of strands with the sailor (helices) on one side and m^7GTP ligand at the base of the ensemble. m^7GTP binding site is stabilized by a hydrophobic cluster of four phenylalanines. Structures of apo- and m^7GTP-bound flu A cap-binding domains are generally similar (Guilligay et al. 2008; Liu et al. 2013). Binding of m^7GTP is stabilized by partial aromatic stacking and salt bridges. Notably, H432 and K339/R355/N439 form salt bridges with α- and γ-phosphate groups, respectively. N439 is on a so-called 424-loop. The binding of m^7-GTP is not rigid, and deviation in its orientation is allowed (Guilligay et al. 2008; Tsurumura et al. 2013).

Flu B cap-binding domain was found to bind not only methylated cap but also unmethylated ones or even GDP substrates. It appears that in flu B, certain degree of conformational variations can be tolerated, leading to differential side chain orientations which allow interaction with a wide range of substrates (Wakai et al. 2011; Liu et al. 2015a; Xie et al. 2016).

PB2 627-Domain (538–680) and NLS-Domain (690–759)

Although the holo-heterotrimeric polymerase complex structure is already available, various X-ray structures of these domains harboring mutations are still valuable sources for the investigation into host determinants. These structures include PDB: 2VY7, 2VY8, 3KC6, 3KHW, 2GMO (solution structure of NLS-domain). Structures composing of both domains are also available (PDB: 2VY6, 3CW4) (Tarendeau et al. 2007; Tarendeau et al. 2008; Kuzuhara et al. 2009; Yamada et al. 2010).

The first two-thirds of 627-domain is composed of eight helices (six α-helices and two 3_{10} helices), while the remaining one-third mainly consists of turns and strands. The domain is compactly folded so that the helices reside on one side of the molecule, while the strands occupy the other side, resembling a blanket sheet covering on the bundle of helices. The 627-domain and NLS-domain are linked by a flexible linker. This linker was found unstructured and buried in the interface between 627-domain and the NLS-domain.

The NLS-domain itself adopts a globular fold with three antiparallel β-strands in the middle hydrophobic core, surrounded by unstructured loops and three helices. In pdb structure 2GMO, the bipartite NLS (738–755) is held close to this globular fold by a putative salt bridge D701-R753. The bipartite NLS shows different secondary structures. When 627-NLS domains were expressed alone, the bipartite NLS is unstructured. The bipartite NLS nevertheless adopts a well-formed helix as shown

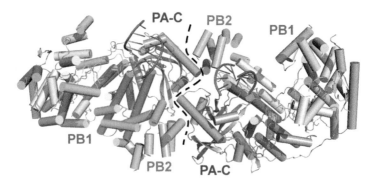

Fig. 5.8 Dimer structure of RdRP revealed by electron microscopy (PDB: 3J9B). Cartoon representation of RdRP dimer structure determined by electron microscopy. PA, PB1, and PB2 are colored in pink, pale blue, and pale green, respectively. The interacting interface is located at PA-C and PB2 N-terminal region

in apo flu C, or promoter bound flu A or flu B heterotrimeric complexes. This helical conformation may be implicated during replication and transcription, but its relevance in the context of nuclear import is not evidenced, since it has been established that PB2 is imported to the nucleus by importin-alpha, upon binding to which the NLS adopts an extended conformation (Pumroy et al. 2015).

Oligomerization of RdRP

RdRp has been shown to be capable of forming higher order oligomer. PB1 and PB2 were proposed to be the interacting site for oligomerization (Jorba et al. 2008). Consistent to this finding, a newer study showed that while the PA-PB1 heterodimer forms homogeneous monomeric and stable particle, the addition of truncated PB2 (residues 1–250) would result in the formation of dimerized RdRP particles (Swale et al. 2016).

The RdRP oligomer could be disrupted by the addition of vRNA. When vRNA is present, the RdRP oligomer will be dissociated into a monomeric RdRP form (Resa-Infante et al. 2010; Swale et al. 2016). This finding is contested by another study, where RdRP with a truncated PB2 (PB2 1–130) is capable of forming tetramer when vRNA or cRNA promoter is added to the dimerized RdRP. Furthermore, a further truncated PB2 (1–86) could not support either dimerization or tetramerization. This suggested that residues 86–130 might be crucial for the oligomerization of RdRP (Chang et al. 2015).

The EM structure of RdRP oligomer reveals two interacting interface (Fig. 5.8). The first interface is stronger, involving PA-C and the PB1-C/PB2-N binding interface, presumably covering PB2 residues 86–130. The second interface is more flexible. The PB1 finger domain is predicted to be involved in this interaction (Chang et al. 2015).

Structure of RNP RdRP Complex and NP

Early electron microscopy studies showed that RNP has a rod-shaped structure. The RNP strand exhibits a double-helical arrangement, with a terminal loop at one end, folding back on itself (Pons et al. 1969; Compans et al. 1972). The RdRP is located at the other end of the rod structure (Murti et al. 1988). In the absence of RdRP, vRNA can form complex with NP resembling the structure of native RNP (Yamanaka et al. 1990), while RNase-treated RNP retains its native conformation (Ruigrok and Baudin 1995). These suggest that the RNP conformation is primarily maintained by NP. A RNA chain wraps on the exterior of the NP helix. The 5′ and 3′ ends of the RNA forms a panhandle structure and are in contact with the polymerase. This architecture is consistent with the observation that influenza vRNA on the RNP is susceptible to RNase cleavage. Early studies estimated 15–20 nucleotides per NP molecules. Individual vRNPs are 10–15 nm in diameter and 30–120 nm long depending on the length of the gene segment (Noda et al. 2006).

Electron microscopy 3D reconstruction of mini-RNP (RNP with short vRNA-like genome) reveals a NP ring structure attached to the RdRP complex. RdRP interacts with two NP molecules at two sites (Martín-Benito et al. 2001; Coloma et al. 2009). This is consistent with previous interaction study, where NP was found to interact with PB1 and PB2 (Biswas et al. 1998, Poole et al. 2004). Since vRNA extends from the RdRP and folds back onto itself, where the extremities of both ends are bound by RdRP, it is natural that two molecules of NP, one near the 5′ end and the other one near the 3′ end, interact with RdRP at two locations.

3D Cryo-EM structures by Arranz et al. and Moeller et al. revealed structural characteristics of the RNP: (1) NP coiled to a double helix forming major and minor grooves as in a DNA molecule. Intra-strand NP-NP contact is via tail-loop-mediated oligomerization, whereas inter-strand NP molecules also contact at minor groove via a separate interface. (2) Orientation of intra-strand NP leads to a continuous path of positive residues allowing RNA molecules to attach onto the exterior of the RNP. (3) N-terminus of NP containing a nuclear localization signal (NLS1) is exposed on the surface and is accessible to host factors (Arranz et al. 2012; Moeller et al. 2012). Moreover, the structures constructed by Arranz et al. and Moeller et al. also show that two NP molecules attach to the RdRP complex (Fig. 5.9), in agreement with previous finding.

Further to the cryo-EM structures, a NP-NP dimer structure was published which implicates additional binding interface plausibly accounting for inter-strand interactions in RNP context. The dimeric NP interface lies on the "back" of the NP crescent and spans through residues 149–167 and 482–498. (PDB: 4IRY, Ye et al. 2012).

The periodicity of vRNA on NP strand is 20–32 nucleotides on average. The terminal loop comprises 3 to 8 NP molecules (Arranz et al. 2012; Moeller et al. 2012). In Moeller and coworkers' reconstructed structures, an extra NP molecule can be found adjacent to the RdRP complex, in addition to the two NPs identified from previous studies (Fig. 5.9) (Moeller et al. 2012).

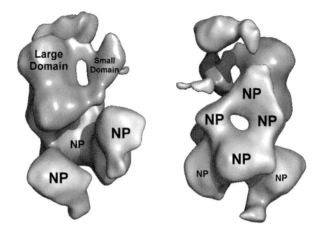

Fig. 5.9 Low-resolution structure of native RNP revealed by cryo-EM. Cryo-EM structure reconstructed to a resolution of 20 Å showing the polymerase end of native RNP. The RdRP heterotrimer is colored in pink. A large domain and a smaller arm domain can be identified. Moeller et al. has regarded the small domain to be the PA-C domain, due to the matching between their sizes and shapes. Comparison with mini-RNP EM structure reveals that the arm domain can undergo conformational change. However, in various high-resolution RdRP structures recently determined, the conformational change of RdRP is caused by substantial rearrangement of PB2 or in situ rotation of PA-N and PB2 cap-binding domain. PA-C domain remains at the same position in all determined crystal structures. Therefore, it is unsure whether the small arm domain, capable of conformational change, represents the PA-C domain. Two NP molecules are shown to interact with the large domain of RdRP, inconsistent with previous studies. An additional mass is observed at the top of the large domain. Moeller et al. suggested that it represents an additional NP molecule

Structure of NP

Flu A, flu B, and flu C NP are 498, 560, and 565 amino acid residues in length, respectively. Flu B NP (BNP) contains an extra N-terminal domain of 72 residues which is not found in Flu A NP. The N-terminus of Flu C NP resembles its Flu A counterpart, but its C-terminus is about 51 residues longer. X-ray structures of ANP and BNP are available (PDB: 2IQH, 2Q06, 3TJ0, 4IRY, 3ZDP) (Ye et al. 2006; Ng et al. 2006; Ng et al. 2012; Ye et al. 2012; Chenavas et al. 2013a). The structure of NP is largely conserved among the Orthomyxoviridae family and is formed predominantly with helices, which fold to form a compact structure. The N-terminal region of BNP evolves to become more extended than the flu A counterpart and has been found to be implicated in a number of functions (Liu et al. 2015b).

The structure of NP has been extensively reviewed in Ng et al. 2009, Chenavas et al. 2013b and Yang et al. 2014. Briefly, the NP molecule comprises a head domain, a body domain, and a protruding tail loop (Fig. 5.10) (Ng et al. 2008; Ye et al. 2006; Ng et al. 2012; Ye et al. 2012; Chenavas et al. 2013a). Flu A H5N1 NP composes of a continuous fold of 19 helices and 7 beta-strands which are interconnected with flexible loop regions. Three regions are especially flexible, namely, the regions 1–20, 75–90, and 390–438, the third of which starts from the end of a β-strand (residue

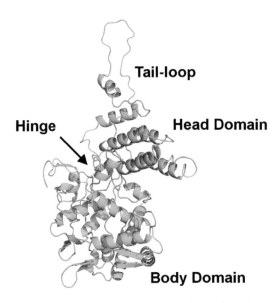

Fig. 5.10 Structure of NP (PDB: 2Q06). The NP molecule consists of a head domain, a body domain, and a tail-loop structure which are crucial for NP-NP homo-oligomerization. Homo-oligomerization is achieved by insertion of the tail loop to the binding groove at the body domain of a neighboring protomer. Conformational flexibility is conferred by the tail-loop and the unstructured regions which form a hinge at the middle of the molecule. This conformational flexibility is important for maintaining vRNP structure especially at the terminal polymerase end and loop end of the vRNP

390) followed by a long unstructured peptide to the tail loop (residues 402–428) and overlaps with a helix-loop-helix starting at residue 420. The overall arrangement of helices and strands renders the molecule a crescent shape, with head and body domains at two ends and a few linkers in the middle part serving like a hinge. As a result, unlike in PA and PB2 wherein distinct domains are linked up by linkers, domains in NP cannot be defined by a single continuous amino acid sequence (Fig. 5.10). In order to form homo-oligomers, tail-loop of a NP molecule is inserted into the binding groove of a neigherboring protomer.

The general features of BNP are similar to ANP. The two molecules shared backbone root-mean-square deviation of only 1.457 Å despite considerably different orientations of the tail loops.

The NP molecule demonstrates certain extent of flexibility. First the tail loop is flanked and followed by unstructured linker regions making it flexible. In fact, the tail loop of H1N1 NP deviates from the H5N1 tail loop for about 75°, while flu B tail loop deviates from H5N1 tail loop for about 60°. This renders tail-loop insertion at a range of angles possible and thus would also allow the formation of rigid (e.g., trimers or rings) and loose oligomers (e.g., higher-order helical oligomers). Second, the head domain and body domain should allow some degree of rocking. This could relate to possible conformational changes upon polymerase binding or during replication and transcription wherein the template RNA is slightly displaced from the NP.

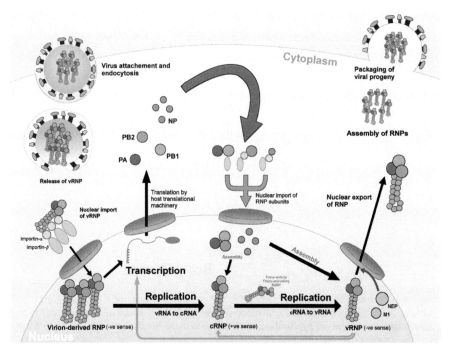

Fig. 5.11 Roles of RNP in influenza life cycle. The segmented genomes of influenza virus are packaged into RNP molecules inside the virion. After virus attachment and endocytosis, the virion-derived RNPs are released into host cytoplasm. With the help of host transport machinery, these vRNPs are transported into nucleus for transcription and replication. Transcription of negative-sense vRNA produces capped mRNA with poly-A tails that is translated by host translational machinery. Newly synthesized RNP subunits are then transported back into nucleus for assembly. Current model strongly suggests that PA-PB1 enters nucleus as a heterodimer complex, while PB2 and NP enter nucleus on their own. On the other hand, replication of negative-sense vRNA produces full length positive-sense replica of the viral genome. cRNA is stabilized by the newly produced RNP subunits, forming an intermediate complex called cRNP. Full length negative-sense vRNA is synthesized using cRNA as template. The cRNA-to-vRNA process has to be mediated by a trans-acting or trans-activating RdRP. The vRNA products are then encapsidated by RdRP and NP, resulting in progeny vRNP. Lastly, these vRNPs are exported from the nucleus, assembled in the cytoplasm and packaged into viral progenies

Functions of RNP

RNP plays a major role in influenza life cycle. vRNP encapsidates the segmented genome of influenza inside virion. It is also involved in the transcription and replication of the viral genome in infected cells. RNP is also capable of nuclear import and export. The overview of RNP functions in influenza life cycle are shown in Fig. 5.11. Its various roles will be described in details below.

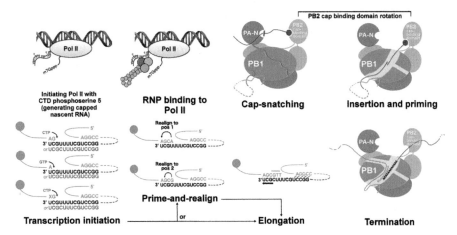

Fig. 5.12 Transcription of viral genome. Influenza RNP requires host Pol II-derived capped RNAs as primer for transcription. Initiating Pol II with phosphorylated serine at position 5 can recruit capping enzymes and add cap structure to its nascent transcript. Influenza RNP can bind specifically to the initiating form of Pol II, causing it to arrest at this stage, and employ cap-snatching mechanism to steal the cap structure from the nascent transcript. The PB2 cap-binding domain can capture the cap structure, while the PA-N endonuclease domain can cleave the RNA at 10–13 nt downstream of the cap, usually after a G or A base. Rotation of PB2 cap-binding domain facilitates the insertion of the cleaved primer into the catalytic cavity of PB1. The 3′ end of cleaved primers can interact with vRNA template at the penultimate or last nucleotide, triggering transcription initiation. After the addition of nucleotides until position 4, the nascent transcript can either undergo realignment or proceed on with elongation. Elongation continues until it stutters over a poly-U track near the 5′ terminus of vRNA

Transcription

Transcription of viral genome results in the production of viral proteins. RNP can steal cap from Pol II transcripts through its cap-snatching mechanism. This is followed by transcription initiation, realignment, and elongation. Poly-A tail is added to the nascent transcript at the end before the mRNA is processed by host translational machinery. The overview of RNP transcription is shown in Fig. 5.12.

Binding of Capped Cellular RNAs

The cap-binding activity resides on the PB2 cap-binding domain. This domain first captures a capped nascent RNA from Pol II and then directs it to the nearby PA-N endonuclease domain. It is unclear whether PA-N captures the capped RNA by chance or it is aided by cellular factors.

The "424-loop" on cap-binding domain was originally found to exert allosteric regulation on PB1 activity (Guilligay et al. 2008). Aligning 2VQZ (flu A m⁷GTP bound form) to the recent flu B mRNA primer bound structure (PDB: 5MSG,

Reich et al. 2017) reveals that this "424-loop" must be displaced outward if a primer is bound instead of m⁷GTP; otherwise the loop will clash with primer RNA. This loop again is in close proximity to the incoming primer RNA within 6 Å. Hence its role in directing correct priming is expected.

On the other hand, as nascent Pol II transcripts are targeted for cap-snatching (Gu et al. 2015; Koppstein et al. 2015), the interaction between RdRP and Pol II is also crucial for cap-binding. In vitro direct binding of RdRP to phosphoserine 5 Pol II CTD has been proven (Engelhardt et al. 2005; Martínez-Alonso et al. 2016). Moreover, a co-crystal structure of RdRP with CTD peptide was determined recently (Lukarska et al. 2017). This structure reveals that the phosphoserine 5 Pol II CTD binds directly to PA-C, where residues K289, R454, K635, and R638 are involved.

Cap-Snatching by Endonuclease Cleavage

The influenza RNP requires a unique cap-snatching mechanism to enable successful transcription of viral proteins. 5′ capped oligonucleotides derived from host RNA are captured, cleaved, and used as primer for initiating transcript elongation. The cap-snatching endonuclease function was originally thought to be residing in PB1 or PB2 subunit (Li et al. 2001). The determination of PA-N structure unquestionably established that the endonuclease function resides in PA (Yuan et al. 2009; Dias et al. 2009).

PA-N contains a PD-(D/E)XK nuclease motif that is conserved in all influenza virus. The negatively charged catalytic site contains H41, E80, D108, and E119 for coordinating the metal ions. Another residue, K134, is proposed to be the catalytic lysine. PA-N endonuclease cleaves captured cellular RNAs at 10–13 nt downstream of the cap structure. Besides, it also displays preference for cleavage after a dinucle-otide CA or a G (Rao et al. 2003; Datta et al. 2013; Sikora et al. 2014; Koppstein et al. 2015). Divalent metal ion is required for endonuclease activity, and the presence of different ions affects its endonuclease activity (Doan et al. 1999; Crepin et al. 2010; Xiao et al. 2014; Kotlarek and Worch 2016).

Initiation, Elongation, and Termination of Viral mRNA

PB1 was originally misidentified as having endonucleolytic actvitiy (Li et al. 2001); it was later revealed that the endonuclease active site resides in PA (Yuan et al. 2009; Dias et al. 2009). The present model proposed that after cap-binding and cleavage, PB2 rotates and inserts the capped primer into the active site of PB1 (Pflug et al. 2014; Reich et al. 2014), where it forms base pairing with the template. Primer with A or G at their 3′ end could form A:U or G:U base pairing with the 3′-ultimate U nucleotide of the viral template, initiating transcription at position 2 of the vRNA template. Besides, cleaved primers with G at the end can also pair with the 3′-pen-ultimate C nucleotide (Koppstein et al. 2015), thus initiating transcription at posi-tion 3. Furthermore, a prime-and-realign phenomenon during mRNA synthesis was observed by several studies (Geerts-Dimitriadou et al. 2011a; Geerts-Dimitriadou et al. 2011b; Sikora et al. 2014; Koppstein et al. 2015). This happens when the nascent transcript slips back and reiterates transcription of the first template nucleotides. Recently, it was identified that the PB1 priming loop is involved in this prime-and-realign mechanism of capped primer, although its importance in mRNA synthesis remains unknown (Oymans and te Velthuis 2017).

PB2 may also play a role in assisting transcription initiation. A recent study by Hara et al. (Hara et al. 2017) showed that R142A abolished both replication and transcription. R142 is in direct contact with PB1 N276 on the palm, forming a hydrogen bond (bat flu A, PDB: 4WSB). In comparison, flu B capped-mRNA primer bound structure (PDB: 5MSG) has two arginines at the equivalent position, forming hydrogen bonds with PB1 (R142/R144, equivalent to K140/R142 in flu A). K145/R146 in flu B (R143/R144 in flu A) is within 6 Å to the incoming capped-mRNA primer. Thus it is highly plausible that the first two positive side chains (R142/R144 in flu B, K140/R142 in flu A) hold the strand by anchoring to PB1 and the following two interact with incoming primers to direct them to PB1 active site. Besides, another study proposed that 627-domain is required for accurate cleavage of capped primer and the transcription initiation at correct position (Nilsson et al. 2017).

After transcription initiation, current model proposes that elongation of mRNA proceeds in a template-dependent manner. Template vRNA and nascent mRNA separate and leave the active site by their respective exit channels. The 5′ promoter of vRNA is anchored to the RdRP. Stuttering occurs when the vRNA reaches its 5′-terminus, resulting in the poly-A tails of the mRNA products (Poon et al. 1999; Poon et al. 2000). The fate of the template vRNA after transcription remains unclear.

Replication

Replication of viral genome results in full copy of viral genome. The replicated viral genomes can be encapsidated into new vRNPs and packaged into viral progenies. Influenza virus replicates its genome in two steps: (1) copying a negative-sense vRNA into an intermediate positive-sense cRNA and (2) copying the positive-sense cRNA back into a negative-sense vRNA. In both steps, RNP performs de novo synthesis of nucleotides. It involves the de novo synthesis of a pppApG dinucleotide, followed by elongation and read through of the entire template. Furthermore, regarding to the second step of replication, a *trans*-activating model and a *trans*-acting model were proposed. Besides, a realignment mechanism has also been observed during the second step of replication. The overview of vRNA-to-cRNA replication is shown in Fig. 5.13, while the overview of cRNA-to-vRNA replication is shown in Fig. 5.14.

Fig. 5.13 Replication of vRNA to cRNA. (a) PB1 priming loop facilitates de novo synthesis of pppApG dinucleotide using the vRNA as template at position 1 and 2 of the promoter. (b) Elongation of nascent transcript is accompanied by the breaking of interactions between 5′ and 3′ vRNA at the panhandle base pairing region. (c) As the nascent transcript leaves the catalytic cavity, it is stabilized by free RdRP and NP. Unlike transcription, the 5'-vRNA promoter has to be released in order to ensure a full read through of the viral genome. It is unclear when the release of 5'-vRNA promoter takes place. (d) The newly synthesized cRNA is assembled into a cRNP

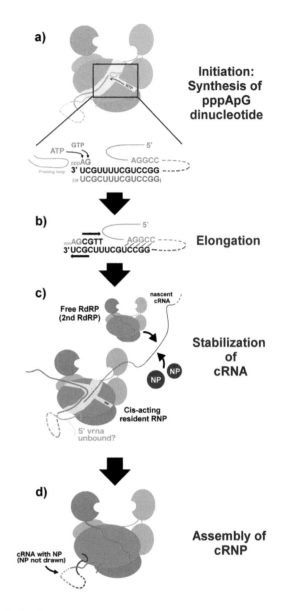

De Novo Synthesis and PB1 Priming Loop

The cRNA/vRNA replication process can be initiated de novo, meaning it can be achieved without primers. De novo replication is initiated by the synthesis of pppApG dinucleotide (Deng et al. 2006). Different mechanisms of pppApG production were observed between cRNA and vRNA synthesis. When vRNA is used as template, primer-independent initiation occurs at the 3′-ultimate position. After the synthesis of pppApG using position 1 and 2 of vRNA, elongation will follow

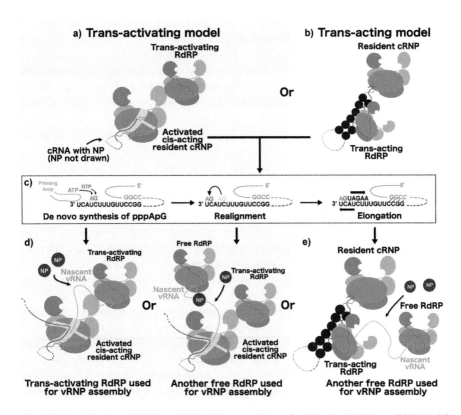

Fig. 5.14 Replication of cRNA to vRNA. The second step of replication (cRNA to vRNA) is different from the first step (vRNA to cRNA) as it requires the assistance of a trans-acting/*trans*-activating RdRP. Two models have been proposed. (a) *Trans*-activating model: the resident cRNP is responsible for polymerase activity. But this activity has to be activated by a trans-activating RdRP. (b) *Trans*-acting model: another RdRP is responsible for polymerase activity in *trans*. The *trans*-acting RdRP binds to the cRNP and replicates the cRNA template on the resident cRNP. (c) Replication is initiated by the synthesis of pppApG using positions 4 and 5 of the cRNA 3′ terminus. The pppApG dinucleotide is then realigned to positions 1 and 2, followed by elongation of nascent transcript. (d) In the trans-activating model, either the trans-activating RdRP encapsidates the newly synthesized vRNA and forms a progeny vRNP or another free RdRP is needed for the encapsidation. (e) In the trans-acting model, a free RdRP is required for the encapsidation of the nascent vRNA transcript

immediately at position 3 (Deng et al. 2006). This process is termed terminal initiation. Another study suggested that terminal initiation with vRNA starts at position 2 (instead of positions 1 and 2), thus proposing that an additional purine nucleotide has to be added to the 3′-terminus of vRNA at position 1 by host factors in order to ensure the replication of full length viral genome (Zhang et al. 2010). On the other hand, when cRNA is used as a template, pppApG is synthesized using positions 4 and 5 of the template. This pppApG will then be realigned to positions 1 and 2, followed by elongation in a template-dependent manner (Deng et al. 2006; Zhang et al. 2010). This process is termed internal initiation. One study showed that terminal initiation is significantly faster than the internal initiation that involves a primer-and-realign mechanism (Reich et al. 2017).

A recent finding showed that the synthesis of the first three nucleotides from vRNA templates can compensate for the breaking of base pairing at the panhandle region. In contrast, the synthesis of nucleotides from cRNA 3' extremity cannot compensate for the energy requirement, thus a prime-and-realign mechanism is needed (Reich et al. 2017).

Kinetic study showed that synthesis of pppApG is a rate-limiting step. Addition of ApG dinucleotide or uncapped primer ending with AG-3' significantly accelerate the level of RNA synthesis. Capped primer accelerates the rate of reaction even further, indicating RNA with 5' cap enhances efficiency (Reich et al. 2017).

Both the initiation of RNA synthesis and the prime-and-realign mechanism are closely related to the priming loop of PB1. RdRps capable of de novo synthesis usually contain a loop structure that can act as stacking platform for incoming NTP to catalyze the formation of the first phosphodiester bond (Butcher et al. 2001; Tao et al. 2002; Caillet-Saguy et al. 2014; Appleby et al. 2015). In the full structure of RdRP, this priming element can be identified in PB1. PB1 residues 641–657 form a conserved antiparallel β-loop that resembles HCV priming loop (Pflug et al. 2014).

Recent study confirmed that this priming loop is crucial for terminal initiation of de novo synthesis using vRNA as template. The priming loop does not affect internal initiation when cRNA is used as template. This loop might have to undergo conformational change and play a role in determining the rate of priming, initiation, and elongation. P651 on the loop was discovered to interact with the initiating NTP during de novo initiation (te Velthuis et al. 2016). By deletion study, the priming loop was observed to stimulate realignment during vRNA synthesis and suppresses internal extension. In contrast, during primer-dependent mRNA synthesis, priming loop was shown to suppress prime-and-realign mechanism (Oymans and te Velthuis 2017).

Trans-Acting or Trans-Activating Cis-Acting RNP

Using a genetic trans-complementation experiment, it was discovered that, with a template-bound replication-defective (i.e., PB2 R142A or R130A) RNP, even though it is not capable of synthesizing new RNA by itself, progeny vRNA could be produced in the presence of free RdRPs that have normal replication function. This indicates that the replication of viral genome occurs in *trans*. In contrast, the inability of mRNA synthesis by a template-bound transcription-defective (i.e., PB2 E361N) RNP cannot be rescued by a *trans* RdRP, indicating that mRNA synthesis occurs in *cis* (Jorba et al. 2009).

Previous study has demonstrated that the virion-derived vRNP is able to produce both mRNA and cRNA (Vreede et al. 2007); it is at least possible for the RNP to synthesize cRNA in *cis*. Thus the trans-acting model suggests that during the cRNP-to-vRNP phase, free RdRP first interacts with template-bound cRNP, presumably through the RdRP. The RdRP then translocates to the 3'-end of the cRNA and initiates replication in *trans*. Yet another free RdRP is required for the encapsidation of the newly synthesized vRNA. It is then assembled into a progeny vRNP (Jorba et al. 2009). This *trans*-acting model is consistent with two other findings: (1) oligomerization of RdRP can be observed (Jorba et al. 2008; Chang et al. 2015; Swale et al. 2016), and (2) electron micrographs of negatively stained RNPs reveal branched RNP structures (Moeller et al. 2012).

Recent study has shown that purified vRNP can perform mRNA and cRNA synthesis in vitro, while purified cRNP cannot synthesize vRNA either de novo or in the presence of an ApG primer. Only when purified RdRP is added to cRNP, synthesis of vRNA can be observed (York et al. 2013). This further supports the notion that free, trans-acting RdRP is required for successful cRNA-to-vRNA synthesis.

York et al. did not support the *trans*-acting model. Instead, a *trans*-activating model was proposed. The *trans*-activating model suggests that the *trans* RdRP only activates the resident RdRP, but the actual synthesis of vRNA is carried out in *cis*. This is supported by the finding that even an RdRP with nonfunctional polymerase activity could produce newly synthesized vRNA in conjunction with the resident cRNP (York et al. 2013). Therefore, the resident cRNP is responsible for RNA synthesis instead of the *trans* RdRP. In this model, it is possible that the *trans*-activating RdRP also fulfills the role of binding the nascent 5' vRNA. This allows the *trans*-activating RdRP to assemble into a progeny vRNP (te Velthuis and Fodor 2016).

Stabilization of cRNAs

Stabilization of replicative intermediate cRNAs has been reported to be important for the replication of viral genome. A recent study reported that the PB2 627-domain is required for the stabilization of cRNA (Nilsson et al. 2017). On the other hand, the RNA binding capability of NP is also essential for stabilizing newly synthesized cRNA, preventing the replicative intermediate being degraded by cellular nuclease (Vreede et al. 2004; Vreede and Brownlee 2007). The stabilization of cRNA has been proposed as a mechanism to regulate the transition of transcription in early infection to replication in late infection (Vreede et al. 2004; Vreede and Brownlee 2007). The exact mechanism of cRNA stabilization by PB2 and NP remains unclear. However, the 627-domin is located close to the exit channel of nascent RNA during replication. It is plausible that this domain is involved in guiding incoming NP to newly synthesized vRNA for encapsidation. In fact, evidence supports that PB2 627-domain and NP interact with each other directly. Ng et al. demonstrated that this interaction is related to the region around R150 on NP, while Hsia et al. further showed that a peptide from NP spanning residues S145 to G185 is sufficient to cause chemical shift perturbation on PB2 627-domain, with PB2 D605 and V606 playing vital roles in the interaction (Ng et al. 2012; Hsia et al. 2018).

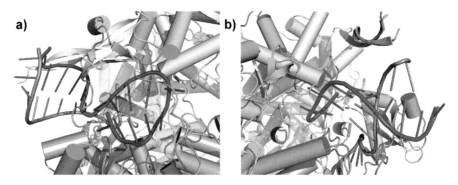

Fig. 5.15 Structure of vRNA promoter and its interaction with RdRP (PDB: 4WSB). (a) 5'-vRNA promoter (colored in red) is shown to interact with both PA and PB1 (colored pink and pale blue, respectively). 5' Terminus of vRNA forms a hook structure, inconsistent with the previously proposed corkscrew model. PA-C RNA-binding groove is involved in binding 5' vRNA, and the PA arch encircles both the 5' vRNA and a PB1 β-ribbon. The 5' vRNA is also situated near the NTP entry channel (located at the right side of the diagram). (b) 3' vRNA promoter (colored in blue) is shown to interact with all three subunits (PA, PB1, and PB2 colored in pink, pale blue, and pale green, respectively). The 3' extremity of vRNA is not shown in this structure. The long and flexible PB1 NLS β-ribbon is shown to be in close proximity with both 5' and 3' promoters

vRNA & cRNA Promoter Binding

Previous studies indicated that PB1 is highly involved in promoter binding (González and Ortín 1999a; González and Ortín 1999b; Jung and Brownlee 2006). Full RdRP structure confirmed the importance of PB1 in promoter binding. PB1 interacts with both 5' and 3' of vRNA and cRNA (Figs. 5.15a and 5.15b) (Pflug et al. 2014; Reich et al. 2014; Thierry et al. 2016). 5' promoter region is held by both PA and PB1 (Fig. 5.15a), while 3' promoter region interacts with all three subunits (Fig. 5.15b). A corkscrew conformation of the viral promoter could be observed, confirming previous studies and models (Flick et al. 1996; Hobom and Flick 1999; Brownlee and Sharps 2002; Tomescu et al. 2014).

PB1 N-terminal H32, T34, and Y38 residues are involved in the interaction with the 5' vRNA. Besides, a PB1 β-hairpin (residues 353–370; bat strain residues) inserts through the PA arch and interacts with the 5' vRNA, where R365 forms multivalent interactions with the RNA backbone (Pflug et al. 2014). On the other hand, PB1 interacts with the 5'-3' duplex region with the NLS-containing β-ribbon (residues 177–212) and C-terminal region of PB1 (residues 672–676). From the structure of Flu A RdRP, it is unclear how R571 and R572 mentioned by Li et al. could affect 5' vRNA binding (Pflug et al. 2014; Li et al. 1998).

Although PB1 plays an important role in the binding of the 5' and 3' viral promoters, PA and PB2 are also essential.

PA-PB1 heterodimer has been shown to involve in 5'-promoter binding (Lee et al. 2002; Deng et al. 2005). Recently, the Kd between PA-PB1 dimer and 5'-vRNA promoter was determined to be as low as 0.2–0.4 nM, whereas their interaction to

3' promoter is much weaker (Swale et al. 2016). The structure of PA-C reveals a putative RNA-binding groove. Highly conserved basic residues (K328, K539, R566, and K574) are found in the groove (Obayashi et al. 2008; He et al. 2008). The elucidation of full RdRp structure confirms that the putative RNA-binding groove indeed interacts with RNA. It binds specifically to the 5' viral promoter (Fig. 5.15a). The first 10 nucleotides of the 5' promoter form a hook that is sandwiched in a pocket formed on one side by strands β17–β18 and β20 of PA. The other side is formed by the PA arch, with a PB1 β-hairpin (bat influenza residues 353–370) inserting through the arch. The conserved basic residues located in the RNA groove form polar interactions with the 5' hook (Pflug et al. 2014). Co-crystal of flu B RdRP with viral promoters shows that both 5'-vRNA promoter and 5' cRNA promoter bind to PA in a highly similar mode (Thierry et al. 2016).

Also, it was recently discovered that RanBP5 regulates the binding of 5'-vRNA promoter to the PA-PB1 dimer. The formation of PA-PB1-RanBP5 heterotrimer prevents the binding of 5' promoter to PA-PB1 dimer (Swale et al. 2016).

For PB2, its N-terminal region is in close contact with the promoter via residues 36–49 as revealed in the full RdRP structure (Pflug et al. 2014). Agreeing with this, during polymerase assembly, the binding of PA-PB1 dimer to 3' vRNA necessitates the presence of PB2-N (Swale et al. 2016).

Hara et al. identified R124 in PB2-N and established by cross-linking experiment that this is important for both vRNA and cRNA promoter binding (Hara et al. 2017). Structurally the side-chain conformation of this residue is conserved in all available polymerase structures, either apo or promoter bound. In capped-mRNA primer bound flu B polymerase structure, it is apparent that R124 lies within the N2-linker, pointing toward incoming primer or template vRNA. However, its direct interaction with RNA is unlikely since several loops and helices on PB2 cause steric hindrances. It is thus conceivable that R124 assumes a structural role related to the lid domain instead of directly in contact with RNA.

Furthermore, two configurations of 3' vRNA promoter have been observed, one represents the pre-initiation state, while the other one resembles the initiation state (Fig. 5.16). In the pre-initiation state, the 3' vRNA promoter extremity interacts with PB1 at the exterior of RdRP; in the initiation-resembling state, it is inserted into the PB1 catalytic cavity (Reich et al. 2014; Reich et al. 2017).

Conformational Change of RdRP Triggered by Promoter Binding

Promoter binding has been implicated in the activation of cap-binding and endonuclease activity (Lee et al. 2003; Rao et al. 2003). This observation is supported by recent discovery of two conformations of RdRP bound by 5'-vRNA promoter and 5' cRNA promoter (Fig. 5.17) (Thierry et al. 2016).

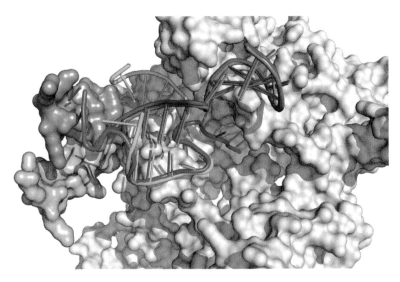

Fig. 5.16 Two configurations of 3' vRNA promoter. The structure of flu B 3' vRNA promoter in pre-initiation state (PDB, 4WRT; red, 5' promoter; cyan, 3' promoter) and initiation-resembling state (PDB, 5MSG; pale red, 5' promoter; blue, 3' promoter) is shown as cartoon representation. PB1 is shown as surface diagram (colored in pale blue). In pre-initiation state, the 3' extremity of vRNA lies at the exterior of PB1 and interacts with its NLS-containing β-ribbon. In initiation-resembling state, the 3' extremity of vRNA inserts into the catalytic cavity of PB1 through a narrow template entry channel. The conformation of 5' vRNA remains the same in both states, whereas the PB1 NLS-containing β-ribbon shows substantial displacement. In the pre-initiation state, the ribbon (colored in pink) is "pulled" toward the 3' vRNA extremities. In the initiation-resembling state, the same ribbon (colored in green) moves further away from the promoter

In an apo (resting) state (flu C, PDB: 5D98), PB2 subunit exists in a compact state wherein the lid domain and the cap-binding domain exhibit extensive contact with PB1. Entry of capped mRNA is obscured. 627-domain is situated out of the protein core. NLS sequence forms a helix packed against PA-N domain, presumably resulting in blocking of the endonuclease activity. This conformation is similar to a cRNA promoter-bound form (flu B, PDB: 5EPI) with two obvious differences in the latter state: (1) although NLS-peptide remains bound to PA-N, the whole NLS-PA-N is rotated by around 90°; (2) the 627-domain is protruded out with a partially unfolded cap-627 linker, now further away from the protein core, leading to shifts of the lid domain and the cap-binding domain. The overall conformation becomes more extended. The orientation of 627-domain as such should register biological significance, since it is now in close proximity of the vRNA exit channel.

In situ rotation of cap-binding domain was observed in vRNA promoter bound flu A and flu B structures (PDB: 4WSB and 4WSA) (Fig. 5.18). This is particularly important for the cap-binding domain to direct the primer to PA-N domain for cleavage. It is currently unknown whether the binding of capped mRNA alone is sufficient to cause the drastic domain rearrangement. It is likely that more than one conformation may exist in equilibrium when in solution, as Thierry et al. suggested.

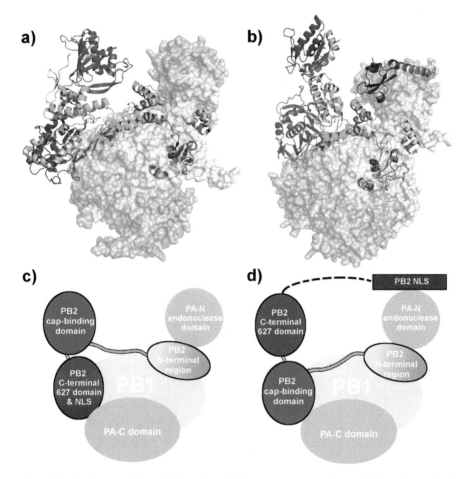

Fig. 5.17 Radical repacking of PB2 subunit. PB2 was shown to undergo radical conformational change in two crystal structures of flu B RdRP (PDB: 4FMZ & 5EPI). (a) The vRNA-bound form (PDB: 4FMZ) resembles the RdRP structure of bat influenza and flu B RdRP interacting with both 5′ and 3′ of vRNA (PDB: 4WSB & 4WSA). In this form, the PB2 cap-binding domain is situated at the top of the U-structure and faces the PA-N domain. Whereas the PB2 627-domain is located near PA-C and the palm domain of PB1, while the NLS domain is not visible in the structure. (b) In the cRNA-bound form (PDB: 5EPI), PB2 cap-binding domain and PB2 627 domain switch their positions. The cap-binding domain is situated near the palm domain of PB1, while the 627-domain moves nearer to the PA-N domain. Furthermore, the PB2 NLS helix becomes visible and is shown to interact directly with PA-N. The apo form of flu C RdRP (PDB: 5D98) resembles the cRNA-bound form. (c) and (d) Schematic diagrams of RdRP corresponding to (a) and (b)

Drastic rearrangement of the 627-domain is registered, albeit the polymerase shape becomes compact again. The overall relocation concerns lid domain, cap-binding domain, cap-627 linker, and the 627-domain so that (1) cap-627 linker is now closely packed against lid domain and (2) 627-domain is now situated next to the "back" of cap-binding domain, leaving the primer entry site open. Endonuclease domain is no

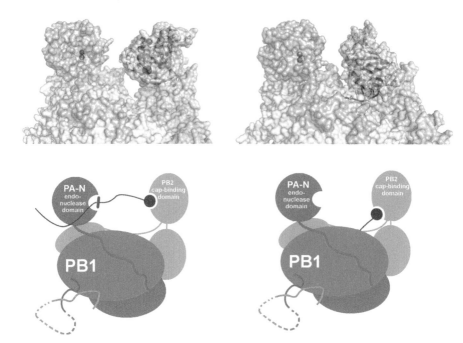

Fig. 5.18 Rotation of cap-binding domain. In situ rotation of cap-binding domain was observed in several crystal structures (PDB: 4WSB, 4WSA, and 5MSG). The cap-binding cavity of PB2 can either point toward PA-N for cap-snatching, or it can point toward PB1 catalytic cavity for priming and transcription initiation

longer bound by NLS-peptide. Indeed the electron density for the NLS-domain is not registered in any of available vRNA promoter-bound structures.

However, no significant structural change could be observed near the 5′ promoter binding site on both PA and PB1, thus the exact mechanism of conformational change and the activation of cap-snatching mechanism remains obscure.

Lastly, the difference between the tails of mRNA and cRNA products are affected by the binding of 5′ promoter to RdRP. Transcription of vRNA requires the 5′ promoter to associate tightly to RdRP, this steric constraint would hinder elongation and cause stuttering at the poly-U stretch, resulting in a poly-A tail in the mRNA product (Pritlove et al. 1998; Poon et al. 1998). On the other hand, replication of vRNA requires the release of 5′ promoter in order to allow read through of the entire genome. The exact mechanism of this switch is not clear, but PA arch and the PA RNA-binding groove might play a role due to their close proximity to 5′ promoter.

Conclusion

The RNP complex plays multiple crucial roles in influenza virus life cycle. In this article, the functions of RNP in transcription and replication are described in detail. It is worthwhile to note that much research has also been done on the role of RNP in host specificity and pathogenicity. Furthermore, several interactome studies have revealed the extensive interaction network of RNP with host factors. Lastly, RNP is also a promising target for drug discovery; numerous small molecules or peptides have been found to disrupt RNP assembly and activity.

From the perspective of structural research, there are several important problems remained unsolved. Firstly, although high-resolution structures of RdRP have been determined, the exact structure of RNP complex remains unknown. NP has long been identified to interact with RdRP (presumably through PB1 and PB2), yet its exact mode of binding has not been revealed. Secondly, it has been observed that RdRP undergoes substantial conformational change (particularly PB2). However, the cause and regulation of this phenomenon is not understood. Moreover, it would be fruitful to understand the effect of such conformational change on the activity of RNP. One would conjecture such structural change might play a role in the regulation between replication and transcription. Thirdly, several regions in the RdRP were unrecognized before the elucidation of full RdRP structure, e.g., PA arch, PB1 β-ribbon (residues 352–398 of bat flu A), and the 550-loop of PA. These regions are mostly uncharacterized at present, and their functions are not known. Fourthly, the actual initiation state of RdRP is not yet resolved. Base pairing cannot be observed in the PB1 catalytic cavity in all present structures. Structural determination of the initiation state could provide us with valuable insight on the mechanism of nucleotide synthesis by RdRP. Fifthly, regardless of the *trans*-acting model or *trans*-activating model, both involve dimerization or oligomerization of RdRP. In depth characterization of their modes of interaction would be much helpful and would give us understanding on the role and mechanism of the *trans* RdRP in RNA synthesis. Finally, numerous residues on the RdRP have been discovered to affect host specificity. However, their exact mechanism remains unclear and the existing structure of RdRP could not explain their phenotypes.

Although many host factors have been identified as interacting partners with RdRP or RNP, most of these interactions remain uncharacterized. Only the binding mode of Pol II CTD domain with RdRP and importin-α with NP and PB2 were structurally determined (Pumroy et al. 2015; Nakada et al. 2015; Lukarska et al. 2017). The characterization of interaction between host factors and RNP and structural determination of these complexes could give us a full picture on the function of RNP.

Altogether, previous research has provided us with profound understanding on the role of influenza RNP. However, some "black boxes" still remain at present, and efforts in clarifying the full mechanism of RNP, its role in host specificity, and its interactions with host factors are much needed. These basic scientific knowledge on RNP could assist us in further understanding the central process of viral growth and for developing measures to combat the virus infection.

References

Appleby TC, Perry JK, Murakami E et al (2015) Structural basis for RNA replication by the hepatitis C virus polymerase. Science 347:771–775

Area E, Martín-Benito J, Gastaminza P et al (2004) 3D structure of the influenza virus polymerase complex: localization of subunit domains. Proc Natl Acad Sci U S A 101:308–313

Arranz R, Coloma R, Chichón FJ et al (2012) The structure of native influenza virion ribonucleoproteins. Science 338:1634–1637

Biswas SK, Nayak DP (1994) Mutational analysis of the conserved motifs of influenza a virus polymerase basic protein 1. J Virol 68:1819–1826

Brownlee GG, Sharps JL (2002) The RNA polymerase of influenza a virus is stabilized by interaction with its viral RNA promoter. J Virol 76:7103–7113

Butcher SJ, Grimes JM, Makeyev EV et al (2001) A mechanism for initiating RNA-dependent RNA polymerization. Nature 410:235–240

Caillet-Saguy C, Lim SP, Shi P-Y et al (2014) Polymerases of hepatitis C viruses and flaviviruses: structural and mechanistic insights and drug development. Antivir Res 105:8–16

Chang S, Sun D, Liang H et al (2015) Cryo-EM structure of influenza virus RNA polymerase complex at 4.3 Å resolution. Mol Cell 57:925–935

Chenavas S, Crépin T, Delmas B et al (2013a) Influenza virus nucleoprotein: structure, RNA binding, oligomerization and antiviral drug target. Future Microbiol 8:1537–1545

Chenavas S, Estrozi LF, Slama-Schwok A et al (2013b) Monomeric nucleoprotein of influenza a virus. PLoS Pathog 9:e1003275

Coloma R, Valpuesta JM, Arranz R et al (2009) The structure of a biologically active influenza virus ribonucleoprotein complex. PLoS Pathog 5:e1000491

Compans RW, Content J, Duesberg PH (1972) Structure of the ribonucleoprotein of influenza virus. J Virol 10:795–800

Crepin T, Dias A, Palencia A et al (2010) Mutational and metal binding analysis of the endonuclease domain of the influenza virus polymerase PA subunit. J Virol 84:9096–9104

Datta K, Wolkerstorfer A, Szolar OHJ et al (2013) Characterization of PA-N terminal domain of influenza a polymerase reveals sequence specific RNA cleavage. Nucleic Acids Res 41:8289–8299

Deng T, Sharps J, Fodor E, Brownlee GG (2005) In vitro assembly of PB2 with a PB1-PA dimer supports a new model of assembly of influenza a virus polymerase subunits into a functional trimeric complex. J Virol 79:8669–8674

Deng T, Vreede FT, Brownlee GG (2006) Different de novo initiation strategies are used by influenza virus RNA polymerase on its cRNA and viral RNA promoters during viral RNA replication. J Virol 80:2337–2348

Dias A, Bouvier D, Crépin T et al (2009) The cap-snatching endonuclease of influenza virus polymerase resides in the PA subunit. Nature 458:914–918

Doan L, Handa B, Roberts NA, Klumpp K (1999) Metal ion catalysis of RNA cleavage by the influenza virus endonuclease. Biochemistry 38:5612–5619

DuBois RM, Slavish PJ, Baughman BM et al (2012) Structural and biochemical basis for development of influenza virus inhibitors targeting the PA endonuclease. PLoS Pathog 8:e1002830

Eisfeld AJ, Neumann G, Kawaoka Y (2014) At the centre: influenza a virus ribonucleoproteins. Nat Rev Microbiol 13:28–41

Engelhardt OG, Smith M, Fodor E (2005) Association of the influenza a virus RNA-dependent RNA polymerase with cellular RNA polymerase II. J Virol 79:5812–5818

Flick R, Neumann G, Hoffmann E et al (1996) Promoter elements in the influenza vRNA terminal structure. RNA 2:1046–1057

Geerts-Dimitriadou C, Goldbach R, Kormelink R (2011a) Preferential use of RNA leader sequences during influenza a transcription initiation in vivo. Virology 409:27–32

Geerts-Dimitriadou C, Zwart MP, Goldbach R, Kormelink R (2011b) Base-pairing promotes leader selection to prime in vitro influenza genome transcription. Virology 409:17–26

González S, Ortín J (1999a) Characterization of influenza virus PB1 protein binding to viral RNA: two separate regions of the protein contribute to the interaction domain. J Virol 73:631–637

González S, Ortín J (1999b) Distinct regions of influenza virus PB1 polymerase subunit recognize vRNA and cRNA templates. EMBO J 18:3767–3775

González S, Zürcher T, Ortín J (1996) Identification of two separate domains in the influenza virus PB1 protein involved in the interaction with the PB2 and PA subunits: a model for the viral RNA polymerase structure. Nucleic Acids Res 24:4456–4463

Gu W, Gallagher GR, Dai W et al (2015) Influenza a virus preferentially snatches noncoding RNA caps. RNA 21:2067–2075

Guilligay D, Tarendeau F, Resa-Infante P et al (2008) The structural basis for cap binding by influenza virus polymerase subunit PB2. Nat Struct Mol Biol 15:500–506

Guu TSY, Dong L, Wittung-Stafshede P, Tao YJ (2008) Mapping the domain structure of the influenza a virus polymerase acidic protein (PA) and its interaction with the basic protein 1 (PB1) subunit. Virology 379:135–142

Hara K, Kashiwagi T, Hamada N, Watanabe H (2017) Basic amino acids in the N-terminal half of the PB2 subunit of influenza virus RNA polymerase are involved in both transcription and replication. J Gen Virol 98:900–905

He X, Zhou J, Bartlam M et al (2008) Crystal structure of the polymerase PAC–PB1N complex from an avian influenza H5N1 virus. Nature 454:1123–1126

Hengrung N, El Omari K, Serna Martin I et al (2015) Crystal structure of the RNA-dependent RNA polymerase from influenza C virus. Nature 527:114–117

Hobom G, Flick R (1999) Interaction of influenza virus polymerase with viral RNA in the "corkscrew" conformation. J Gen Virol 80:2565–2572

Hsia H-P, Yang Y-H, Szeto W-C, Nilsson BE, Lo C-Y, Ng AK-L, Fodor E, Shaw P-C, Menéndez-Arias L (2018) Amino acid substitutions affecting aspartic acid 605 and valine 606 decrease the interaction strength between the influenza virus RNA polymerase PB2 '627' domain and the viral nucleoprotein. PLOS ONE 13(1):e0191226

Jorba N, Area E, Ortin J (2008) Oligomerization of the influenza virus polymerase complex in vivo. J Gen Virol 89:520–524

Jorba N, Coloma R, Ortín J (2009) Genetic trans-complementation establishes a new model for influenza virus RNA transcription and replication. PLoS Pathog 5:e1000462

Jung TE, Brownlee GG (2006) A new promoter-binding site in the PB1 subunit of the influenza a virus polymerase. J Gen Virol 87:679–688

Koppstein D, Ashour J, Bartel DP (2015) Sequencing the cap-snatching repertoire of H1N1 influenza provides insight into the mechanism of viral transcription initiation. Nucleic Acids Res 43:5052–5064

Kotlarek D, Worch R (2016) New insight into metal ion-driven catalysis of nucleic acids by influenza PA-Nter. PLoS One 11:e0156972

Kuzuhara T, Kise D, Yoshida H et al (2009) Structural basis of the influenza a virus RNA polymerase PB2 RNA-binding domain containing the pathogenicity-determinant lysine 627 residue. J Biol Chem 284:6855–6860

Lee MTM, Bishop K, Medcalf L et al (2002) Definition of the minimal viral components required for the initiation of unprimed RNA synthesis by influenza virus RNA polymerase. Nucleic Acids Res 30:429–438

Lee M-TM, Klumpp K, Digard P, Tiley L (2003) Activation of influenza virus RNA polymerase by the 5′ and 3′ terminal duplex of genomic RNA. Nucleic Acids Res 31:1624–1632

Li M-L, Ramirez BC, Krug RM (1998) RNA-dependent activation of primer RNA production by influenza virus polymerase: different regions of the same protein subunit constitute the two required RNA-binding sites. EMBO J 17:5844–5852

Li M-L, Rao P, Krug RM (2001) The active sites of the influenza cap-dependent endonuclease are on different polymerase subunits. EMBO J 20:2078–2086

Liu Y, Qin K, Meng G et al (2013) Structural and functional characterization of K339T substitution identified in the PB2 subunit cap-binding pocket of influenza a virus. J Biol Chem 288:11013–11023

Liu M, Lam MK-H, Zhang Q et al (2015a) The functional study of the N-terminal region of influenza B virus nucleoprotein. PLoS One 10:e0137802

Liu Y, Yang Y, Fan J et al (2015b) The crystal structure of the PB2 cap-binding domain of influenza B virus reveals a novel cap recognition mechanism. J Biol Chem 290:9141–9149

Lukarska M, Fournier G, Pflug A et al (2017) Structural basis of an essential interaction between influenza polymerase and pol II CTD. Nature 541:117–121

Martín-Benito J, Area E, Ortega J et al (2001) Three-dimensional reconstruction of a recombinant influenza virus ribonucleoprotein particle. EMBO Rep 2:313–317

Martínez-Alonso M, Hengrung N, Fodor E (2016) RNA-free and ribonucleoprotein-associated influenza virus polymerases directly bind the Serine-5-phosphorylated carboxyl-terminal domain of host RNA polymerase II. J Virol 90:6014–6021

Moeller A, Kirchdoerfer RN, Potter CS et al (2012) Organization of the influenza virus replication machinery. Science 338:1631–1634

Moen SO, Abendroth J, Fairman JW et al (2014) Structural analysis of H1N1 and H7N9 influenza a virus PA in the absence of PB1. Sci Rep 4:5944

Murti KG, Webster RG, Jones IM (1988) Localization of RNA polymerases on influenza viral ribonucleoproteins by immunogold labeling. Virology 164:562–566

Nakada R, Hirano H, Matsuura Y (2015) Structure of importin-alpha bound to a non-classical nuclear localization signal of the influenza a virus nucleoprotein. Sci Rep 5:15055

Ng AK-L, Zhang H, Tan K et al (2008) Structure of the influenza virus a H5N1 nucleoprotein: implications for RNA binding, oligomerization, and vaccine design. FASEB J 22:3638–3647

Ng AK-L, Wang J-H, Shaw P-C (2009) Structure and sequence analysis of influenza a virus nucleoprotein. Sci China Ser C Life Sci 52:439–449

Ng AK-L, Lam MK-H, Zhang H et al (2012) Structural basis for RNA binding and homo-oligomer formation by influenza B virus nucleoprotein. J Virol 86:6758–6767

Nilsson BE, te Velthuis AJW, Fodor E (2017) Role of the PB2 627 domain in influenza a virus polymerase function. J Virol 91:e02467-16–e02e02467

Noda T, Sagara H, Yen A et al (2006) Architecture of ribonucleoprotein complexes in influenza a virus particles. Nature 439:490–492

Obayashi E, Yoshida H, Kawai F et al (2008) The structural basis for an essential subunit interaction in influenza virus RNA polymerase. Nature 454:1127–1131

Ohtsu Y, Honda Y, Sakata Y et al (2002) Fine mapping of the subunit binding sites of influenza virus RNA polymerase. Microbiol Immunol 46:167–175

Oymans J, te Velthuis A (2017) Correct And efficient initiation of viral RNA synthesis by the influenza A virus RNA polymerase bioRvix 138487

Perez DR, Donis RO (2001) Functional analysis of PA binding by influenza a virus PB1: effects on polymerase activity and viral infectivity. J Virol 75:8127–8136

Pflug A, Guilligay D, Reich S, Cusack S (2014) Structure of influenza a polymerase bound to the viral RNA promoter. Nature 516:355–360

Pons MW, Schulze IT, Hirst GK, Hauser R (1969) Isolation and characterization of the ribonucleoprotein of influenza virus. Virology 39:250–259

Poon LL, Pritlove DC, Fodor E, Brownlee GG (1999) Direct evidence that the poly(A) tail of influenza A virus mRNA is synthesized by reiterative copying of a U track in the virion RNA template. J Virol 73:3473–3476

Poon LL, Fodor E, Brownlee GG (2000) Polyuridylated mRNA synthesized by a recombinant influenza virus is defective in nuclear export. J Virol 74:418–427

Pritlove DC, Poon LL, Fodor E et al (1998) Polyadenylation of influenza virus mRNA transcribed in vitro from model virion RNA templates: requirement for 5′ conserved sequences. J Virol 72:1280–1286

Pumroy RAA, Ke S, Hart DJJ et al (2015) Molecular determinants for nuclear import of influenza a PB2 by importin α isoforms 3 and 7. Structure 23:374–384

Rao P, Yuan W, Krug RM (2003) Crucial role of CA cleavage sites in the cap-snatching mechanism for initiating viral mRNA synthesis. EMBO J 22:1188–1198

Reguera J, Gerlach P, Cusack S (2016) Towards a structural understanding of RNA synthesis by negative strand RNA viral polymerases. Curr Opin Struct Biol 36:75–84

Reich S, Guilligay D, Pflug A et al (2014) Structural insight into cap-snatching and RNA synthesis by influenza polymerase. Nature 516:361–366

Reich S, Guilligay D, Cusack S (2017) An in vitro fluorescence based study of initiation of RNA synthesis by influenza B polymerase. Nucleic Acids Res 45:3353–3368

Resa-Infante P, Recuero-Checa MA, Zamarreño N et al (2010) Structural and functional characterization of an influenza virus RNA polymerase-genomic RNA complex. J Virol 84:10477–10487

Ruigrok RW, Baudin F (1995) Structure of influenza virus ribonucleoprotein particles. II. Purified RNA-free influenza virus ribonucleoprotein forms structures that are indistinguishable from the intact influenza virus ribonucleoprotein particles. J Gen Virol 76(Pt 4):1009–1014

Sikora D, Rocheleau L, Brown EG, Pelchat M (2014) Deep sequencing reveals the eight facets of the influenza a/HongKong/1/1968 (H3N2) virus cap-snatching process. Sci Rep 4:6181

Sugiyama K, Obayashi E, Kawaguchi A et al (2009) Structural insight into the essential PB1-PB2 subunit contact of the influenza virus RNA polymerase. EMBO J 28:1803–1811

Swale C, Monod A, Tengo L et al (2016) Structural characterization of recombinant IAV polymerase reveals a stable complex between viral PA-PB1 heterodimer and host RanBP5. Sci Rep 6:24727

Tao Y, Farsetta DL, Nibert ML, Harrison SC (2002) RNA synthesis in a cage-structural studies of reovirus polymerase lambda3. Cell 111:733–745

Tarendeau F, Boudet J, Guilligay D et al (2007) Structure and nuclear import function of the C-terminal domain of influenza virus polymerase PB2 subunit. Nat Struct Mol Biol 14:229–233

Tarendeau F, Crepin T, Guilligay D et al (2008) Host determinant residue lysine 627 lies on the surface of a discrete, folded domain of influenza virus polymerase PB2 subunit. PLoS Pathog 4:e1000136

Thierry E, Guilligay D, Kosinski J et al (2016) Influenza polymerase can adopt an alternative configuration involving a radical repacking of PB2 domains. Mol Cell 61:125–137

Tomescu AI, Robb NC, Hengrung N et al (2014) Single-molecule FRET reveals a corkscrew RNA structure for the polymerase-bound influenza virus promoter. Proc Natl Acad Sci U S A 111:E3335–E3342

Torreira E, Schoehn G, Fernández Y et al (2007) Three-dimensional model for the isolated recombinant influenza virus polymerase heterotrimer. Nucleic Acids Res 35:3774–3783

Tsurumura T, Qiu H, Yoshida T et al (2013) Conformational polymorphism of m7GTP in crystal structure of the PB2 middle domain from human influenza a virus. PLoS One 8:e82020

te Velthuis AJW, Fodor E (2016) Influenza virus RNA polymerase: insights into the mechanisms of viral RNA synthesis. Nat Rev Microbiol 14:479–493

te Velthuis AJW, Robb NC, Kapanidis AN et al (2016) The role of the priming loop in influenza a virus RNA synthesis. Nat Microbiol 1:16029

Vreede FT, Brownlee GG (2007) Influenza virion-derived viral ribonucleoproteins synthesize both mRNA and cRNA in vitro. J Virol 81:2196–2204

Vreede FT, Jung TE, Brownlee GG (2004) Model suggesting that replication of influenza virus is regulated by stabilization of replicative intermediates. J Virol 78:9568–9572

Wakai C, Iwama M, Mizumoto K, Nagata K (2011) Recognition of cap structure by influenza B virus RNA polymerase is less dependent on the methyl residue than recognition by influenza a virus polymerase. J Virol 85:7504–7512

Wunderlich K, Juozapaitis M, Ranadheera C et al (2011) Identification of high-affinity PB1-derived peptides with enhanced affinity to the PA protein of influenza a virus polymerase. Antimicrob Agents Chemother 55:696–702

Xiao S, Klein ML, LeBard DN et al (2014) Magnesium-dependent RNA binding to the PA endonuclease domain of the avian influenza polymerase. J Phys Chem B 118:873–889

Xie L, Wartchow C, Shia S et al (2016) Molecular basis of mRNA cap recognition by influenza B polymerase PB2 subunit. J Biol Chem 291:363–370

Yamada S, Hatta M, Staker BL et al (2010) Biological and structural characterization of a host-adapting amino acid in influenza virus. PLoS Pathog 6:e1001034

Yamanaka K, Ishihama A, Nagata K (1990) Reconstitution of influenza virus RNA-nucleoprotein complexes structurally resembling native viral ribonucleoprotein cores. J Biol Chem 265:11151–11155

Yang Y, Tang Y-S, Shaw P-C (2014) Structure and function of nucleoprotein from Orthomyxoviruses. Biodesign 2:91–99

Ye Q, Krug RM, Tao YJ (2006) The mechanism by which influenza a virus nucleoprotein forms oligomers and binds RNA. Nature 444:1078–1082

Ye Q, Guu TSY, Mata DA et al (2012) Biochemical and structural evidence in support of a coherent model for the formation of the double-helical influenza a virus ribonucleoprotein. MBio 4:e00467–e00412

York A, Hengrung N, Vreede FT et al (2013) Isolation and characterization of the positive-sense replicative intermediate of a negative-strand RNA virus. Proc Natl Acad Sci 110(45):E4238

Yuan P, Bartlam M, Lou Z et al (2009) Crystal structure of an avian influenza polymerase PAN reveals an endonuclease active site. Nature 458:909–913

Zhang S, Wang J, Wang Q, Toyoda T (2010) Internal initiation of influenza virus replication of viral RNA and complementary RNA in Vitro. J Biol Chem 285:41194–41201

Zhao C, Lou Z, Guo Y et al (2009) Nucleoside monophosphate complex structures of the endonuclease domain from the influenza virus polymerase PA subunit reveal the substrate binding site inside the catalytic center. J Virol 83:9024–9030

Chapter 6
Structural Homology Between Nucleoproteins of ssRNA Viruses

Mikel Valle

Introduction

Almost 40 years ago, pioneering crystallographic studies on two icosahedral viruses (Abad-Zapatero et al. 1980; Harrison et al. 1978) revealed a clear structural similarity between their capsid proteins. The resemblance was not expected since the proteins that build the capsids of the two viruses have no sequence homology between them; however they display the same so-called jelly roll fold. Additional early works on viral structures supported the notion of structural homology between capsid proteins and an evolutionary divergence of the viruses from common ancestors (Rossmann et al. 1983). Currently, the classical taxonomy of viruses by their genomic features (Baltimore 1971) is challenged by a structure-based classification (Abrescia et al. 2012). In the latter, the viral universe is segregated into four major lineages (PRD1/Adenovirus-like, Picornavirus-like, HK97-like, and BTV-like), and new groups have been recently proposed (Nasir and Caetano-Anolles 2017). A drawback of these new structure-based classifications is that only icosahedral viruses are clearly grouped, and helical and non-icosahedral enveloped viruses lie outside the described lineages. The strong spatial restrictions for capsid proteins in icosahedral arrangement seem to limit their structural variation; thus, the similarities between them are kept at recognition levels. Nucleoproteins from viruses that do not construct well-ordered icosahedral particles exhibit larger structural variability, and their relationships are harder to reveal. Nevertheless, the latest structural studies on nucleoproteins and virions from non-icosahedral viruses substantiate new homologies between viral groups with different morphologies. This chapter is focused on a recently described structural homology between the nucleoproteins from several families of ssRNA viruses that infect eukaryotes.

M. Valle (✉)
Molecular Recognition and Host-Pathogen Interactions, Center for Cooperative Research in Biosciences, CIC bioGUNE, Derio, Spain
e-mail: mvalle@cicbiogune.es

© Springer Nature Singapore Pte Ltd. 2018
J. R. Harris, D. Bhella (eds.), *Virus Protein and Nucleoprotein Complexes*,
Subcellular Biochemistry 88, https://doi.org/10.1007/978-981-10-8456-0_6

Eukaryotic ssRNA Viruses

Among viruses that infect eukaryotic organisms, RNA viruses are the most abundant and diverse, especially the ones with (+)ssRNA genomes. It seems that the compartments in the cytoplasm provide a rich niche where RNA replication complexes are constructed via interactions with proteins and membranes from the hosts (Nagy and Pogany 2011). In the last ICTV release of virus classification (Adams et al. 2017), (+)ssRNA viruses are distributed in 3 orders and 22 unassigned families (Fig. 6.1, that includes only viral families relevant for this chapter), while the less populated group of (−)ssRNA viruses contains 1 order and 4 unassigned families (in this last ICTV release, the previously unassigned family *Bunyaviridae* is redistributed in several families within the new order *Bunyavirales*, but this chapter keeps the name of this family to easily refer to previous works). A tentative phylogeny of eukaryotic (+)ssRNA viruses has been proposed based on the sequence homology between their RNA-dependent RNA polymerases (RdRp), the only common gene to all the families, and the structure of their viral genomes (Koonin 1991). This phylogeny distinguishes three superfamilies: alphavirus-like, picornavirus-like, and flavivirus-like. On the other hand, (−)ssRNA viruses, whose RdRp differ significantly from the ones of the (+)ssRNA groups, are segregated in the order *Mononegavirales* (which includes eight families with monopartite genomes) and in several unassigned families with segmented genomes (Fig. 6.1) (Koonin et al. 2015).

Flexible Filamentous Plant Viruses

Flexible filamentous plant viruses are plant pathogens that contain a monopartite (+)ssRNA genome protected by hundreds of copies of their coat protein (CP) arranged in helical mode (Kendall et al. 2008). Their infective particles are long (several hundreds of nm) and thin (10–15 nm diameter) flexible filaments. They are transmitted by mechanical contact or by arthropod vectors and cause severe economic impact in agriculture. Currently there are more than 380 species (Adams et al. 2017) grouped in four families: *Alphaflexiviridae* (50 species, where genus *Potexvirus* has 35 representatives), *Betaflexiviridae* (89 species, genus *Carlavirus* includes 47 different viruses), *Closteroviridae* (49 species), and *Potyviridae* (195 species, and genus *Potyvirus* includes 160). All those viruses display a very similar architecture for their non-enveloped virions, although some genus within family *Closteroviridae* have segmented genomes.

Most of the flexuous filamentous plant virus groups belong to the alphavirus-like superfamily (Fig. 6.1). They share a closely related RdRp, a capping enzyme, and the superfamily 1 helicase gene (Koonin and Dolja 1993). However, family *Potyviridae* fits in the picornavirus-like superfamily following the RdRp-based phylogeny, the expression and processing of a polyprotein, and the presence of a

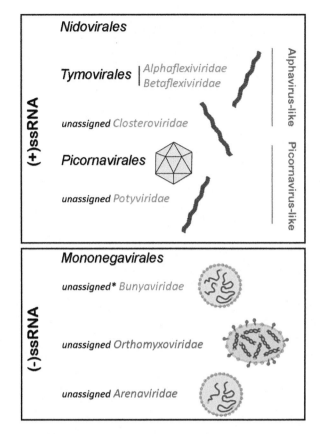

Fig. 6.1 Groups of ssRNA eukaryotic viruses. Some of the orders and families of eukaryotic ssRNA viruses are shown grouped accordingly to the polarity of their ssRNA. Only the viral families relevant for this chapter are included, together with cartoons that represent the architecture of their virions or infective particles. The names of the different families are seen in green (plant infecting viruses), red (animal infecting viruses), or orange (family with plant and animal viruses). (*) *Bunyaviridae* family is currently reassigned in the order Bunyavirales (see main text)

genome-linked VPg protein. Potyviruses are clear outsiders within the picornavirus-like superfamily where the icosahedral capsid made of proteins with the jelly roll fold is abundant; however potyviruses display helical and filamentous virions. It is thought that a common CP gene for flexible filamentous viruses has been transferred and finally shared by all the families (Koonin et al. 2015).

Enveloped and Segmented (−)ssRNA Viruses

Families Orthomyxoviridae (e.g., influenza virus), Bunyaviridae (e.g., Rift Valley fever virus or RVFV), and Arenaviridae (e.g., Lassa fever virus) have been sometimes grouped within the order Multinegavirales, i.e., enveloped viruses with segmented (−)ssRNA genomes or sNSV (segmented negative-strand viruses). They present genomes divided into two (*Arenaviridae*), three (*Bunyaviridae*), and six to eight (*Orthomyxoviridae*) fragments. These subgenomic segments have complementary ends and form circular nucleocapsids (Raju and Kolakofsky 1989; Hsu et al. 1987) together with nucleoproteins and the viral polymerase. The ribonucleoprotein complexes of arenaviruses and bunyaviruses are rather flexible and unstructured, but in influenza they construct double-helical nucleoproteins (Arranz et al. 2012). For all the representatives of this tentative order, the genomic material is protected inside an envelope coming from the membrane of the infected cell. Most of the viruses within these three families infect animals, but the genus *Tospovirus* (e.g., tomato spotted wilt virus or TSWV, family *Bunyaviridae*) is a plant-infecting group that can multiplicate within the arthropod vector (usually thrips) leading to persistent vector transmission (Kormelink et al. 2011).

Structure of Flexible Filamentous Plant Viruses

Initial structural studies of flexuous filamentous plant viruses revealed their common overall architecture (Kendall et al. 2008). Apart from possible differences at their ends (for instance, the presence of VPg linked to the 5′ genomic end in potyviruses), low-resolution X-ray fiber diffraction data and cryoEM 3D maps showed filaments of 120–130 Å diameter constructed by CPs arranged in helical mode, with about nine subunits per turn. The studies were carried out with soyben mosaic virus (SMV), a potyvirus, and three potexviruses (family *Alphaflexiviridae*), potato virus X (PVX), papaya mosaic virus (PapMV), and narcissus mosaic virus (NMV) (Kendall et al. 2013; Yang et al. 2012; Kendall et al. 2008). The flexible nature of the virions precluded atomic resolved data, and the virions were depicted following a right-handed helical arrangement, as observed for rod-shaped rigid viruses such as tobacco mosaic virus or TMV (Namba and Stubbs 1986). In recent years, by using single-particle based helical reconstruction of cryoEM data, several virions have been characterized at higher structural detail: bamboo mosaic virus (BaMV), a potexvirus resolved at 5.6 Å resolution (DiMaio et al. 2015); pepino mosaic virus (PepMV), another potexvirus solved at 3.9 Å resolution (Agirrezabala et al. 2015); and watermelon mosaic virus (WMV), a potyvirus solved at 4.0 Å resolution (Zamora et al. 2017). The availability of structures for flexible filamentous plant viruses from different families (*Alphaflexiviridae* and *Potyviridae*) allows for direct comparison (Fig. 6.2).

The three described virions (BaMV, PepMV, and WMV) display almost identical helical arrangement, with about 34.5–35 Å of helical pitch and 8.8 subunits per turn in left-handed helices (Fig. 6.2a–b). The CPs show a core domain rich in alpha helices and two long arms at both ends of the protein. The assembly of the CPs is mostly mediated by flexible N- and C-terminal arms, in a way that slight relative movements between CP subunits are allowed (Fig. 6.2c–d), and this is the structural

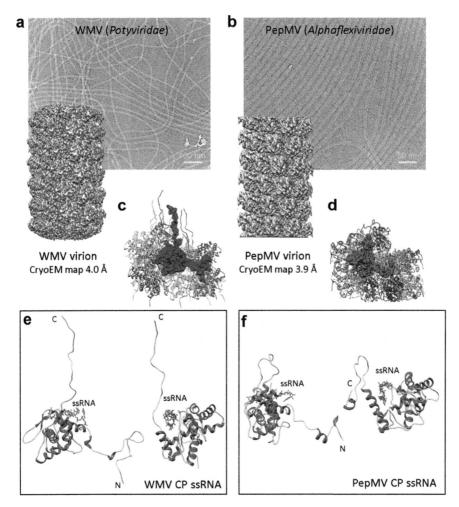

Fig. 6.2 Structure of two flexible filamentous plant viruses belonging to different families. (**a**) CryoEM micrograph field of a WMV (family *Potyviridae*) sample, together with the rendering of the cryoEM map (EMD-3785) for the virion depicted blue (Zamora et al. 2017). (**b**) An electron micrograph field for a sample with PepMV (family *Alphaflexiviridae*) virions is shown, and the corresponding cryoEM map (EMD-3236) rendered in red (Agirrezabala et al. 2015). (**c**) Representation of the atomic models (pdb code 5ODV) of several CPs from WMV as seen in the virion. The atomic coordinates are seen in ribbons with different blue colors for each subunit. One of the CP monomers is depicted as a solid surface. (**d**) Similar depiction for the atomic models of CPs subunits of PepMV (pdb code 5FN1) shown in red. (**e**) Two views of the atomic model of the CP from WMV including a fragment of ssRNA. (**f**) The atomic coordinates for the CP from PepMV are depicted in similar orientations as in (e)

Fig. 6.3 Comparison of
the CP structure from
WMV and PepMV. The
ribbon representation for
WMV CP (pdb code
5ODV (Zamora et al.
2017)) is depicted in
rainbow colors. The atomic
structure for PepMV CP
(pdb code 5FN1,
(Agirrezabala et al. 2015))
is seen in gray ribbons.
The 3D alignment between
both structures was
performed in Matras
(Kawabata 2003). The
numbers indicate the
residue number at the N-
and C-terminal ends of
both atomic coordinates

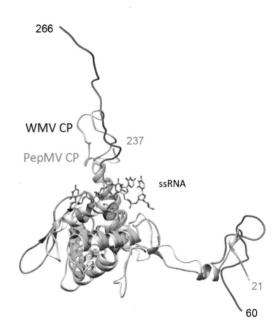

basis for the flexible nature of the virions (Zamora et al. 2017; DiMaio et al. 2015; Agirrezabala et al. 2015). Essentially, the C-terminal arm contributes to the oligomerization between CP subunits at different turns of the helix, i.e., the axial or longitudinal assembly. In all the cases, a final segment of the N-terminal end of the CPs is missing in the atomic models due to its high flexibility. A significant difference is seen in the role of this N-terminal arm. While in potexviruses the N-terminal of one CP interacts with the next subunit in side-by-side contact (Fig. 6.2d) (DiMaio et al. 2015; Agirrezabala et al. 2015), in the potyvirus, a longer N-terminal segment bridges the next subunit in the helix, and by a sharp turn, also interacts with another CP copy at adjacent turn (Fig. 6.2c) (Zamora et al. 2017), displaying a dual role supporting side-by-side and axial polymerization.

Structural Homology Between CPs from Flexible Filamentous Plant Viruses

Remarkably, despite the low sequence identity between CPs from two different families (*Potyviridae* and *Alphaflexiviridae*), their 3D fold is almost identical (Fig. 6.2e–f) with rmsd values at the core of the protein (excluding flexible N- and C-terminal arms) bellow 3 Å (Zamora et al. 2017), and all the essential alpha-helical elements of their structure superimpose (Fig. 6.3). Thus, at least for these two families, their CPs are clear structural homologues, which suggest that a gene transfer occurred at some time between families that are distant with regard to other genetic elements and characteristics.

Conserved RNA-Binding Site

In both, potexviruses and potyviruses, the CP in the virion binds to five nucleotides of the ssRNA in a very similar mode (Zamora et al. 2017; DiMaio et al. 2015; Agirrezabala et al. 2015). Although the density for the ssRNA in those cryoEM maps of virions is an average of RNA segments with different compositions, the signal attributed to the nucleic acid is alike in the three available density maps. The higher-resolution studies (Zamora et al. 2017; Agirrezabala et al. 2015) showed that one out of the five nucleotides bound by each CP fits in a binding pocket (nucleotide labeled as U_4 in Fig. 6.4). Essentially, several residues from the CP interact with consecutive phosphates backbone groups, and the nucleoside in between goes deep into the binding pocket (Fig. 6.4b).

Direct comparison of the atomic models for WMV (Fig. 6.4c) and PepMV (Fig. 6.4d) reveals that three amino acids that participate in the ssRNA-binding pocket are at the same position in the CP of both viruses (Zamora et al. 2017) (Fig. 6.4e). Furthermore, these serine (S), arginine (R), and aspartic (D) acid residues are universally conserved along the four families of flexible filamentous plant viruses (Fig. 6.5) (Zamora et al. 2017; Dolja et al. 1991), with the exception of two potexviruses (bamboo mosaic virus and foxtail mosaic virus) where the conserved arginine is substituted by histidine. Despite the lack of structures for CPs from other families, the high conservation of invariant amino acids suggests that the CPs from flexuous filamentous plant viruses display the same fold and contain a highly conserved RNA-binding site.

Architecture of Enveloped Viruses with Segmented (−)ssRNA Genomes

The members of sNSV have a common overall design that includes the presence of an envelope that protects a variable number of ribonucleoproteins (RNPs) inside (Fig. 6.6). The envelope is taken from the host cell membrane by budding (Lyles 2013) and contains viral glycoproteins that have different roles during the viral cycle. The shape of the virions ranges from pleomorphic (*Arenaviridae*) to spherical and/or elongated (*Orthomyxoviridae* and *Bunyaviridae*). Most of the representatives infect animals, with the exception of plant-infecting tospoviruses (*Bunyaviridae*). Several bunyaviruses and arenaviruses are present in rodents and arthropods and occasionally infect humans in outbreaks of hemorrhagic fever and encephalitis-related diseases. Family Orthomyxoviridae includes well-known influenza viruses that have a large impact in human health. Influenza representatives

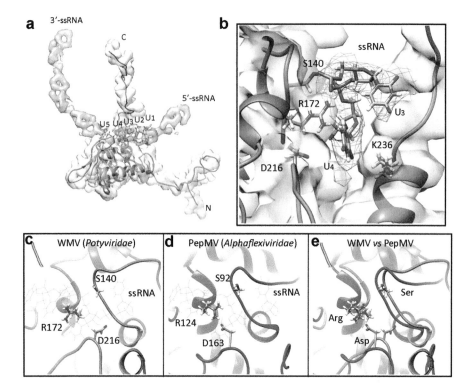

Fig. 6.4 Conserved ssRNA-binding pocket. (**a**) Semi transparent depiction of one CP subunit from WMV (gray) segmented from the cryoEM map of the virion (EMD-3785 (Zamora et al. 2017)), together with the density attributed to the path of the ssRNA (in red) and the derived atomic model (pdb code 5ODV). (**b**) Close-up view of the ssRNA-binding pocket in the CP of WMV with some of the amino acids highlighted. (**c**), (**d**), and (**e**) show the regions that participate in the RNA binding pockets of the CP from WMV (**c**), PepMV (**d**), and a comparison between them (**e**). Three key and conserved amino acids are seen

infect birds and mammals and are transmitted by aerosols between humans. The nucleoproteins of sNSV are associated to the genomic segments and the RdRp in nucleocapsids of dissimilar morphologies (Ruigrok et al. 2011). These nucleoproteins are mainly helical globular with a positively charged groove for RNA binding (Reguera et al. 2014), but no structural homology has been described between nucleoproteins of different families.

Large part of the structural information in *Orthomyxoviridae* family has been obtained for influenza A virus. Nucleoproteins of influenza virus polymerize through the insertion of a loop into the neighboring subunit (Ye et al. 2006). In the constructed RNPs, the ssRNA is in closed conformation, and the viral RdRp binds both RNA ends. CryoET analysis of influenza virions showed the presence of helical RNPs inside the virus (Fig. 6.6a), and cryoEM of isolated RNPs revealed a double-helical architecture with two antiparallel strands of nucleoproteins (Fig. 6.6b) (Arranz et al. 2012). For bunyaviruses (RVFV is used as a representa-

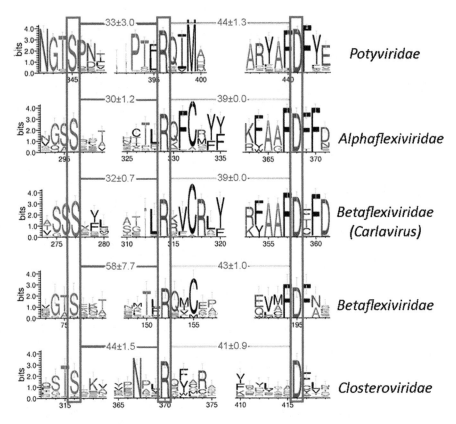

Fig. 6.5 Conservation of amino acids in the RNA-binding pocket along the families of flexible filamentous plant viruses. Consensus sequence logos (Crooks et al. 2004) for CPs from different families of flexuous filamentous plant viruses. The conserved invariant amino acids (Ser or S, Arg or R, and Asp or D) are highlighted in red boxes

tive), loose and flexible RNPs (Raymond et al. 2012) are seen protected inside a spherical shell of glycoproteins (Fig. 6.6c) inserted in the enveloping membrane (Huiskonen et al. 2009; Freiberg et al. 2008). Crystallographic structures of nucleoproteins from bunyaviruses with and without RNA have shown several oligomeric states, from tetramers to hexamers (Fig. 6.6d), where side-by-side interaction between subunits is mediated by N- and/or C-terminal arms (Zhou et al. 2013). The number of ssRNA nucleotides bound by each nucleoprotein subunit can vary from 7 as seen for RVFV (Raymond et al. 2012) up to 11 for orthobunyaviruses (Reguera et al. 2013; Niu et al. 2013; Dong et al. 2013; Ariza et al. 2013), one of the genus in the family Bunyaviridae. The structure of their native RNPs is not well known, but it seems to be rather flexible, ranging from loose and unstructured as in RVFV (Raymond et al. 2012) to different levels of helical arrangement as in La Crosse orthobunyavirus (Reguera et al. 2013). The members of the family Arenaviridae present unique nucleoproteins. This way, the nucleoprotein of Lassa virus

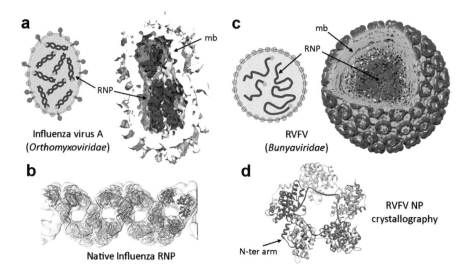

Fig. 6.6 Morphology and organization of segmented (−)ssRNA viruses. (**a**) Rendering of a segmented cryoelectron tomogram for influenza A virus (image courtesy of J. Martín-Benito) (Arranz et al. 2012) and a cartoon that summarizes the general features of the virion. (**b**) Semi transparent view of the cryoEM map for native influenza RNPs (EMD2205) with fitted atomic coordinates for its nucleoproteins (pdb code 4BBL) (Arranz et al. 2012). (**c**) Display that includes the representation of the cryoEM map for RVFV (EMD-5124 (Sherman et al. 2009)) and a schematic cartoon of the viral architecture. (**d**) Crystallographic structure of the hexameric form of the nucleoprotein from RVFV assembled with ssRNA (pdb code 4H5O (Raymond et al. 2012))

(*Arenaviridae*) displays an additional C-terminal domain with exonuclease activity that seems to be involved in immune suppression (Hastie et al. 2011; Qi et al. 2010).

Structural Homology between Nucleoproteins of Eukaryotic ssRNA Viruses

It is clear that flexible filamentous plant viruses (at least two of the families) display high structural homology between their CPs, which are also nucleoproteins (Zamora et al. 2017). Despite the abundant structural information about nucleoproteins of sNSV, no structural homology was detected by direct comparison between atomic coordinates of nucleoproteins from different families (Ruigrok et al. 2011). However, using the atomic models of CPs from flexuous filamentous plant viruses as structural targets, structural similarities with several nucleoproteins of sNSV emerged (Zamora et al. 2017; Agirrezabala et al. 2015). The core region of CPs shows structural homology with nucleoproteins of representatives from families *Bunyaviridae* (Zamora et al. 2017; Agirrezabala et al. 2015) and *Orthomyxoviridae* (Zamora et al. 2017) (Fig. 6.7). The structures share similar topology where alpha-helical secondary structure elements are easily aligned. Both N- and C-terminal

Fig. 6.7 Structural homology between nucleoproteins of eukaryotic ssRNA viruses. The panels show ribbon representations for the core regions of nucleoproteins from different viruses in rainbow color mode (left side). At the right side, the core of the nucleoprotein subunit is seen green, and the N- and C-terminal extensions are red and yellow, respectively. Other subunits that interact with the colored ones are depicted gray to illustrate their oligomerization. The atomic coordinates are (**a**) WMV CP and ssRNA (pdb code 5ODV (Zamora et al. 2017)); (**b**) PepMV CP and ssRNA, (pdb code 5FN1 (Agirrezabala et al. 2015)); (**c**) in the left side, a single influenza virus A nucleoprotein subunit, (pdb code 3ZDP (Chenavas et al. 2013)), and at the right side, two interacting nucleoproteins (pdb code 2IQH (Ye et al. 2006)); (**d**) RVFV nucleoprotein in complex with ssRNA, (pdb code 4H5O (Raymond et al. 2012)); (**e**) La Crosse virus nucleoprotein and ssRNA, (pdb code 4BHH(Reguera et al. 2013); (**f**) and TSWV nucleoprotein in complex with ssRNA, (pdb code 5IP2 (Komoda et al. n.d.). The numbers indicate the residue number at the N- and C-terminal ends of the atomic coordinates

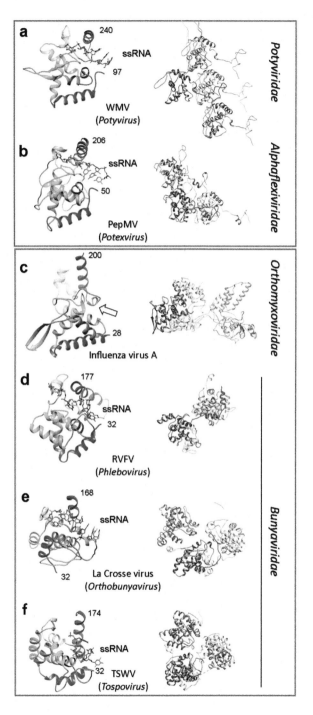

ends are in similar positions, and the grooves for ssRNA binding are at the same location within the nucleoproteins. For the nucleoprotein of influenza virus, there is not any available atomic structure in complex with ssRNA, but the proposed binding site (Ye et al. 2006) aligns well with that of the other nucleoproteins (indicated by an arrow in Fig. 6.7c). The estimated probability that all these nucleoproteins (the ones displayed in Fig. 6.7 and related) belong to the same fold is above 90% (Zamora et al. 2017; Kawabata 2003). In essence, there is a clear fragment of about 150 residues that shares the same fold in nucleoproteins of different ssRNA viruses, and each of these proteins has additional regions of variable length at N- and C-terminal ends. There is no significant structural homology with nucleoproteins from family *Arenaviridae*, although as mentioned earlier, their nucleoproteins have adopted an additional domain and might have diverged from a similar fold.

The oligomerization of these nucleoproteins takes place through interactions via N- and C-terminal arms (right panels in Fig. 6.7), although the final multimeric RNPs have different arrangements (from loose to full helical). Remarkably, the N-terminal arm oligomerization modes between nucleoproteins in potexviruses (family *Alphaflexiviridae*) and phlebovirues (family *Bunyaviridae*) are comparable and use the same groove in the neighboring subunit to receive the N-terminal arm (Agirrezabala et al. 2015). Nucleoproteins of influenza viruses are a clear exception, since a folded and large C-terminal region (depicted yellow in the right panel of Fig. 6.7c) contributes to oligomerization by the insertion of an internal loop in the adjacent subunit (Arranz et al. 2012; Ye et al. 2006).

Evolutionary Implications

Structural homology between proteins is usually understood as an indication of common evolutionary origin (homology) rather than the product of convergent evolution (analogy). This is based on the observation that the structure of proteins is more conserved than their sequence of amino acids (Illergard et al. 2009) and that convergent evolution of protein domains is a rare event (Gough 2005). In the current matter, the structural similarity between nucleoproteins is further supported by their role as viral proteins that bind and protect ssRNA genomes. It can be presumed that the genes of nucleoproteins from flexible filamentous plant viruses and from at least two families of sNSV share a common ancestor (Fig. 6.8). The homology between these two sets of viruses was not anticipated until atomic structures for the first group (the flexuous plant viruses) were available (Zamora et al. 2017; Agirrezabala et al. 2015). This suggests that the structure of flexible filamentous plant virus nucleoproteins displays a fold closer to a common ancestor protein and that their homology with nucleoproteins of sNSV is still recognized. However, nucleoproteins from sNSV are diverse and show lower levels of structural homology between them. This is an indication of a higher structural divergence within sNSV nucleoproteins.

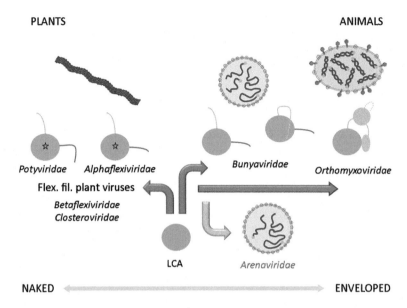

Fig. 6.8 Landscape for putative evolution of nucleoproteins in ssRNA viruses. The two sets of viruses are segregated in two main groups, naked and plant viruses (left) and enveloped and animal viruses (right). Their nucleoproteins are represented by a green circle, and the N- and C-terminal extensions are depicted red and yellow, respectively. In the nucleoprotein (or CP) of flexible filamentous plant virus, the star indicates the conserved RNA-binding site. LCA: last common ancestor

The evolution of RNA viruses is hard to unveil, and in the current scenario, we do not known how the nucleoprotein gene has spread along several families of ssRNA viruses. Some viral evolutionary mechanism such as cross-species transmission (Geoghegan et al. 2017) and transfer of genes between virus and host (Aiewsakun and Katzourakis 2015) have recently been acknowledged as frequent events. For instance, CP sequences from potato virus Y (PVY, a potyvirus) have been found in the genomes of grapevines, probably after nonhomologous recombination with retrotransposable elements (Tanne and Sela 2005). In the same line, genomic sequences from bunyaviruses and orthomyxoviruses have been found as endogenous viral elements in insects and crustaceans (Theze et al. 2014; Ballinger et al. 2013; Katzourakis and Gifford 2010). Importantly, RNA sequencing works have found a large genomic diversity of RNA viruses and related sequences in invertebrates (Shi et al. 2016; Li et al. 2015). The phylogenetic analysis of these viromes reveals frequent recombination, gene transfer events, and genetic reassortments. Invertebrates, specially insects, play a central role as vectors for several of the ssRNA viruses discussed in this chapter and might have provided a niche for an evolutionary explosion of eukaryotic RNA viruses (Koonin et al. 2015).

Regardless of the evolutionary mechanisms that transferred the nucleoprotein gene, there are some general trends that might explain the current diversity of morphologies in these viral families. Naked and filamentous forms are linked to plant-infecting viruses (Fig. 6.8), while enveloped viruses are essentially animal pathogens, with the exception of tospoviruses (within the groups discussed in this chapter). There is a clear relationship between the presence of a cell wall in the host cell and the lack of viral envelope (Buchmann and Holmes 2015). Also, the need to cross the narrow plasmodesmata between plant cells during infection favors filamentous versus spherical virions in plant viruses (Hong and Ju 2017). It is possible that the naked nucleoproteins from flexuous plant viruses undergo functional restrictions that limit their structural variation and they conserve a very close fold and a specific RNA-binding site. Nucleoproteins from snSV, however, are protected inside the membranous envelope and have explored a wider structural landscape and several oligomerization strategies.

Acknowledgments This work was supported by grant BFU2015-66326-P from the Spanish Ministry of Economy and Competitiveness (MINECO). Also the Severo Ochoa Excellence Accreditation (SEV-2016-0644) by MINECO is acknowledged.

References

Abad-Zapatero C, Abdel-Meguid SS, Johnson JE, Leslie AG, Rayment I, Rossmann MG, Suck D, Tsukihara T (1980) Structure of southern bean mosaic virus at 2.8 A resolution. Nature 286(5768):33–39

Abrescia NG, Bamford DH, Grimes JM, Stuart DI (2012) Structure unifies the viral universe. Annu Rev Biochem 81:795–822. https://doi.org/10.1146/annurev-biochem-060910-095130

Adams MJ, Lefkowitz EJ, King AMQ, Harrach B, Harrison RL, Knowles NJ, Kropinski AM, Krupovic M, Kuhn JH, Mushegian AR, Nibert M, Sabanadzovic S, Sanfacon H, Siddell SG, Simmonds P, Varsani A, Zerbini FM, Gorbalenya AE, Davison AJ (2017) Changes to taxonomy and the international code of virus classification and nomenclature ratified by the international committee on taxonomy of viruses (2017). Arch Virol 162:2505. https://doi.org/10.1007/s00705-017-3358-5

Agirrezabala X, Mendez-Lopez E, Lasso G, Sanchez-Pina MA, Aranda M, Valle M (2015) The near-atomic cryoEM structure of a flexible filamentous plant virus shows homology of its coat protein with nucleoproteins of animal viruses. Elife 4:e11795. https://doi.org/10.7554/eLife.11795

Aiewsakun P, Katzourakis A (2015) Endogenous viruses: connecting recent and ancient viral evolution. Virology 479-480:26–37. https://doi.org/10.1016/j.virol.2015.02.011

Ariza A, Tanner SJ, Walter CT, Dent KC, Shepherd DA, Wu W, Matthews SV, Hiscox JA, Green TJ, Luo M, Elliott RM, Fooks AR, Ashcroft AE, Stonehouse NJ, Ranson NA, Barr JN, Edwards TA (2013) Nucleocapsid protein structures from orthobunyaviruses reveal insight into ribonucleoprotein architecture and RNA polymerization. Nucleic Acids Res 41(11):5912–5926. https://doi.org/10.1093/nar/gkt268

Arranz R, Coloma R, Chichon FJ, Conesa JJ, Carrascosa JL, Valpuesta JM, Ortin J, Martin-Benito J (2012) The structure of native influenza virion ribonucleoproteins. Science 338(6114):1634–1637. https://doi.org/10.1126/science.1228172

Ballinger MJ, Bruenn JA, Kotov AA, Taylor DJ (2013) Selectively maintained paleoviruses in Holarctic water fleas reveal an ancient origin for phleboviruses. Virology 446(1–2):276–282. https://doi.org/10.1016/j.virol.2013.07.032

Baltimore D (1971) Expression of animal virus genomes. Bacteriol Rev 35(3):235–241

Buchmann JP, Holmes EC (2015) Cell walls and the convergent evolution of the viral envelope. Microbiol Mol Biol Rev 79(4):403–418. https://doi.org/10.1128/MMBR.00017-15

Chenavas S, Estrozi LF, Slama-Schwok A, Delmas B, Di Primo C, Baudin F, Li X, Crepin T, Ruigrok RW (2013) Monomeric nucleoprotein of influenza A virus. PLoS Pathog 9(3):e1003275. https://doi.org/10.1371/journal.ppat.1003275

Crooks GE, Hon G, Chandonia JM, Brenner SE (2004) WebLogo: a sequence logo generator. Genome Res 14(6):1188–1190. https://doi.org/10.1101/gr.849004

DiMaio F, Chen CC, Yu X, Frenz B, Hsu YH, Lin NS, Egelman EH (2015) The molecular basis for flexibility in the flexible filamentous plant viruses. Nat Struct Mol Biol 22(8):642–644. https://doi.org/10.1038/nsmb.3054

Dolja VV, Boyko VP, Agranovsky AA, Koonin EV (1991) Phylogeny of capsid proteins of rod-shaped and filamentous RNA plant viruses: two families with distinct patterns of sequence and probably structure conservation. Virology 184(1):79–86

Dong H, Li P, Bottcher B, Elliott RM, Dong C (2013) Crystal structure of Schmallenberg orthobunyavirus nucleoprotein-RNA complex reveals a novel RNA sequestration mechanism. RNA 19(8):1129–1136. https://doi.org/10.1261/rna.039057.113

Freiberg AN, Sherman MB, Morais MC, Holbrook MR, Watowich SJ (2008) Three-dimensional organization of Rift Valley fever virus revealed by cryoelectron tomography. J Virol 82(21):10341–10348. https://doi.org/10.1128/JVI.01191-08

Geoghegan JL, Duchene S, Holmes EC (2017) Comparative analysis estimates the relative frequencies of co-divergence and cross-species transmission within viral families. PLoS Pathog 13(2):e1006215. https://doi.org/10.1371/journal.ppat.1006215

Gough J (2005) Convergent evolution of domain architectures (is rare). Bioinformatics 21(8):1464–1471. https://doi.org/10.1093/bioinformatics/bti204

Harrison SC, Olson AJ, Schutt CE, Winkler FK, Bricogne G (1978) Tomato bushy stunt virus at 2.9 A resolution. Nature 276(5686):368–373

Hastie KM, Kimberlin CR, Zandonatti MA, MacRae IJ, Saphire EO (2011) Structure of the Lassa virus nucleoprotein reveals a dsRNA-specific 3′ to 5′ exonuclease activity essential for immune suppression. Proc Natl Acad Sci U S A 108(6):2396–2401. https://doi.org/10.1073/pnas.1016404108

Hong JS, Ju HJ (2017) The plant cellular systems for plant virus movement. Plant Pathol J 33(3):213–228. https://doi.org/10.5423/PPJ.RW.09.2016.0198

Hsu MT, Parvin JD, Gupta S, Krystal M, Palese P (1987) Genomic RNAs of influenza viruses are held in a circular conformation in virions and in infected cells by a terminal panhandle. Proc Natl Acad Sci U S A 84(22):8140–8144

Huiskonen JT, Overby AK, Weber F, Grunewald K (2009) Electron cryo-microscopy and single-particle averaging of Rift Valley fever virus: evidence for GN-GC glycoprotein heterodimers. J Virol 83(8):3762–3769. https://doi.org/10.1128/JVI.02483-08

Illergard K, Ardell DH, Elofsson A (2009) Structure is three to ten times more conserved than sequence--a study of structural response in protein cores. Proteins 77(3):499–508. https://doi.org/10.1002/prot.22458

Katzourakis A, Gifford RJ (2010) Endogenous viral elements in animal genomes. PLoS Genet 6(11):e1001191. https://doi.org/10.1371/journal.pgen.1001191

Kawabata T (2003) MATRAS: A program for protein 3D structure comparison. Nucleic Acids Res 31(13):3367–3369

Kendall A, McDonald M, Bian W, Bowles T, Baumgarten SC, Shi J, Stewart PL, Bullitt E, Gore D, Irving TC, Havens WM, Ghabrial SA, Wall JS, Stubbs G (2008) Structure of flexible filamentous plant viruses. J Virol 82(19):9546–9554. https://doi.org/10.1128/JVI.00895-08

Kendall A, Bian W, Maris A, Azzo C, Groom J, Williams D, Shi J, Stewart PL, Wall JS, Stubbs G (2013) A common structure for the potexviruses. Virology 436(1):173–178. https://doi.org/10.1016/j.virol.2012.11.008

Komoda K, Narita M, Yamashita K, Tanaka I, Yao M (n.d.) Tomato spotted wilt tospovirus nucleocapsid protein-ssRNA complex. pdb code 5IP2

Koonin EV (1991) The phylogeny of RNA-dependent RNA polymerases of positive-strand RNA viruses. J Gen Virol 72(Pt 9):2197–2206. https://doi.org/10.1099/0022-1317-72-9-2197

Koonin EV, Dolja VV (1993) Evolution and taxonomy of positive-strand RNA viruses: implications of comparative analysis of amino acid sequences. Crit Rev Biochem Mol Biol 28(5):375–430. https://doi.org/10.3109/10409239309078440

Koonin EV, Dolja VV, Krupovic M (2015) Origins and evolution of viruses of eukaryotes: the ultimate modularity. Virology 479-480C:2–25. https://doi.org/10.1016/j.virol.2015.02.039

Kormelink R, Garcia ML, Goodin M, Sasaya T, Haenni AL (2011) Negative-strand RNA viruses: the plant-infecting counterparts. Virus Res 162(1–2):184–202. https://doi.org/10.1016/j.virusres.2011.09.028

Li CX, Shi M, Tian JH, Lin XD, Kang YJ, Chen LJ, Qin XC, Xu J, Holmes EC, Zhang YZ (2015) Unprecedented genomic diversity of RNA viruses in arthropods reveals the ancestry of negative-sense RNA viruses. Elife 4:e05378. https://doi.org/10.7554/eLife.05378

Lyles DS (2013) Assembly and budding of negative-strand RNA viruses. Adv Virus Res 85:57–90. https://doi.org/10.1016/B978-0-12-408116-1.00003-3

Nagy PD, Pogany J (2011) The dependence of viral RNA replication on co-opted host factors. Nat Rev Microbiol 10(2):137–149. https://doi.org/10.1038/nrmicro2692

Namba K, Stubbs G (1986) Structure of tobacco mosaic virus at 3.6 A resolution: implications for assembly. Science 231(4744):1401–1406

Nasir A, Caetano-Anolles G (2017) Identification of capsid/coat related protein folds and their utility for virus classification. Front Microbiol 8:380. https://doi.org/10.3389/fmicb.2017.00380

Niu F, Shaw N, Wang YE, Jiao L, Ding W, Li X, Zhu P, Upur H, Ouyang S, Cheng G, Liu ZJ (2013) Structure of the Leanyer orthobunyavirus nucleoprotein-RNA complex reveals unique architecture for RNA encapsidation. Proc Natl Acad Sci U S A 110(22):9054–9059. https://doi.org/10.1073/pnas.1300035110

Qi X, Lan S, Wang W, Schelde LM, Dong H, Wallat GD, Ly H, Liang Y, Dong C (2010) Cap binding and immune evasion revealed by Lassa nucleoprotein structure. Nature 468(7325):779–783. https://doi.org/10.1038/nature09605

Raju R, Kolakofsky D (1989) The ends of La Crosse virus genome and antigenome RNAs within nucleocapsids are base paired. J Virol 63(1):122–128

Raymond DD, Piper ME, Gerrard SR, Skiniotis G, Smith JL (2012) Phleboviruses encapsidate their genomes by sequestering RNA bases. Proc Natl Acad Sci U S A 109(47):19208–19213. https://doi.org/10.1073/pnas.1213553109

Reguera J, Malet H, Weber F, Cusack S (2013) Structural basis for encapsidation of genomic RNA by La Crosse Orthobunyavirus nucleoprotein. Proc Natl Acad Sci U S A 110(18):7246–7251. https://doi.org/10.1073/pnas.1302298110

Reguera J, Cusack S, Kolakofsky D (2014) Segmented negative strand RNA virus nucleoprotein structure. Curr Opin Virol 5:7–15. https://doi.org/10.1016/j.coviro.2014.01.003

Rossmann MG, Abad-Zapatero C, Murthy MR, Liljas L, Jones TA, Strandberg B (1983) Structural comparisons of some small spherical plant viruses. J Mol Biol 165(4):711–736

Ruigrok RW, Crepin T, Kolakofsky D (2011) Nucleoproteins and nucleocapsids of negative-strand RNA viruses. Curr Opin Microbiol 14(4):504–510. https://doi.org/10.1016/j.mib.2011.07.011

Sherman MB, Freiberg AN, Holbrook MR, Watowich SJ (2009) Single-particle cryo-electron microscopy of Rift Valley fever virus. Virology 387(1):11–15. https://doi.org/10.1016/j.virol.2009.02.038

Shi M, Lin XD, Tian JH, Chen LJ, Chen X, Li CX, Qin XC, Li J, Cao JP, Eden JS, Buchmann J, Wang W, Xu J, Holmes EC, Zhang YZ (2016) Redefining the invertebrate RNA virosphere. Nature 540:539. https://doi.org/10.1038/nature20167

Tanne E, Sela I (2005) Occurrence of a DNA sequence of a non-retro RNA virus in a host plant genome and its expression: evidence for recombination between viral and host RNAs. Virology 332(2):614–622. https://doi.org/10.1016/j.virol.2004.11.007

Theze J, Leclercq S, Moumen B, Cordaux R, Gilbert C (2014) Remarkable diversity of endogenous viruses in a crustacean genome. Genome Biol Evol 6(8):2129–2140. https://doi.org/10.1093/gbe/evu163

Yang S, Wang T, Bohon J, Gagne ME, Bolduc M, Leclerc D, Li H (2012) Crystal structure of the coat protein of the flexible filamentous papaya mosaic virus. J Mol Biol 422(2):263–273. https://doi.org/10.1016/j.jmb.2012.05.032

Ye Q, Krug RM, Tao YJ (2006) The mechanism by which influenza a virus nucleoprotein forms oligomers and binds RNA. Nature 444(7122):1078–1082. https://doi.org/10.1038/nature05379

Zamora M, Mendez-Lopez E, Agirrezabala X, Cuesta R, Lavin JL, Sanchez-Pina MA, Aranda M, Valle M (2017) Potyvirus virion structure shows conserved protein fold and RNA binding site in ssRNA viruses. Sci Adv 3(9):eaao2182. https://doi.org/10.1126/sciadv.aao2182

Zhou H, Sun Y, Guo Y, Lou Z (2013) Structural perspective on the formation of ribonucleoprotein complex in negative-sense single-stranded RNA viruses. Trends Microbiol 21(9):475–484. https://doi.org/10.1016/j.tim.2013.07.006

Chapter 7
Zika Virus Envelope Protein and Antibody Complexes

Lianpan Dai, Qihui Wang, Hao Song, and George Fu Gao

Introduction

Zika virus (ZIKV) is an arthropod-borne human pathogen, belonging to *Flavivirus* genus in *Flaviviridae* family (Knipe et al. 2013). Other members in this genus include dengue virus (DENV), yellow fever virus (YFV), West Nile virus (WNV), Japanese encephalitis virus (JEV), tick-borne encephalitis virus (TBEV), etc. (Knipe et al. 2013). ZIKV was first identified in 1947 from Zika forest in Uganda (Dick et al. 1952). A group of British scientists investigated the YFV there and occasionally isolated the ZIKV in a "sentinel" monkey (Yun and Lee 2017). This isolate was named as MR766 strain and is the prototype of the African lineage of ZIKV (Dick et al. 1952). One year later, ZIKV was isolated from mosquito in the same region (Dick et al. 1952). In the following years, mounting evidences support that mosquito is the major vector for ZIKV transmission. Due to the serological

L. Dai (✉) · H. Song
Research Network of Immunity and Health (RNIH), Beijing Institutes of Life Science,
Chinese Academy of Sciences, Beijing, China
e-mail: dailp@biols.ac.cn; songh@biols.ac.cn

Q. Wang
CAS Key Laboratory of Microbial Physiological and Metabolic Engineering, Institute of
Microbiology, Chinese Academy of Sciences, Beijing, China
e-mail: wangqihui@im.ac.cn

G. F. Gao (✉)
Research Network of Immunity and Health (RNIH), Beijing Institutes of Life Science,
Chinese Academy of Sciences, Beijing, China

CAS Key Laboratory of Pathogenic Microbiology and Immunology, Institute of
Microbiology, Chinese Academy of Sciences, Beijing, China

National Institute for Viral Disease Control and Prevention, Chinese Center for Disease
Control and Prevention (China CDC), Beijing, China
e-mail: gaof@im.ac.cn

© Springer Nature Singapore Pte Ltd. 2018
J. R. Harris, D. Bhella (eds.), *Virus Protein and Nucleoprotein Complexes*,
Subcellular Biochemistry 88, https://doi.org/10.1007/978-981-10-8456-0_7

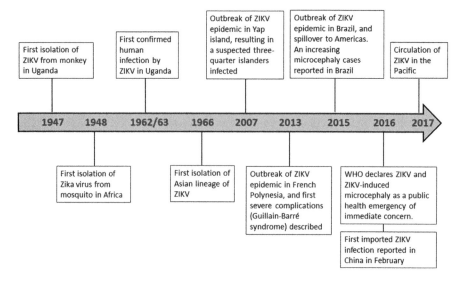

Fig. 7.1 Timeline of important events for Zika virus

cross-reactivity with other closely related flaviviruses like DENV, ZIKV infections were frequently misdiagnosed (Baud et al. 2017). The first confirmed human infection of ZIKV was documented in Uganda during 1962–1963 (Simpson 1964). The timeline of important events for ZIKV is shown in Fig. 7.1

Epidemics

Before 2000, ZIKV infections were only sporadically reported in tropical regions of Africa and Asia (Wikan and Smith 2016). In the new millennium, ZIKV diseases were endemic in the Pacific Islands and Southeast Asia (Wikan and Smith 2016). In 2007, the first epidemic of ZIKV emerged in Yap Island in the Western Pacific (Wikan and Smith 2016). An estimate of up to 75% of the islanders were infected (Duffy et al. 2009). Six years later in 2013, a larger outbreak occurred in French Polynesia located in the Southern Pacific (Cao-Lormeau et al. 2014). Approximately 30,000 residents were supposed to be infected (Cao-Lormeau et al. 2014). Coincidentally, an increased incidence of Guillain-Barré syndrome (GBS) was documented during the same time with ZIKV epidemic in French Polynesia, the first to link the ZIKV infection to GBS (Dos Santos et al. 2016; Musso et al. 2014). The biggest explosion of Zika disease occurred in Brazil from 2015, causing more than a million of suspected ZIKV infections (Bogoch et al. 2016; Wikan and Smith 2016; Garcia Serpa Osorio-de-Castro et al. 2017). Catastrophic clinical consequences were observed in fetuses infected with ZIKV during pregnancy in Brazil (Brito and Cordeiro 2016; Kleber de Oliveira et al. 2016; Baud et al. 2017). During the ZIKV outbreak in Brazil, the incidence of microcephaly in newborns dramatically elevated. Zika disease then spread to other countries in Latin America and the USA (Baud et al. 2017). As a result, the World Health Organization (WHO) announced the explosive microcephaly, neurological disorders, and their associations with ZIKV infection to be a public health emergency of immediate concern for 9 months (Gulland 2016; Baud et al. 2017). In 2016, ZIKV was even imported into China, the most populous country, becoming a serious challenge for the society (Li et al. 2016).

Pathogenesis

Historically, the clinical syndrome caused by ZIKV infection is mild. Although ZIKV has been isolated for more than half a century, many of its distinguishing features were not uncovered until recently. ZIKV can cause congenital Zika syndrome in newborns and GBS in adults. Infection of fetuses during pregnancy has been associated with placental insufficiency, microcephaly, cerebral calcifications, and miscarriage (Miner et al. 2016; Rasmussen et al. 2016). ZIKV transmission is mainly through mosquitoes but can also be transmitted via sex contact (D'Ortenzio et al. 2016; Hills et al. 2016). ZIKV was found to persist in human semen and sperm for several months (Deckard et al. 2016). By applying the mouse model, we and others have demonstrated that ZIKV infection leads to testis damage and male infertility (Govero et al. 2016; Ma et al. 2016). In sum, ZIKV can break through barriers from the blood to the brain, placenta, testes, and eyes, respectively, all the four immune privileged organs (Miner and Diamond 2017).

Zika Virus Genome and Viral Proteins

Like other flaviviruses, such as DENV, JEV, YFV, WNV, and TBEV, ZIKV is an enveloped, single-stranded, positive-sense RNA virus (Knipe et al. 2013). The RNA genome of ZIKV is translated into a long polypeptide in the cytoplasm of the infected cells directly. The polypeptide is further cleaved and processed by host and viral proteases into three structural proteins (premembrane (prM), envelope (E), capsid (C)) during and posttranslation, which form the virus particle, and seven nonstructural proteins (NS1, NS2A, NS2B, NS3, NS4A, NS4B, and NS5), which perform essential functions in genome replication, polyprotein processing, and manipulation of cellular processes for viral advantage (Shi and Gao 2017) (Fig. 7.2). Like other flaviviruses, replication of ZIKV genome takes place in the endoplasmic reticulum (ER) membrane-associated replication complexes (RC), including seven NS proteins and host factors and with NS3 and NS5 residing in the functional center.

NS1, a glycoprotein, exists as a membrane-associated homodimer after translocation into the ER lumen and is necessary during the replication of viral genome

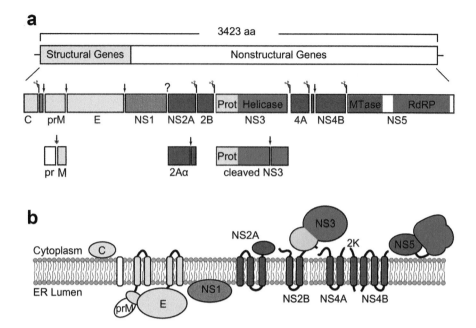

Fig. 7.2 Genome organization of ZIKV and the processed proteins. (**a**) Polyprotein processing and cleavage products. ZIKV has a single-stranded positive-sense genome of approximately 11 Kb. The boxes below the genome indicate precursors and mature proteins generated by the proteolytic processing cascade. (**b**) Polyprotein topology in the membrane. The proposed membrane orientation of the ZIKV proteins is shown. The figure is modified from Fig. 7.1 in Shi and Gao (2017)

and virion maturation (Shi and Gao 2017; Avirutnan et al. 2010; Chung et al. 2006a). Infected cells also secrete NS1 into the extracellular space as a hexameric lipoprotein particle (sNS1) (Xu et al. 2016). The sNS1 lipoprotein could elicit not only protective antibodies but also autoantibodies contributing to dengue shock syndrome (DSS). Further, sNS1 is involved in pathogenesis by activating innate immunity cells, leading to DSS, and is involved in immune evasion by interacting with different components of the complement system (Avirutnan et al. 2010; Chung et al. 2006a). NS1 also represents the major antigenic marker for viral infection (Young et al. 2000). The NS2A, NS2B, NS4A, and NS4B are small, hydrophobic membrane proteins that are required for virus assembly and play an important role in the inhibition of the interferon (IFN) response. NS3 and NS5 are two proteins with different enzymatic activities: NS3 protein encodes for viral serine protease (active only together with NS2B cofactor), helicase, nucleoside triphosphatase, and RNA triphosphatase. NS5, the largest and most conserved viral protein, encodes for a methyltransferase (MTase) at the N-terminal, while C-terminal encodes for the RNA-dependent RNA polymerase. Small molecule inhibitors targeting the enzyme activities of NS3 and NS5 have been developed, and antiviral intervention targeting NS1 protein also shows attractive prospects (Kang et al. 2017; Lei et al. 2016; Duan et al. 2017; Chan et al. 2017).

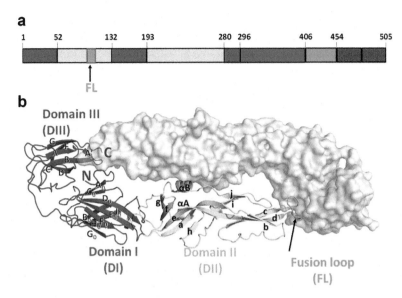

Fig. 7.3 Overall structure of the ZIKV E protein at its pre-fusion state. (a) Schematic diagram of domain organization for ZIKV E protein: DI (red), DII (yellow), and DIII (blue). A 48-residue stem region is colored cyan. The transmembrane domain is colored dark blue. (b) Dimer structure of ZIKV E (PDB:5JHM). One protomer is shown as surface and the other is shown as cartoon. The three extracellular domains (DI, DII, and DIII) are marked in red, yellow, and blue, respectively, while the fusion loop (FL) is colored in green. Secondary structural elements are labeled

E Protein: Structures and Function

E Protein Structure in Pre-Fusion State

Flavivirus E protein mediates virus entry and membrane fusion and contains the putative receptor-binding sites for host cells, though no known host receptor has ever been identified to date (Dai et al. 2016b; Bhardwaj et al. 2001; Chu et al. 2005; Lee and Lobigs 2000). Therefore, E protein is the major target for neutralizing antibodies. E protein of ZIKV belongs to the classical type II fusion protein. Unlike the type I fusion protein which anchors in the virus surface as spikes, 180 copies of E protomers lay flat on the envelope of the virion forming 90 head-to-tail homodimers in its pre-fusion state (Dai et al. 2016a). The ectodomain of each E protomer consists of three structurally distinct domains rich in β-sheets: a β-barrel domain I (DI), which locates in the middle to link the other two domains; a finger like domain II (DII), in which tip contains the hydrophobic fusion loop (FL); and a C-terminus domain III (DIII), which shows an IgC-like fold (Fig. 7.3). This pre-fusion structure of ZIKV E protein reassembles the E protein of other flavivirus. Ninety E-dimers compact together as the icosahedral symmetry and display "herringbone"-like arrangement. Every three E-dimers array parallel to form a raft structure. Therefore, each virion is covered by 30 rafts. In this pre-fusion state, each of the FL of DII is covered by DIII in the neighboring protomer, leading to the hydrophobic FL unexposed. This protection of FL can stabilize the E protein in its pre-fusion conformation and prevents the premature membrane fusion during virion maturation.

E Protein Structures During Membrane Fusion

During the flavivirus infection, virus attaches the cell membrane and is internalized into the target cell for fusion via endocytosis (Knipe et al. 2013). The structures of ZIKV E protein during membrane fusion have not been reported. According to the structure-known E proteins from other flavivirus, such as DENV and TBEV, the acidic pH condition in later endosome triggers the conformational transition for the E protein on the virus surface (Modis et al. 2004; Bressanelli et al. 2004). During this process, DII rotates more than 20 degrees, leading to the exposure of FL and subsequently dissociation of E-dimers. Meanwhile, DIII swings around 60–70 degrees toward DI, and the E protomer stands on the virus surface from lying state (Bressanelli et al. 2004; Modis et al. 2004). Along with that, E protomers are rearranged to form 60 trimers, and the FLs are exposed on top in a head-to-head manner. This exposure is instable and will lead to the insertion of FLs into the membrane of endosome via hydrophobic interaction, initiating the membrane fusion process (Fig. 7.4a–c). E protein in this state is known as the fusion intermediate. In the proximity of virus membrane, the stem region of E protein which is composed of two α-helices connects with the transmembrane domain. When FL is inserted into the endosomal membrane, the two α-helices of stem region relocate toward the FL and subsequently pull closer the membranes between the virus and endosome. The E proteins then form the trimeric structure at post-fusion state (Fig. 7.5). Consequently, these two membranes fuse together and release the viral RNA into the cytoplasm of the host cell. Antibodies can neutralize virus in multiple stages by blocking virus attachment to host cells or by interfering with virus fusion at post-attachment steps (Fig. 7.6, left panel).

E Protein During Virus Assembly

E proteins are incorporated into virion at late stage of virus life cycle (Knipe et al. 2013). The newly assembled virus progenies firstly enter into the lumen of endoplasmic reticulum (ER) to form the immature virions. In this stage, DII of E protein forms the binding partner with prM as a prM/E heterodimer (Li et al. 2008). Every 3 heterodimers gather together to form a trimeric protrusion. Each virion is anchored by 60 such protrusions (Fig. 7.4d). Subsequently, immature virions enter into *trans*-Golgi network for virus maturation. During this process, prM is cleaved

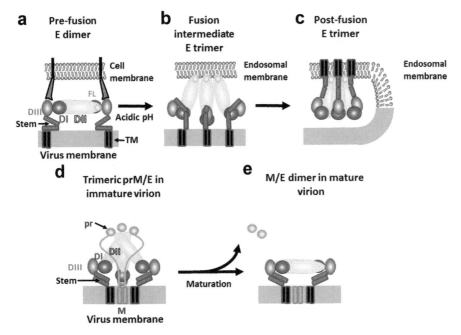

Fig. 7.4 **Schematic demonstration of conformational changes of E protein during virus entry, fusion, and maturation (modified from Fig. 7.1a in (Heinz and Stiasny 2017)).** (a) Pre-fusion E-dimer binds to putative receptor on the cell surface, and the virus is internalized to the endosome. (b) At acidic pH condition, E-dimer dissociates and rearranges to form fusion intermediate E trimer with the fusion loops inserting into the endosomal membrane. This movement will pull the membranes of virus and endosome in proximity. (c) E proteins form the post-fusion trimer after the membrane fusion. (d) Trimeric prM/E heterodimer in immature virion. (e) During the virus maturation, pr is cleaved and dissociates from E protein to generate the M/E-dimer in mature virion

by furin-like protease intracellularly to yield pr peptide and M protein. In the acidic surrounding of *trans*-Golgi network, the pr still sticks to DII of E protein to protect the hydrophobic FL. When virion is released into the extracellular space in neutral pH condition, pr is dissociated from the E protein (Fig. 7.4e). In this circumstance, 60 prM/E trimeric protrusions are rearranged to form 90 E homodimers, generating the smooth surface on mature virion in pre-fusion state. Nevertheless, this process is always inefficient, and the virus usually contains uncut prM, thereof prM/E trimers. The egress of these uncut prM to extracellular space will lead to the reversal of prM/E to spiky structure of immature trimer (Plevka et al. 2011). As a result, the immature or partially immature virions ("mosaic" particles) are generated (Plevka et al. 2011). In this regard, some crypic epitopes in mature virion (e.g., FL epitopes) are exposed in the immature particles or "mosaic" particles for antibody recognition. Thereby, antibodies can achieve protective efficacy through either virus neutralization or Fcγ receptor and complement-dependent effector (Vogt et al. 2011).

Fig. 7.5 Structural model of the ZIKV E protein at its post-fusion state. Trimer structure of DENV-2 E (PDB:1OK8) is shown as the model for ZIKV E. Two protomers are shown as surface, and one is shown as cartoon. The domains are colored as in Fig. 7.3. The protruding FLs are marked in green circle

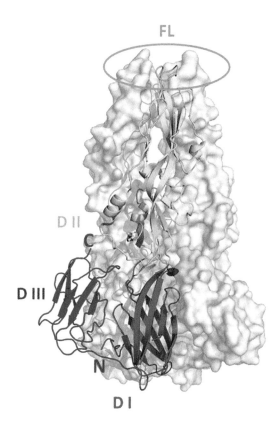

Complexity of the Antigenic Surface

Like other flaviviruses, the dynamic of the antigenic surface of ZIKV complicates the recognition of (monoclonal antibody) MAbs to virus. DENV-2 is genetically close to ZIKV. Increasing the temperature to 37 °C and beyond, the smooth surface of DENV-2 becomes spiky, and the E proteins are rearranged to expose some hidden epitopes for MAb engagement (Fibriansah et al. 2013; Zhang et al. 2013). In contrast, ZIKV particles are demonstrated highly thermal stable. They keep the smooth envelope even in temperature up to 40 °C and thereby occlude the cryptic binding sites for MAbs (Kostyuchenko et al. 2016). Besides, flavivirus can change their conformation of E protein via "breathing" (Dowd et al. 2014; Kuhn et al. 2015). Virus "breathing" describes the conformational fluctuation of E proteins on the virus particle, which alters the antigenic surface of the virion for MAbs neutralization (Dowd et al. 2011). This indicates the dynamic nature of E proteins on the virus surface. When virus is "breathing," some hidden epitopes can be intermittently exposed, leaving the chance for antibody recognition.

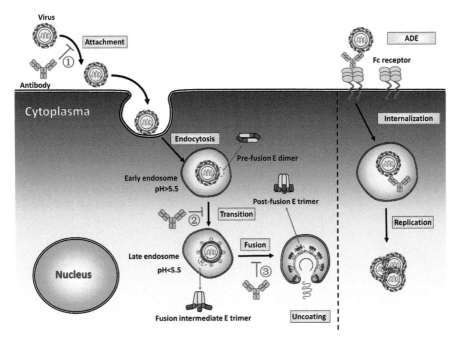

Fig. 7.6 Schematic demonstration of the proposed mechanism for MAbs neutralizing flavivirus at multistages and ADE. The left panel shows the life cycle of flavivirus. MAbs can neutralize virus entry and fusion by (1) causing steric hindrance for virus attachment, (2) locking the E protein in pre-fusion state to prevent conformational transition, and (3) causing steric hindrance for virus fusion. The right panel shows the ADE of flavivirus infection. Fc receptor-bearing cells engage the antibody-bound virus. The virus is then internalized into the target cell and proliferates

Overview of Neutralizing MAbs Against ZIKV

In flavivirus infections, humoral immunity represents an important component of the host response (Knipe et al. 2013; Pierson et al. 2008). Administration of polyclonal or purified MAbs against either E or NS1 protein helps to clear the viruses (Dai et al. 2016b; Dowd and Pierson 2011; Chung et al. 2006b; Schlesinger et al. 1993). Although ZIKV NS1 MAbs are also characterized and subjected for diagnosis (Stettler et al. 2016), their efficacy in ZIKV clearance remains to be determined. In contrast, the E-specific MAbs display the inhibition to ZIKV infection in vitro and protect mice against lethal ZIKV challenge (Dai et al. 2016a; Zhao et al. 2016; Barba-Spaeth et al. 2016; Stettler et al. 2016; Wang et al. 2016; Sapparapu et al. 2016; Robbiani et al. 2017). In addition, MAbs targeting E protein markedly reduce tissue pathology, decrease vertical transmission, and prevent ZIKV-induced microcephaly in a mice model (Sapparapu et al. 2016), emphasizing the therapeutic potential of E MAbs in preventing ZIKV-related damages. Accordingly, in this chapter, we further review the MAbs targeting E protein on their epitopes and probably neutralizing mechanisms.

Table 7.1 Summary for the ZIKV MAbs with clear epitopes

Name	Origin	Epitope[a]	Neutralization in vitro (μg/mL)[b]	Efficacy in vivo	References
2A10G6	Mice	FL	249	Yes	(Dai et al. 2016a; Deng et al. 2011)
ZV-67	Mice	DIII	0.143~0.511	Yes	(Zhao et al. 2016)
Z006	Human	DIII	0.002	ND	(Robbiani et al. 2017)
Z3L1	Human	DI, DII	0.17~0.24	Yes	(Wang et al. 2016)
Z20	Human	DII, DII	0.37~0.89	Yes	(Wang et al. 2016)
C8[c]	Human	DII, DIII	0.0023~0.0143	ND	(Barba-Spaeth et al. 2016)
A11[c]	Human	DII, DI, DIII	0.0759~0.1356	ND	(Barba-Spaeth et al. 2016)
ZIKV-117	Human	DII, DII	0.0054~0.0249	Yes	(Hasan et al. 2017; Sapparapu et al. 2016)
Z23	Human	DIII, DI	0.37~0.56	Yes	(Wang et al. 2016)
C10	Human	DII, DI, DII, DIII, DII	0.0020~0.0095	ND	(Barba-Spaeth et al. 2016)

[a]Some MAbs recognize the tertiary/quaternary epitopes. Here the epitopes in black and red represent the domains from either protomer in one dimer, while the orange one indicates the epitope located in the neighboring dimer

[b]The numbers listed here represent the concentration of indicated MAb used to inhibit 50% of ZIKV

[c]Dengue cross-reactive MAbs

ND not determined

Since the latest outbreak of ZIKV infection, hundreds of ZIKV E MAbs have been developed from either humans or mice within a short period of time (Robbiani et al. 2017; Zhao et al. 2016; Dai et al. 2016a; Barba-Spaeth et al. 2016; Sapparapu et al. 2016; Wang et al. 2016; Stettler et al. 2016). To date, there are 10 MAbs with clear epitopes uncovered through structural studies, which are summarized in Table 7.1. Herein, four were previously isolated from either mice immunized with DENV (2A10G6) or DENV patients (C8, C10, and A11) exhibiting cross-protection or cross-neutralizing activities against ZIKV infection (Dai et al. 2016a; Barba-Spaeth et al. 2016). One (ZV-67) is generated in mice infected with ZIKV (Zhao et al. 2016), and the five left were isolated from ZIKV patients, either by human hybridoma (ZIKV-117) (Sapparapu et al. 2016) or sequencing the antigen-specific memory B cells (Z006, Z3L1, Z23 and Z20) (Robbiani et al. 2017; Wang et al. 2016). Except for 2A10G6, the other 9 MAbs conveyed moderate-to-high neutralizing activities, with the 50% inhibition of virus infection at a concentration lower than 1 μg/mL (Table 7.1). Furthermore, 6 MAbs have been tested for the protection efficacies in mice model and conferred mice protection against lethal ZIKV challenge (Table 7.1) (Zhao et al. 2016; Dai et al. 2016a; Sapparapu et al. 2016; Wang et al. 2016).

As summarized in Table 7.1, the epitopes of MAbs that neutralize ZIKV infection vary a lot. Most MAbs function through binding to either the single protomer or the two protomers in one dimer. Additionally, two MAbs (Z23 and C10) even recognize the epitopes across the neighboring dimers (Wang et al. 2016; Zhang et al. 2016). In this part, the E MAbs are classified based on the protomer they recognized, and their probable neutralizing mechanisms will be discussed.

Single Protomer-Binding MAbs

As introduced above, E protein contains three domains, namely, DI, DII, and DIII. Previous structural studies on flaviviruses suggested that all three domains contain independent neutralizing epitopes (Dai et al. 2016b). Taking MAb 5H2 as an example of DI-targeting MAb, this MAb covers the DI of serotype 4 of DENV (DENV-4) and is supposed to interfere with the fusogenic trimer formation during membrane fusion, thereby neutralizing virus infection (Cockburn et al. 2012). Due to the similar structure as well as similar conformational rearrangement of E protein during membrane fusion among flaviviruses, MAbs that cover the similar regions as MAb 5H2 should also neutralize ZIKV infection, although no DI-targeting ZIKV-neutralizing MAbs have been characterized to date. Further efforts are needed to isolate the DI MAbs.

In flaviviruses, FL resides in DII and represents as the dominant epitope during flavivirus infection (Knipe and Howley 2013). Previous studies on patients or infected mice indicated that the FL MAbs account for about 50% of E MAbs (Oliphant et al. 2006; Dejnirattisai et al. 2010). Similarly, about half of the E-targeting MAbs elicited by ZIKV infection recognize FL (Sapparapu et al. 2016; Wang et al. 2016; Stettler et al. 2016). Because of the high conservation of FL among flaviviruses, FL MAbs usually cross-react to a broad-spectrum of flaviviruses. Murine MAb 2A10G6 is a broadly neutralizing antibody against flaviviruses and showed cross-protection against four serotypes of DENV, WNV, YFV, and ZIKV (Dai et al. 2016a; Deng et al. 2011). Through structural studies, MAb 2A10G6 is displayed to engage ZIKV E protein on the top of the distal end of the DII, covering the key FL residues, such as W101 and F108 (Dai et al. 2016a). In addition, it utilizes the bc loop of E protein to further stabilize the antibody binding (Fig. 7.7). During membrane fusion, the virus inserts its FL into the endosomal membrane to facilitate membrane fusion. FL-targeting MAbs are likely to cause steric hindrance and inhibit E insertion into the endosomal membrane and then inhibit virus infection. However, due to the unique structural features of ZIKV, this virus displays higher thermal stability (Kostyuchenko et al. 2016), leading to less "breathing" of the E protein and a less solvent-accessible FL. In line with that, FL MAbs, such as MAb 2A10G6, hardly bind to the virions to inhibit virus infection, which explains the much lower neutralizing activities of FL MAbs against ZIKV infection than against other flaviviruses. Nevertheless, MAb 2A10G6 confers partial protection against lethal ZIKV challenge in mice model (Dai et al. 2016a).

DIII is speculated to contain binding sites for host receptors for flavivirus attachment and entry (Lee and Lobigs 2000; Mandl et al. 2000; Bhardwaj et al. 2001; Chu et al. 2005; Watterson et al. 2012). Moreover, during membrane fusion, DIII

2A10G6

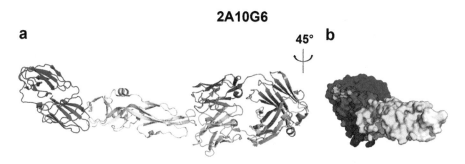

Fig. 7.7 **Epitope for FL MAb.** The three extracellular domains (DI, DII, and DIII) are marked in red, yellow, and blue, respectively, while the fusion loop (FL) is colored in green. The heavy chain and light chain of MAb are indicated in cyan and orange, respectively. (a) Overall structure of 2A10G6 complexed with E protein. (b) Footprint for the epitope of 2A10G6 on E protein and the interacted residues are marked in white (cutoff at 4.5 Å). The PDB identifier for the epitope analysis in 2A10G6 is 5JHL

undergoes large conformational transition as mentioned above (Mukhopadhyay et al. 2005; Shi and Gao 2017). Thus, DIII MAbs may prevent virus infection through multiple ways. Many of the most potent neutralizing MAbs identified to date recognize DIII, and therefore DIII is usually clipped out as vaccine formulation against Flavivirus infection (Heinz and Stiasny 2012; Chu et al. 2007; Ramanathan et al. 2009; Chavez et al. 2010). ZIKV DIII has been used as the "bait" to fish the specific neutralizing MAbs in patients (MAb Z006) (Robbiani et al. 2017) and to stimulate DIII-specific response for MAb isolation (MAb ZV-67) (Zhao et al. 2016). Although these two DIII-targeting MAbs were identified through different ways in different species, complex structures indicate that both MAbs bind to the lateral region (LR) on DIII (Fig. 7.8). LR is also the neutralizing target for other flaviviruses, such as DENV and WNV, suggesting that this region is an epitope hotspot for high neutralizing MAbs. In detail, both MAbs recognize the DIII loops connecting I_0A, BC, and DE. However, MAb Z006 also engages CD loop and FG loop, while MAb ZV-67 interacts with strand A and G (Fig. 7.8). Due to the differences in epitopes, Z006 cross-binds to serotype 1 of DENV (DENV-1), whereas the other is ZIKV-specific.

Aside from the epitopes in discrete domains, MAbs might associate with region across domains in one protomer, which is exemplified by MAb Z3L1 (Fig. 7.9) (Wang et al. 2016). This MAb is hooked from ZIKV patient by E protein and engages the DI, DII, and the hinge region linking the two domains. The crystal structure of Z3L1-E indicates that this MAb interacts with E protein at pre-fusion state, including the β-strands of D_0, E_0, and F_0, the 150-loop in DI, the loops connecting strands of fg and hI in DII, and the αBI_0 in DI–DII hinge region (Fig. 7.9). Based on the structural studies on E protein from DENV-2, the DII undergoes rotation with respect to DI by more than 20 degrees to transit from pre- to post-fusion state (Mukhopadhyay et al. 2005; Shi and Gao 2017). Accordingly, MAb Z3L1 is likely to provide hindrance for this transition.

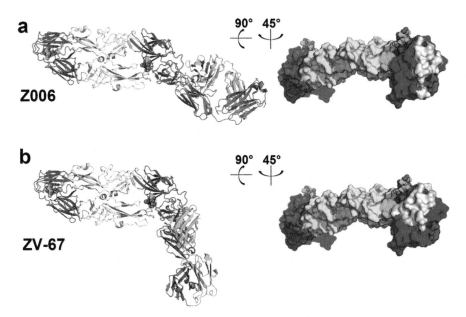

Fig. 7.8 Epitopes in DIII. Domains in E protein are colored as in Fig. 7.7. The cyan and orange represent the heavy chain and light chain, respectively. The left panel displays the overall structures of Z006 or ZV-67 complexed with E protein, respectively, while the right panel exhibits the epitopes of the indicated MAb. (**a**) Z006; (**b**) ZV-67 (cutoff at 4.5 Å). The PDB identifier for the epitope analysis in Z006 is 5VIG and for ZV67 is 5KVG, respectively

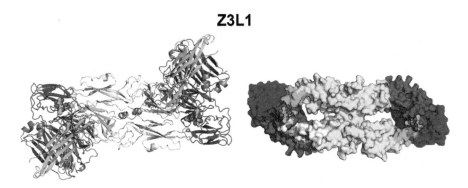

Fig. 7.9 Epitope of MAb Z3L1. Domains in E protein are colored as in Fig. 7.7. The cyan and orange represent the heavy chain and light chain, respectively. The *left panel* displays the overall structure of Z3L1 complexed with E protein, while the *right panel* exhibits the epitopes of MAb Z3L1 (cutoff at 4.5 Å). The PDB identifier for the epitope analysis in Z3L1 is 5GZN

E-Dimer-Dependent MAbs

Recently, a new class of neutralizing MAbs with high potency have been character-ized against DENV infection (Rouvinski et al. 2015). They bind to the epitopes across two protomers in one dimer (EDE) and subsequently lock the E protein at pre-fusion structure, preventing the essential conformational rearrangement needed for virus membrane fusion. Interestingly, two groups of EDE (EDE1 and EDE2) MAbs cross-neutralize ZIKV/DENV infections. MAbs C8 and A11, representative of EDE1 and EDE2, respectively, recognize a serotype-invariant site at the E-dimer interface in both ZIKV and DENV1–4 (Rouvinski et al. 2015; Barba-Spaeth et al. 2016). The epitopes include the exposed main chain of FL and the two conserved glycan chains (N67 and N153 glycans), which are supposed to be the binding site for viral glycoprotein prM during virus maturation (Barba-Spaeth et al. 2016). MAb A11 mainly engages the DII in one protomer, including the loops between ij and bc as well as FL, and also interacts with the other protomer through K316 in DIII and the V153 together with the carbohydrates at N154 in DI (Fig. 7.10a). Compared with MAb A11, more residues in E protein contribute to the association with MAb C8 (Fig. 7.10b). Aside from the three loops in the DII that MAb A11 binds to, MAb C8 also engages strands of b and d in this domain. With respect to the other pro-tomers, the strands of B,E, the loop linking AA′ in DIII, and the residue of R2 in DI contribute to the MAb recognition. However, the carbohydrates in 150-loop are not involved in the interactions with MAb C8, similar as observed in that of MAb C8 with DENV-2 E protein (Fig. 7.10b). The conserved epitopes of MAbs A11 and C8 in both ZIKV and DENV indicate the Achilles' heel for vaccine development and design of therapeutic MAbs.

In addition to the MAbs identified in DENV patients, EDE MAbs are also char-acterized in ZIKV patients. Two independent groups isolated MAbs Z20 and ZIKV-117 from two ZIKV patients, respectively (Wang et al. 2016; Sapparapu et al. 2016). Interestingly, both MAbs bind to the DII across E-dimer, as displayed in Fig. 7.10c, d. MAb Z20 mainly binds to the ab loop, bc loop, hi loop, and ij loop and strands of a, b, d, i, and j placed on one protomer. At the same time, this MAb also engages loops of gf and αBI_0 located in the other DII. Cryo-EM structure of Fab-ZIKV indi-cates that, like MAb Z20, MAb ZIKV-117 also binds to both DII in one E-dimer. However, a few varied residues in the epitopes between these two MAbs lead to their different specificities. ZIKV-117 only recognizes ZIKV E protein, while MAb Z20 exerts weak cross-neutralization against DENV.

Neighboring E Cross-Linking MAbs

Another class of MAbs bind to regions across neighboring E-dimers. As demon-strated by cryo-EM structures of MAb-virion, both C10 and Z23 bind to the two adjacent E-dimers (Wang et al. 2016; Zhang et al. 2016). Z23 mainly engages the LR

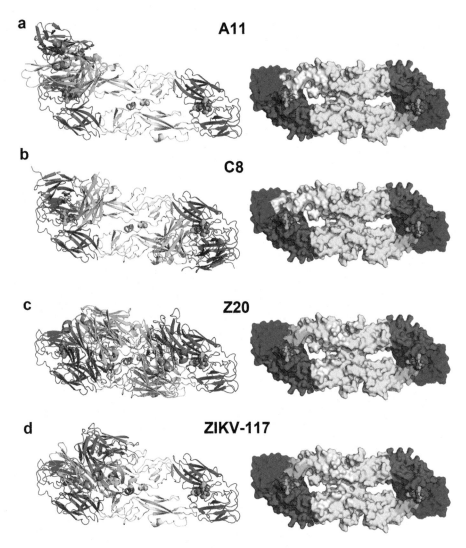

Fig. 7.10 Epitopes for E-dimer-dependent MAbs. Domains in E protein are colored as in Fig. 7.7. Cyan and orange represent the heavy chain and light chain, respectively. The left panel displays the overall structures of A11, C8, Z20, or ZIKV-117 complexed with E protein, respectively, while the right panel exhibits the epitopes of the indicated MAb. (**a**) A11, (**b**) C8, (**c**) Z20, and (**d**) ZIKV-117 (cutoff at 4.5 Å). The PDB identifier for the epitope analysis in A11 is 5LCV, for C8 is 5LBS, for Z20 is 5GZO, and for ZIKV-117 is 5UHY

in DIII in one protomer, similar to MAb Z006 and ZV-67. In addition, this MAb also inserts into the cleft between two adjacent E-dimers and then contacts with DI in the other E-dimer. Through this kind of binding, MAb Z23 is likely to prevent ZIKV infection at multiple stages. First, Z23 is supposed to block viral attachment to host receptors, since DIII is likely to contain binding sites for host receptor. Second, the

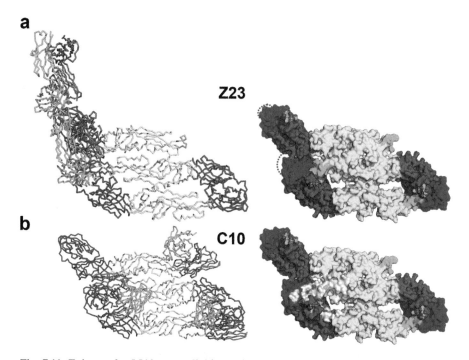

Fig. 7.11 **Epitopes for MAbs cross-linking adjacent E-dimers. Domains in E protein are colored as in Fig. 7.7.** The cyan and orange represent the heavy chain and light chain, respectively. The left panel displays the overall structures of Z23 or C10 complexed with ZIKV virion, respectively, while the right panels exhibit the epitopes of the indicated MAb (cutoff at 4.5 Å). (**a**) Z23. Due to the low resolution of ZIKV/Z23 complex (9.6Å) as determined by cryo-electron microscopy (cryo-EM), the detailed epitope could not be distinguished and is highlighted with an ellipse as indicated by arrow. (**b**) C10. The epitope of the MAb is marked in white on the surface structure as indicated by arrows. Carbohydrates are indicated as spheres. The PDB identifier for the epitope analysis in Z23 is 5GZR and for C10 is 5H37, respectively

engagement of the adjacent two E-dimers will stabilize the dimer conformations and inhibit trimer formation. Third, the insertion of MAb Z23 into the cleft may cause steric hindrance for DIII conformational change, which is necessary for fusogenic trimers formation (Fig. 7.11a). Cryo-EM structures of C10-ZIKV at different pH provide more direct evidence to support that this MAb prevents infection through locking the E protein at dimer conformations and inhibiting membrane fusion (Zhang et al. 2016). MAb C10 covers the region of FL on DII of one protomer of the virion and makes contact with DI, DII, and DIII on the other protomer. Simultaneously, MAb C10 also binds to N52 in the neighboring E-dimer (Fig. 7.11b). It is displayed that the E protein layer of the ZIKV-C10 complex remains at a similar radius at pH 6.5 as the uncomplexed ZIKV at pH 8.0, unlike uncomplexed ZIKV at pH 6.5 (Zhang et al. 2016). This indicates that through binding to the inter-dimer interface, MAb C10 locks E in the pre-fusion dimer conformations, which inhibits domain reorganization.

Antibody-Dependent Enhancement of Infection (ADE)

ADE describes the increase of virus infection when virus-reactive antibodies are insufficient to neutralize virus or they are in sub-neutralizing concentrations (Beltramello et al. 2010). In this regard, antibodies can bind efficiently to virus but do not reach the neutralization threshold. Thereby, these antibodies help virus uptake by cells expressing Fc or complement receptors, subsequently promoting the virus infectivity (Fig. 7.6, right panel). Due to the cross-reactivity, plasma immunity to DENV and WNV is found to drive ADE of infection with ZIKV. In this circumstance, epitope accessibility plays an important role in shaping the neutralizing activity of antibody and ADE. FL-targeting MAbs comprise a significant fraction of flavivirus-elicited humoral response. FL is buried by the DI and DIII from the neighboring E protomer, occluding the binding sites for antibody recognition. FL MAbs can barely recognize the mature virion via virus "breathing"; FL MAbs have to be in high concentration to reach the neutralizing threshold. These MAbs are usually at high risk of ADE of flavivirus infection. In a recent vaccine development against ZIKV, when key residues in the epitopes for FL MAb binding are mutated, vaccine can induce antisera with dramatically decreased ADE (Richner et al. 2017). Besides, LALA mutation in Fc region of MAb, which abrogates the binding to the Fc receptor, can also eliminate the ADE of flavivirus infection (Fibriansah et al. 2015).

Perspective

ZIKV is a re-emerged pathogen which has been underestimated for more than half a century. ZIKV infections lead to catastrophic consequences for the newborn. So far, no prophylactics and therapeutics are available. Therefore, development of vaccine and protective antibody is urgently demanded. In a recent report, MAb ZIKV-117 was demonstrated to be able to abrogate the maternal-fetal transmission of ZIKV, suggesting a promising therapeutics for clinical applications. However, due to the high mutation rate of RNA virus, MAbs "cocktail" targeting different epitopes will be required to prevent virus evasion. Delineating ZIKV E protein and the MAbs bound to epitopes by structural studies dissect the molecular basis for antibody neutralization. The ten structure-known MAb-E proteins or virions pinpoint the vulnerable sites of ZIKV. These structural information uncovers the hotspots for potent neutralizing MAbs and will undoubtedly facilitate the structure-based MAb design.

Aside from that, FL-targeting antibodies dominate the flavivirus-induced humoral response, including ZIKV, DENV, and WNV. FL MAbs are usually cross-reactive and display weakly neutralizing activities. In fact, they are proposed to contribute mainly to the ADE of flavivirus infection. Structural analysis of 2A10G6 bound to ZIKV E reveals the classical binding mode of FL MAb. The interactions uncovered in this structure are valuable for the vaccine design in that it indicates key residues in FL for the ADE which should be removed or mutated (Richner et al. 2017). Besides, based on the indication from complex structure of EDE and FL Mabs bound to ZIKV E, new vaccines can be designed by E protein engineering to lock or mutate the epitope for FL MAb but maintain that for EDE MAb.

References

Avirutnan P, Fuchs A, Hauhart RE, Somnuke P, Youn S, Diamond MS, Atkinson JP (2010) Antagonism of the complement component C4 by flavivirus nonstructural protein NS1. J Exp Med 207(4):793–806. https://doi.org/10.1084/jem.20092545

Barba-Spaeth G, Dejnirattisai W, Rouvinski A, Vaney MC, Medits I, Sharma A, Simon-Loriere E, Sakuntabhai A, Cao-Lormeau VM, Haouz A, England P, Stiasny K, Mongkolsapaya J, Heinz FX, Screaton GR, Rey FA (2016) Structural basis of potent Zika-dengue virus antibody cross-neutralization. Nature 536(7614):48–53. https://doi.org/10.1038/nature18938

Baud D, Gubler DJ, Schaub B, Lanteri MC, Musso D (2017) An update on Zika virus infection. Lancet 390:2099. https://doi.org/10.1016/S0140-6736(17)31450-2

Beltramello M, Williams KL, Simmons CP, Macagno A, Simonelli L, Quyen NT, Sukupolvi-Petty S, Navarro-Sanchez E, Young PR, de Silva AM, Rey FA, Varani L, Whitehead SS, Diamond MS, Harris E, Lanzavecchia A, Sallusto F (2010) The human immune response to dengue virus is dominated by highly cross-reactive antibodies endowed with neutralizing and enhancing activity. Cell Host Microbe 8(3):271–283. https://doi.org/10.1016/j.chom.2010.08.007

Bhardwaj S, Holbrook M, Shope RE, Barrett AD, Watowich SJ (2001) Biophysical characterization and vector-specific antagonist activity of domain III of the tick-borne flavivirus envelope protein. J Virol 75(8):4002–4007. https://doi.org/10.1128/JVI.75.8.4002-4007.2001

Bogoch II, Brady OJ, Kraemer MUG, German M, Creatore MI, Kulkarni MA, Brownstein JS, Mekaru SR, Hay SI, Groot E, Watts A, Khan K (2016) Anticipating the international spread of Zika virus from Brazil. Lancet 387(10016):335–336. https://doi.org/10.1016/S0140-6736(16)00080-5

Bressanelli S, Stiasny K, Allison SL, Stura EA, Duquerroy S, Lescar J, Heinz FX, Rey FA (2004) Structure of a flavivirus envelope glycoprotein in its low-pH-induced membrane fusion conformation. EMBO J 23(4):728–738. https://doi.org/10.1038/sj.emboj.7600064

Brito CA, Cordeiro MT (2016) One year after the Zika virus outbreak in Brazil: from hypotheses to evidence. Rev Soc Bras Med Trop 49(5):537–543. https://doi.org/10.1590/0037-8682-0328-2016

Cao-Lormeau VM, Roche C, Teissier A, Robin E, Berry AL, Mallet HP, Sall AA, Musso D (2014) Zika virus, French polynesia, south pacific, 2013. Emerg Infect Dis 20(6):1085–1086. https://doi.org/10.3201/eid2006.140138

Chan JF, Chik KK, Yuan S, Yip CC, Zhu Z, Tee KM, Tsang JO, Chan CC, Poon VK, Lu G, Zhang AJ, Lai KK, Chan KH, Kao RY, Yuen KY (2017) Novel antiviral activity and mechanism of bromocriptine as a Zika virus NS2B-NS3 protease inhibitor. Antivir Res 141:29. https://doi.org/10.1016/j.antiviral.2017.02.002

Chavez JH, Silva JR, Amarilla AA, Moraes Figueiredo LT (2010) Domain III peptides from flavivirus envelope protein are useful antigens for serologic diagnosis and targets for immunization. Biologicals : journal of the International Association of Biological Standardization 38(6):613–618. https://doi.org/10.1016/j.biologicals.2010.07.004

Chu JJ, Rajamanonmani R, Li J, Bhuvanakantham R, Lescar J, Ng ML (2005) Inhibition of West Nile virus entry by using a recombinant domain III from the envelope glycoprotein. J Gen Virol 86(Pt 2):405–412. https://doi.org/10.1099/vir.0.80411-0

Chu J-HJ, Chiang C-CS, Ng M-L (2007) Immunization of Flavivirus West Nile recombinant envelope domain III protein induced specific immune response and protection against West Nile
virus infection. J Immunol 178(5):2699–2705. https://doi.org/10.4049/jimmunol.178.5.2699

Chung KM, Liszewski MK, Nybakken G, Davis AE, Townsend RR, Fremont DH, Atkinson JP,
Diamond MS (2006a) West Nile virus nonstructural protein NS1 inhibits complement activation by binding the regulatory protein factor H. Proc Natl Acad Sci U S A 103(50):19111–
19116. https://doi.org/10.1073/pnas.0605668103

Chung KM, Nybakken GE, Thompson BS, Engle MJ, Marri A, Fremont DH, Diamond MS (2006b)
Antibodies against West Nile virus nonstructural protein NS1 prevent lethal infection through
fc gamma receptor-dependent and -independent mechanisms. J Virol 80(3):1340–1351. https://
doi.org/10.1128/Jvi.80.3.1340-1351.2006

Cockburn JJB, Sanchez MEN, Goncalvez AP, Zaitseva E, Stura EA, Kikuti CM, Duquerroy S,
Dussart P, Chernomordik LV, Lai CJ, Rey FA (2012) Structural insights into the neutralization
mechanism of a higher primate antibody against dengue virus. EMBO J 31(3):767–779. https://
doi.org/10.1038/emboj.2011.439

Dai L, Song J, Lu X, Deng YQ, Musyoki AM, Cheng H, Zhang Y, Yuan Y, Song H, Haywood
J, Xiao H, Yan J, Shi Y, Qin CF, Qi J, Gao GF (2016a) Structures of the Zika virus envelope protein and its complex with a flavivirus broadly protective antibody. Cell Host Microbe
19(5):696–704. https://doi.org/10.1016/j.chom.2016.04.013

Dai L, Wang Q, Qi J, Shi Y, Yan J, Gao GF (2016b) Molecular basis of antibody-mediated neutralization and protection against flavivirus. IUBMB Life 68(10):783–791. https://doi.org/10.1002/
iub.1556

Deckard DT, Chung WM, Brooks JT, Smith JC, Woldai S, Hennessey M, Kwit N, Mead P (2016)
Male-to-male sexual transmission of Zika virus--Texas, January 2016. MMWR Morb Mortal
Wkly Rep 65(14):372–374. https://doi.org/10.15585/mmwr.mm6514a3

Dejnirattisai W, Jumnainsong A, Onsirisakul N, Fitton P, Vasanawathana S, Limpitikul
W, Puttikhunt C, Edwards C, Duangchinda T, Supasa S, Chawansuntati K, Malasit P,
Mongkolsapaya J, Screaton G (2010) Cross-reacting antibodies enhance dengue virus infec-
tion in humans. Science 328(5979):745–748. https://doi.org/10.1126/science.1185181

Deng Y-Q, Dai J-X, Ji G-H, Jiang T, Wang H-J, H-o Y, Tan W-L, Liu R, Yu M, Ge B-X, Zhu Q-Y,
Qin ED, Guo Y-J, Qin C-F (2011) A broadly flavivirus cross-neutralizing monoclonal antibody
that recognizes a novel epitope within the fusion loop of E protein. PloS One 6(1):e16059.
https://doi.org/10.1371/journal.pone.0016059

Dick GW, Kitchen SF, Haddow AJ (1952) Zika virus. I. Isolations and serological specificity.
Trans R Soc Trop Med Hyg 46(5):509–520

D'Ortenzio E, Matheron S, Yazdanpanah Y, de Lamballerie X, Hubert B, Piorkowski G, Maquart
M, Descamps D, Damond F, Leparc-Goffart I (2016) Evidence of sexual transmission of Zika
virus. N Engl J Med 374(22):2195–2198. https://doi.org/10.1056/NEJMc1604449

Dos Santos T, Rodriguez A, Almiron M, Sanhueza A, Ramon P, de Oliveira WK, Coelho GE,
Badaro R, Cortez J, Ospina M, Pimentel R, Masis R, Hernandez F, Lara B, Montoya R,
Jubithana B, Melchor A, Alvarez A, Aldighieri S, Dye C, Espinal MA (2016) Zika virus and
the Guillain-Barre syndrome - case series from seven countries. N Engl J Med 375(16):1598–
1601. https://doi.org/10.1056/NEJMc1609015

Dowd KA, Pierson TC (2011) Antibody-mediated neutralization of flaviviruses: a reductionist
view. Virology 411(2):306–315. https://doi.org/10.1016/j.virol.2010.12.020

Dowd KA, Jost CA, Durbin AP, Whitehead SS, Pierson TC (2011) A dynamic landscape for antibody binding modulates antibody-mediated neutralization of West Nile virus. PLoS Pathog
7(6):e1002111. https://doi.org/10.1371/journal.ppat.1002111

Dowd KA, Mukherjee S, Kuhn RJ, Pierson TC (2014) Combined effects of the structural heterogeneity and dynamics of flaviviruses on antibody recognition. J Virol 88(20):11726–11737.
https://doi.org/10.1128/JVI.01140-14

Duan W, Song H, Wang H, Chai Y, Su C, Qi J, Shi Y, Gao GF (2017) The crystal structure of Zika
virus NS5 reveals conserved drug targets. EMBO J 36(7):919–933. https://doi.org/10.15252/
embj.201696241

Duffy MR, Chen TH, Hancock WT, Powers AM, Kool JL, Lanciotti RS, Pretrick M, Marfel M, Holzbauer S, Dubray C, Guillaumot L, Griggs A, Bel M, Lambert AJ, Laven J, Kosoy O, Panella A, Biggerstaff BJ, Fischer M, Hayes EB (2009) Zika virus outbreak on Yap Island, Federated States of Micronesia. N Engl J Med 360(24):2536–2543. https://doi.org/10.1056/NEJMoa0805715

Fibriansah G, Ng TS, Kostyuchenko VA, Lee J, Lee S, Wang J, Lok SM (2013) Structural changes in dengue virus when exposed to a temperature of 37 degrees C. J Virol 87(13):7585–7592. https://doi.org/10.1128/JVI.00757-13

Fibriansah G, Ibarra KD, Ng TS, Smith SA, Tan JL, Lim XN, Ooi JS, Kostyuchenko VA, Wang J, de Silva AM, Harris E, Crowe JE, Jr., Lok SM (2015) DENGUE VIRUS. Cryo-EM structure of an antibody that neutralizes dengue virus type 2 by locking E protein dimers. Science 349 (6243):88–91. doi:https://doi.org/10.1126/science.aaa8651

Garcia Serpa Osorio-de-Castro C, Silva Miranda E, Machado de Freitas C, Rochel de Camargo K Jr, Cranmer HH (2017) The Zika virus outbreak in Brazil: knowledge gaps and challenges for risk reduction. Am J Public Health 107(6):960–965. https://doi.org/10.2105/AJPH.2017.303705

Govero J, Esakky P, Scheaffer SM, Fernandez E, Drury A, Platt DJ, Gorman MJ, Richner JM, Caine EA, Salazar V, Moley KH, Diamond MS (2016) Zika virus infection damages the testes in mice. Nature 540(7633):438–442. https://doi.org/10.1038/nature20556

Gulland A (2016) Zika virus is a global public health emergency, declares WHO. BMJ 352:i657

Hasan SS, Miller A, Sapparapu G, Fernandez E, Klose T, Long F, Fokine A, Porta JC, Jiang W, Diamond MS, Crowe JE Jr, Kuhn RJ, Rossmann MG (2017) A human antibody against Zika virus crosslinks the E protein to prevent infection. Nat Commun 8:14722. https://doi.org/10.1038/ncomms14722

Heinz FX, Stiasny K (2012) Flaviviruses and flavivirus vaccines. Vaccine 30(29):4301–4306. https://doi.org/10.1016/j.vaccine.2011.09.114

Heinz FX, Stiasny K (2017) The antigenic structure of Zika virus and its relation to other flaviviruses: implications for Infection and immunoprophylaxis. Microbiol Molecular Biol Rev 81(1). https://doi.org/10.1128/MMBR.00055-16

Hills SL, Russell K, Hennessey M, Williams C, Oster AM, Fischer M, Mead P (2016) Transmission of Zika virus through sexual contact with travelers to areas of ongoing transmission - continental United States, 2016. MMWR Morb Mortal Wkly Rep 65(8):215–216. https://doi.org/10.15585/mmwr.mm6508e2

Kang C, Keller TH, Luo D (2017) Zika Virus Protease: An Antiviral Drug Target. Trends Microbiol 25:797. https://doi.org/10.1016/j.tim.2017.07.001

Kleber de Oliveira W, Cortez-Escalante J, De Oliveira WT, do Carmo GM, Henriques CM, Coelho GE, Araujo de Franca GV (2016) Increase in reported prevalence of microcephaly in infants born to women living in areas with confirmed Zika virus transmission during the first trimester of pregnancy - Brazil, 2015. MMWR Morb Mortal Wkly Rep 65(9):242–247. https://doi.org/10.15585/mmwr.mm6509e2

Knipe DM, Howley PM (2013) Fields virology, 6th edn. Wolters Kluwer/Lippincott Williams & Wilkins Health, Philadelphia

Knipe DM, Howley PM, Cohen JI, Griffin DE, Lamb RA, Martin MA, Racanielo VR, Be R (2013) Fields virology, 6th edn. Lippincott Williams & Wilkins, Philadelphia

Kostyuchenko VA, Lim EX, Zhang S, Fibriansah G, Ng TS, Ooi JS, Shi J, Lok SM (2016) Structure of the thermally stable Zika virus. Nature 533(7603):425–428. https://doi.org/10.1038/nature17994

Kuhn RJ, Dowd KA, Beth Post C, Pierson TC (2015) Shake, rattle, and roll: impact of the dynamics of flavivirus particles on their interactions with the host. Virology 479-480:508–517. https://doi.org/10.1016/j.virol.2015.03.025

Lee E, Lobigs M (2000) Substitutions at the putative receptor-binding site of an encephalitic flavivirus alter virulence and host cell tropism and reveal a role for glycosaminoglycans in entry. J Virol 74(19):8867–8875. https://doi.org/10.1128/jvi.74.19.8867-8875.2000

Lei J, Hansen G, Nitsche C, Klein CD, Zhang L, Hilgenfeld R (2016) Crystal structure of Zika virus NS2B-NS3 protease in complex with a boronate inhibitor. Science 353(6298):503–505. https://doi.org/10.1126/science.aag2419

Li L, Lok SM, Yu IM, Zhang Y, Kuhn RJ, Chen J, Rossmann MG (2008) The flavivirus precursor membrane-envelope protein complex: structure and maturation. Science 319(5871):1830–1834. https://doi.org/10.1126/science.1153263

Li J, Xiong Y, Wu W, Liu X, Qu J, Zhao X, Zhang S, Li J, Li W, Liao Y, Gong T, Wang L, Shi Y, Xiong Y, Ni D, Li Q, Liang M, Hu G, Li D (2016) Zika virus in a traveler returning to China from Caracas, Venezuela, February 2016. Emerg Infect Dis 22(6):1133–1136. https://doi.org/10.3201/eid2206.160273

Ma W, Li S, Ma S, Jia L, Zhang F, Zhang Y, Zhang J, Wong G, Zhang S, Lu X, Liu M, Yan J, Li W, Qin C, Han D, Qin C, Wang N, Li X, Gao GF (2016) Zika virus causes testis damage and leads to male infertility in mice. Cell 167(6):1511–1524. e1510. https://doi.org/10.1016/j.cell.2016.11.016

Mandl CW, Allison SL, Holzmann H, Meixner T, Heinz FX (2000) Attenuation of tick-borne encephalitis virus by structure-based site-specific mutagenesis of a putative flavivirus receptor binding site. J Virol 74(20):9601–9609

Miner JJ, Diamond MS (2017) Zika virus pathogenesis and tissue tropism. Cell Host Microbe 21(2):134–142. https://doi.org/10.1016/j.chom.2017.01.004

Miner JJ, Cao B, Govero J, Smith AM, Fernandez E, Cabrera OH, Garber C, Noll M, Klein RS, Noguchi KK, Mysorekar IU, Diamond MS (2016) Zika virus infection during pregnancy in mice causes placental damage and fetal demise. Cell 165(5):1081–1091. https://doi.org/10.1016/j.cell.2016.05.008

Modis Y, Ogata S, Clements D, Harrison SC (2004) Structure of the dengue virus envelope protein after membrane fusion. Nature 427(6972):313–319. https://doi.org/10.1038/nature02165

Mukhopadhyay S, Kuhn RJ, Rossmann MG (2005) A structural perspective of the flavivirus life cycle. Nat Rev Microbiol 3(1):13–22. https://doi.org/10.1038/nrmicro1067

Musso D, Nilles EJ, Cao-Lormeau VM (2014) Rapid spread of emerging Zika virus in the Pacific area. Clin Microbiol Infect 20(10):O595–O596. https://doi.org/10.1111/1469-0691.12707

Oliphant T, Nybakken GE, Engle M, Xu Q, Nelson CA, Sukupolvi-Petty S, Marri A, Lachmi BE, Olshevsky U, Fremont DH, Pierson TC, Diamond MS (2006) Antibody recognition and neutralization determinants on domains I and II of West Nile virus envelope protein. J Virol 80(24):12149–12159. https://doi.org/10.1128/JVI.01732-06

Pierson TC, Fremont DH, Kuhn RJ, Diamond MS (2008) Structural insights into the mechanisms of antibody-mediated neutralization of flavivirus infection: implications for vaccine development. Cell Host Microbe 4(3):229–238. https://doi.org/10.1016/j.chom.2008.08.004

Plevka P, Battisti AJ, Junjhon J, Winkler DC, Holdaway HA, Keelapang P, Sittisombut N, Kuhn RJ, Steven AC, Rossmann MG (2011) Maturation of flaviviruses starts from one or more icosahedrally independent nucleation centres. EMBO Rep 12(6):602–606. https://doi.org/10.1038/embor.2011.75

Ramanathan MP, Kutzler MA, Kuo YC, Yan J, Liu H, Shah V, Bawa A, Selling B, Sardesai NY, Kim JJ, Weiner DB (2009) Coimmunization with an optimized IL15 plasmid adjuvant enhances humoral immunity via stimulating B cells induced by genetically engineered DNA vaccines expressing consensus JEV and WNV E DIII. Vaccine 27(32):4370–4380. https://doi.org/10.1016/j.vaccine.2009.01.137

Rasmussen SA, Jamieson DJ, Honein MA, Petersen LR (2016) Zika virus and birth defects--reviewing the evidence for causality. N Engl J Med 374(20):1981–1987. https://doi.org/10.1056/NEJMsr1604338

Richner JM, Himansu S, Dowd KA, Butler SL, Salazar V, Fox JM, Julander JG, Tang WW, Shresta S, Pierson TC, Ciaramella G, Diamond MS (2017) Modified mRNA vaccines protect against Zika virus infection. Cell 168(6):1114–1125. e1110. https://doi.org/10.1016/j.cell.2017.02.017

Robbiani DF, Bozzacco L, Keeffe JR, Khouri R, Olsen PC, Gazumyan A, Schaefer-Babajew D, Avila-Rios S, Nogueira L, Patel R, Azzopardi SA, Uhl LFK, Saeed M, Sevilla-Reyes EE, Agudelo M, Yao KH, Golijanin J, Gristick HB, Lee YE, Hurley A, Caskey M, Pai J, Oliveira T, Wunder EA Jr, Sacramento G, Nery N Jr, Orge C, Costa F, Reis MG, Thomas NM, Eisenreich T, Weinberger DM, de Almeida ARP, West AP Jr, Rice CM, Bjorkman PJ, Reyes-Teran G, Ko AI, MacDonald MR, Nussenzweig MC (2017) Recurrent potent human neutralizing antibodies to Zika virus in Brazil and Mexico. Cell 169(4):597–609. e511. https://doi.org/10.1016/j.cell.2017.04.024

Rouvinski A, Guardado-Calvo P, Barba-Spaeth G, Duquerroy S, Vaney MC, Kikuti CM, Navarro Sanchez ME, Dejnirattisai W, Wongwiwat W, Haouz A, Girard-Blanc C, Petres S, Shepard WE, Despres P, Arenzana-Seisdedos F, Dussart P, Mongkolsapaya J, Screaton GR, Rey FA (2015) Recognition determinants of broadly neutralizing human antibodies against dengue viruses. Nature 520(7545):109–113. https://doi.org/10.1038/nature14130

Sapparapu G, Fernandez E, Kose N, Bin C, Fox JM, Bombardi RG, Zhao H, Nelson CA, Bryan AL, Barnes T, Davidson E, Mysorekar IU, Fremont DH, Doranz BJ, Diamond MS, Crowe JE (2016) Neutralizing human antibodies prevent Zika virus replication and fetal disease in mice. Nature 540(7633):443–447. https://doi.org/10.1038/nature20564

Schlesinger JJ, Foltzer M, Chapman S (1993) The fc portion of antibody to yellow-fever virus-Ns1 is a determinant of protection against Yf encephalitis in mice. Virology 192 (1):132-141. DOI:DOI. https://doi.org/10.1006/viro.1993.1015

Shi Y, Gao GF (2017) Structural biology of the Zika virus. Trends Biochem Sci 42(6):443–456. https://doi.org/10.1016/j.tibs.2017.02.009

Simpson DI (1964) Zika virus infection in man. Trans R Soc Trop Med Hyg 58:335–338

Stettler K, Beltramello M, Espinosa DA, Graham V, Cassotta A, Bianchi S, Vanzetta F, Minola A, Jaconi S, Mele F, Foglierini M, Pedotti M, Simonelli L, Dowall S, Atkinson B, Percivalle E, Simmons CP, Varani L, Blum J, Baldanti F, Cameroni E, Hewson R, Harris E, Lanzavecchia A, Sallusto F, Corti D (2016) Specificity, cross-reactivity and function of antibodies elicited by Zika virus infection. Science 353(6301):823–826. https://doi.org/10.1126/science.aaf8505

Vogt MR, Dowd KA, Engle M, Tesh RB, Johnson S, Pierson TC, Diamond MS (2011) Poorly neutralizing cross-reactive antibodies against the fusion loop of West Nile virus envelope protein protect in vivo via Fcgamma receptor and complement-dependent effector mechanisms. J Virol 85(22):11567–11580. https://doi.org/10.1128/JVI.05859-11

Wang Q, Yang H, Liu X, Dai L, Ma T, Qi J, Wong G, Peng R, Liu S, Li J, Li S, Song J, Liu J, He J, Yuan H, Xiong Y, Liao Y, Li J, Yang J, Tong Z, Griffin BD, Bi Y, Liang M, Xu X, Qin C, Cheng G, Zhang X, Wang P, Qiu X, Kobinger G, Shi Y, Yan J, Gao GF (2016) Molecular determinants of human neutralizing antibodies isolated from a patient infected with Zika virus. Sci Transl Med 8(369):369ra179. https://doi.org/10.1126/scitranslmed.aai8336

Watterson D, Kobe B, Young PR (2012) Residues in domain III of the dengue virus envelope glycoprotein involved in cell-surface glycosaminoglycan binding. J Gen Virol 93(Pt 1):72–82. https://doi.org/10.1099/vir.0.037317-0

Wikan N, Smith DR (2016) Zika virus: history of a newly emerging arbovirus. Lancet Infect Dis 16(7):e119–e126. https://doi.org/10.1016/S1473-3099(16)30010-X

Xu X, Song H, Qi J, Liu Y, Wang H, Su C, Shi Y, Gao GF (2016) Contribution of intertwined loop to membrane association revealed by Zika virus full-length NS1 structure. EMBO J 35(20):2170–2178. https://doi.org/10.15252/embj.201695290

Young PR, Hilditch PA, Bletchly C, Halloran W (2000) An antigen capture enzyme-linked immunosorbent assay reveals high levels of the dengue virus protein NS1 in the sera of infected patients. J Clin Microbiol 38(3):1053–1057

Yun SI, Lee YM (2017) Zika virus: An emerging flavivirus. J Microbiol 55(3):204–219. https://doi.org/10.1007/s12275-017-7063-6

Zhang X, Sheng J, Plevka P, Kuhn RJ, Diamond MS, Rossmann MG (2013) Dengue structure differs at the temperatures of its human and mosquito hosts. Proc Natl Acad Sci U S A 110(17):6795–6799. https://doi.org/10.1073/pnas.1304300110

Zhang S, Kostyuchenko VA, Ng TS, Lim XN, Ooi JS, Lambert S, Tan TY, Widman DG, Shi J, Baric RS, Lok SM (2016) Neutralization mechanism of a highly potent antibody against Zika virus. Nat Commun 7:13679. https://doi.org/10.1038/ncomms13679

Zhao H, Fernandez E, Dowd KA, Speer SD, Platt DJ, Gorman MJ, Govero J, Nelson CA, Pierson TC, Diamond MS, Fremont DH (2016) Structural basis of Zika virus-specific antibody protection. Cell 166(4):1016–1027. https://doi.org/10.1016/j.cell.2016.07.020

Chapter 8
The Retrovirus Capsid Core

Wei Zhang, Luiza Mendonça, and Louis L. Mansky

The Retrovirus Capsid Core Is a Functional Unit in the Virus Life Cycle

The *Retroviridae* represents a family of enveloped single-strand positive-sense RNA viruses that utilize reverse transcription and genome integration to infect cells (Hu and Hughes 2012; Pan et al. 2013). After entering a host cell, viral RNA is reverse transcribed into DNA by the virus-encoded reverse transcriptase (RT). The double-stranded DNA copy of the viral genome is then integrated permanently into the host cell chromosome, after which the cell's transcription and translation machinery is utilized to express the viral genes and produce progeny virus particles. From studying human immunodeficiency virus type 1 (HIV-1), it is known that the retrovirus capsid core takes part at several critical steps during the early phase of infection. For example, the core protects the viral genome from the cellular innate immune response (Mortuza et al. 2008; Li et al. 2016). The core retains RT and serves as a closed container for regulating reverse transcription. And the core facilitates transport of the viral genome and associated proteins to the nucleus, enabling the reverse transcribed DNA to enter the nucleus for the integration process (Peng et al. 2014). The structural stability of the core has a significant impact on the infectivity of the virus such that over-stabilization or premature disassembly of the core results in reduction or loss of infectivity.

Studies of retrovirus structural proteins and particles using X-ray crystallography, nuclear magnetic resonance (NMR) spectroscopy, cryo-electron microscopy (cryo-EM), cryo-electron tomography (cryo-ET), and dynamic simulation (Kingston et al. 2000; Briggs et al. 2004; Bailey et al. 2012; Zhao et al. 2013; Cao et al. 2015;

W. Zhang (✉) · L. Mendonça · L. L. Mansky
Institute for Molecular Virology, Department of Diagnostic and Biological Sciences, University of Minnesota, Minneapolis, MN, USA
e-mail: zhangwei@umn.edu; luiza@umn.edu; mansky@umn.edu

© Springer Nature Singapore Pte Ltd. 2018
J. R. Harris, D. Bhella (eds.), *Virus Protein and Nucleoprotein Complexes*, Subcellular Biochemistry 88, https://doi.org/10.1007/978-981-10-8456-0_8

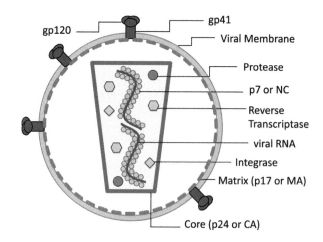

Fig. 8.1 Schematic representation of the cross section of a mature lentivirus particle. The virus particle exterior has a viral membrane and the envelope glycoproteins, i.e., gp41 and gp120. Matrix (MA) proteins are associated along the inner leaflet of the viral membrane. The capsid (CA) protein core encloses the nucleocapsid (NC) protein-coated viral RNA genome, as well as the viral enzymes, i.e., protease, reverse transcriptase, and integrase

Mattei et al. 2016; Meissner et al. 2017) have revealed the various structural characteristics of the retrovirus capsid core and their functional roles in the virus life cycle. The functionality of the retroviral capsid core depends on several physical and chemical properties that support a metastable structure and enable interactions with virus-specific host cellular factors. First, the core is a closed shell with sufficient volume to house two viral RNA strands plus several indispensable viral enzymes, including protease, RT, and integrase (IN) (Fig. 8.1). Second, the protein shell of the core is a selective barrier that is permeable to essential components needed for reverse transcription, including salts, water, and nucleotides (Jacques et al. 2016; Perilla et al. 2017). Third, the surface of the capsid core carries chemical groups that can bind infection-promoting host factors and facilitate nuclear targeting (Peng et al. 2014; Liu et al. 2016). Fourth, the metastable capsid shell disassembles following reverse transcription to release the DNA intermediate and integration complex into the nucleus (Schaller et al. 2011; Bichel et al. 2013).

The Retrovirus Capsid Core Has a Pleomorphic Structure

The *Retroviridae* family consists of two subfamilies: *Orthoretrovirinae* (including six genera: *Alpharetrovirus*, *Betaretrovirus*, *Gammaretrovirus*, *Deltaretrovirus*, *Epsilonretrovirus*, and *Lentivirus*) and *Spumaretrovirinae* (including spumaretrovirus). Orthoretroviruses assemble first as immature particles on cell membranes and mature into infectious particles after budding (Sundquist and Krausslich 2012; Christopher and Peijun 2013). The immature particle is composed mainly of the viral membrane, envelope proteins, Gag/Gag-Pol/Gag-Pro-Pol polyproteins, and diploid retroviral genome. Through Gag-membrane and Gag-Gag interactions, Gag polyproteins arrange themselves into an incomplete paracrystalline organization at the inner leaflet of the viral membrane (Yeager et al. 1998; Briggs et al. 2004; Briggs et al. 2006; Wright et al. 2007; Briggs et al. 2009; Maldonado et al. 2016). During maturation, Gag is cleaved by the viral protease into three major proteins, matrix (MA), capsid (CA), and nucleocapsid (NC), plus other virus-specific spacer peptides and subdomains. Among these, MA remains bound to the viral membrane, and a subset of CA proteins, typically involving thousands of copies, assembles into a capsid shell, which has a single layer of CA lattice that encapsulates the NC-RNA ribonucleoprotein complexes, viral replication enzymes (RT and IN), and dNTPs and tRNAs that serve as primers during reverse transcription. Using cryo-EM and cryo-ET, it was observed that the shapes of the capsid cores of retroviruses differ among genera (Zhang et al. 2015). For example, lentiviruses, such as HIV-1, typically have a conical core, whereas the cores of other retroviruses, such as human T-cell leukemia virus type 1 (HTLV-1), are mostly ovoid or polyhedral. The size and shape of the cores are heterogeneous even in the same virus genus. The pleomorphic nature of the retroviral capsid core suggests there is substantial flexibility in the curvature of the protein shell when CA proteins interconnect with one another to form the lattice structure. In contrast to other retroviruses, however, spumaretrovirus Gag protein is not processed into the classical orthoretroviral MA, CA, and NC subunits and has its own unique process for particle morphogenesis (Enssle et al. 1996; Hamann et al. 2014).

Two-Domain Capsid Protein: A Major Building Block of the Core in Orthoretroviruses

Despite there being little similarity in their amino acid sequences, various, 25-kDa orthoretrovirus CA proteins share conserved secondary and tertiary structures, as determined by X-ray crystallography and NMR spectroscopy (Khorasanizadeh et al. 1999; Campos-Olivas et al. 2000; Kingston et al. 2000; Cornilescu et al. 2001; Tang et al. 2002; Mortuza et al. 2004; Bartonova et al. 2008; Macek et al. 2009; Schur et al. 2015). CA contains two independently folded α-helix-rich domains that are connected by a flexible linker. The N-terminal domain of CA (NTD) is made up mostly of six or seven α-helices, while the C-terminal domain (CTD) contains four α-helices (Fig. 8.2a). The region around the residues in helix 8 (H8) and part of the loop region connecting it to helix 7 (H7) is known as the major homology region (MHR) and is highly conserved among all retroviruses (Wills and Craven 1991). The two-domain configuration of CA gives the protein substantial conformational flexibility when forming protein-protein interactions in both immature and mature particles. CA participates in lattice formation as a structural component of Gag polyprotein; CA-CA interactions within the Gag lattice are significantly constrained by both its N-terminal and C-terminal protein components. The N-terminal

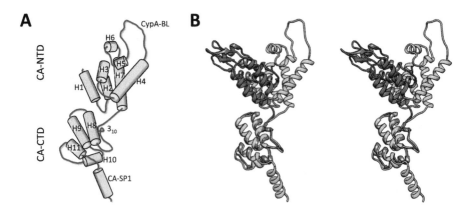

Fig. 8.2 HIV-1 CA protein structure. (**a**) Diagram of a CA molecule, indicating the sequence of the helices in the NTD and CTD domains. CypA-BL represents the cyclophilin A-binding loop. CA-SP1 represents the CA-SP1 helix present in an immature Gag lattice structure. The major homology region is composed of H8 and part of the loop between H7 and H8. (**b**) Structures of the HIV-1 CA molecule in immature (cyan, PDB, 5 L93) and mature (orange, PDB, 5MCX) particles, aligned according to CA-CTDs

end of CA is usually MA or a spacer peptide (e.g., p12 in murine leukemia virus (MLV)) situated between MA and CA. The C-terminal end of CA is either NC or a spacer such as sp1 in lentiviruses or sp. in *Rous sarcoma virus* (RSV). In a mature particle, proteolytic cleavage liberates CA and promotes formation of the lattice structure of the capsid core, where the two-domain CA proteins assume another conformation (Fig. 8.2b) to form different sets of protein-protein interactions. The two-domain configuration of CA likely contributes the structural plasticity underlying the varied curvature of the protein shell in the mature virus that results in the pleomorphism of retroviral capsid cores.

Cleavage at the C-Terminus of CA Marks a Turning Point in Representative Orthoretrovirus Maturation

Cryo-ET studies of in vitro-assembled tubular arrays of MPMV CA-NC proteins (Bharat et al. 2012), RSV CA-SP proteins (Jaballah et al. 2017), XMRV (a chimeric MLV) immature CA-NC proteins (Hadravová et al. 2012), and immature HIV-1 lattice (Schur et al. 2015; Mattei et al. 2016) demonstrated that within the retroviral immature Gag lattice, CA organizes into a paracrystalline hexagonal structure along the inner leaflet of the viral membrane. In immature HIV-1, CA hexamers are stabilized by six-helix bundles of CA-SP1 domains (Wright et al. 2007; Briggs et al. 2009). The center-to-center distance between the hexamers within the Gag lattice is ~80 Å. The Gag hexamer in an immature HIV-1 particle has the shape of a

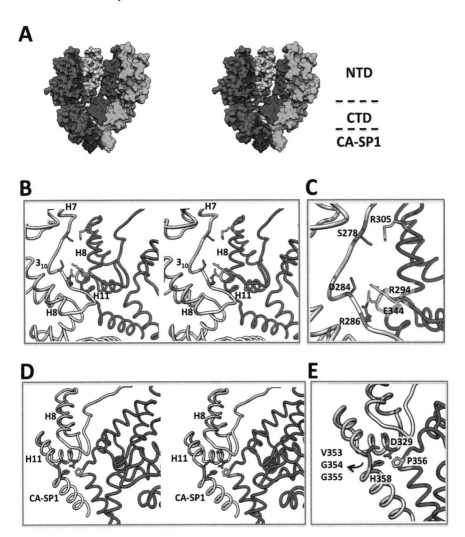

Fig. 8.3 Structure and molecular interactions that stabilize the immature HIV1-CA lattice (Schur et al. 2016). (**a**) Surface representation of the four CA molecules on the back side of a Gag hexamer.(**b**) Stereo diagram showing the L-shaped linker region between H7 and H8, which is stabilized by H8 of a six fold-related CA molecule. (**c**) Magnified view of panel b showing residues involved in NTD-CTD interactions. (**d**) Stereo diagram showing a network of interactions stabilizing the CA-SP1 helix bundles. (**e**) Magnified view of panel d showing residues involved at the interaction site. The molecular model is a reproduction of structure PDB 5 L93

wineglass with the NTD forming the tope of the glass, the CTD at the bottom, and the CA-SP1 six-helix bundle as the stem of the glass (Fig. 8.3a) (Schur et al. 2016). Examined from outside a Gag hexamer with the viral membrane at the top, CA NTD and CTD appear to have left-handed twist organizations, while the CA-SP1 bundle has a right-handed twist.

The HIV-1 immature Gag lattice and CA-SP1 helix bundle are well represented at the atomic level by two reported structures (Schur et al. 2016, Wagner et al. 2016b). The first is the Gag lattice structure in a native HIV-1 immature particle determined at 3.9-Å resolution using cryo-ET and subtomogram averaging methods (Schur et al. 2016). The other is the crystal structure of the HIV-1 CA CTD-SP1 domains (Wagner et al. 2016b). As shown in Fig. 8.3b–c in the immature Gag lattice, the MHR region, including H8 and the strand between H8 and the 3_{10} helix, interacts with both H8 and H11 in a neighboring CA molecule within the same CA hexamer. This configuration enables H8 of one CA molecule to interact with the NTD (H7) and CTD (H8) of a neighboring CA molecule. This network of interactions constrains the orientation of the CA NTD relative to the CTD and stabilizes CA in the immature conformation. Remarkably, even without the NTD domains, the HIV-1 CA CTD-SP1 protein forms a molecular organization that is exactly the same as in the immature HIV-1 Gag lattice. The crystal structure of CTD-SP1 (Wagner et al. 2016b) can be superimposed on the Gag lattice in a native virus (Schur et al. 2016), which suggests the contacts between the CTD and CA-SP1 domains are the primary structural determinants of the immature Gag lattice.

All orthoretroviral immature Gag lattice structures reported to date share similar molecular organizations in their CTD regions. Based on two atomic resolution structures of CA-SP1, there are two critical sets of molecular contacts in the CTD and SP1 regions (Schur et al. 2016; Wagner et al. 2016b). The first are the contacts among the helices that form the helical bundle organization. The second are the contacts critical for stabilizing the CA-SP1 helices, which involve interactions with the β-turn that stretch from H11 and the junction helix, i.e., residues V353, G354, and G355 (Fig. 8.3d–e) (Wagner et al. 2016b). While both the helical bundle and the β-turn elements are present in HIV-1, the RSV CA-SP1 has only the helix bundle, and MPMV has only the β-turn region (Wagner et al. 2016b). These structures confirm that CA-SP1 interaction among CTDs causes the CA CTDs to orient in a way that brings H8 into close contact with both the MHR and H7 of a neighboring CA molecule and enables CA to assume a conformation favorable for the immature lattice. During HIV-1 maturation, the last proteolytic cleavage occurs in the middle of the CA-SP1 helix. Thereafter, the CA proteins reorganize into a hexamer with a different conformation that is characterized in the mature particle. The proteolytic cleavage in the middle of the CA-SP1 region thus functions as a molecular switch that transforms an immature Gag lattice into a mature lattice.

The Retrovirus Capsid Core Assumes a Fullerene Structure and Is Permeable to Nucleotides

The protein shell of a retrovirus capsid core is modeled as a closed fullerene structure – a continuous curved lattice of CA hexamers interspersed with 12 CA pentamers (Fig. 8.4a) (Ganser et al. 1999; Zhao et al. 2013; Perilla and Gronenborn 2016).

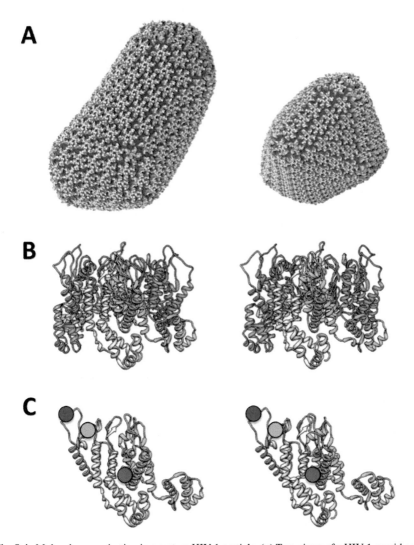

Fig. 8.4 Molecular organization in a mature HIV-1 particle. (**a**) Two views of a HIV-1 capsid core generated through all-atom fitting of CA into a cryo-ET reconstruction map (Zhao et al. 2013, Perilla and Gronenborn 2016). Figure courtesy of Dr. Juan Perilla. (**b**) Ribbon diagram of a hexamer structure determined by cryo-ET (PDB 5MCX) (Mattei et al. 2016). (**c**) Ligand-binding sites, blue, CypA (Gamble et al. 1996); green, CPIPB (Goudreau et al. 2013), and putative binding site for MxB (Fricke et al. 2014), red, CPSF6, NUP153, and PF174 (Bhattacharya et al. 2014, Price et al. 2014)

The spatial distribution of the CA hexamers and pentamers determines the size and shape of the capsid core. For example, the conical HIV-1 capsid has seven pentamers at the wide end of the core and five pentamers at the narrow end. While retrovirus CA hexamers have been detected in many 2D (Pornillos et al. 2009; Bailey et al. 2012) and 3D crystals (Pornillos et al. 2009; Gres et al. 2015) and helical tubes (Li et al. 2000; Zhao et al. 2013), it has been a challenge to resolve the structure of CA pentamers and prove their existence in the mature retrovirus capsid core. The pentameric structures of retrovirus CA proteins have been studied in the icosahedral capsid formed by a RSV mutant CA (Cardone et al. 2009) and in cross-linked HIV-1 CA crystals (Pornillos et al. 2011) through dynamic modeling (Zhao et al. 2013; Grime et al. 2016) and cryo-ET reconstruction of intact HIV-1 virions (Mattei et al. 2016). Determination of the native pentameric and hexameric HIV-1 CA structure at sub-nanometer resolution in intact virions revealed the presence and molecular arrangement of CA pentamers in the HIV-1 capsid core (Mattei et al. 2016).

In each capsomer within a retrovirus capsid core shell, the CA NTD takes part in intra-capsomer (intra-hexameric and intra-pentameric) interactions and forms the inner ring of the capsomer, which is 40 Å thick and 90 Å in diameter. The CA CTD forms an outer ring by making connections among the capsomers (Mattei et al. 2016). An examination of a capsomer from the side, with the NTD at the top, reveals that both the NTD and CTD are arranged with left-handed twists (Fig. 8.4b). NTDs interact with CTDs of neighboring molecules within the same capsomer. In contrast to the arrangement in immature particles, where NTD H1 and H2 are located at local twofold and threefold axes of the hexagonal Gag lattice, within a mature particle hexamer, H1–H3 are situated (or H1 and H2 in the pentamer) toward the center of the capsomer and form the 18-helix (or 10 helices in a pentamer) bundle and a central pore (Cardone et al. 2009; Pornillos et al. 2009; Zhang et al. 2015; Mattei et al. 2016). Within the pentamer, however, only H1 and H2 contribute to NTD-NTD interactions within the same pentamer, leaving the CPSF6 binding site open and accessible (Mattei et al. 2016).

The retrovirus capsid core has an irregular shape and presents a CA protein shell with changing curvature (Perilla and Gronenborn 2016). In HIV-1, the narrow end of the conical capsid core has greater curvature than the wide end of the core. The protein shell is also curved in the direction perpendicular to the longitudinal axis of the core. Studies of the native HIV-1 capsid (Byeon et al. 2009; Zhao et al. 2013; Mattei et al. 2016) and 2D crystallography (Pornillos et al. 2009; Bailey et al. 2012) show that several factors contribute to the curvature of the HIV-1 core, including the relative orientations of the NTD and CTD within each CA molecule, variability in the in-plane tilt at the local twofold and threefold axes of CTDs in the hexamers, and insertion of pentamers. The presence of pentamers offers sharp turns in the lattice (Mattei et al. 2016). Several host cell factors have been shown to interact with the CA capsomers (Fig. 8.4c) and accommodate the changing curvature of the capsid shell (Stremlau et al. 2004, Yap et al. 2004; Liu et al. 2016).

Another prominent structural feature of the HIV-1 core is the central channel of the capsomer, which can switch between open and closed conformations under different pH conditions (Jacques et al. 2016). At a basic pH, the pore is in a closed state, while at an acidic pH, the pore is open and forms a channel 8 Å in diameter, which is sufficient to allow transit of a dNTP molecule while excluding larger molecules. It was shown that a highly conserved arginine residue at CA position 18 (R18, CA numbering) acts as the channel gate (Jacques et al. 2016). Residue R18 is located at the N-terminal β-hairpin of CA, which sits right at the sixfold symmetry axis. In a recent 6.8-Å subtomogram-averaged structure of HIV-1 capsomers (Mattei et al. 2016), the central pore formed by NTDs is occupied by electron densities attributable to triphosphates that neutralize the positive charges and stabilize the core. In HIV-1, it has been demonstrated that protonation of a histidine residue at position 12 leads to movement of the N-terminal β-hairpin in CA, which influences the state and size of the pore (Jacques et al. 2016).

Capsid Interaction with Host Factors Interferes with Capsid Stability and Restrict Infectivity

After entering the cell, the retrovirus capsid protein shell protects the viral genome from attacks by cellular factors, provides space for reverse transcription, and facilitates nuclear entry. The majority of retrovirus genera, including *Deltaretrovirus* (Bieniasz et al. 1995), *Gammaretrovirus* (Yamashita and Emerman 2004) and *Spumaretrovirus* (Bieniasz et al. 1995), accesses the host cell chromosomes via mitosis to achieve stable integration. Lentiviruses, such as HIV-1, however, can efficiently deliver a DNA copy of the virus genome through the nuclear pore complex and replicate in certain types of nondividing cells such as macrophages (Hilditch and Towers 2014). Alpharetroviruses are also able to infect nondividing cells but are far less efficient than lentiviruses (Hatziioannou and Goff 2001; Katz et al. 2002). Betaretroviruses are usually thought to be unable to infect nonproliferating cells, but a MMTV-based vector was recently shown to transduce non-dividing cells (Konstantoulas and Indik 2014). A HIV-1 chimera in which the CA sequence was swapped for that of MLV (gammaretrovirus) lacks the ability to infect nonproliferating cells, indicating that the capsid is a major determinant in this capacity (Yamashita and Emerman 2004).

The capsid protein shell of HIV-1 undergoes uncoating prior nuclear entry. During uncoating, the shell disassembles and releases the newly synthesized double-stranded DNA (Ambrose and Aiken 2014; Campbell and Hope 2015). The processes of reverse transcription, trafficking to the nucleus and nuclear entry, all require the capsid to have a metastable structure. Over-stabilization of the core structure or premature uncoating results in reduction or loss of infectivity. A panel of cell- and virus-specific cellular proteins that recognize the quaternary structure of the retroviral capsid core has been identified. These proteins include CypA, TRIM5α, MxB, CPSF6, and NUP153 (Ambrose and Aiken 2014; Campbell and Hope 2015). Each of these cellular factors has its role in the normal cell life cycle. During retrovirus infection, however, they can promote or restrict viral replication, depending on the specific cell type or organism. Given how crucial the optimal stability of the capsid is to the viral infection, modulating capsid stability makes an attractive target for antiviral strategies (Shi et al. 2011).

CypA

Cyclophilin A (CypA) is omnipresent in the cytoplasm of eukaryotic cells. It is a cellular peptidylprolyl isomerase that catalyzes the isomerization of peptide bonds at proline residues and facilitates protein folding. CypA can be packed into HIV-1 virions, but it was confirmed that only target cell CypA is relevant for HIV-1 infectivity (Sokolskaja and Luban 2006). In certain nonhuman primate cells, inhibition of the CA-CypA interaction with the immunosuppressive drug cyclosporin A (CsA) (Ke et al. 1994) dramatically increases viral infectivity (Towers et al. 2003). The crystal structure of the HIV-1 CA NTD in complex with human CypA reveals a sequence-specific interaction between residues 85–93 of CA (known as the CypA-binding loop) (Fig. 8.2a) and the active site of the enzyme (Gamble et al. 1996). A cryo-EM reconstruction of CypA in complex with the assembled HIV-1 CA tubular structure at 8 Å resolution suggests one CypA molecule can interact with two CA subunits and bridge two neighboring CA hexamers along the direction of the highest capsid surface curvature. CypA binding is therefore believed to stabilize CA assemblies with high surface curvature (Liu et al. 2016). The precise function of CypA in HIV-1 replication remains unknown, however. Capsid binding to CypA has been shown to promote HIV-1 infection in human cells, though the same interaction enhances the anti-HIV-1 restriction activity in certain nonhuman primate cells (Ikeda et al. 2004). Because the curvature of the conical HIV-1 capsid varies, within specific cellular environments, CypA could introduce subtle changes in the CA conformation, thereby promoting HIV-1 virion core detection by another host restriction factor such as simian TRIM5α, which can interact with the HIV-1 core through innate pattern recognition (Towers et al. 2003; Sokolskaja and Luban 2006; Towers 2007) enhance antiviral activity by the small HIV-1 inhibitor PF-3450074 (PF74) (Shi et al. 2011) or protect HIV-1 from recognition by the Ref-1 restriction factor (Towers et al. 2003; Sokolskaja and Luban 2006; Towers 2007).

TRIM5α

The alpha isoform of tripartite motif protein 5 (TRIM5α) is a host restriction factor that combats retrovirus infection by inducing premature disassembly of the viral capsid core (Black and Aiken 2010), leading to unproductive reverse transcription and activation of downstream innate immune responses. TRIM5α is composed of a tripartite motif that includes a RING, B-box 2, a coiled-coil domain, and a C-terminal B30.2/SPRY domain (Fig. 8.5a). The coiled-coil domain mediates formation of homodimers by full-length TRIM5α molecules (Sanchez et al. 2014). The B30.2/SPRY domain determines the binding specificity of the viral capsid (Yang et al. 2012) and destabilizes the HIV-1 capsid structure. Mixing rhesus macaque coiled-coil SPRY with the tubular structure of HIV-1 CA and purified cores resulted into

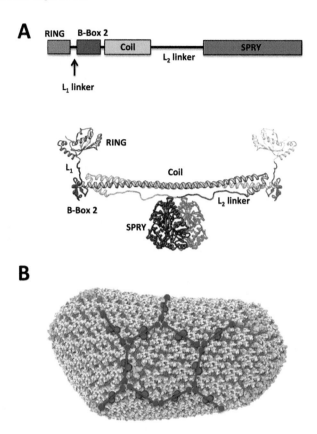

Fig. 8.5 Structure of Trim5α and its interactions with the HIV-1 capsid. (**a**) Schematic representations of the Trim5α molecule domain organization and the full-length antiparallel protein dimer (Wagner et al. 2016a, b). The ribbon diagram of the Trim5α dimer is created by combining the structures for isolated domains: RING (4TKP) (Yudina et al. 2015), B-Box-2, coiled-coil (4NT3) (Goldstone et al. 2014), and SPPY (4B3N) (Yang et al. 2012). (**b**) Schematic representation of an HIV-1 capsid core bound by a TRIM5α hexagonal net (Li et al. 2016). The blue stick with two red dots represents one Trim5α dimer molecule

disruption of the tube and core structures at the inter-hexamer interfaces, releasing fragments of protofilaments consisting of CA hexamers (Zhao et al. 2011). The B-box 2 domain is involved in higher-order assembly of TRIM5α homodimers and their interactions with the retrovirus capsid core. The RING domain functions as an E3 ubiquitin ligase that catalyzes synthesis of polyubiquitin chains linked through a lysine residue on the target protein and promotes innate immune responses (Yudina et al. 2015). Dimerization of TRIM5α RING domains enhances the E3 ubiquitin ligase activity of TRIM5α and contributes to its antiretroviral functionality (Yang et al. 2012).

The anti-HIV-1 activity of TRIM5α is enhanced by its high-order oligomerization. Disruption of the higher-order structure of TRIM5α impairs its restriction capacity. The high-order structure of TRIM5α dimers is organized as a hexagonal net in solution (Ganser-Pornillos et al. 2011; Li et al. 2016) (Fig. 8.5b), and this net structure is able to bind to the assembled capsid cores (Fig. 8.5c). On the other hand, TRIM5α does not significantly interact with monomeric or soluble CA protein. A cryo-ET study (Li et al. 2016) revealed the mesh decoration of TRIM5α on disulfide-cross-linked HIV-1 capsid cores. It was found that the hexagonal nets formed open and flexible hexameric rings, with the SPRY domains centered on the edges and the B-box and RING domains at the vertices. It is believed that only a few molecules are needed to cover the surface of the capsid (Li et al. 2016) and fulfill its restriction function.

Interactions between the TRIM5α net and the HIV-1 capsid core are dependent on the flexibility of the SPRY domain and the entire network. First, the crystal structure showed that the variable loop regions of TRIM5α PRY/SPRY, which carry crucial amino acid residues for capsid interaction and antiviral specificity, are flexible. This flexibility enables residues in these regions to interact with viral capsids of different shapes. Second, the hexagonal nets formed by TRIM5α dimers have differing arm lengths and angles (Li et al. 2016; Wagner et al. 2016a), enabling formation of paracrystalline structures that can interact with assembled cores with different surface curvatures. The flexibility of the TRIM5α assembly may come from the flexing of the coiled-coil domains and/or the hinge between B-box 2 and the coiled-coil domains. In summary, individual TRIM5α molecules interact only weakly with CA proteins. However, dimerization, higher-order oligomerization, and the conformational plasticity of the net structure all enhance pattern recognition by TRIM5α, enabling it to interact with diverse and pleomorphic retroviral capsids and fulfill its restriction activity.

MxB

Myxovirus resistance protein 2 (MxB) is an interferon-induced HIV-1 restriction factor (Kane et al. 2013). MxB is a guanosine triphosphatase (GTPase) that belongs to the dynamin superfamily, although its antiviral function is independent of its GTPase activity (Kane et al. 2013). The molecule recognizes and binds to the viral core when it is near the host cell nucleus. Immunofluorescent staining revealed that wild-type MxB forms cytoplasmic puncta localized to the nuclear rim (Kane et al. 2013; Fribourgh et al. 2014). MxB inhibits uncoating of the HIV-1 capsid, a process downstream of reverse transcription, and inhibits import into the nucleus (Bhattacharya et al. 2014). The MxB molecule is composed of a nuclear localization signal (NLS), GTPase domain, a stalk region, and a leucine zipper motif or bundle signaling element (BSE) domain (Fig. 8.6a). The N-terminus of MxB, which includes the NLS, is important for MxB-capsid interaction (Kane et al. 2013). The crystal structure of MxB lacking its N-terminal region (Fig. 8.6b) adopts an extended antiparallel dimer structure through interactions between the stalk domains (Fribourgh et al. 2014; Xu et al. 2015). The same dimeric MxB organization is also

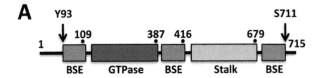

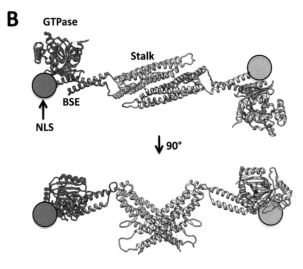

Fig. 8.6 Domain organization of MxB and its crystallographic dimer structure (Fribourgh et al. 2014). (**a**) Schematic representation of an MxB molecule. Arrows in the schematic denote the first and last visible residues in the structure in (**b**). (**b**) Top and side views of the dimeric MxB structure. Individual domains of one molecule are colored as in (**a**). The N-terminal region nuclear localization signal (NLS), which is not included in the crystal structure, is represented as a red sphere

seen in the 4.6-Å cryo-EM structure of the helical assembly of full-length MxB fused with a N-terminal maltose-binding protein (MBP) (Alvarez et al. 2017). Dimerization of MxB was shown to be important for its subcellular localization, binding to the HIV-1 capsid, and viral restriction (Fribourgh et al. 2014). Higher-order oligomerization is also believed to enhance capsid interaction and inhibition of HIV-1 infection (Alvarez et al. 2017). The MxB N-terminal domain and MBP domain were not resolved in the full-length MxB helical structure, which is suggestive of the flexibility of this region. Interactions between MxB and the virus capsid can be modulated by a small-molecule inhibitor. The benzimidazole-based compound CPIPB prevents the binding of MxB to the HIV-1 capsid (Bhattacharya et al. 2014; Fricke et al. 2014), suggesting that the capsid-binding site for MxB overlaps with the binding pocket for CPIPB, which is located between the base of the Cyp A-binding loop and the loop that connects H6 and H7 in the HIV-1 capsid (Fig. 8.4c) (Goudreau et al. 2013).

CPSF6 and NUP153

The capsid protein and the assembled capsid core of HIV-1 provide critical functionality for transporting the reverse transcription complex from cytoplasm into the nucleus (Ambrose and Aiken 2014; Campbell and Hope 2015). The nuclear import process is facilitated by the interaction between CA and the assembled capsid core with various nuclear targeting cofactors, including cellular protein cleavage and polyadenylation specific factor 6 (CPSF6), nucleoporin 153 kDa (NUP153), and transportin 3 (TNPO3). CPSF6 is part of an mRNA processing complex that shuttles between the nucleus and cytoplasm (Ruegsegger et al. 1998). It has been proposed that CPSF6 binding to the HIV-1 capsid facilitates recruitment of nuclear import factors such as TNPO3 and nucleoporins (Price et al. 2012). Both structural and functional studies (Bhattacharya et al. 2014; Price et al. 2014) have shown that CPSF6 and NUP153 bind to HIV-1 CA hexamers or the assembled capsid with at least tenfold higher affinity than they bind unassembled CA proteins. The preference of NUP153 and CPSF6 for assembled capsid supports a model wherein the HIV-1 capsid remains largely intact near the nuclear envelope of infected cells. In addition, both ligands bind to the CA hexamer via a phenylalanine-glycine (FG) motif at the common binding pocket site, which is located at the interface of the NTD of one CA molecule and the CTD of a neighboring molecule in the same CA hexamer (Fig. 8.4c). Evidently, the binding sites of NUP153 and CPSF6 also overlap with that of the antiviral compound PF74 (Bhattacharya et al. 2014). It was proposed that PF74 inhibits HIV-1 infection at a low dose by competing for cofactor binding and at a high dose by irreversibly blocking reverse transcription, possibly as a result of irreversible changes to the capsid ultrastructure.

Summary

During the early phase of the retroviral life cycle, the capsid core plays critical but not fully defined roles in viral DNA synthesis and nuclear entry. The CA protein, which is the major structural component of the capsid core, undergoes two key irreversible conformational changes which link the capsid core to two other steps in retroviral replication. First, during virus maturation, CA is liberated from Gag by the viral protease and forms the capsid core that has a metastable and permeable structure – features imperative for efficient reverse transcription upon infection of permissive target cells. Subsequently, the process of capsid core disassembly facilitates translocation of the viral DNA to the nucleus of the infected cell. The study of the molecular mechanisms for these conformational changes and their interrelationships with other steps in the retroviral life cycle is an active area of research.

In contrast to non-enveloped viruses where the capsid-host interactions generally occur outside or at the cell periphery, retroviral capsid cores interact with host cell factors in the cytoplasm of the infected cell. Host cellular factors can either bind the core at the exposed loops of the CA proteins or at the regions formed between neighboring CA molecules within the same capsomer. Host cell factors may also recognize the varied capsid core curvature. The study of virus-host cell interactions via CA-restriction factor interactions has implications to the mechanisms of host innate immunity to retroviral infection and is a highly promising area of structural virology. Studies of the retroviral capsid core yield important insights into the identification of targets for antiretroviral intervention.

Acknowledgment This work is supported by NIH grants R01 GM098550 and R01 GM124279. We thank Dr. Juan Perilla for providing the figures that show the all-atom fitting model of CA in a HIV-1 capsid core structure.

References

Alvarez FJD, He S, Perilla JR, Jang S, Schulten K, Engelman AN, Scheres SHW, Zhang P (2017) CryoEM structure of MxB reveals a novel oligomerization interface critical for HIV restriction. Sci Adv 3(9):e1701264

Ambrose Z, Aiken C (2014) HIV-1 uncoating: connection to nuclear entry and regulation by host proteins. Virology 454-455:371–379

Bailey GD, Hyun JK, Mitra AK, Kingston RL (2012) A structural model for the generation of continuous curvature on the surface of a retroviral capsid. J Mol Biol 417(3):212–223

Bartonova V, Igonet S, Sticht J, Glass B, Habermann A, Vaney MC, Sehr P, Lewis J, Rey FA, Krausslich HG (2008) Residues in the HIV-1 capsid assembly inhibitor binding site are essential for maintaining the assembly-competent quaternary structure of the capsid protein. J Biol Chem 283(46):32024–32033

Bharat TA, Davey NE, Ulbrich P, Riches JD, de Marco A, Rumlova M, Sachse C, Ruml T, Briggs JA (2012) Structure of the immature retroviral capsid at 8 a resolution by cryo-electron microscopy. Nature 487(7407):385–389

Bhattacharya A, Alam SL, Fricke T, Zadrozny K, Sedzicki J, Taylor AB, Demeler B, Pornillos O, Ganser-Pornillos BK, Diaz-Griffero F, Ivanov DN, Yeager M (2014) Structural basis of HIV-1 capsid recognition by PF74 and CPSF6. Proc Natl Acad Sci U S A 111(52):18625–18630

Bichel K, Price AJ, Schaller T, Towers GJ, Freund SM, James LC (2013) HIV-1 capsid undergoes coupled binding and isomerization by the nuclear pore protein NUP358. Retrovirology 10:81

Bieniasz PD, Weiss RA, McClure MO (1995) Cell cycle dependence of foamy retrovirus infection. J Virol 69(11):7295–7299

Black LR, Aiken C (2010) TRIM5alpha disrupts the structure of assembled HIV-1 capsid complexes in vitro. J Virol 84(13):6564–6569

Briggs JA, Johnson MC, Simon MN, Fuller SD, Vogt VM (2006) Cryo-electron microscopy reveals conserved and divergent features of gag packing in immature particles of Rous sarcoma virus and human immunodeficiency virus. J Mol Biol 355(1):157–168

Briggs JA, Riches JD, Glass B, Bartonova V, Zanetti G, Krausslich HG (2009) Structure and assembly of immature HIV. Proc Natl Acad Sci U S A 106(27):11090–11095

Briggs JA, Watson BE, Gowen BE, Fuller SD (2004) Cryoelectron microscopy of mouse mammary tumor virus. J Virol 78(5):2606–2608

Byeon IJ, Meng X, Jung J, Zhao G, Yang R, Ahn J, Shi J, Concel J, Aiken C, Zhang P, Gronenborn AM (2009) Structural convergence between Cryo-EM and NMR reveals intersubunit interactions critical for HIV-1 capsid function. Cell 139(4):780–790

Campbell EM, Hope TJ (2015) HIV-1 capsid: the multifaceted key player in HIV-1 infection. Nat Rev Microbiol 13(8):471–483

Campos-Olivas R, Newman JL, Summers MF (2000) Solution structure and dynamics of the Rous sarcoma virus capsid protein and comparison with capsid proteins of other retroviruses. J Mol Biol 296(2):633–649

Cao S, Maldonado JO, Grigsby IF, Mansky LM, Zhang W (2015) Analysis of human T-cell leukemia virus type 1 particles by using cryo-electron tomography. J Virol 89(4):2430–2435

Cardone G, Purdy JG, Cheng N, Craven RC, Steven AC (2009) Visualization of a missing link in retrovirus capsid assembly. Nature 457(7230):694–698

Christopher A, Peijun Z (2013) HIV-1 Maturation. In: Freed EO (ed) Advances in HIV-1 Assembly and Release. Springer, pp 153–166

Cornilescu CC, Bouamr F, Yao X, Carter C, Tjandra N (2001) Structural analysis of the N-terminal domain of the human T-cell leukemia virus capsid protein. J Mol Biol 306(4):783–797

Enssle J, Jordan I, Mauer B, Rethwilm A (1996) Foamy virus reverse transcriptase is expressed independently from the gag protein. Proc Natl Acad Sci U S A 93(9):4137–4141

Fribourgh JL, Nguyen HC, Matreyek KA, Alvarez FJD, Summers BJ, Dewdney TG, Aiken C, Zhang P, Engelman A, Xiong Y (2014) Structural insight into HIV-1 restriction by MxB. Cell Host Microbe 16(5):627–638

Fricke T, White TE, Schulte B, de Souza Aranha DA, Vieira AD, Campbell EM, Brandariz-Nunez A, Diaz-Griffero F (2014) MxB binds to the HIV-1 core and prevents the uncoating process of HIV-1. Retrovirology 11:68

Gamble TR, Vajdos FF, Yoo S, Worthylake DK, Houseweart M, Sundquist WI, Hill CP (1996) Crystal structure of human cyclophilin a bound to the amino-terminal domain of HIV-1 capsid. Cell 87(7):1285–1294

Ganser BK, Li S, Klishko VY, Finch JT, Sundquist WI (1999) Assembly and analysis of conical models for the HIV-1 core. Science 283(5398):80–83

Ganser-Pornillos BK, Chandrasekaran V, Pornillos O, Sodroski JG, Sundquist WI, Yeager M (2011) Hexagonal assembly of a restricting TRIM5alpha protein. Proc Natl Acad Sci U S A 108(2):534–539

Goldstone DC, Walker PA, Calder LJ, Coombs PJ, Kirkpatrick J, Ball NJ, Hilditch L, Yap MW, Rosenthal PB, Stoye JP, Taylor IA (2014) Structural studies of postentry restriction factors reveal antiparallel dimers that enable avid binding to the HIV-1 capsid lattice. Proc Natl Acad Sci U S A 111(26):9609–9614

Goudreau N, Lemke CT, Faucher AM, Grand-Maitre C, Goulet S, Lacoste JE, Rancourt J, Malenfant E, Mercier JF, Titolo S, Mason SW (2013) Novel inhibitor binding site discovery on HIV-1 capsid N-terminal domain by NMR and X-ray crystallography. ACS Chem Biol 8(5):1074–1082

Gres AT, Kirby KA, KewalRamani VN, Tanner JJ, Pornillos O, Sarafianos SG (2015) STRUCTURAL VIROLOGY. X-ray crystal structures of native HIV-1 capsid protein reveal conformational variability. Science 349(6243):99–103

Grime JM, Dama JF, Ganser-Pornillos BK, Woodward CL, Jensen GJ, Yeager M, Voth GA (2016) Coarse-grained simulation reveals key features of HIV-1 capsid self-assembly. Nat Commun 7:11568

Hadravová R, de Marco A, Ulbrich P, Stokrova J, Dolezal M, Pichova I, Ruml T, Briggs JA, Rumlova M (2012) In vitro assembly of virus-like particles of a gammaretrovirus, the murine leukemia virus XMRV. J Virol 86(3):1297–1306

Hamann MV, Mullers E, Reh J, Stanke N, Effantin G, Weissenhorn W, Lindemann D (2014) The cooperative function of arginine residues in the prototype foamy virus gag C-terminus mediates viral and cellular RNA encapsidation. Retrovirology 11:87

Hatziioannou T, Goff SP (2001) Infection of nondividing cells by Rous sarcoma virus. J Virol 75(19):9526–9531

Hilditch L, Towers GJ (2014) A model for cofactor use during HIV-1 reverse transcription and nuclear entry. Curr Opin Virol 4:32–36

Hu WS, Hughes SH (2012) HIV-1 reverse transcription. Cold Spring Harb Perspect Med 2(10)

Ikeda Y, Ylinen LM, Kahar-Bador M, Towers GJ (2004) Influence of gag on human immunodeficiency virus type 1 species-specific tropism. J Virol 78(21):11816–11822

Jaballah SA, Bailey GD, Desfosses A, Hyun J, Mitra AK, Kingston RL (2017) In vitro assembly of the Roeus Sarcoma Virus capsid protein into hexamer tubes at physiological temperature. Sci Rep **7**(1):2913

Jacques DA, McEwan WA, Hilditch L, Price AJ, Towers GJ, James LC (2016) HIV-1 uses dynamic capsid pores to import nucleotides and fuel encapsidated DNA synthesis. Nature 536(7616):349–353

Kane M, Yadav SS, Bitzegeio J, Kutluay SB, Zang T, Wilson SJ, Schoggins JW, Rice CM, Yamashita M, Hatziioannou T, Bieniasz PD (2013) MX2 is an interferon-induced inhibitor of HIV-1 infection. Nature 502(7472):563–566

Katz RA, Greger JG, Darby K, Boimel P, Rall GF, Skalka AM (2002) Transduction of interphase cells by avian sarcoma virus. J Virol 76(11):5422–5434

Ke H, Mayrose D, Belshaw PJ, Alberg DG, Schreiber SL, Chang ZY, Etzkorn FA, Ho S, Walsh CT (1994) Crystal structures of cyclophilin a complexed with cyclosporin A and N-methyl-4-[(E)-2-butenyl]-4,4-dimethylthreonine cyclosporin A. Structure 2(1):33–44

Khorasanizadeh S, Campos-Olivas R, Clark CA, Summers MF (1999) Sequence-specific 1H, 13C and 15N chemical shift assignment and secondary structure of the HTLV-I capsid protein. J Biomol NMR 14(2):199–200

Kingston RL, Fitzon-Ostendorp T, Eisenmesser EZ, Schatz GW, Vogt VM, Post CB, Rossmann MG (2000) Structure and self-association of the Rous sarcoma virus capsid protein. Structure 8(6):617–628

Konstantoulas CJ, Indik S (2014) Mouse mammary tumor virus-based vector transduces nondividing cells, enters the nucleus via a TNPO3-independent pathway and integrates in a less biased fashion than other retroviruses. Retrovirology 11:34

Li S, Hill CP, Sundquist WI, Finch JT (2000) Image reconstructions of helical assemblies of the HIV-1 CA protein. Nature 407(6802):409–413

Li YL, Chandrasekaran V, Carter SD, Woodward CL, Christensen DE, Dryden KA, Pornillos O, Yeager M, Ganser-Pornillos BK, Jensen GJ, Sundquist WI (2016) Primate TRIM5 proteins form hexagonal nets on HIV-1 capsids. Elife 5

Liu C, Perilla JR, Ning J, Lu M, Hou G, Ramalho R, Himes BA, Zhao G, Bedwell GJ, Byeon IJ, Ahn J, Gronenborn AM, Prevelige PE, Rousso I, Aiken C, Polenova T, Schulten K, Zhang P (2016) Cyclophilin a stabilizes the HIV-1 capsid through a novel non-canonical binding site. Nat Commun 7:10714

Macek P, Chmelik J, Krizova I, Kaderavek P, Padrta P, Zidek L, Wildova M, Hadravova R, Chaloupkova R, Pichova I, Ruml T, Rumlova M, Sklenar V (2009) NMR structure of the N-terminal domain of capsid protein from the mason-pfizer monkey virus. J Mol Biol 392(1):100–114

Maldonado JO, Cao S, Zhang W, Mansky LM (2016) Distinct morphology of human T-cell leukemia virus type 1-like particles. Virus 8(5)

Mattei S, Glass B, Hagen WJ, Krausslich HG, Briggs JA (2016) The structure and flexibility of conical HIV-1 capsids determined within intact virions. Science 354(6318):1434–1437

Meissner ME, Mendonca LM, Zhang W, Mansky LM (2017) Polymorphic nature of human T-cell leukemia virus type 1 particle cores as revealed through characterization of a chronically infected cell line. J Virol 91(16):e00369

Mortuza GB, Dodding MP, Goldstone DC, Haire LF, Stoye JP, Taylor IA (2008) Structure of B-MLV capsid amino-terminal domain reveals key features of viral tropism, gag assembly and core formation. J Mol Biol 376(5):1493–1508

Mortuza GB, Haire LF, Stevens A, Smerdon SJ, Stoye JP, Taylor IA (2004) High-resolution structure of a retroviral capsid hexameric amino-terminal domain. Nature 431(7007):481–485

Pan X, Baldauf HM, Keppler OT, Fackler OT (2013) Restrictions to HIV-1 replication in resting CD4+ T lymphocytes. Cell Res 23(7):876–885

Peng K, Muranyi W, Glass B, Laketa V, Yant SR, Tsai L, Cihlar T, Muller B, Krausslich HG (2014) Quantitative microscopy of functional HIV post-entry complexes reveals association of replication with the viral capsid. Elife 3:e04114

Perilla JR, Gronenborn AM (2016) Molecular architecture of the retroviral capsid. Trends Biochem Sci 41(5):410–420

Perilla JR, Zhao G, Lu M, Ning J, Hou G, Byeon IL, Gronenborn AM, Polenova T, Zhang P (2017) CryoEM structure refinement by integrating NMR chemical shifts with molecular dynamics simulations. J Phys Chem B 121(15):3853–3863

Pornillos O, Ganser-Pornillos BK, Kelly BN, Hua Y, Whitby FG, Stout CD, Sundquist WI, Hill CP, Yeager M (2009) X-ray structures of the hexameric building block of the HIV capsid. Cell 137(7):1282–1292

Pornillos O, Ganser-Pornillos BK, Yeager M (2011) Atomic-level modelling of the HIV capsid. Nature 469(7330):424–427

Price AJ, Fletcher AJ, Schaller T, Elliott T, Lee K, KewalRamani VN, Chin JW, Towers GJ, James LC (2012) CPSF6 defines a conserved capsid interface that modulates HIV-1 replication. PLoS Pathog 8(8):e1002896

Price AJ, Jacques DA, McEwan WA, Fletcher AJ, Essig S, Chin JW, Halambage UD, Aiken C, James LC (2014) Host cofactors and pharmacologic ligands share an essential interface in HIV-1 capsid that is lost upon disassembly. PLoS Pathog 10(10):e1004459

Ruegsegger U, Blank D, Keller W (1998) Human pre-mRNA cleavage factor Im is related to spliceosomal SR proteins and can be reconstituted in vitro from recombinant subunits. Mol Cell 1(2):243–253

Sanchez JG, Okreglicka K, Chandrasekaran V, Welker JM, Sundquist WI, Pornillos O (2014) The tripartite motif coiled-coil is an elongated antiparallel hairpin dimer. Proc Natl Acad Sci U S A 111(7):2494–2499

Schaller T, Ocwieja KE, Rasaiyaah J, Price AJ, Brady TL, Roth SL, Hue S, Fletcher AJ, Lee K, KewalRamani VN, Noursadeghi M, Jenner RG, James LC, Bushman FD, Towers GJ (2011) HIV-1 capsid-cyclophilin interactions determine nuclear import pathway, integration targeting and replication efficiency. PLoS Pathog 7(12):e1002439

Schur FK, Hagen WJ, Rumlova M, Ruml T, Muller B, Krausslich HG, Briggs JA (2015) Structure of the immature HIV-1 capsid in intact virus particles at 8.8 a resolution. Nature 517(7535):505–508

Schur FK, Obr M, Hagen WJ, Wan W, Jakobi AJ, Kirkpatrick JM, Sachse C, Krausslich HG, Briggs JA (2016) An atomic model of HIV-1 capsid-SP1 reveals structures regulating assembly and maturation. Science 353(6298):506–508

Shi J, Zhou J, Shah VB, Aiken C, Whitby K (2011) Small-molecule inhibition of human immunodeficiency virus type 1 infection by virus capsid destabilization. J Virol 85(1):542–549

Sokolskaja E, Luban J (2006) Cyclophilin, TRIM5, and innate immunity to HIV-1. Curr Opin Microbiol 9(4):404–408

Stremlau M, Owens CM, Perron MJ, Kiessling M, Autissier P, Sodroski J (2004) The cytoplasmic body component TRIM5alpha restricts HIV-1 infection in old world monkeys. Nature 427(6977):848–853

Sundquist WI, Krausslich HG (2012) HIV-1 assembly, budding, and maturation. Cold Spring Harb Perspect Med 2(7):a006924

Tang C, Ndassa Y, Summers MF (2002) Structure of the N-terminal 283-residue fragment of the immature HIV-1 gag polyprotein. Nat Struct Biol 9(7):537–543

Towers GJ (2007) The control of viral infection by tripartite motif proteins and cyclophilin a. Retrovirology 4:40

Towers GJ, Hatziioannou T, Cowan S, Goff SP, Luban J, Bieniasz PD (2003) Cyclophilin a modulates the sensitivity of HIV-1 to host restriction factors. Nat Med 9(9):1138–1143

Wagner JM, Roganowicz MD, Skorupka K, Alam SL, Christensen D, Doss G, Wan Y, Frank GA, Ganser-Pornillos BK, Sundquist WI, Pornillos O (2016a) Mechanism of B-box 2 domain-mediated higher-order assembly of the retroviral restriction factor TRIM5alpha. Elife 5

Wagner JM, Zadrozny KK, Chrustowicz J, Purdy MD, Yeager M, Ganser-Pornillos BK, Pornillos O (2016b) Crystal structure of an HIV assembly and maturation switch. Elife 5

Wills JW, Craven RC (1991) Form, function, and use of retroviral gag proteins. AIDS 5(6):639–654

Wright ER, Schooler JB, Ding HJ, Kieffer C, Fillmore C, Sundquist WI, Jensen GJ (2007) Electron cryotomography of immature HIV-1 virions reveals the structure of the CA and SP1 Gag shells. EMBO J 26(8):2218–2226

Xu B, Kong J, Wang X, Wei W, Xie W, Yu XF (2015) Structural insight into the assembly of human anti-HIV dynamin-like protein MxB/Mx2. Biochem Biophys Res Commun 456(1):197–201

Yamashita M, Emerman M (2004) Capsid is a dominant determinant of retrovirus infectivity in nondividing cells. J Virol 78(11):5670–5678

Yang H, Ji X, Zhao G, Ning J, Zhao Q, Aiken C, Gronenborn AM, Zhang P, Xiong Y (2012) Structural insight into HIV-1 capsid recognition by rhesus TRIM5alpha. Proc Natl Acad Sci U S A 109(45):18372–18377

Yap MW, Nisole S, Lynch C, Stoye JP (2004) Trim5alpha protein restricts both HIV-1 and murine leukemia virus. Proc Natl Acad Sci U S A 101(29):10786–10791

Yeager M, Wilson-Kubalek EM, Weiner SG, Brown PO, Rein A (1998) Supramolecular organization of immature and mature murine leukemia virus revealed by electron cryo-microscopy: implications for retroviral assembly mechanisms. Proc Natl Acad Sci U S A 95(13):7299–7304

Yudina Z, Roa A, Johnson R, Biris N, de Souza Aranha DA, Vieira VT, Reszka N, Taylor AB, Hart PJ, Demeler B, Diaz-Griffero F, Ivanov DN (2015) RING dimerization links higher-order assembly of TRIM5alpha to synthesis of K63-linked Polyubiquitin. Cell Rep 12(5):788–797

Zhang W, Cao S, Martin JL, Mueller JD, Mansky LM (2015) Morphology and ultrastructure of retrovirus particles. AIMS Biophys 2(3):343–369

Zhao G, Ke D, Vu T, Ahn J, Shah VB, Yang R, Aiken C, Charlton LM, Gronenborn AM, Zhang P (2011) Rhesus TRIM5alpha disrupts the HIV-1 capsid at the inter-hexamer interfaces. PLoS Pathog 7(3):e1002009

Zhao G, Perilla JR, Yufenyuy EL, Meng X, Chen B, Ning J, Ahn J, Gronenborn AM, Schulten K, Aiken C, Zhang P (2013) Mature HIV-1 capsid structure by cryo-electron microscopy and all-atom molecular dynamics. Nature 497(7451):643–646

Chapter 9
Nucleoprotein Intermediates in HIV-1 DNA Integration: Structure and Function of HIV-1 Intasomes

Robert Craigie

Introduction

Integration of a DNA copy of the viral genome into the host DNA is an essential step in the replication cycle of HIV-1 and other retroviruses (Coffin et al. 1977). Once integrated, the viral DNA is replicated along with cellular DNA during each cycle of cell division, and populations of long-lived cells with integrated proviruses have obstructed efforts to cure AIDS; although viral replication can now be effectively suppressed by antiviral drugs, these infected cells are a reservoir from which virus reemerges if treatment is interrupted. Biochemical studies have shown that retroviral DNA integration occurs by a mechanism that is shared by a class of DNA mobile genetic elements that are ubiquitous in both prokaryotes and eukaryotes and by retrotansposons. Although the mechanism of DNA integration is closely related among these classes of elements, the source of the DNA to be integrated differs. In the case of DNA transposons, the transpose encoded by the transposon excises the transposon from its original location in the genome and inserts it into a new location. Retrotransposons must first transcribe an RNA copy of their genome which then undergoes an intermediate step of reverse transcription within the same cell to make the DNA copy that is then integrated at a new site. Retroviruses have evolved the additional step of packaging the transcribed viral RNA in the form of a virion that is budded from the infected cell. The virion subsequently infects another cell where reverse transcription and DNA integration occur. Transposons, retrotansposons, and retroviruses share the common feature that, prior to integration, the two ends of the mobile DNA are tightly associated with the enzyme that catalyzes the DNA integration reaction. This protein is called transposase in the case of transposons and integrase in the case of retrotransposons and retroviruses. In retroviruses

R. Craigie (✉)
Laboratory of Molecular Biology, National Institute of Diabetes and Digestive and Kidney Diseases, National Institutes of Health, Bethesda, MD, USA
e-mail: robertc@niddk.nih.gov

© Springer Nature Singapore Pte Ltd. 2018
J. R. Harris, D. Bhella (eds.), *Virus Protein and Nucleoprotein Complexes*,
Subcellular Biochemistry 88, https://doi.org/10.1007/978-981-10-8456-0_9

and retrotransposons, terminal CA dinucleotides are joined to target DNA, whereas terminal sequences are more divergent among transposons. Complexes between integrase and viral DNA are collectively termed intasomes. Retroviral intasomes undergo a series of transitions between initial formation and catalysis of the DNA cutting and joining steps of DNA integration. Here, we focus on our current knowledge of the structure and function of HIV-1 intasomes, with reference to related systems as required to put this knowledge in context. First, we review key discoveries that led to the recent breakthroughs with high-resolution structural studies of HIV-1 intasomes.

Mechanism of DNA Integration

The Preintegration Complex (PIC)

The establishment of an in vitro system for retroviral DNA integration by Brown and colleagues in 1987 (Brown et al. 1987) was a pivotal step in biochemical studies of retroviral DNA integration. They discovered that cytoplasmic extracts of cells infected with Moloney murine leukemia virus (MoMLV) supported integration of the viral DNA made by reverse transcription into an exogenously added plasmid DNA in vitro. Importantly, the integrated viral DNA was flanked by a 4 bp repeat of target DNA at the site of integration, a hallmark of correct MoMLV DNA integration. Integration activity sedimented as a very large nucleoprotein complex with an S value of approximately 160S (Bowerman et al. 1989); for comparison, the S value of eukaryotic ribosomes is 80S. These complexes have been termed preintegration complexes (PICs). HIV-1 DNA was subsequently shown to form part of similarly large preintegration complexes (Ellison et al. 1990; Farnet and Haseltine 1990). In addition to viral DNA, PICs contain many proteins derived from the infecting virion and cellular proteins acquired from the cytoplasm of the infected cell (Farnet and Haseltine 1991; Bukrinsky et al. 1993; Li et al. 2001). The organization and composition of PICs is still poorly defined because their low abundance in extracts of infected cells limits the types of studies that can be attempted. It is likely that components of the PIC are jettisoned between initial formation after reverse transcription in the cytoplasm and transport to the nucleus for integration into cellular DNA. However, functional studies of integration activity clearly show that integrase must be tightly associated with the viral DNA ends because integrase is the enzyme that catalyzes integration (see below) and integration activity is retained upon treating PICs with greater than 0.5 M NaCl and separating them from loosely bound protein factors by gel filtration or sedimentation. These complexes of integrase associated with the viral DNA ends (intasomes) form a substructure of the PIC and are the topic of this chapter. Since the low abundance of PICs in infected cells precludes structural studies, recent progress in this area has built upon biochemical studies of intasomes assembled from purified components in vitro.

DNA Cutting and Joining Steps

In vitro DNA integration assays using PICs isolated from infected cells provided the first definitive evidence that the immediate precursor DNA for integration is linear. At the time, retroviral DNA was known to exist in both linear and circular forms, and the circular form was a strong candidate based on analogy with the extensively studied bacteriophage lambda integration system. Analysis of the in vitro product of DNA integration revealed that only the 3' ends of the viral DNA are joined to target DNA at the site of integration. The unjoined 5' ends of the viral DNA extended two nucleotides beyond the CA terminal nucleotides on the joined viral DNA strands, consistent with a linear viral precursor, and not four nucleotides as would be expected if the viral DNA were circularized prior to integration (Fujiwara and Mizuuchi 1988; Brown et al. 1989). The DNA cutting and joining steps deduced from these studies are depicted in Fig. 9.1. In the first step, 3' end processing, two nucleotides are generally removed from the 3' ends of the initially blunt-ended linear viral DNA (Fig. 9.1b). A target DNA is recruited (Fig. 9.1c), and then in the second step, DNA strand transfers and the hydroxyl groups at each 3' end of the viral DNA attack a pair of phosphodiester bonds in the target DNA. The sites of attack on the two target DNA strands are separated by five nucleotides in the case of HIV-1.

In the integration intermediate (Fig. 9.1d), only the 3' ends of the viral DNA are joined to target DNA, and the 5' ends of the viral DNA remain unjoined with a 5 bp overhang. Completion of integration requires removal of each overhanging dinucleotide, filling in the single-strand gaps between viral and target DNA by a polymerase and ligation. The enzymes that carry out these steps have not been determined, but the reactions can be readily carried out in vitro by the known activities of cellular nucleases, polymerases, and ligases (Yoder and Bushman 2000).

Integrase

Discovery of Integrase

Early evidence for a key role of a virally encoded enzyme in retroviral DNA integration came from genetic experiments (Panganiban and Temin 1984; Donehower and Varmus 1984; Schwartzberg et al. 1984). Mutations that abolished integration, but allowed infection to proceed normally upon completion of the viral DNA synthesis by reverse transcription, mapped to 3' end of the viral *pol* gene. This part of *pol* includes a protein that we now know catalyzes the key steps of DNA integration and is called integrase (IN). Mutations that disrupt the termini of the viral DNA resulted in an essentially identical phenotype, suggesting that this protein acts on the viral DNA ends to promote integration, even before any biochemical evidence was obtained (Panganiban and Temin 1983; Colicelli and Goff 1985; Roth et al. 1989). However, biochemical studies of integrase actually began much earlier when

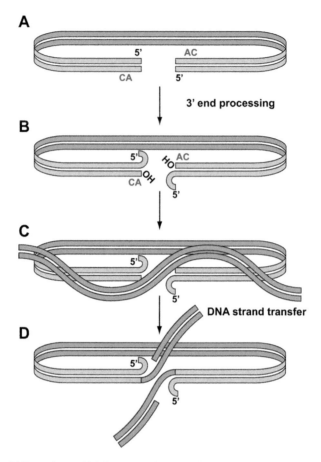

Fig. 9.1 The DNA cutting and joining steps of retroviral DNA integration. The viral DNA made by reverse transcription is linear and blunt ended (Fig. 9.1a). In the first step, 3′ end processing, two nucleotides are removed from each 3′ end of the viral DNA (Fig. 9.1b). This exposes the terminal CA dinucleotide (colored in red) that is conserved among all retroviruses. In the second step, DNA strand transfer, the 3′ hydroxyl groups at the two viral DNA ends attack a pair phosphodiester bonds on each strand of target DNA. In the case of HIV-1, five base pairs separate the sites of attack on the two target DNA strands. In the resulting integration intermediate (Fig. 9.1d), the 3′ ends of the viral DNA are covalently joined to target DNA, while the 5′ ends of the viral DNA and the 3′ ends of the target DNA are unjoined. The viral integrase protein catalyzes both 3′ end processing and DNA strand transfer. Cellular proteins complete the integration process by removing the overhanging dinucleotide at the 5′ ends of the viral DNA, filling in the 5 bp single-strand connection and ligation of the 5′ ends of the viral DNA to the 3′ ends of the unjoined target DNA strand. This figure is reproduced from Fig. 1 of Craigie (2012)

Grandgenett Grandgenett et al. (1978) and colleagues identified a protein from avian sarcoma leucosis virus (ASLV) (p33) that had a DNA-nicking activity on substrate corresponding to the circularized form of ASLV DNA. This protein was later identified as the ASLV integrase protein.

Early In Vitro Assay Systems

The finding that detergent-disrupted MoMLV virions could support in vitro integration of linear DNA with termini that mimic the viral DNA ends (Fujiwara and Craigie 1989) suggested that integrase might be sufficient to carry out the DNA cutting and joining steps of integration. This was confirmed by assays using purified integrase protein of MoLV (Craigie et al. 1990), ASLV (Katzman et al. 1989) (Katz et al. 1990) and HIV-1 (Sherman and Fyfe 1990; Bushman et al. 1990; Engelman et al. 1991) as the only protein factor. The first assay systems used a linear "mini-viral DNA" substrate with phage lambda DNA as the target for integration (Fujiwara and Craigie 1989; Bushman et al. 1990). Integration events were scored by in vitro packaging of the lambda DNA, infection of E. coli, and selection for a marker present on the mini-viral DNA substrate. Subsequent assays used oligonucleotides mimicking the ends of the viral DNA as the viral DNA substrate; oligonucleotides also served as the target DNA for integration, and the products were detected by gel electrophoresis (Fig. 9.2). In the first step of the reaction (3′ end processing), two nucleotides are cleaved from the 3′ end of the viral DNA substrate to expose the CA-3′ ends that are to be joined to target DNA (Fig. 9.2a). In the next step (DNA strand transfer), the hydroxyl groups at the termini exposed by the 3′ end processing reaction attack a pair of phosphodiester bonds in the target DNA (Fig. 9.2b). Integration can occur at essentially any position along the target DNA, so the products are heterogeneous in length. In the case of HIV-1, the products of DNA strand transfer produced in such assays resulted largely from integration of a single viral DNA end into a single strand of target DNA rather than two viral DNA ends integrated into each target DNA strand with the sites of integration separated by 5 bp as occurs in the normal DNA strand transfer reaction. A separate in vitro activity, termed disintegration, represents the chemical reversal of DNA strand transfer (Chow et al. 1992). Although not thought to occur during virus infection, disintegration activity provided useful information on the domain organization of integrase protein (see below).

Chemistry of DNA Integration

The development of simple assay systems with purified components and a physical readout ignited detailed biochemical studies of integration and laid the foundation to the development of high-throughput screens for inhibitors of integrase. Many

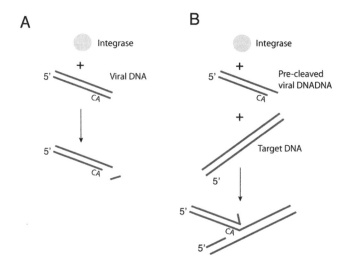

Fig. 9.2 HIV-1 integration assay with oligonucleotide DNA substrates. (**a**) 3′ end processing. In the presence of a divalent metal ion, integrase cleaves two nucleotides from the 3′ ends of the viral DNA (shown in red) to expose the terminal CA-$_{OH}$ that, in the next step, attacks a phosphodiester bond in the target DNA. (**b**) DNA strand transfer. In the presence of a divalent metal ion and an oligonucleotide lacking the dinucleotide that is normally removed in the 3′ end processing reaction, integrase catalyzes a phosphoryl transfer reaction that covalently joins viral to target DNA (shown in blue). Cleaved oligonucleotides produced by the 3′end processing reaction can also go on to carry out DNA strand transfer. The sites of integration into target DNA are random to a first approximation, although there are sequence preferences. This in vitro assay is a "loosened up" version of the fully authentic reaction in which a pair of viral DNA ends are joined to target DNA on opposite strands. Integrase, which functions as a multimer, is depicted as a circle for simplicity

DNA recombinases catalyze recombination by cleaving DNA and conserving bond energy by forming a covalent intermediate with one of the cleaved DNA ends. However, in the case of retroviral DNA integration, the 3′ hydroxyl of the DNA strand to be joined is exposed prior to the joining reaction. This raised the possibility that DNA strand transfer might occur by a one-step transesterification reaction in which phosphodiester bond cleavage and formation of the new phosphodiester bond occur in one step. This was tested by incorporating chiral phosphorothioate in the target DNA and monitoring the change in chirality after the reaction. If a covalent protein-DNA intermediate is involved, two reaction steps would be required, and the chirality would be retained, whereas a single-step transesterification would result in inversion of chirality. The chirality was found to invert, showing that integration occurred by a one-step transesterification mechanism (Engelman et al. 1991). In vitro, the product of 3′ end processing exists as both a linear dinucleotide and a circular dinucleotide in which the 3′ end of the viral DNA itself bends back and attacks the scissile phosphodiester bond. This enabled the stereochemical course of the 3′ end processing reaction to also be followed by putting a chiral phosphorothioate at the site of cleavage. The chirality inverted during the course of this

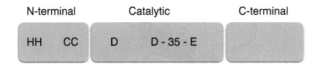

Fig. 9.3 The three domains of HIV-1 integrase. The catalytic core domain is highly conserved in retroviral integrases, retrotransposon integrases, and transposases of many DNA transposons. It includes a triad of conserved acidic residues, the D,D-35-E motif, that are key residues of the active site. The N-terminal domain is conserved among retroviral and retrotransposon integrases and contains a pair of conserved His and Cys residues, the HHCC motif, that coordinate a Zn^{2+} ion. The C-terminal domain is less well conserved than the other two domains

reaction as well, demonstrating that 3′ end processing also occurs by a one-step mechanism (Engelman et al. 1991).

Domain Organization and Early Mutagenesis Studies

Partial proteolysis studies revealed that HIV-1 integrase is comprised of three domains (Engelman and Craigie 1992)(Fig. 9.3). Protein sequence alignment revealed three amino residues, D64, D116, and E152 that are absolutely conserved among retroviral integrases, retrotransposon integrases, and some classes of transposases. Mutation of any of these residues abolished 3′ processing and DNA strand transfer activity (Kulkosky et al. 1992; Engelman and Craigie 1992; van Gent et al. 1992; Leavitt et al. 1993). These studies, together with inversion of chirality for both DNA strand transfer and 3′ end processing, suggested that both reactions share a common chemical mechanism and are catalyzed by the same active site. The residues of the D,D–35-E motif were proposed to coordinate divalent metal ions for catalysis (Kulkosky et al. 1992), and this role was subsequently confirmed by structural studies. The N-terminal domain is well conserved among retroviral integrases and contains a pair of His and Cys residues, the HHCC motif, that binds zinc (Zheng et al. 1996). The C-terminal domain is the least conserved of the three domains.

HIV-1 Integrase Domain Structures

Catalytic Domain Structure

Structural studies of HIV-1 integrase and other retroviral integrases have been impeded by the propensity of the protein to aggregate, and early attempts at crystallization were unsuccessful. This led investigators to focus their efforts on individual domains. The catalytic core domain of HIV-1 integrase, like the full-length proteins, aggregates in solution. However, the mutation F185 K in the catalytic domain was found to result in a protein that is much more soluble than the wild-type catalytic

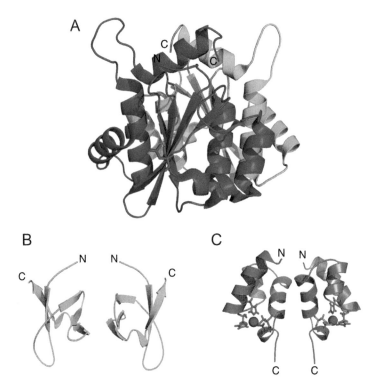

Fig. 9.4 Structures of the domains of HIV-1 integrase. (**a**) The catalytic domain (pdb: 1ITG) is dimeric with an extensive mainly hydrophobic interface of 1300 Å². The monomers have a fold consisting of a five-stranded β-sheet surrounded by helices, a fold that was first observed in RNAase H and subsequently in many polynucleotidyl transferases. The active site residues D64 and D116 are shown in red. The third residue of the D,D35, E motif, E152 is disordered in this structure. The active sites are located on opposite sides of the nearly spherical dimer. (**b**) The C-terminal domain (pdb: 1IHV). The C-terminal domain is a dimer in solution with each subunit comprised of a five-stranded β-barrel resembling an SH3 domain. (**c**) The N-terminal domain (pdb: 1WJV). It is a four-helix bundle with Zn²⁺ (red spheres) coordinated to His12, His16, Cys40, and Cys43. The structure is dimeric with a predominantly hydrophobic interface

domain and retains disintegration activity in vitro (Jenkins et al. 1995). The domain with this mutation was successfully crystallized, and the structure was solved (Dyda et al. 1994). The structure contained a five-stranded beta sheet flanked by helices (Fig. 9.4a), with two of the previously identified catalytic residues in close proximity, while the third catalytic residue was on a disordered loop. The structure established that integrase belongs to a superfamily of polynucleotidyl transferases that at the time included ribonuclease H (Yang et al. 1990) and the Holliday junction resolvase RuvC (Ariyoshi et al. 1994) but has subsequently expanded to include many additional enzymes. Unfortunately, the structure provided little information on how integrase interacts with DNA. The two active sites in the structure, which were located on opposite sides of the nearly spherical dimer, were separated by approximately 30 Å. This spacing is incompatible with the 5 bp spacing of the sites of joining of the two viral DNA ends to two strands of target DNA and suggested that the active multimer must be larger than a dimer.

C-Terminal Domain Structure

The structure of the C-terminal domain was solved by NMR (Lodi et al. 1995). The protein was a dimer in solution with each subunit comprised of a five-stranded β-barrel with a topology similar to the Src-homology 3 (SH3) domain (Fig. 9.4b).

N-Terminal Domain Structure

The N-terminal domain structure was also determined by NMR (Cai et al. 1997). The structure was dimeric with each subunit comprising a four-helix bundle with zinc tetrahedrally coordinated to the conserved residues of the HHCC motif, H12, H16, C40, and C43 (Fig. 9.4c). The two subunits in the dimer are arranged parallel to each other, and the interface is mainly hydrophobic. The N-terminal domain of HIV-2 integrase has an essentially similar structure (Eijkelenboom et al. 1997).

Multi-domain Structures

The first multi-domain integrase structure to be solved was the catalytic domain together with the C-terminal domain (Chen et al. 2000a). The structure was a Y-shaped dimer (Fig. 9.5a). The catalytic domain formed the only dimer interface, and this interface was essentially the same as in in the isolated catalytic domain dimer. The C-terminal domains were monomeric, 55 Å apart and linked to the catalytic domain by alpha helices. The second multi-domain structure to be determined was the N-terminal together with catalytic domain (Wang et al. 2001). This structure is dimeric with the same catalytic domain dimer interface as observed in earlier structures (Fig. 9.5b). Although dimerized, the N-terminal dimer interface was different from that of the solution structure of the N-terminal domain alone. The lack of density for the linker connecting the two domains did not allow unambiguous assignment of the connectivity between the N-terminal and catalytic domains. The missing linker can easily span the distance in the arrangement shown at the top of Fig. 9.5b. The connectivity shown below cannot be excluded, although it would require extreme stretching of the linker. However, the N-terminal domain is located in a similar flanking position in the structure of the N-terminal together with the catalytic domain of HIV-2 integrase (Hare et al. 2009; Jaskolski et al. 2009).

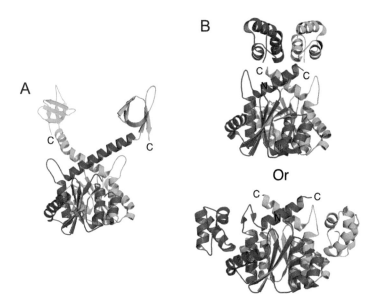

Fig. 9.5 Two-domain structures of HIV-1 integrase. (**a**) Catalytic together with C-terminal domain (pdb: 1EX4). The catalytic domain shares the same dimer interface as in the structure of the isolated catalytic domain. The C-terminal helix of the isolated catalytic domain structure extends through the linker region to position the C-terminal domains far from the catalytic domain dimer, and the dimer interface seen in the solution structure of the isolated C-terminal domain is not present. The catalytic domains and linkers are colored in cyan and magenta and the C-terminal domains in green. (**b**) N-terminal together with the catalytic domain (pdb: 1K6Y). The absence of density in the linker region makes the assignment of connectivity ambiguous. The connectivity shown at the top is easily accommodated by the length of the missing linker region. The alternative connectivity shown below is possible, but the linker would be required to adopt an extremely stretched conformation. However, it is worth noting that this connectivity has been observed in the corresponding structure of HIV-2 integrase (Hare et al. 2009). The subunits are colored magenta and cyan (**b**)

The structures of the isolated domains and two-domain structures provided high-resolution information on the individual domains but provided few clues as how viral DNA assembles with integrase in the intasome.

Intasomes

Intasomes

Intasome is a collective term for stable nucleoprotein complexes on the retroviral DNA integration reaction pathway (Fig. 9.6) and is analogous to transpososomes that mediate transposition of a class of DNA transposons. The first intasome on the reaction pathway is the stable synaptic complex (SSC) which comprises a pair of

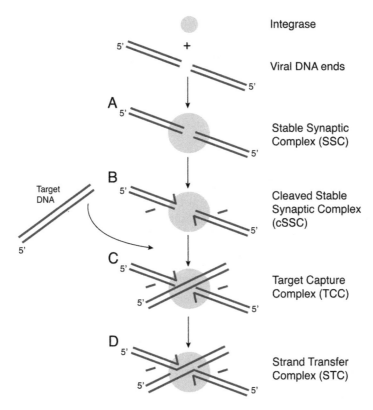

Fig. 9.6 Retroviral DNA integration is mediated by a series of intasomes along the reaction pathway. (**a**) Integrase, together with a pair of viral DNA ends, in the presence of divalent metal ions, forms a stable synaptic complex (SSC), the first intasome on the reaction pathway. (**b**) Within the SSC, integrase cleaves two nucleotides from each 3′ end of the viral DNA to form the cleaved stable synaptic complex (cSSC). (**c**) The cSSC then captures a target DNA molecule into which the viral DNA ends are to be integrated to form the target capture complex (TCC). This initial association of the cSSC with target DNA is non-covalent. (**d**) The hydroxyl groups at the 3′ ends of the viral DNA attack a pair of phosphodiester bonds in the target DNA to accomplish DNA strand transfer and form the strand transfer complex (STC). Completion of integration requires disassembly of the STC, removal of the two overhanging nucleotides from the 5′ ends of the viral DNA, filling in of the single-strand connections between viral and target DNA by a polymerase, and ligation. These latter steps are carried out by cellular enzymes

viral DNA ends bridged by integrase. Within the SSC, two nucleotides are removed from the 3′ ends of the viral DNA to form the cleaved stable synaptic complex (cSSC). The cSSC then non-covalently captures a target DNA to form the target capture complex (TCC). Finally, within the TCC, a pair of transesterification reactions occurs to form the strand transfer complex (STC).

Biochemical Studies of HIV-1 Intasomes

The first indication that the ends of the viral DNA are stably bridged by integrase in the preintegration complex (PIC) came with functional studies of Moloney murine leukemia virus PICs. PICs incubated at high ionic strength and separated from released protein factors by sedimentation or gel filtration retain integration activity, suggesting that integrase remains associated with the viral DNA ends (Lee and Craigie 1994).

Early biochemical studies HIV-1 DNA integration used reaction conditions in which most of the integration products resulted from "half-site" integration of a single viral DNA end into one strand of target DNA. Under these conditions, highly stable complexes of integrase with viral DNA ends were not detected. However, under improved reaction conditions that support concerted integration of pairs of viral DNA ends (Hindmarsh and Leis 1999; Sinha et al. 2002; Li and Craigie 2005; Sinha and Grandgenett 2005), highly stable complexes of integrase and viral DNA ends (SSC intasomes) are formed (Li et al. 2006; Pandey et al. 2007). Although the overall efficiency of concerted two-end DNA integration in vitro is low, the limiting step is formation of intasomes; once assembled essentially all the intasomes successfully complete DNA strand transfer to form STCs (Li et al. 2006).

Overcoming Obstacles to High-Resolution Structural Studies of HIV-1 Intasomes

Although the structures of all the individual domains of HIV-1 integrase were solved by the mid-1990s, it would be almost two decades before the first HIV-1 intasome structures were determined. The early biochemical assays that resulted in mainly single-site integration appear to be a "loosened-up" reaction system in which the chemical steps of integration occur normally, but stable intasomes are bypassed. Improvements in in vitro assay systems were required for intasomes to be detected biochemically. The biggest obstacle to structural studies was the tendency of integrase and intasomes to aggregate in solution. This problem was partly overcome by the finding that fusing the DNA-binding domain of *Sulfolobus solfataricus* chromosomal protein Sso7d to the N-terminus of HIV-1 integrase resulted in a protein that was not only hyperactive in vitro but also assembled intasomes with greatly improved solubility properties (Li et al. 2014). The fusion protein is also active with oligonucleotide viral DNA substrate, whereas the wild-type protein requires several hundred additional base pairs of non-specific DNA for integration activity and assembly of intasomes (Li and Craigie 2005; Li et al. 2006). The reasons for these differences are not understood. Importantly, the Sso7d fusion IN is active in the virus, albeit with a somewhat lower efficiency than wild-type HIV-1 IN (Li et al. 2014).

Fig. 9.7 Schematic depiction of the PFV intasome. The PFV intasome is a tetramer in which only the inner two protomers of the tetramer (*blue* and *tan*) interact with viral DNA (*red*). The N-terminal extension domain (NED), N-terminal (NTD), and C-terminal domains of the outer protomers are disordered

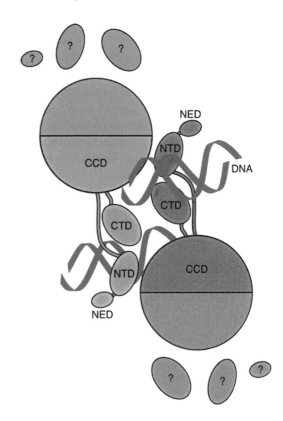

Structure of Prototype Foamy Virus (PFV) Intasomes

A major breakthrough in structural studies of intasomes came when the Cherepanov group solved the crystal structures of the prototype foamy virus (PFV) SSC and STC intasomes (Hare et al. 2010a; Maertens et al. 2010). PFV is a relative of HIV-1 with relatively low sequence similarity but shares predicted secondary structures and key active site residues (Valkov et al. 2009). It also has an additional small domain at the N-terminus, the N-terminal extension domain (NED), that is not present in HIV-1 integrase. PFV integrase is much more active than HIV-1 integrase in vitro, and, unlike HIV-1 integrase, both PFV integrase and PFV intasomes are highly soluble. The PFV SSC structure is a tetramer comprised of a dimer of dimers (Fig. 9.7). Each dimer has essentially the same catalytic dimer as in the isolated HIV-1 integrase catalytic domain structure. All the interactions with viral DNA occur with the domains of the inner subunits within the tetramer. The N-terminal domains of the two inner subunits are exchanged and extend to the active site region of each other. The NED, N-terminal, and C-terminal domains of the outer subunits are disordered in the structure. DNA-protein interactions play a dominant role and act as the "glue" that stabilizes the intasome. Such a structure could not have been predicted from the structures of the isolated integrase domains. Structures of PFV intasomes that have been determined importantly include all complexes on the reaction pathway (SSC, cSSC, TCC, and STC) (Hare et al. 2010a; Maertens et al. 2010; Hare et al. 2012) as well as the cSSC in complex with DNA strand transfer inhibitors (Hare et al. 2010a, b). The structure of the cSSC in complex with compounds such as raltegravir elucidated the mechanism of this class of inhibitor. Binding of the inhibitor displaces the attacking 3′ hydroxyl from the active site. Although the PFV structures are excellent for modeling the interaction of HIV-1 integrase inhibitors, sequence differences necessitate structures of HIV-1 intasomes in complex with inhibitors to understand the detailed molecular interactions of HIV-1 intasomes with inhibitors and the structural basis for clinical drug resistance. This is especially true for understanding resistance mutations away from the immediate vicinity of the active site, where differences in the two proteins make modeling less reliable.

Structure of HIV-1 Intasomes and Implications

Structure of HIV-1 Intasomes

Fusion of Sso7d to the N-terminus of HIV-1 IN (Sso7d-IN) was a key step that facilitated the first HIV-1 intasome structures. The other strategy that contributed was the assembly of HIV-1 STC intasomes on branched DNA substrate that mimics the product of DNA strand transfer. It had previously been shown that PFV STCs assembled on such a DNA substrate were identical to STCs made through the normal forward reaction pathway (Yin et al. 2012). HIV-1 STCs aggregate less than HIV-1 SSCs and were therefore chosen for the first structural studies of HIV-1 intasomes. The assembly efficiency is relatively low, and multiple purification steps were required to obtain material that was suitable for structural studies. After a final step of gel filtration, the intasomes appeared relatively homogeneous as judged by the gel filtration profile, although the peak exhibited a slight hump suggesting the presence of more than one species (Li et al. 2014). Attempts to crystallize these intasomes failed. With hindsight, crystallization was unlikely to work as we know multiple species are present, even after purification.

The STC intasomes required the presence of high ionic strength (0.5 M NaCl) and glycerol in the buffer to prevent aggregation. Such conditions have generally been regarded as unfavorable for cryo-electron microscopy (cryoEM). However, recent advances in cryoEM techniques partially negate these factors, and cryoEM was highly successful in determining the structure of HIV-1 intasomes even with less than optimal buffer conditions (Passos et al. 2017).

The initial studies focused on the smaller species with the erroneous assumption that larger species were simply aggregates of the same basic unit. The density map resolved to 3.5 Å to 4.5 Å, with the highest resolution being in the core of the structure around the active site. The structure was tetrameric and essentially the same as the previously determined PFV intasome structure (Fig. 9.8). The intasome com-

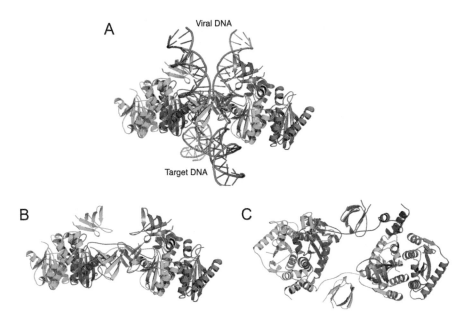

Fig. 9.8 Structure of the tetrameric HIV-1 STC intasome (pdb: 5U1C). Most of the contacts between protein and DNA are made by the inner protomers, colored in magenta and cyan. The C-terminal domains of the outer protomers (colored green and orange) interact with viral DNA, while the N-terminal domains of the outer protomers are disordered. The overall architecture is similar to that of PFV intasomes except for the additional contact of the C-terminal domain of the outer protomers with viral DNA. (**b**) The same view with the DNA removed. (**c**) As in B, it is viewed from the top. The large empty space in the middle that is occupied by DNA (not shown) highlights the critical role of protein-DNA interactions in holding the intasome together

prised a dimer of dimers. Like the PFV intasome, most of the interactions with DNA are mediated by the inner protomer, and the N-terminal domains of each inner protomer extend out to the active site region of the other inner protomer. Unlike the PFV intasome, the C-terminal domains of the outer protomers were resolved and contribute to the interaction with viral DNA. The N-terminal domains of the outer protomers were disordered as in PFV intasomes. The target DNA is distorted from B form, as is the target DNA in PFV STC intasomes, with five bp between the active sites as expected from the target site duplication. The Sso7d domain that was added to the N-terminus of integrase to improve the solubility of the intasomes was disordered.

In addition to tetrameric STC intasomes, cryoEM revealed higher-order species, the best resolved of which contained 12 integrase protomers. This higher-order STC intasome has the same arrangement of domains interacting with DNA as in the tetrameric STC. However, some of these "positionally conserved" domains are contributed by additional subunits in the higher-order STC (Fig. 9.9).

Tetrameric STC   Higher-order STC

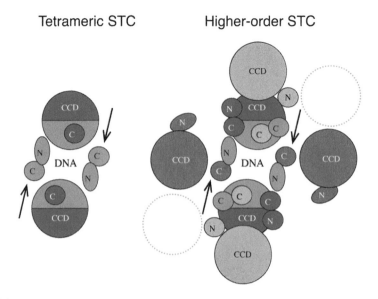

Fig. 9.9 Schematic representation of the domain arrangements in the HIV-1 tetrameric and higher-order STC intasomes. The domains of the inner protomers are colored light blue and the outer protomers in dark blue. Arrows show the positionally conserved C-terminal domains. In the tetrameric structure, they are contributed by the inner protomers, whereas in the higher-order structure, they are contributed by flanking dimers. Dotted circles represent regions of poorly defined density. This figure is reproduced from supplementary Fig. S16 of Passos et al. (2017)

Reinterpreting Earlier Domain Structures with Hindsight of the HIV-1 STC Intasome Structure

Earlier structures of individual domains and two-domain structures of HIV-1 integrase provided few clues as to intasome structure. Although many modeling studies were attempted based on these structures, none of these came close to predicting the functional DNA-integrase structure. This is hardly surprising given the role of protein-DNA interactions in organizing the intasome, and none of the earlier structures contained DNA. The catalytic domain dimer interface was the only interface consistently observed - in the structures of retroviral integrases in the absence of DNA. The isolated HIV-1 integrase N-terminal domain was a dimer in solution (Cai et al. 1997). In the context of the N-terminal together with the catalytic domain, although the N-terminal domain dimerized, the interface was entirely different from that observed with the isolated N-terminal domain; additionally, the N-terminal domain dimer was packed against the catalytic domain (Wang et al. 2001). A similar ambiguity was seen with the C-terminal domain. Whereas the isolated C-terminal domain was a dimer in solution (Lodi et al. 1995), in the structure of the catalytic together with the C-terminal domain, the C-terminal domain was monomeric and separated from the catalytic domain by an alpha-helical linker (Chen et al. 2000a). To complicate the picture further, in the structure of the catalytic domain together with the

C-terminal domain of simian immunodeficiency virus (SIV) integrase, only one of the C-terminal domains was discernable, and it was packed against the catalytic domain (Chen et al. 2000b). The variety of domain arrangements seen in these structures was generally thought to reflect the flexibility of the linkers between domains and not necessarily be directly relevant to intasome architecture. It is striking that many of these interfaces seen in the absence of DNA are present in the HIV-1 STC intasome structures. For example, in the higher-order STC, positionally conserved domains share the interface observed in the structure of the N-terminal together with a catalytic domain. Strikingly, in the higher-order STC, the linker between the catalytic and C-terminal domain is helical and extends away from the catalytic domain as in the structure of the catalytic together with C-terminal domain in the absence of DNA. Two of the C-terminal domains also dimerize exactly as in the NMR structure of the isolated C-terminal domain. All these structures together highlight the plasticity of integrase in using a common set of interfaces to assemble different structures.

Comparison with Intasome Structures of Other Retroviruses

The PFV SSC was the first intasome structure to be solved for any retrovirus (Hare et al. 2010a). Given the similarity of PFV integrase to other retroviral integrases, it seemed reasonable to assume that other retroviral intasomes would have a similar organization. However, in PFV intasomes the C-terminal domains of the protomers involved in interactions with the viral and target DNA extend away from the catalytic domain of one inner protomer toward- the catalytic domain of the other inner protomer (Fig. 9.7). This is possible because PFV integrase has a long linker between the catalytic and C-terminal domain. This represented a conundrum because the catalytic to C-terminal domain linkers of some retroviral integrases are too short to accommodate the same arrangement of domains as in the PFV tetrameric intasomes (Fig. 9.10). The solution became apparent when the structures of *Rous sarcoma virus* (RSV) (Yin et al. 2016) and mouse mammary tumor virus (MMTV) intasomes (Ballandras-Colas et al. 2016) were determined. The intasomes had the same arrangement of domains around the viral DNA as PFV intasomes, but the positionally conserved C-terminal domains were contributed by an additional pair of flanking dimers in an octameric intasome. Maedi visna virus (MVV) intasomes were found to be hexadecameric (Ballandras-Colas et al. 2017) but again had the same set of positionally conserved domains interacting with a DNA. Structures of both tetrameric and hexadecameric intasomes have been determined for HIV-1. The HIV-1 hexadecamer has a weak density in positions of the extra protomers in the MVV hexadecamer; it is therefore possible that this HIV-1 higher-order intasome is a hexadecamer like MVV, with some protomers disordered. Biophysical studies also suggest the presence of additional higher-order intasome species, the structures of which are yet to be determined. At the time of writing, the functional significance of multiple species of HIV-1 intasomes is unclear. However, they all

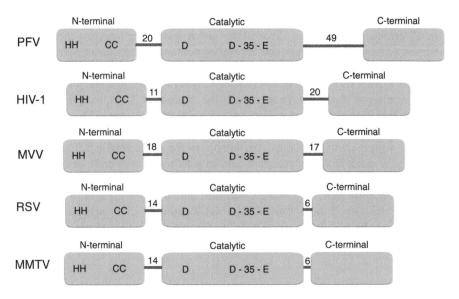

Fig. 9.10 Comparison of the linker lengths in the following retroviral integrases: HIV-1, PFV, maedi visna virus (MVV), *Rous sarcoma virus* (RSV), and mouse mammary tumor virus (MMTV). Whereas the N-terminal to catalytic domain linkers exhibit little variation in length, the catalytic to C-terminal domain linkers range from 49 residues in PFV to 6 residues in RSV and MMTV. Relative linker lengths are drawn to scale.The NED of PFV integrase is omitted for clarity

use essentially the same protein-protein and protein-DNA interfaces at the core of the structure but differ in the protomers that contribute the positionally conserved domains.

Beyond Intasomes

HIV-1 intasomes are sufficient for catalysis of the chemical steps of HIV-1 DNA integration in the absence of other protein factors. Although outside the scope of this chapter, it would be remiss to not mention the important role of other viral and cellular protein factors in the overall DNA integration process. Two of the steps that clearly involve recruitment of such factors are nuclear entry (reviewed in (Matreyek and Engelman 2013)) and targeting of intasomes to specific regions of chromatin (reviewed in (Debyser et al. 2015)). One of the best understood cellular factors in HIV-1 DNA integration is lens epithelium-derived growth factor p75 splice variant (LEDGF). LEDGF contains a domain that binds HIV-1 integrase and a domain that binds chromatin. Intasomes are therefore preferentially targeted to regions of chromatin that are enriched for LEDGF, and this is a major mechanism behind the preferential integration of HIV-1 in transcriptionally active chromatin. Roles for many other cellular factors in retroviral DNA integration can be expected to be uncovered.

Perspectives and Drug Discovery

The first clinically approved HIV-1 inhibitors were developed using lead compounds identified by high-throughput screens independent of any structural information on HIV-1 integrase. They were found to target cSSC intasomes rather than free integrase protein. The structure of the PFV cSSC in complex with inhibitors showed that they clearly work by displacing the 3'-OH which is poised to attack the target DNA away from the active site. Sequence similarity allows modeling of the HIV-1 in the intasome active site based on the PFV intasome structure (Krishnan et al. 2010; Johnson et al. 2013). However, sequence divergence makes modeling away from the immediate vicinity of the active site less reliable. Structures of HIV-1 cSSC intasomes will be required to understand the detailed mechanism of action of HIV-1 integrase inhibitors and how mutations can confer drug resistance.

Acknowledgments We thank Alan Engelman and Wei Yang for valuable input to the manuscript. This work was supported by the Intramural Program of the National Institute of Diabetes and Digestive Diseases of the National Institutes of Health and by the Intramural AIDS Targeted Antiviral Program of the Office of the Director of the NIH.

References

Ariyoshi M, Vassylyev DG, Iwasaki H, Nakamura H, Shinagawa H, Morikawa K (1994) Atomic structure of the RuvC resolvase: a holliday junction-specific endonuclease from E. Coli. Cell 78:1063–1072

Ballandras-Colas A, Browne M, Cook NJ, Dewdney TG, Domeler B, Cherepanov P, Lyumkis D, Engelman AN (2016) Cryo-EM reveals a novel octameric integrase structure for betaretroviral intasome function. Nature 530(7590):358. https://doi.org/10.1038/nature16955

Ballandras-Colas A, Maskell DP, Serrao E, Locke J, Swuec P, Jonsson SR, Kotecha A, Cook NJ, Pye VE, Taylor IA, Andresdottir V, Engelman AN, Costa A, Cherepanov P (2017) A supramolecular assembly mediates lentiviral DNA integration. Science 355(6320):93–95. https://doi.org/10.1126/science.aah7002

Bowerman B, Brown PO, Bishop JM, Varmus HE (1989) A nucleoprotein complex mediates the integration of retroviral DNA. Genes Dev 3:469–478

Brown PO, Bowerman B, Varmus HE, Bishop JM (1987) Correct integration of retroviral DNA in vitro. Cell 49:347–356

Brown PO, Bowerman B, Varmus HE, Bishop JM (1989) Retroviral integration: structure of the initial covalent product and its precursor, and a role for the viral IN protein. Proc Natl Acad Sci U S A 86:2525–2529

Bukrinsky MI, Sharova N, McDonald TL, Pushkarskaya T, Tarpley WG, Stevenson M (1993) Association of integrase, matrix, and reverse transcriptase antigens of human immunodeficiency virus type 1 with viral nucleic acids following acute infection. Proc Natl Acad Sci U S A 90:6125–6129

Bushman FD, Fujiwara T, Craigie R (1990) Retroviral DNA integration directed by HIV integration protein in vitro. Science 249:1555–1558

Cai M, Zheng R, Caffrey M, Craigie R, Clore GM, Gronenborn AM (1997) Solution structure of the N-terminal zinc binding domain of HIV-1 integrase. Nat Struct Biol 4:567–577

Chen JCH, Krucinski J, Miercke LJW, Finer-Moore JS, Tang AH, Leavitt AD, Stroud RM (2000a) Crystal structure of the HIV-1 integrase catalytic core and C- terminal domains: a model for viral DNA binding. Proc Natl Acad Sci U S A 97:8233–8238

Chen ZG, Yan YW, Munshi S, Li Y, Zugay-Murphy J, Xu B, Witmer M, Felock P, Wolfe A, Sardana V, Emini EA, Hazuda D, Kuo LC (2000b) X-ray structure of simian immunodeficiency virus integrase containing the core and C-terminal domain (residues 50-293) - an initial glance of the viral DNA binding platform. J Mol Biol 296:521–533

Chow SA, Vincent KA, Ellison V, Brown PO (1992) Reversal of integration and DNA splicing mediated by integrase of human immunodeficiency virus. Science 255:723–726

Coffin JM, Hughes SH, Varmus HE (1977) Retroviruses. Cold Spring Harbor Laboratory Press, New York

Colicelli J, Goff SP (1985) Mutants and pseudorevertants of Moloney murine leukemia virus with alterations at the integration site. Cell 42:573–580

Craigie R (2012) The molecular biology of HIV integrase. Future Virol 7(7):679–686. https://doi. org/10.2217/fvl.12.56

Craigie R, Fujiwara T, Bushman F (1990) The IN protein of Moloney murine leukemia virus processes the viral DNA ends and accomplishes their integration in vitro. Cell 62:829–837

Debyser Z, Christ F, De Rijck J, Gijsbers R (2015) Host factors for retroviral integration site selection. Trends Biochem Sci 40(2):108–116. https://doi.org/10.1016/j.tibs.2014.12.001

Donehower LA, Varmus HE (1984) A mutant murine leukemia virus with a single missense codon in pol is defective in a function affecting integration. Proc Natl Acad Sci U S A 81:6461–6465

Dyda F, Hickman AB, Jenkins TM, Engelman A, Craigie R, Davies DR (1994) Crystal structure of the catalytic domain of HIV-1 integrase: similarity to other polynucleotidyl transferases. Science 266:1981–1986

Eijkelenboom AP, van den Ent FM, Vos A, Doreleijers JF, Hård K, Tullius TD, Plasterk RH, Kaptein R, Boelens R (1997) The solution structure of the amino-terminal HHCC domain of HIV-2 integrase: a three-helix bundle stabilized by zinc. Curr Biol 7:739–746

Ellison V, Abrams H, Roe T, Lifson J, Brown P (1990) Human immunodeficiency virus integration in a cell-free system. J Virol 64:2711–2715

Engelman A, Craigie R (1992) Identification of conserved amino acid residues critical for human immunodeficiency virus type 1 integrase function in vitro. J Virol 66:6361–6369

Engelman A, Mizuuchi K, Craigie R (1991) HIV-1 DNA integration: mechanism of viral DNA cleavage and DNA strand transfer. Cell 67:1211–1221

Farnet CM, Haseltine WA (1990) Integration of human immunodeficiency virus type 1 DNA in vitro. Proc Natl Acad Sci U S A 87:4164–4168

Farnet CM, Haseltine WA (1991) Determination of viral proteins present in the human immunodeficiency virus type 1 preintegration complex. J Virol 65:1910–1915

Fujiwara T, Craigie R (1989) Integration of mini-retroviral DNA: a cell-free reaction for biochemical analysis of retroviral integration. Proc Natl Acad Sci U S A 86:3065–3069

Fujiwara T, Mizuuchi K (1988) Retroviral DNA integration: structure of an integration intermediate. Cell 54:497–504

Grandgenett DP, Vora AC, Schiff RD (1978) A 32,000-dalton nucleic acid-binding protein from avian retrovirus cores possesses DNA endonuclease activity. Virology 89:119–132

Hare S, Gupta SS, Valkov E, Engelman A, Cherepanov P (2010a) Retroviral intasome assembly and inhibition of DNA strand transfer. Nature 464:232–236

Hare S, Maertens GN, Cherepanov P (2012) 3'-processing and strand transfer catalysed by retroviral integrase in crystallo. EMBO J 31(13):3020–3028. https://doi.org/10.1038/emboj.2012.118

Hare S, Shun MC, Gupta SS, Valkov E, Engelman A, Cherepanov P (2009) A novel co-crystal structure afords the design of gain-of-function lentiviral integrase mutants in the presence of modified PSIP1/LEDGF/p75. PLoS Pathog 5(1):e1000259. https://doi.org/10.1371/journal. ppat.1000259

Hare S, Vos AM, Clayton RF, Thuring JW, Cummings MD, Cherepanov P (2010b) Molecular mechanisms of retroviral integrase inhibition and the evolution of viral resistance. Proc Natl Acad Sci U S A 107(46):20057–20062. https://doi.org/10.1073/pnas.1010246107

Hindmarsh P, Leis J (1999) Reconstitution of concerted DNA integration with purified components. In: Advances in virus research, Vol 52, vol 52. Advances in virus research, pp 397–410

Jaskolski M, Alexandratos JN, Bujacz G, Wlodawer A (2009) Piecing together the structure of retroviral integrase, an important target in AIDS therapy. FEBS J 276(11):2926–2946. https://doi.org/10.1111/j.1742-4658.2009.07009.x

Jenkins TM, Hickman AB, Dyda F, Ghirlando R, Davies DR, Craigie R (1995) Catalytic domain of human immunodeficiency virus type 1 integrase: identification of a soluble mutant by systematic replacement of hydrophobic residues. Proc Natl Acad Sci U S A 92:6057–6061

Johnson BC, Metifiot M, Ferris A, Pommier Y, Hughes SH (2013) A homology model of HIV-1 integrase and analysis of mutations designed to test the model. J Mol Biol 425(12):2133–2146. https://doi.org/10.1016/j.jmb.2013.03.027

Katz RA, Merkel G, Kulkosky J, Leis J, Skalka AM (1990) The avian retroviral IN protein is both necessary and sufficient for integrative recombination in vitro. Cell 63:87–95

Katzman M, Katz RA, Skalka AM, Leis J (1989) The avian retroviral integration protein cleaves the terminal sequences of linear viral DNA at the in vivo sites of integration. J Virol 63:5319–5327

Krishnan L, Li XA, Naraharisetty HL, Hare S, Cherepanov P, Engelman A (2010) Structure-based modeling of the functional HIV-1 intasome and its inhibition. Proc Natl Acad Sci U S A 107(36):15910–15915. https://doi.org/10.1073/pnas.1002346107

Kulkosky J, Jones KS, Katz RA, Mack JP, Skalka AM (1992) Residues critical for retroviral integrative recombination in a region that is highly conserved among retroviral/retrotransposon integrases and bacterial insertion sequence transposases. Mol Cell Biol 12:2331–2338

Leavitt AD, Shiue L, Varmus HE (1993) Site-directed mutagenesis of HIV-1 integrase demonstrates differential effects on integrase functions in vitro. J Biol Chem 268:2113–2119

Lee MS, Craigie R (1994) Protection of retroviral DNA from autointegration: involvement of a cellular factor. Proc Natl Acad Sci U S A 91:9823–9827

Li L, Olvera JM, Yoder KE, Mitchell RS, Butler SL, Lieber M, Martin SL, Bushman FD (2001) Role of the non-homologous DNA end joining pathway in the early steps of retroviral infection. EMBO J 20(12):3272–3281

Li M, Craigie R (2005) Processing the viral DNA ends channels the HIV-1 integration reaction to concerted integration. J Biol Chem 280:29334–29339

Li M, Jurado KA, Lin S, Engelman A, Craigie R (2014) Engineered hyperactive Integrase for concerted HIV-1 DNA integration. PLoS One 9(8). https://doi.org/10.1371/journal.pone.0105078

Li M, Mizuuchi M, Burke TR, Craigie R (2006) Retroviral DNA integration: reaction pathway and critical intermediates. EMBO J 25(6):1295–1304

Lodi PJ, Ernst JA, Kuszewski J, Hickman AB, Engelman A, Craigie R, Clore GM, Gronenborn AM (1995) Solution structure of the DNA binding domain of HIV-1 integrase. Biochemistry 34:9826–9833

Maertens GN, Hare S, Cherepanov P (2010) The mechanism of retroviral integration from X-ray structures of its key intermediates. Nature 468(7321):326–329. https://doi.org/10.1038/nature09517

Matreyek KA, Engelman A (2013) Viral and cellular requirements for the nuclear entry of retroviral Preintegration nucleoprotein complexes. Viruses-Basel 5(10):2483–2511. https://doi.org/10.3390/v5102483

Pandey KK, Bera S, Zahm J, Vora A, Stillmock K, Hazuda D, Grandgenett DP (2007) Inhibition of human immunodeficiency virus type I concerted integration by strand transfer inhibitors which recognize a transient structural intermediate. J Virol 81(22):12189–12199. https://doi.org/10.1128/jvi.02863-06

Panganiban AT, Temin HM (1983) The terminal nucleotides of retrovirus DNA are required for integration but not virus production. Nature 306:155–160

Panganiban AT, Temin HM (1984) The retrovirus pol gene encodes a product required for DNA integration: identification of a retrovirus int locus. Proc Natl Acad Sci U S A 81:7885–7889

Passos DO, Li M, Yang RB, Rebensburg SV, Ghirlando R, Jeon Y, Shkriabai N, Kvaratskhelia M, Craigie R, Lyumkis D (2017) Cryo-EM structures and atomic model of the HIV-1 strand transfer complex intasome. Science 355(6320):89–92. https://doi.org/10.1126/science.aah5163

Roth MJ, Schwartzberg PL, Goff SP (1989) Structure of the termini of DNA intermediates in the integration of retroviral DNA: dependence on IN function and terminal DNA sequence. Cell 58:47–54

Schwartzberg P, Colicelli J, Goff SP (1984) Construction and analysis of deletion mutations in the pol gene of Moloney murine leukemia virus: a new viral function required for productive infection. Cell 37:1043–1052

Sherman PA, Fyfe JA (1990) Human immunodeficiency virus integration protein expressed in Escherichia Coli possesses selective DNA cleaving activity. Proc Natl Acad Sci U S A 87:5119–5123

Sinha S, Grandgenett DP (2005) Recombinant human immunodeficiency virus type 1 integrase exhibits a capacity for full-site integration in vitro that is comparable to that of purified preintegration complexes from virus-infected cells. J Virol 79(13):8208–8216

Sinha S, Pursley MH, Grandgenett DP (2002) Efficient concerted integration by recombinant human immunodeficiency virus type 1 integrase without cellular or viral cofactors. J Virol 76(7):3105–3113

Valkov E, Gupta SS, Hare S, Helander A, Roversi P, McClure M, Cherepanov P (2009) Functional and structural characterization of the integrase from the prototype foamy virus. Nucleic Acids Res 37(1):243–255. https://doi.org/10.1093/nar/gkn938

van Gent DC, Groeneger AA, Plasterk RH (1992) Mutational analysis of the integrase protein of human immunodeficiency virus type 2. Proc Natl Acad Sci U S A 89:9598–9602

Wang JY, Ling H, Yang W, Craigie R (2001) Structure of a two-domain fragment of HIV-1 integrase: implications for domain organization in the intact protein. EMBO J 20(24):7333–7343

Yang W, Hendrickson WA, Crouch RJ, Satow Y (1990) Structure of ribonuclease H phased at 2 a resolution by MAD analysis of the selenomethionyl protein. Science 249:1398–1405

Yin Z, Lapkouski M, Yang W, Craigie R (2012) Assembly of prototype foamy virus strand transfer complexes on product DNA bypassing catalysis of integration. Protein Sci 21(12):1849–1857

Yin Z, Shi K, Banerjee S, Pandey KK, Bera S, Grandgenett DP, Aihara H (2016) Crystal structure of the Rous sarcoma virus intasome. Nature 530 (7590):362–+. https://doi.org/10.1038/nature16950

Yoder KE, Bushman FD (2000) Repair of gaps in retroviral DNA integration intermediates. J Virol 74(23):11191–11200. https://doi.org/10.1128/jvi.74.23.11191-11200.2000

Zheng R, Jenkins TM, Craigie R (1996) Zinc folds the N-terminal domain of HIV-1 integrase, promotes multimerization, and enhances catalytic activity. Proc Natl Acad Sci U S A 93:13659–13664

Chapter 10
Oligomerization of Retrovirus Integrases

Duane P. Grandgenett and Hideki Aihara

Introduction

Retroviruses have the unique capacity to stably integrate their viral cDNA genome into the host chromosomal DNA which is a necessary step in their replication cycle (Fig. 10.1). The viral cDNA is produced by reverse transcription of the viral RNA genome, while the integration of the linear viral DNA genome is catalyzed by the viral integrase (IN). The stably integrated provirus is transcribed by cellular RNA polymerase II that results in the synthesis of viral RNA for translation and incorporation into assembling virus particles at the cell membrane. Immature virus particles are released which are subjected to viral proteolysis producing mature viral particles initiating new rounds of infection (Engelman and Cherepanov 2014; Skalka 2014).

Rous sarcoma virus (RSV) IN is derived from viral protease-mediated cleavage of the viral pGag-Pol precursor (Pr180) in immature particles resulting in the dimeric reverse transcriptase (RT) containing two subunits (p95 and p63) and the catalytic active dimeric IN in infectious virions (Fig. 10.2). Note that a single copy of the IN protein is still associated with the RT. A similar pathway exists for human immunodeficiency virus type-1 (HIV-1) Pr160 that produces RT with no covalently attached IN but only free IN. Related pathways for production of RT and IN in various retroviruses are similar to HIV-1. There are ~150 IN molecules per virus particle. IN plays multifunctional roles in the retrovirus life cycle besides integration

D. P. Grandgenett
Saint Louis University Health Sciences Center, Department of Microbiology and
Immunology, Institute for Molecular Virology, Doisy Research Center, St. Louis, MO, USA
e-mail: duane.grandgenett@health.slu.edu

H. Aihara (✉)
Department of Biochemistry, Molecular Biology and Biophysics, University of Minnesota,
Minneapolis, MN, USA
e-mail: aihar001@umn.edu

© Springer Nature Singapore Pte Ltd. 2018   211
J. R. Harris, D. Bhella (eds.), *Virus Protein and Nucleoprotein Complexes*,
Subcellular Biochemistry 88, https://doi.org/10.1007/978-981-10-8456-0_10

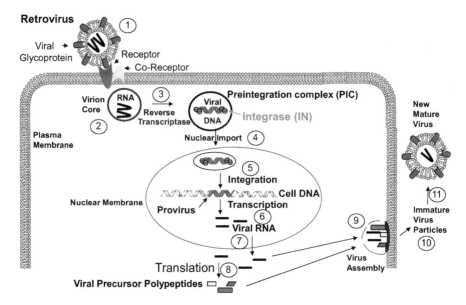

Fig. 10.1 Overview of the retrovirus replication cycle. The infection of a cell by a retrovirus is initiated by the binding of the viral glycoprotein to a specific cellular receptor sometimes, requiring the participation of a co-receptor (step 1). The virus membrane fuses with the cell membrane which results in the entry of the virus core particle into the cytoplasm (step 2). Incomplete uncoating of the core shell facilitates the reverse transcription of the viral plus strand RNA into double-stranded linear viral DNA (~10 kbp)(step 3) which results in the cytoplasmic preintegration complex (PIC). Integrase (IN) in the PIC cleaves a dinucleotide from the 3' OH ends of the viral DNA. The PIC is transported into the nucleus via a nuclear membrane pore (step 4). IN within the PIC coordinates the integration (step 5) of the viral 3' OH ends DNA into the host cell DNA resulting in a permanently insert viral DNA, termed the provirus. Transcription (step 6) by cellular RNA polymerase II yields different size viral plus strand RNAs (step 7) which are translated into different structural and nonstructural proteins (step 8). The non-translated viral RNA and the translated viral proteins are transported to the plasma membrane for virus assembly (step 9). The viral-particle budding step (step 10) releases an immature noninfectious virus particle which undergoes a viral protease maturation step (step 11) that results in new infectious viral particles

including reverse transcription, nuclear import, interactions with cellular protein cofactors for integration, and virus particle maturation (Fig. 10.1) (Skalka 2014; Grandgenett et al. 2015).

During integration, the retrovirus IN proteins are capable of mediating two successive reactions. The first reaction is the 3' OH processing of the viral DNA blunt ends releasing a dinucleotide adjacent to the conserved CA motif at the unique U5 and U3 ends of the long terminal repeats (LTR) (Fig. 10.3a). These processing steps occur in the cytoplasmic preintegration complex (PIC)(Fig. 10.1) that is temporally and spatially separated from the concerted integration of the 3' OH recessed ends into host chromosomes (Fig. 10.3b, c). The concerted insertion of the two viral DNA ends by IN occurs on opposite strands of the target DNA producing a stagger cut that results in a host site duplication specific for different retroviruses (Fig. 10.3c) (Lesbats et al. 2016).

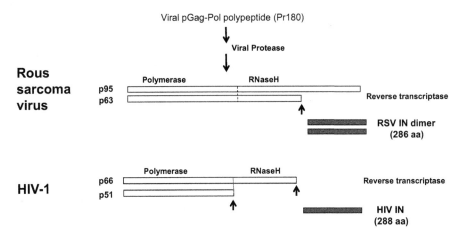

Fig. 10.2 Forms of mature IN and reverse transcriptase in virus particles. The proteolysis of the Rous sarcoma virus (RSV) precursor pGag-Pol polypeptide (Pr180) by the viral protease occurs in virus particles. There are several proteolytic events necessary prior to formation of the shown mature reverse transcriptase and separated IN. The reverse transcriptase maintains one copy of the IN protein on one of its subunits (p95). IN purified from avian retrovirus particles is a dimer. The pGag-Pol precursor for HIV-1 is Pr160. The mature reverse transcriptase and IN in HIV-1 virus particles are shown. IN in HIV-1 virus particles is mainly a monomer in the presence of a reducing agent but forms oligomers in nonreducing conditions

In this chapter, we discuss the oligomerization properties of different retrovirus INs that are necessary for 3' OH processing and concerted integration activities in vitro. The minimal oligomeric form of IN necessary for 3' OH processing is a dimer. The minimal oligomeric form of IN for concerted integration of the two viral DNA ends into a target DNA substrate is a tetramer. Higher-order octameric, dodecameric, and hexadecameric IN structures are also found associated with two viral DNA ends for concerted integration into a target DNA substrate.

Domain Structure of Retroviral IN

Similar domain organization is shared among INs from five different retrovirus genera structurally characterized to date, except for the presence of an additional N-terminal extension domain (NED) on prototype foamy virus (PFV) and murine leukemia virus (MLV) IN (Fig. 10.4). In brief, the NED of PFV IN binds viral DNA in the minor groove (Hare et al. 2010a). The N-terminal domain (NTD) fold is stabilized by a zinc-finger (HHCC), and it binds viral DNA in the major groove in a sequence-specific fashion through a helix-turn-helix motif (Bushman et al. 1993; Zheng et al. 1996; Cai et al. 1997). The NTD functions *in trans*, with respect to the catalytic domain, i.e., NTD binds to the viral DNA juxtaposed to that engaged by the catalytic domain it is connected to. This arrangement helps hold the two viral

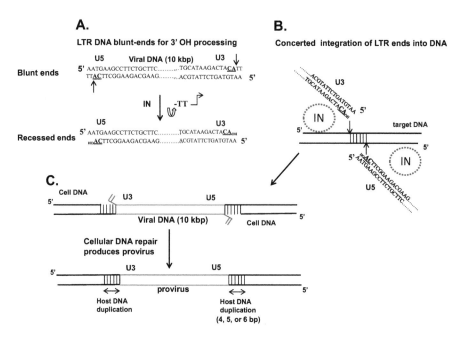

Fig. 10.3 IN 3'OH processing activity and concerted integration of 10 kbp viral DNA into cellular DNA. A. The unique U3 and U5 long terminal repeat (LTR) sequences are located at the termini of viral ~10 kbp blunt-ended DNA. The position of the invariant CA dinucleotide (bolded and underlined) is near the end of the DNA. The nucleotide sequence of the RSV LTRs is shown. The 3' OH processing activity of IN releases the $pTpT_{OH}$ dinucleotide from each DNA end. B. IN forms a specific IN-DNA complex with the 3' OH recessed linear DNA ends (LTR ends shown only) for the covalent insertion of the viral ends into the target DNA producing a 6 bp stagger for RSV and MMTV, 5 bp for HIV-1, and 4 bp for PFV and MLV. C. The staggered cell DNA ends are repaired by cellular proteins which result in the 4, 5, or 6 bp duplications of the target site DNA flanking the covalently inserted provirus

DNA ends together in an IN-DNA complex that promotes concerted integration. The catalytic core domain (CCD) has the RNaseH fold and contains a triad of invariant acidic amino acids where the last two are separated by 35 residues and is called the D,D(35)E motif (Kulkosky et al. 1992), responsible for both the 3' OH processing and concerted integration of the viral DNA ends. The C-terminal domain (CTD) is composed of β-strand barrels resembling SH3 domains (Eijkelenboom et al. 1995). The CTD is the least conserved domain with respect to amino acid sequences among different species and is functionally the most diverse domain which includes the ability to bind both viral and target DNAs, promote IN-IN interactions, bind host factors, and affect assembly of IN-DNA complexes (see below). The secondary structure elements of the three common domains (NTD, CCD, and CTD) of IN are very similar for all retroviruses, illustrated for RSV IN (Fig. 10.5), even though there is minimal amino acid sequence identity (~20 to 25%) between the different viruses.

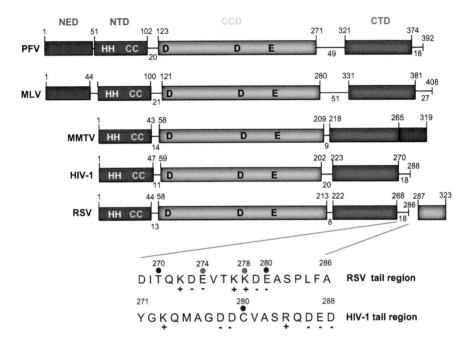

Fig. 10.4 IN domain organization from different retroviruses. Drawings for the IN domains of spumavirus prototype foamy virus (PFV), gammaretrovirus murine leukemia virus (MLV), betaretrovirus mouse mammary tumor virus (MMTV), lentivirus human immunodeficiency virus type 1 (HIV-1), and alpharetrovirus Rous sarcoma virus (RSV) are shown. The top amino acid numbers define the beginning and end of each domain, while the lower numbers delineate the linker size that connects the domains. The N-terminal extension domain (NED) is dark brown, the N-terminal domain (NTD) is purple, the catalytic core domain (CCD) is blue, and the C-terminal domain (CTD) is red. Most of the individual domain structures or combinations of two domains have been defined at the atomic level for the purified retrovirus INs. The CTD of each IN possesses a tail region at their COOH terminus varying in length from 18 residues to 57 residues. The amino acid sequences of the RSV and HIV-1 tail regions are shown. A 37 amino acid protein fragment (green) cleaved from the RSV IN in virus particles has no known biological function

Oligomerization of RSV IN from Virions and Recombinant IN Proteins in the Absence of DNA

The oligomerization of retrovirus IN subunits is necessary for both the 3' OH processing and concerted integration of linear viral DNA (Engelman et al. 1993; Bao et al. 2003; Engelman and Cherepanov 2014; Skalka 2014; Lesbats et al. 2016). The rest of this chapter will concentrate on biological, biochemical, and structural studies that defined the multimerization properties of alpharetrovirus INs. These viruses including RSV and other strains of avian sarcoma/leukemia viruses (ASLV) possess the same or very similar IN amino acid sequences, except as noted. All of the avian retrovirus INs are 286 residues in length (approximately 32,000 Da). As needed, properties of other retrovirus INs will be compared with RSV IN to provide

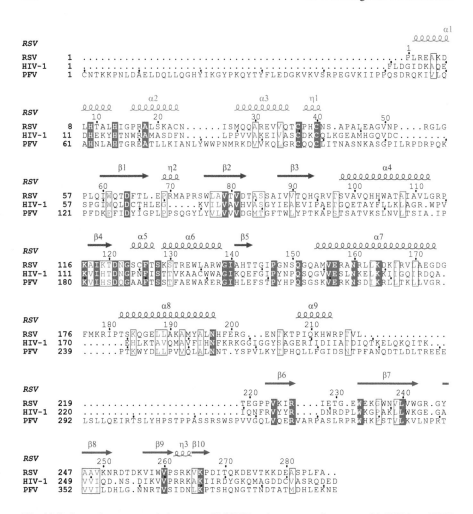

Fig. 10.5 Secondary structural elements of RSV IN and sequence alignment with HIV-1 and PFV IN. The secondary structure elements of RSV IN of the NTD (1–44) (green), CCD (58–213) (red), and CTD (222–269) (blue) are shown. These structural elements of IN were determined from X-ray diffraction data of IN in the presence of DNA (Yin et al. 2016) which are similar to those found in the absence of DNA (Shi et al. 2013). The residue numbers at the top are for RSV IN, marked with a dot for every ten residues. The alignment reveals highly conserved amino acids with HIV-1 and PFV IN, marked in boxed bright red. MMTV IN also possesses these same conserved secondary structures and amino acids (Ballandras-Colas et al. 2016). The secondary structure of the NED for PFV (Hare et al. 2010a) and MLV (Crowe et al. 2016) have been determined

clarity of presentation. Another chapter (Craigie's Chapter) in this book is dedicated to defining the oligomerization properties of HIV-1 IN.

Retrovirus IN has been purified and characterized from avian myeloblastosis virus (AMV) (Grandgenett et al. 1978) and RSV (Prague A strain) (Knaus et al. 1984). IN purified from these virus particles is dimeric and has a molecular weight under native conditions of ~64 kDa determined by glycerol gradient sedimentation. The successful purification of IN from other retrovirus particles has not been reported.

Solution studies using purified recombinant full-length IN from various strains of alpharetroviruses have also demonstrated that IN is dimeric in solution (McCord et al. 1998; Coleman et al. 1999; Moreau et al. 2004; Bojja et al. 2011). Small-angle X-ray scattering (SAXS) and protein cross-linking studies of ASLV IN in solution suggests that the NTD domain of one dimer interacts via a linker to contact the CCD and CTD of the second dimer, called a "reaching dimer" model that produces a closed tetrameric structure (Bojja et al. 2011; Andrake and Skalka 2015). As mentioned, the dimeric status of IN is necessary for 3' OH processing and the tetrameric form for concerted integration activities (Andrake and Skalka 2015). Studies of recombinant full-length RSV IN and several C-terminally truncated forms of IN, which have modifications in the C-terminal tail region following the well-ordered I269 residue at the end of the β-sheet in the CTD (Figs. 10.4 and 10.5) (Yin et al. 2016) demonstrate that IN dimers are capable of producing tetramers and octamers in the presence of viral or viral/target DNA substrates that is necessary for concerted integration (Shi et al. 2013; Pandey et al. 2014; Yin et al. 2016; Pandey et al. 2017) (see below). The results suggest that the three domains of IN and their connecting linkers play critical roles for important IN-IN interactions as well as for IN-viral DNA complex formation.

Site-directed mutagenesis, protein-protein cross-linking, and solution alteration studies of RSV IN have provided some insights into these important interactions to maintain its dimeric state. Early mutagenesis studies utilizing separately purified domains highly suggested that both the CCD and CTD possess self-association properties (Andrake and Skalka 1995). Constructed IN (49–286 residues) lacking the NTD is dimeric (Bojja et al. 2011; Shi et al. 2013), while IN (1–207 residues) is monomeric (Bojja et al. 2011). In solution protein-protein cross-linking coupled with mass spectrometry studies has revealed numerous interactions between each domain of an IN dimer (Bojja et al. 2011). However, monomers, dimers, and tetramers are also observed in solution depending on the pH, salt, and protein concentrations of wild-type (wt) or C-terminally truncated IN (1–270 residues) (Jones et al. 1992; Coleman et al. 1999; Moreau et al. 2004; Shi et al. 2013).

Single amino acid mutational analysis of IN has identified several residues that are critical to maintain dimer status necessary for concerted integration activity (Moreau et al. 2004; Bojja et al. 2011; Shi et al. 2013). For example, modification of W259 in the CTD of IN promotes monomer formation that eliminates enzymatic activities (Bojja et al. 2011; Shi et al. 2013). Other mutations in the CTD modulate activities but do not affect its dimeric status (Moreau et al. 2004). W259 of RSV IN appears to be critical to maintain the dimer interface as well as being involved in multiple interactions near the viral DNA 5' end necessary for concerted integration activity (Yin et al. 2016) (Fig. 10.6) (see later structural details also). A similar possibility exists, with PFV IN (T363) (Hare et al. 2010a) that appears to play similar roles as RSV IN W259. Interpretations of single or multiple mutations in IN are complicated by the fact that individual subunits play different roles within a protein

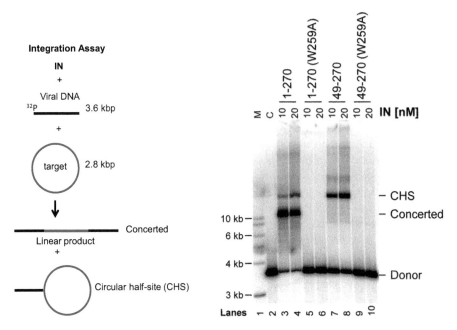

Fig. 10.6 C-terminal truncated IN 1–270 and the critical role of the amino acid residue W259 in RSV IN concerted integration activity. RSV IN constructs 1–270 and 49–270, and their W259A mutants, were assayed at the indicated concentrations for concerted and circular half-site integration (CHS) activities (right), illustrated on the left. The viral 3' OH recessed DNA (3.6 kbp) was labeled with ^{32}P at its 5' end, and the supercoiled target DNA (2.8 kbp) was unlabeled. The concerted and CHS integration products and donor substrate are indicated on the far right size of the gel. Molecular weight markers are in lane 1 while lane 2 contains no IN

complex containing multiple IN subunits. Likewise, this is the case for all retrovirus IN-DNA complexes containing multiple IN subunits as well as the Mu transposase-DNA complexes (Montano et al. 2012).

Atomic Structure of Retrovirus IN Domains and Their Functions

The crystal structures of HIV-1 (Dyda et al. 1994) and RSV (Bujacz et al. 1995) CCDs demonstrated that these active sites are very similar to the superfamily of polynucleotidyl transferases that includes the RT RNaseH (Nowotny 2009). Briefly, the CCD features a five-stranded beta sheet flanked by α-helices on both sides. The active site on one edge of the β-sheet harbors the conserved carboxylate residues of the D,D(35)E motif, Asp64, Asp121, and Glu157 that coordinate two Mg^{2+} ions essential for catalysis (Fig. 10.4). CCD dimerizes symmetrically via the interface on the opposite side of the domain. The CCD domain of RSV IN has been crystallized numerous times with nearly identical dimeric structures but is modified by pH and

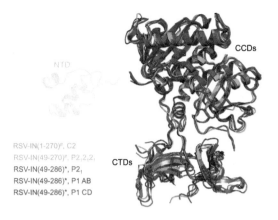
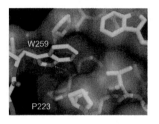

RSV-IN(1-270)#, C2
RSV-IN(49-270)#, P2₁2₁2₁
RSV-IN(49-286)*, P2₁
RSV-IN(49-286)*, P1 AB
RSV-IN(49-286)*, P1 CD

Fig. 10.7 Native conformation of the RSV IN dimer. Crystal structures of two- or three-domain RSV IN constructs determined in various crystal forms show a conserved "native" conformation for the CCD-CTD dimer. NTD was poorly ordered in the IN (1–270) structure, most likely due to the solubility-enhancing F199 K amino acid substitution in CCD that disrupts the hydrophobic NTD-CCD interface. W259 is buried in the asymmetric CTD-CTD interface, docked into a hydrophobic pocket as well as hydrogen-bonded to Pro223 backbone carbonyl group. As discussed later, this native conformation of IN dimer represents its catalytic conformation in the intasome assembly. The central role of W259 in stabilizing the native conformation explains severe defects of IN W259A mutants in integration reactions (Fig. 10.6). The structures marked by (*) are from (Yang et al. 2000), and those marked by (#) are from (Shi et al. 2013)

certain mutations (Bujacz et al. 1995; Bujacz et al. 1997; Lubkowski et al. 1998; Ballandras et al. 2011). Notably, the wt Rous-associated virus type-1 (RAV-1) IN differs with the conserved avian retrovirus CCD sequence at A182 (Ballandras et al. 2011). The RAV-1 IN sequence is T182 which causes a rearrangement of the dimer interface suggesting flexibility to help explain the multifunctionally of retroviral INs. Atomic structures of many other single- and two-domain fragments (CCD + CTD, NTD + CCD) of RSV, HIV-1, and PFV IN were determined and are listed in Li et al. (Li et al. 2011). The structures of each domain of all retrovirus INs, including MLV (Crowe et al. 2016; Guan et al. 2017) and MMTV (Ballandras-Colas et al. 2016), possess similar conserved structural features that are also associated with RSV IN.

The X-ray structure of the CCD-CTD of RSV IN (residues 49–286) (Yang et al. 2000) and near full-length IN (1–270) (Shi et al. 2013) revealed very similar structures despite different crystal packing interactions (Fig. 10.7). The NTD domain of IN 1–270 was poorly ordered and was not resolved. The dimeric CCDs interact with each other through the conserved symmetric dimerization interface observed in other retrovirus INs. The two CTDs dimerize through an asymmetric interface and are not related by a twofold rotational symmetry. Trp259 mentioned above is buried in the CTD dimer interface, playing a critical role in stabilizing the IN dimer.

Functional analysis was performed to determine the role of the NTD and CTD for 3' OH processing and concerted integration (Fig. 10.3). The NTD and CTD are not necessary for 3' OH processing but are essential along with the CCD for concerted integration. IN 1–270 is fully capable of both reactions as full-length IN 1–286

(Fig. 10.6) (Shi et al. 2013). IN 1–214 is completely inactive for concerted or circular half-site (CHS) integration, while IN 49–270 was capable of only promoting the CHS integration reaction, where only one end of the viral DNA is inserted into the target substrate (Shi et al. 2013) (Fig. 10.6). The results suggested that the NTD domain plays a critical role in arranging the two viral DNA ends in a stable synaptic complex (SSC) necessary for concerted integration, which has been rationalized by the swapping of the NTD observed in the PFV (Hare et al. 2010a; Yin et al. 2012), RSV (Yin et al. 2016), MMTV (Ballandras-Colas et al. 2016), HIV-1 (Passos et al. 2017), and maedi-visna virus (MVV) (Ballandras-Colas et al. 2017) IN-DNA complex structures. The CCD-CTD structure obtained for the native dimer form of RSV IN (Yang et al. 2000; Shi et al. 2013) was suggested to be the functional confirmation poised for viral DNA binding and catalysis, which was confirmed by the crystal structure of the RSV strand transfer complex (STC) containing a branched viral/target DNA substrate (Yin et al. 2016). This explains the significant CHS integration activity of the RSV IN CCD-CTD fragment. In the higher-order IN-DNA complex, the asymmetrically associated CTDs of RSV IN further interact with each other and with the NTD of the catalytic IN subunit to cross-link between the two viral DNAs. X-ray analysis allowed a detail description of how eight IN molecules played varying roles in holding the RSV STC together (Yin et al. 2016) (see below).

PFV and MLV IN share similar structural features including the NED and the much longer linkers between the domains, especially those connecting CCD and CTD that distinguish them from RSV, HIV-1, and MMTV IN (Fig. 10.4). Recently, X-ray analysis of the combined NED and NTD regions of MLV (IN 1–105) was determined (Guan et al. 2017) and compared with these same PFV domains (Hare et al. 2010a). Not surprisingly, these two domains from each virus were very similar for both the binding of Zn^{++} and their ability to bind DNA, determined with PFV IN and modeled with MLV IN (Guan et al. 2017). The IN-DNA interactions of the NED improve stability of PFV IN-viral DNA complexes (Hare et al. 2010a). Some differences are noted between the NED and NTD of MLV IN which defines their orientations in comparison to PFV IN. The large size flexible linkers associated with these two INs play a dominant role in shaping of the PFV IN-DNA complexes in contrast to those observed with shorter linkers associated with RSV IN (Yin et al. 2016), MMTV IN (Ballandras-Colas et al. 2016), lentiviruses HIV-1 IN (Passos et al. 2017), and MVV IN (Ballandras-Colas et al. 2017), allowing PFV IN to form the intasome with only four monomers.

Biological Roles for Oligomerization of Retrovirus IN In Vivo

The biological roles for IN forming multimers with viral DNA in the cytoplasmic PICs were first detected with MLV and HIV-1. Analysis of isolated MLV (Wei et al. 1997; Wei et al. 1998) and HIV-1 (Chen et al. 1999) PICs from virus-infected cells demonstrated that IN protected the viral DNA termini (~20 bp) and extended regions of the viral LTR DNA ends up to ~200 bp. In contrast with reconstructed concerted integration reactions using IN and viral DNAs, DNase I protection footprint analysis demonstrated that ~20 bp and 30 bp of the viral DNA ends on >1.0 kbp viral

DNA substrate were protected by RSV IN (Vora and Grandgenett 2001; Vora et al. 2004) and HIV-1 IN (Bera et al. 2009; Pandey et al. 2010; Pandey et al. 2011), respectively. A smaller protective footprint (~16 bp) was also observed with HIV-1 IN (Li et al. 2006). Small DNA oligonucleotides (ODN) from 18 to 21 bp in length containing viral DNA sequences are also optimal for assembly of the RSV SSC containing IN tetramers (Pandey et al. 2014) or IN octamers (Pandey et al. 2017). These results suggest that the oligomerization properties of IN in vitro are likely different than some of those observed in vivo. In support of this possibility, single-particle analysis of HIV-1 PICs in infected cells showed that nuclear entry of the HIV-1 PIC is associated with a reduction of IN molecules, while association of the PIC with chromatin-associated LEDGF/p75 increased IN oligomerization properties (Borrenberghs et al. 2016; Quercioli et al. 2016). Host factor LEDGF/p75 directs the HIV-1 PIC efficiently to active transcription sites for integration (Lesbats et al. 2016; Lusic and Siliciano 2017) where the site selection positions effect influence HIV-1 latency in infected individuals (Chen et al. 2017). Further investigative methods are needed to fully understand the role of IN oligomerization in its many demonstrated multifunctional roles in vivo, including reverse transcription that IN has in the replication of retroviruses in cells (Grandgenett et al. 2015).

IN purified from AMV and RSV virus particles is dimeric under reducing conditions (Grandgenett et al. 1978; Knaus et al. 1984). HIV-1 IN in virus particles is a monomer under reducing conditions but is a monomer, dimer, and trimer or tetramer under non-reducing conditions (Petit et al. 1999; Bischerour et al. 2003). This oligomerization property of IN in virus particles may be associated with the role HIV-1 IN has in the maturation of the internal core structure in virus particles (Fig. 10.1). Surprisingly, this important discovery for the IN structural role was revealed by non-active site inhibitors directed against the LEDGF/p75 binding site at the CCD interface of HIV IN (Christ et al. 2010; Balakrishnan et al. 2013; Desimmie et al. 2013; Feng et al. 2013; Jurado et al. 2013; Gupta et al. 2016). The inhibitors produced defective noninfectious virus particles with abnormal organization of the viral RNA and nucleocapsid producing deformed and empty core structures. These IN inhibitors, termed LEDGINs or allosteric IN inhibitors (ALLINIs), are in various phases of clinical development for HIV-1/AIDS treatment (Feng et al. 2015). Whether the ability of HIV-1 IN to bind to specific regions of the viral RNA is related to the oligomerization properties of IN or connected to virus particle maturation is unknown (Kessl et al. 2016). The biological roles for oligomerization of other retrovirus INs in virus-infected cells besides HIV-1 IN are not well understood.

RSV SSC Are Kinetically Stabilized by HIV-1 Strand Transfer Inhibitors

The active site of retrovirus INs catalyzes the same 3' OH processing, and concerted integration activities imply that their each respective viral DNA and active residues are positioned in the CCD to promote these reactions. There is also an invariant CA dinucleotide on the catalytic strand at the 3' OH processing site of retrovirus DNAs

(Fig. 10.3). The atomic structure of PFV IN active site in the PFV SSC (Hare et al. 2010a) containing two viral DNA ends also supports this conclusion. The positioning of HIV-1 STIs in the active site results in the displacement of the CA dinucleotide out of the site thus, producing an inactive PFV SSC (Hare et al. 2010b; Hare et al. 2012). STIs are interfacial inhibitors that form stable interfaces with the viral CA dinucleotide and specific IN residues including the D,D(35)E motif, displacing the terminal dA nucleotide which prevents strand transfer activity (Pommier et al. 2005; Hare et al. 2010b; Pommier et al. 2015). STIs have varying capacities to kinetically "trap" HIV-1 SSC (Pandey et al. 2007; Pandey et al. 2010) produced by HIV-1 IN that is related to their rate of dissociation from the active site. HIV-1 STIs also differentially inhibit the replication of lentiviruses, alphaviruses (avian), gammaretroviruses (murine), and betaretroviruses (Mason-Pfizer monkey virus) (Koh et al. 2011).

As previously stated, full-length wt RSV IN (1–286 residues) (Fig. 10.4) and various C-terminal IN truncations are dimeric and fully capable of 3' OH processing and concerted integration activities (Shi et al. 2013). Under appropriate assay conditions, AMV and RSV IN can self-assemble unto viral DNA substrates for DNA-binding analysis (Grandgenett et al. 1978; Knaus et al. 1984; Peletskaya et al. 2011), for fluorescence resonance energy transfer (FRET) analysis to investigate interactions of viral DNA ends in IN-DNA complexes where the 5' DNA ends are labeled with Cy3 and Cy5 (Bera et al. 2005), for enzymatic activities (Andrake and Skalka 1995; Moreau et al. 2000; Zhou et al. 2001; Moreau et al. 2002; Andrake and Skalka 2015), and for DNase I protection footprint analysis (Vora and Grandgenett 2001).

To further study the assembly mechanisms of RSV IN onto viral DNA in-solution, a new assembly procedure was necessary to assess the oligomeric state of IN in complexes capable of concerted integration. Previously, HIV-1 IN was shown to assemble onto an ~1 kbp linear DNA substrate producing the SSC that promoted the concerted integration reaction and could be identified by native agarose electrophoresis (Li et al. 2006). An IN tetramer was shown to be present in these HIV-1 SSCs (Li et al. 2006). Further studies showed that STIs can physically trap the HIV-1 SSC and higher-order SSC resulting in the accumulations of these complexes (Pandey et al. 2007; Pandey et al. 2010). These studies suggested that STIs could also be used to kinetically trap the RSV SSC to study assembly mechanisms. STIs effectively inhibit the concerted integration activities of RSV IN similarly to HIV-1 IN (Pandey et al. 2014).

RSV IN dimers can self-assemble onto 18 bp to 22 bp 3' OH recessed U3 gain-of-function (GU3) LTR DNA substrates to form a stable complex in the presence of HIV-1 STIs but not in their absence at 4°C (Pandey et al. 2014; Pandey et al. 2017) (Fig. 10.8a). The STIs physically "trap" the RSV SSC produced with IN (1–269 residues) that lack the 18-residue "tail" region (Fig. 10.4). Once formed, the IN-DNA-STI complex can be isolated by size-exclusion chromatography (SEC) in the absence of STI in the running buffer, suggesting that the complex is kinetically trapped. These kinetically stabilized SSCs contain an IN tetramer by Superdex-200 SEC (Fig. 10.8a), protein-protein cross-linking studies, and SEC-multiangle light scattering (SEC-MALS). The efficiency and stability of producing the STI-trapped

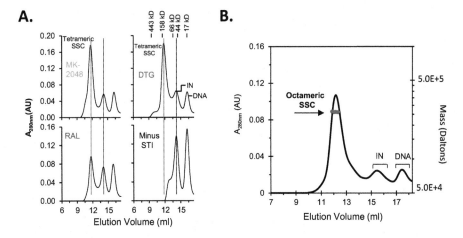

Fig. 10.8 HIV-1 STIs kinetically stabilize RSV SSC containing IN tetramers or octamers. A. RSV STI-trapped SSC assembled with RSV IN (1–269) in the presence of different HIV-1 IN STIs and 3' OH recessed 18 bp GU3 at 4°C were identified by SEC. The STIs MK-2048 (blue) and dolutegravir (DTG) (red) efficiently produce trapped SSCs, while raltegravir (RAL) (green) possesses moderate efficiency. The determined molecular mass by SEC-MALS analysis for the tetrameric intasome is 151,000 ± 2000 Da (Pandey et al. 2014). No intasomes are produced in absence of an STI shown in right lower panel marked minus STI. Elution positions of the STI-trapped intasome (red line), free IN (black line), and free DNA are marked. The elution volume and absorbance are indicated. B. The octameric intasome was produced by RSV IN (1–278) and GU3 DNA substrate in the presence of MK-2048 at 18°C and subjected to SEC-MALS analysis. The molecular mass is 257,000 ± 8000 Da (Pandey et al. 2017)

SSC by each STIs is related to the dissociation rate of the STI from HIV-1 IN-DNA complexes (Grobler et al. 2009; Hightower et al. 2011). Additional mass spectrometry studies demonstrated that the STIs are physically associated with the SEC purified RSV SSC (Pandey et al. 2014).

The oligomerization of RSV IN using the same GU3 substrate is increased to an octameric state when the C-terminal truncated IN (1–278 residues) (Fig. 10.4) is used to produce an STI-trapped SSC at 18°C (Fig. 10.8b) (Pandey et al. 2017). The octameric IN structure is also observed with wt IN (1–286), while IN (1–274) inefficiently forms both the tetrameric and octameric structures. RSV IN (1–269) lacks the ability to produce the SSC containing IN octamers at 18°C. The ends of the viral DNAs are in close proximity in both SSCs as measured by FRET (Pandey et al. 2017). The results suggest that the "tail" region plays a key role in the oligomerization of IN on viral DNA ends with IN (1–278) being the most effective protein to assemble SSC containing octamers. The "tail" region of HIV-1 IN is not critical for viral replication but enhances the functions of IN with increasing efficiency in accordance with its length (Dar et al. 2009; Mohammed et al. 2011). Future structural and functional studies will provide insights into understanding the oligomerization properties of RSV and HIV-1 IN and their associations with viral and viral/target DNA substrates.

X-Ray Structural Analysis of RSV IN in the Presence of a Branched Viral/Target DNA Substrate

For definition again, the SSC contains two viral DNA ends held together by IN capable of concerted integration, while the strand transfer complex (STC) contains a branched DNA structure in which the two viral DNA ends are joined with each strand of the target DNA. This branched DNA mimics the reaction product produced by the concerted integration of the two viral DNA ends into a target DNA substrate (Fig. 10.3b). Intasome is a general term for all of these different retrovirus nucleoprotein complexes involved in retrovirus integration (Engelman and Cherepanov 2017).

In-solution conformational studies have established the precursor IN subunit requirements for assembling intasomes of five different retroviruses. Their architectures were determined by X-ray analysis of their crystal structures or by cryo-electron microscopy (EM). A monomer is the precursor for the PFV SSC containing an IN tetramer (Hare et al. 2010a; Gupta et al. 2012); a dimer precursor is required for the RSV STC (Shi et al. 2013; Yin et al. 2016) and MMTV SSC (Ballandras-Colas et al. 2016), both structures of which contain IN octamers; and a variety of multimeric states from monomers to octamers have been characterized (Lee et al. 1997; Pandey et al. 2011; Passos et al. 2017) for the lentivirus HIV-1 STC containing a tetramer, dodecamer, or a hexadecamer (Passos et al. 2017) and presumably also for the lentivirus MVV SSC and STC that contains an IN hexadecamer (Ballandras-Colas et al. 2017). Despite the variation in oligomeric states of the precursor INs, the commonality associated with all five retrovirus intasomes is the presence of a conserved intasome core (CIC) (Ballandras-Colas et al. 2017) composed of a IN tetramer (a dimer of dimers) responsible for 3' OH processing and the concerted integration reaction.

The groundbreaking X-ray crystallographic studies of PFV IN provided the long-sought high-resolution structural information on retroviral IN-DNA complexes (Hare et al. 2010a, b; Maertens et al. 2010; Hare et al. 2012). The tetrameric PFV intasome consists of a pair of IN dimers, where each IN dimer comprises the catalytic (inner) and the non-catalytic (outer) subunit (Fig. 10.9). The inner and outer IN subunits dimerize via the conserved symmetric CCD dimer interface as described above. However, for the rest of the molecules, the IN dimer is structurally asymmetric, and accordingly, they have distinct functions within the intasome. The inner IN subunit assumes an extended conformation where all four structural domains are linearly arranged spatially in the order of NED, NTD, CTD, and CCD from one end to the other. All four domains of the inner IN molecule are engaged in critical DNA interactions as well as protein-protein contacts to hold the complex together. Importantly, the NED and NTD bind to the minor and major grooves, respectively, of a viral DNA molecule juxtaposed to the viral DNA bound by the CCD of the same IN subunit. This reciprocally domain-swapped arrangement helps ensure concerted integration of both viral DNA ends, and it is a conserved feature in CIC of all retroviral intasome structures characterized to date (see below).

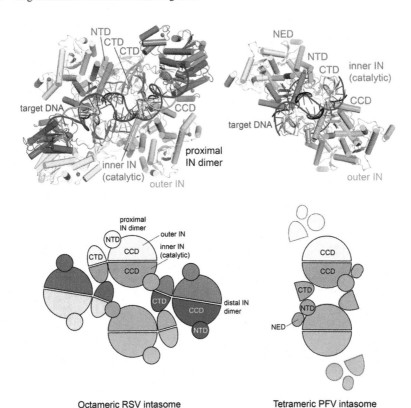

Fig. 10.9 Comparison between the RSV and PFV intasomes. In the octameric RSV intasome (STC) (left panel), CTDs of the distal IN dimers fill the gap between NTD and CCD of the inner catalytic IN subunit. These interactions in turn tether the distal IN CCDs, which serve as the target DNA-binding surface. In the tetrameric PFV intasome (SSC) (right panel), CTD of the inner catalytic IN subunit plays the corresponding role, which is enabled by the longer inter-domain linker connecting CCD and CTD. The NED, NTD, and CTD of the outer PFV IN molecules are disordered and were not resolved in the crystal structure (Hare et al. 2010a)

The CTD plays a key role in bridging between the two viral DNA molecules, by filling the space between NTD and CCD. CTD also serves as the platform for target DNA-binding. The outer IN subunits play a role in target DNA capturing by making direct interaction with nucleosomes (Maskell et al. 2015).

Work on PFV IN was highly instrumental in understanding basic mechanisms of integration and allowed modeling of tetrameric intasome structures for other retroviruses (Krishnan et al. 2010). MLV IN, for instance, shares a similar domain organization with PFV IN and thus likely forms a similar tetrameric intasome. However, it remained unknown how the smaller three-domain INs lacking NED oligomerize to form intasomes. The first intasome structures for INs of this type were solved for the betaretrovirus MMTV and alpharetrovirus RSV (Figs. 10.9 and 10.10) systems by cryo-EM and X-ray crystallography, respectively (Ballandras-Colas et al. 2016; Yin et al. 2016), to address the question. As retroviral INs have a tendency to

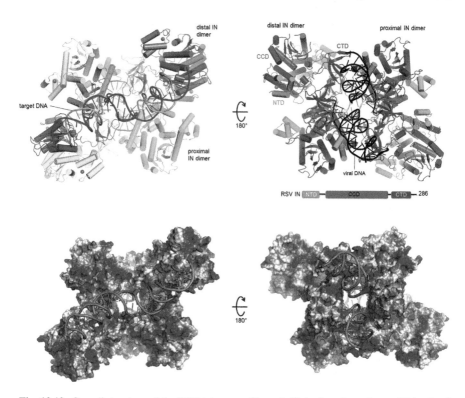

Fig. 10.10 Overall structure of the RSV intasome. (Upper left) A view along the twofold axis of the complex from the target DNA side. Eight IN molecules are colored differently. The gray spheres represent zinc ions bound in NTD. (Upper right) A view along the twofold axis from the viral DNA side, with the three structural domains of IN (NTD, CCD, CTD) colored, respectively, in green, red, and blue colors as indicated by the schematic diagram below the structure. Two IN molecules within each IN dimer are colored slightly differently. (Bottom panels) Electrostatic surface potential (blue, positive; red, negative) is shown to highlight the broad positively charged patches involved in DNA-binding. The orientations correspond to that in the panel right above each image

aggregate (Li and Craigie 2009) and non-specifically associate with nucleic acids (Knaus et al. 1984), the task of preparing homogeneous IN-DNA complexes for structural studies posed a considerable challenge. To overcome this problem, both these studies employed similar biochemical procedures in which purified IN protein and DNA substrates are first mixed in a high ionic strength condition to prevent non-specific association, and the IN-DNA complex is formed by slow dialysis into a low ionic strength condition. Although the specific IN-DNA complex thus formed has limited solubility and mostly precipitates, the precipitated complex could be re-solubilized by addition of salt or further dialysis into conditions containing organic solvents and then purified by size-exclusion chromatography. The final solution condition in which the complex is isolated does not support IN-DNA complex formation, i.e., the purified complex is removed from the binding equilibrium and is kinetically trapped. This is similar to the STI-trapped RSV SC described above, where the complex tightly bound to STI is stable for hours in the absence of STI in the surrounding solution. The extreme biochemical stability of these complexes may reflect the highly stable nature of PIC in viral infection.

Crystallization of the RSV intasome (STC) was achieved using a high concentration (3.2 ~ 4.0 M) of sodium formate as the precipitant (Yin et al. 2016). Conditions with such high ionic strengths are rarely used successfully for non-covalent protein-DNA complexes. However, the PFV intasome was similarly crystallized in a high concentration (1.35 M) of ammonium sulfate (Hare et al. 2010a), highlighting again the high stability of retroviral intasome complexes. The crystal structure of the RSV intasome refined at 3.8 Å resolution showed a novel octameric IN assembly with viral and target DNA molecules. The RSV intasome has both conserved, and unique structural features in comparison to the previously reported PFV and the more recently reported lentiviral intasome structures as discussed below. The cryo-EM structure of the betaretroviral MMTV intasome (SSC) complex at approximately 5–6 Å resolution (CIC structure was approximately 4 Å) showed a similar octameric assembly of IN, but with some important variations.

The RSV STC assembled on the branched DNA substrate, which carries two 'GU3' gain-of-function viral terminal sequence (Vora et al. 2004) and a stretch of target DNA, shows an overall twofold symmetric shape mirroring the symmetrical nature of the concerted integration. The octameric RSV IN includes four IN dimers, two copies each of proximal and distal dimers that take different conformations (Fig. 10.10). As is the case with the PFV intasome, the proximal IN dimer consists of an inner subunit that serves the catalytic role and engages the viral/target DNA junction and an outer subunit associated via the conserved CCD dimer interface (Fig. 10.11). Two juxtaposed proximal IN dimers, each bound to a viral DNA terminus, form a tetramer by swapping NTD of the inner catalytic subunit as observed in the PFV IN tetramer. However, unlike with PFV IN, the inner and outer IN molecules also interact tightly with each other through the asymmetric CTD interface, where both CTDs are positioned adjacent to the CCDs and make viral DNA contacts. Thus, the CTDs bind viral DNA in *cis* whereas the NTD of the inner IN subunit binds viral DNA in *trans*. The NTDs of the outer IN subunits are not exchanged, bound to the CCD surface via a hydrophobic interface involving Phe199 of CCD. Overall the RSV IN tetramer including a dimer of the proximal IN dimers mimics the PFV IN tetramer, except for the unique positioning of the CTDs and their viral DNA contacts, which partially replaces the viral DNA contacts made by the PFV IN NED.

As mentioned above, in the PFV intasome, the CTD of the catalytic inner subunits bridges between the two viral DNA ends to complete the domain-swapped IN tetramer (Hare et al. 2010a; Yin et al. 2016). In the octameric RSV intasome, this critical role is assumed by CTDs of the distal RSV IN dimers, contributed in *trans* (in an inter molecular fashion; Fig. 10.12). This CTD interaction tethers the distal (flanking) IN dimers to the core IN tetramer. Both CTDs of the distal IN dimers make viral DNA contacts; thus, RSV intasome includes a total of eight

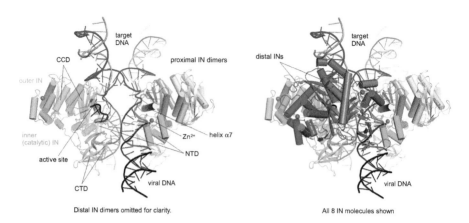

Fig. 10.11 Interactions within the RSV intasome. The left panel shows two proximal IN dimers associated with the viral and target DNAs. Note that the inner catalytic subunits (green/beige) swap NTDs, which bind to viral DNAs in *trans*. The NTD is connected to CCD via an extended linker that traverses the core IN tetramer. The backbone residues of the catalytic triad are colored red to mark the active site. In the left panel, the distal IN dimers are omitted for clarity. In the right panel, all eight IN molecules are shown. The distal INs fit in the gap between NTD and CCD of the catalytic IN subunit

CTDs, interacting with various positions of viral DNA substrates in four distinct fashions to help hold them together. The CCDs of the distal IN dimer do not have any catalytic role and instead serve as the platform for target DNA-binding. Reflecting their distinct roles and interactions within the intasome, the proximal and distal IN dimers show distinct conformations that differ in the relative configurations between the dimerized CCDs and CTDs (Fig. 10.13). The "canted" CCD-CTD dimer with the parallel β-sheet-like conformation of the linker segments was observed in multiple DNA-free RSV IN structures (Yang et al. 2000; Yin et al. 2016) (Fig. 10.7). It was noted that this "native" conformation appears to be poised for binding a viral DNA and presenting its terminus to the active site, based on the positioning of viral DNA with respect to CCD in the PFV intasome structure (Shi et al. 2013). The RSV intasome crystal structure indeed showed that the proximal IN dimer takes this native conformation to engage the viral DNA substrate. The alternative CCD-CTD conformation of the distal IN dimers, which had not been observed for free IN structures, is required for all IN molecules to fit in the complex without clashes (Yin et al. 2016). As discussed above, the distal IN dimers are likely recruited during the intasome assembly after the proximal IN dimers form the core tetramer with two viral DNA molecules. Thus, this unique IN dimer conformation may be achieved through an induced fit mechanism, upon association of a free IN dimers through the C-terminal interactions to complete the octameric assembly.

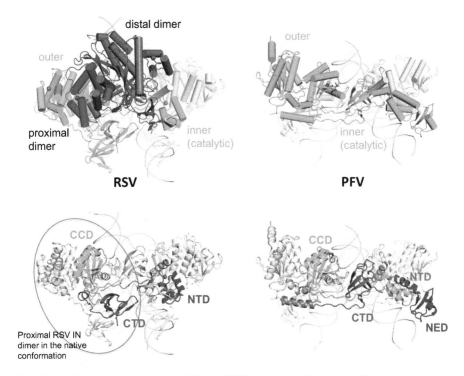

Fig. 10.12 Comparison between the RSV and PFV intasomes (side views). Views perpendicular to the twofold axis of the complexes. For RSV, the distal IN dimer was omitted in the lower panel for clarity. The inner catalytic PFV IN molecule takes an extended conformation with its four structural domains linearly aligned in the order of NED, NTD, CTD, and CCD. The catalytic IN molecules are colored in a gradient of blue to red from the N- to C-terminus, respectively. Note the longer CCD-CTD linker of PFV IN (orange), which allows folding back of the polypeptide to deliver CTD for interaction in *cis* (Hare et al. 2010a). The CTDs of the proximal RSV IN dimer instead makes viral DNA contacts to play analogous roles to PFV IN NED

DNA Conformations

The two viral DNA ends are attached to the target DNA with the canonical 6 bp spacing for avian retrovirus integration. The viral DNAs branch out from a sharply bent target DNA with their helical axes skew and at an angle of ~60°. The viral DNA shows a significant widening of the minor groove near the viral/target DNA junction, where an α-helix α7 inserts to make sequence-specific contacts. The thymine base from the non-transferred strand opposite the terminal adenine on the transferred strand is displaced by Gln151 from a CCD loop immediately preceding α7 (corresponding to Gln215 in PFV and Gln146 in HIV-1 IN). Thus, three nucleotides from the 5′-terminus of the non-transferred strand are unpaired, interacting with a CTD from distal IN. The α7 helix also harbors one of the catalytic triad residues Glu157. Similar engagement of the DNA minor groove by an α-helix

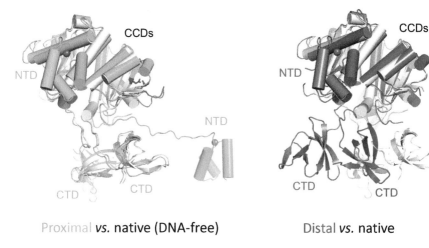

Proximal *vs.* native (DNA-free)   Distal *vs.* native

Fig. 10.13 Two different IN dimer conformations observed in the RSV intasome. On the left, the proximal IN dimer, which consists of the inner catalytic (green) and outer (cyan) IN subunit and forms the core tetramer of the intasome assembly, is superimposed on the native IN dimer (light gray) as shown in Fig. 10.7. The catalytic triad residues and the adjacent loop that plays an important role in viral DNA interaction are colored in red and pink, respectively. On the right, the distal IN dimer CCDs are superimposed on the CCDs of the native IN dimer, showing significant difference of CTD positioning. The asymmetric CTD dimer interface involving W259 is common in both conformations, and the proximal and distal IN dimers differ mainly in the conformation of the CCD-CTD linker segments

harboring a catalytic glutamate residue has been observed for the superfamily of distantly related integrase/transposases, including the cut-and-paste transposase Tn5 (Davies et al. 2000). The swapped NTD from the opposing IN subunit binds in the adjacent major groove for additional sequence recognition. The CCD and NTD of the inner catalytic INs along with CTDs of all eight IN molecules together form a ring-shaped structure to "bundle up" the two viral DNA molecules (Fig. 10.14). The footprint of the RSV IN octamer on each viral DNA is ~20 bp, roughly consistent with previous biochemical data (Vora and Grandgenett 2001; Vora et al. 2004).

The RSV STC crystal contained a target DNA stretch of a total of 38 bp, corresponding to 16 bp on either side of the integration site. The target DNA is bent away from the protein assembly, similar to the target DNAs in transpososome structure (Montano et al. 2012). The DNA structure is particularly distorted in the central 6 bp region between the viral/target junctions, with unstacked bases with a negative roll angle of 57° in the middle. The target DNA in this conformation is stabilized by interactions with the CCD of the inner catalytic IN subunit immediately outside the 6 bp region, including insertion of a short α-helix with Ser124 (corresponding to Ser119 of HIV-1 IN and Ala188 of PFV IN) into the DNA minor groove for possible base contacts (Fig. 10.15). CCDs of the distal IN dimers make less-specific backbone contacts further outside region of the target DNA, which involves a different region of the CCD surface. The sharply bent target DNA conformation helps to drive the

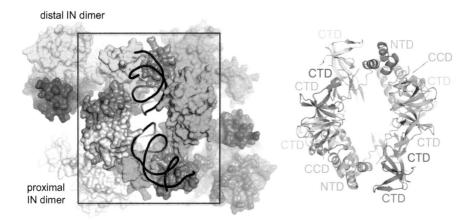

Fig. 10.14 "Bundling" of the two viral DNAs. The two viral DNA ends are encircled by CTDs of all eight INs, and NTD and CCD of the catalytic INs, which form a ring like structure. The two viral DNA molecules in the RSV intasome are positioned close to each other, with the backbone phosphate oxygen atoms at the closest point ~5 Å apart. The highly positively charged surface formed by a network of CTDs and NTDs may alleviate potential electrostatic repulsions between DNA strands

otherwise chemically isoenergetic integration reaction forward, as the "spring loaded" DNA conformation causes the liberated 3'-OH end of the nicked target strand to be misaligned for the reverse disintegration reaction (Chow et al. 1992). In addition to the ~90° bending away from the viral DNAs, the target DNA in the RSV intasome also zigzags in the place perpendicular to the direction of the primary bending. The resulting DNA trajectory has a positive writhe, which is opposite that in a negatively supercoiled nucleosomal DNA (Fig. 10.16). Thus, unlike PFV IN that preferentially integrates into nucleosomal target DNA (Maskell et al. 2015; Kirk et al. 2016), avian retroviral integration may not benefit from nucleosomes in the target DNA. This is consistent with a recent study showing more efficient integration into a naked DNA substrate in vitro by the RSV/ASV integrase (Benleulmi et al. 2015) .

Comparison to the MMTV Intasome

The intasomes from four different genera of retroviruses, ranging in size from the tetramer (spumavirus PFV) to hexadecamer (lentivirus MVV), share a conserved core structure that consists of two IN dimers with CTDs bridging between the swapped NTD and the catalytic CCD. The CTD can either be from the inner catalytic IN contributed in *cis* (as in tetrameric PFV intasome) or from distal/flanking IN molecules contributed in *trans* (as in octameric RSV). The phylogenetically closely related betaretroviral MMTV and alpharetroviral RSV further share the

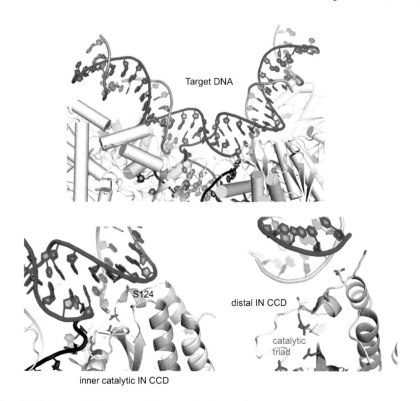

Fig. 10.15 Target DNA contacts. Target DNA in the RSV intasome is bound to the continuous positively charged surface (Fig. 10.9) comprising the CCDs of both proximal and distal IN dimers. The target DNA shows an overall bending of ~90° away from the protein, mostly significantly distorted in the middle. Near the active site harboring the viral/target DNA junction, S124 from a short helix (α5) engages the minor groove. The contacts made by distal IN CCDs are less specific and involve surface-exposed basic side chains and the DNA backbone. The catalytic residues are shown in red sticks in both cases. The distal IN plays no catalytic roles

octameric intasome architecture and show similar positioning of both CTDs that anchor the distal (flanking) IN dimers to the core (Fig. 10.17). However, the two structures also have notable variations in the domain arrangement. CCDs of the flanking MMTV IN dimers are positioned such that the four IN dimers in the MMTV SSC are arranged to form an overall cruciform shape. In RSV STC, the two pairs of diagonally positioned IN dimers are not orthogonally arranged. In addition, CTD of the outer MMTV IN subunit from the core IN dimer is positioned away from DNA, unlike the corresponding RSV IN CTD that forms viral DNA contacts. These differences highlight the diversity and flexibility of retrovirus intasome assemblies. The lentiviral intasome structures (Passos et al. 2017) also show varying domain arrangements outside the core (CIC) and unique positioning and roles of CTD, which is elaborated in another section focused on the HIV-1 integration (Craigie's Chapter).

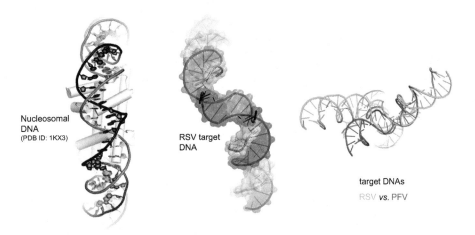

Fig. 10.16 Distorted target DNA conformation in the RSV intasome. Comparison between DNA conformations in a nucleosome (left) and the RSV intasome (middle) and a superposition of target DNAs in RSV and PFV intasomes (right). The trajectory of the RSV target DNA zigzags in the plane perpendicular to the direction of the primary bending as a result of sharp kinks at the integration sites, creating an ~20 Å shift in the helical axis with an overall positive writhe. This is in contrast to the more smoothly bent DNA structures observed in a nucleosome particle or the PFV intasome

Interactions of Retrovirus IN with Host Factors

DNA sequence analysis of multiple retrovirus integration sites in the host genome in retrovirus infected cells clearly established that different retroviruses preferentially integrate their DNAs into specific regions (Schroder et al. 2002; Wu et al. 2003; Mitchell et al. 2004; Kvaratskhelia et al. 2014; Demeulemeester et al. 2015), although most of the host genome is available for integration. HIV-1 shows a preference to integrate into actively transcribed regions, MLV into active promoter and enhancer regions, and ASLV into random regions with little preference for either of the above regions selected by HIV-1 and MLV. But, there is a slight preference for ASLV integration into transcriptional regions in the chicken genome (Barr et al. 2005). Numerous factors like nuclear entry of the PIC, chromatin environments, direct selection of DNA target sequences for local distortability by IN, and interactions of IN with host factors associated with the host chromatin likely all contribute to site selection.

The oligomeric status of IN within the different retrovirus PICs directly interacting with a host protein factor just prior to integration into host chromatin in virus-infected cells is unknown. However, we do know what host factors interact with each IN. LEDGF/p75 is a chromatin-associated protein where the integrase-binding domain (IBD) binds to the dimer interface of HIV-1 IN and tethers the PIC to actively transcribed chromatin (Cherepanov et al. 2003; Cherepanov et al. 2004; Cherepanov et al. 2005; Ciuffi et al. 2005; De Rijck et al. 2010; Benleulmi et al. 2015).

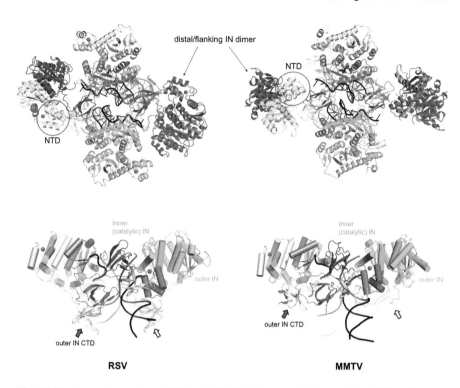

Fig. 10.17 Comparison between RSV and MMTV intasomes. Alpharetrovirus RSV and betaretrovirus MMTV share a conserved octameric intasome assembly, with CTDs of the distal/flanking IN dimers contributed in *trans* bridge between the two viral DNA molecules. These two systems however have significant differences as well, including different positioning of CCDs of the distal/flanking IN dimers (top panels; NTD of corresponding IN molecules are circled to highlight the different orientations of the distal IN CCDs and NTDs) and distinct positioning and interactions involving the outer IN subunit CTD (bottom panels; indicated by arrows). The target DNA present in the RSV intasome/STC crystal structure is not shown. In the lower panels, the distal IN CCDs and NTDs are also omitted for clarity

It is unknown what IN subunit(s) in an HIV-1 intasome that contains dodecamer or hexadecamer of IN (Passos et al. 2017) interacts with LEDGF/p75 for integration into the host genome but most likely exists in the tetrameric CIC structure, although the lentivirus MVV intasome can be modeled with 16 copies of LEDGF/p75 (Ballandras-Colas et al. 2017). The BET proteins (Brd 2,3,4), which are transcriptional co-regulators, are MLV integration cofactors (De Rijck et al. 2013; Gupta et al. 2013; Sharma et al. 2013). The interactions between the BET proteins and MLV IN involve the specific BET ET domain (Crowe et al. 2016) and the C-terminal "tail" of IN (De Rijck et al. 2013; Aiyer et al. 2014; El Ashkar et al. 2014) (Fig. 10.4), specifically the highly conserved 17 residues of MLV starting at W390 and ending at R408, termed ET-binding motif (EBM), in gammaretroviruses. NMR solution structure studies demonstrated a high affinity complex between the ET domain of Brd4 and the EBM of IN, which both in isolation are disordered

(Lin et al. 2008; Aiyer et al. 2014). Other host factors have also been implied to interact with the MLV IN (Studamire and Goff 2010).

Recently, the host FACT complex, which consists of SSRP1 and Spt16, has been shown to directly interact with the CTD of ASLV IN (Winans et al. 2017). The FACT complex is a histone chaperone protein that is involved in DNA replication and facilitates chromatin transcription (Abe et al. 2011; Hsieh et al. 2013; Oliveira et al. 2014; Zhang et al. 2015). The FACT complex stimulated integration activity of avian IN in vitro, and the integration of the viral DNA in virus-infected cells is directly related to the expression levels of FACT complex. It is interesting to speculate that the RSV IN "tail" region may have dual roles in integration: first, promoting the formation of the RSV SSC containing IN octamers (proximal and distal subunits) where the latter subunits support target binding (Yin et al. 2016) and, second, acting as a binding motif for the FACT complex, similar to the "tail" region of MLV IN (Winans et al. 2017). Modeling studies of RSV STC containing octamers suggest that two of the eight tail regions are not associated with the viral DNA (Pandey et al. 2017) possibly allowing them to interact with the FACT complex. Interesting, the viral oncogene vSrc in RSV controls the phosphorylation of S282 in the "tail" region of IN (Fig. 10.4) by a serine kinase in virus particles (Horton et al. 1991; Mumm et al. 1992) which suggests vSrc may influence selection of host integration sites. Further studies are needed to address the precise nature of the interactions of the FACT complex with avian retrovirus IN and specifically the CTD of IN.

RSV IN as Model for HIV-1 IN

As stated previously, all of the retrovirus intasomes contain the common tetrameric CIC. Even though the RSV STC contains this CIC plus two flanking dimers (Yin et al. 2016) and the HIV-1 STC containing this CIC in the predominant tetrameric complex (Passos et al. 2017), these two INs have many commonalities in overall length, linker size connecting domains, and secondary structural features (Figs. 10.4 and 10.5). Further functional similarities have been revealed by site-directed mutagenesis of these INs and viral DNA sequences for activities, their enzymatic sensitivities to HIV-1 STIs both in vitro (Pandey et al. 2014; Pandey et al. 2017) and in vivo (Koh et al. 2011), and the ability of the STIs to kinetically stabilize both the RSV SSC (Pandey et al. 2014; Pandey et al. 2017) (Fig. 10.8) and the HIV-1 SSC with IN tetramers (Bera et al. 2009; Pandey et al. 2010). These apparent functional similarities would appear to mostly map to the active site in the CIC. Further functional studies comparing the two INs and their DNAs by site-directed mutagenesis are discussed below.

The crystal structure of the RSV STC (Yin et al. 2016) has substantiated previous studies of RSV IN and viral DNA requirements and their relationship to HIV-1 IN and its DNA interactions. For example, RSV IN R244 of the inner catalytic subunit of the proximal IN dimer is positioned in the major groove of the viral LTR DNA, closest to G7 of the non-transferred strand. The GC pair at this position is critical for concerted integration of viral DNA by RSV IN (Vora et al. 2004). The equivalent residue E246 of HIV-1 IN interacts with the A7 nucleotide of its non-transferred LTR strand shown by disulfide cross-linking studies (Gao et al. 2001). Mutagenesis of these IN residues on both proteins demonstrated reduced 3' OH processing and strand transfer activities (Lutzke and Plasterk 1998; Peletskaya et al. 2011; Shi et al. 2013). RSV IN W233 is highly conserved in retrovirus INs (Fig. 10.4) (Engelman 1999) and is stacked between R227 and K266. Changing W233 to Glu or Ala but not Phe abolishes binding to the viral DNA LTR sequence and concerted integration activity of RSV IN (Chiu and Grandgenett 2003). Paralleling HIV-1 IN mutations W235E/A/F have corresponding effects on concerted integration activity and virus replication (Chen et al. 1999) suggesting the importance of an aromatic residue at this position in orienting the basic side chains. RSV IN interactions with the viral T66 and R263 bring insights to HIV-1 STI drug-resistance for these same residues of HIV-1 IN (Quashie et al. 2012; Abram et al. 2013). The weakly drug resistance R263K mutation observed in HIV-1 IN and found in patients treated with dolutegravir produces HIV-1 that possesses a significantly reduced replication capacity (Mesplede et al. 2014; Anstett et al. 2015). An R263A mutation in RSV IN affects both its 3' OH processing and concerted integration activities (Shi et al. 2013). The above and future comparison studies of RSV to HIV-1 IN highly suggest that functional studies of RSV IN will provide further insights to HIV-1 IN and the mechanisms associated with STIs.

Conclusions and Perspectives

The oligomerization properties of the different retrovirus INs play dominant roles in the assembly of their intasome structures. The success of unravelling the atomic structures of five different retrovirus intasomes has propelled our understandings of the unique roles that the IN NTDs, CCDs, and CTDs along with their various linker sizes have in assembly of these architectural marvels. Each individual IN subunit plays different multiple roles in the intasome from protein-protein interactions to binding of viral DNA and/or target substrates. The atomic structure of the CIC in all of the retrovirus intasomes has established a firm structural understanding for the concerted integration of the viral DNA genome into the host chromosomes. These understandings will increase our abilities to produce new STIs for drug-resistance HIV-1 IN mutants and possibly development of new inhibitors outside of the active site. The biological roles that these different intasomes assembled in vitro have in the life cycle of each retrovirus are unclear. The higher-order stoichiometry of the IN subunits, particularly associated with the HIV-1 and MVV lentiviral intasomes, is essential to understand to further develop new strategies for prevention of HIV-1/AIDS.

References

Abe T, Sugimura K, Hosono Y, Takami Y, Akita M, Yoshimura A, Tada S, Nakayama T, Murofushi H, Okumura K, Takeda S, Horikoshi M, Seki M, Enomoto T (2011) The histone chaperone facilitates chromatin transcription (FACT) protein maintains normal replication fork rates. J Biol Chem 286(35):30504–30512

Abram ME, Hluhanich RM, Goodman DD, Andreatta KN, Margot NA, Ye L, Niedziela-Majka A, Barnes TL, Novikov N, Chen X, Svarovskaia ES, McColl DJ, White KL, Miller MD (2013) Impact of primary elvitegravir resistance-associated mutations in HIV-1 integrase on drug susceptibility and viral replication fitness. Antimicrob Agents Chemother 57(6):2654–2663

Aiyer S, Swapna GV, Malani N, Aramini JM, Schneider WM, Plumb MR, Ghanem M, Larue RC, Sharma A, Studamire B, Kvaratskhelia M, Bushman FD, Montelione GT, Roth MJ (2014) Altering murine leukemia virus integration through disruption of the integrase and BET protein family interaction. Nucleic Acids Res 42(9):5917–5928

Andrake MD, Skalka AM (1995) Multimerization determinants reside in both the catalytic core and C terminus of avian sarcoma virus integrase. J Biol Chem 270(49):29299–29306

Andrake MD, Skalka AM (2015) Retroviral Integrase: then and Now. Annu Rev Virol 2(1):241–264

Anstett K, Mesplede T, Oliveira M, Cutillas V, Wainberg MA (2015) HIV-1 dolutegravir-resistance substitution R263K cannot co-exist in combination with many classical integrase inhibitor resistance substitutions. J Virol 89:4681–4684

Balakrishnan M, Yant SR, Tsai L, O'Sullivan C, Bam RA, Tsai A, Niedziela-Majka A, Stray KM, Sakowicz R, Cihlar T (2013) Non-catalytic site HIV-1 integrase inhibitors disrupt core maturation and induce a reverse transcription block in target cells. PLoS One 8(9):e74163

Ballandras A, Moreau K, Robert X, Confort MP, Merceron R, Haser R, Ronfort C, Gouet P (2011) A crystal structure of the catalytic core domain of an avian sarcoma and leukemia virus integrase suggests an alternate dimeric assembly. PLoS One 6(8):e23032

Ballandras-Colas A, Brown M, Cook NJ, Dewdney TG, Demeler B, Cherepanov P, Lyumkis D, Engelman AN (2016) Cryo-EM reveals a novel octameric integrase structure for betaretroviral intasome function. Nature 530(7590):358–361

Ballandras-Colas A, Maskell DP, Serrao E, Locke J, Swuec P, Jonsson SR, Kotecha A, Cook NJ, Pye VE, Taylor IA, Andresdottir V, Engelman AN, Costa A, Cherepanov P (2017) A supramolecular assembly mediates lentiviral DNA integration. Science 355(6320):93–95

Bao KK, Wang H, Miller JK, Erie DA, Skalka AM, Wong I (2003) Functional oligomeric state of avian sarcoma virus integrase. J Biol Chem 278(2):1323–1327

Barr SD, Leipzig J, Shinn P, Ecker JR, Bushman FD (2005) Integration targeting by avian sarcoma-leukosis virus and human immunodeficiency virus in the chicken genome. J Virol 79(18):12035–12044

Benleulmi MS, Matysiak J, Henriquez DR, Vaillant C, Lesbats P, Calmels C, Naughtin M, Leon O, Skalka AM, Ruff M, Lavigne M, Andreola ML, Parissi V (2015) Intasome architecture and chromatin density modulate retroviral integration into nucleosome. Retrovirology 12:13

Bera S, Vora AC, Chiu R, Heyduk T, Grandgenett DP (2005) Synaptic complex formation of two retrovirus DNA attachment sites by integrase: a fluorescence energy transfer study. Biochemistry 44(46):15106–15114

Bera S, Pandey KK, Vora AC, Grandgenett DP (2009) Molecular Interactions between HIV-1 integrase and the two viral DNA ends within the synaptic complex that mediates concerted integration. J Mol Biol 389(1):183–198

Bischerour J, Leh H, Deprez E, Brochon JC, Mouscadet JF (2003) Disulfide-linked integrase oligomers involving C280 residues are formed in vitro and in vivo but are not essential for human immunodeficiency virus replication. J Virol 77(1):135–141

Bojja RS, Andrake MD, Weigand S, Merkel G, Yarychkivska O, Henderson A, Kummerling M, Skalka AM (2011) Architecture of a full-length retroviral integrase monomer and dimer, revealed by small angle X-ray scattering and chemical cross-linking. J Biol Chem 286(19):17047–17059

Borrenberghs D, Dirix L, De Wit F, Rocha S, Blokken J, De Houwer S, Gijsbers R, Christ F, Hofkens J, Hendrix J, Debyser Z (2016) Dynamic oligomerization of integrase orchestrates HIV nuclear entry. Sci Rep 6:36485

Bujacz G, Jaskolski M, Alexandratos J, Wlodawer A, Merkel G, Katz RA, Skalka AM (1995) High-resolution structure of the catalytic domain of avian sarcoma virus integrase. J Mol Biol 253(2):333–346

Bujacz G, Alexandratos J, Wlodawer A, Merkel G, Andrake M, Katz RA, Skalka AM (1997) Binding of different divalent cations to the active site of avian sarcoma virus integrase and their effects on enzymatic activity. J Biol Chem 272(29):18161–18168

Bushman FD, Engelman A, Palmer I, Wingfield P, Craigie R (1993) Domains of the integrase protein of human immunodeficiency virus type 1 responsible for polynucleotidyl transfer and zinc binding. Proc Natl Acad Sci U S A 90(8):3428–3432

Cai M, Zheng R, Caffrey M, Craigie R, Clore GM, Gronenborn AM (1997) Solution structure of the N-terminal zinc binding domain of HIV-1 integrase. Nat Struct Biol 4(7):567–577

Chen H, Wei SQ, Engelman A (1999) Multiple integrase functions are required to form the native structure of the human immunodeficiency virus type I intasome. J Biol Chem 274(24):17358–17364

Chen HC, Martinez JP, Zorita E, Meyerhans A, Filion GJ (2017) Position effects influence HIV latency reversal. Nat Struct Mol Biol 24(1):47–54

Cherepanov P, Maertens G, Proost P, Devreese B, Van Beeumen J, Engelborghs Y, De Clercq E, Debyser Z (2003) HIV-1 integrase forms stable tetramers and associates with LEDGF/p75 protein in human cells. J Biol Chem 278(1):372–381

Cherepanov P, Devroe E, Silver PA, Engelman A (2004) Identification of an evolutionarily conserved domain in human lens epithelium-derived growth factor/transcriptional co-activator p75 (LEDGF/p75) that binds HIV-1 integrase. J Biol Chem 279(47):48883–48892

Cherepanov P, Ambrosio AL, Rahman S, Ellenberger T, Engelman A (2005) Structural basis for the recognition between HIV-1 integrase and transcriptional coactivator p75. Proc Natl Acad Sci U S A 102:17308–17313

Chiu R, Grandgenett DP (2003) Molecular and genetic determinants of Rous sarcoma virus integrase for concerted DNA integration. J Virol 77(11):6482–6492

Chow SA, Vincent KA, Ellison V, Brown PO (1992) Reversal of integration and DNA splicing mediated by integrase of human immunodeficiency virus. Science 255(5045):723–726

Christ F, Voet A, Marchand A, Nicolet S, Desimmie BA, Marchand D, Bardiot D, Van der Veken NJ, Van Remoortel B, Strelkov SV, De Maeyer M, Chaltin P, Debyser Z (2010) Rational design of small-molecule inhibitors of the LEDGF/p75-integrase interaction and HIV replication. Nat Chem Biol 6(6):442–448

Ciuffi A, Llano M, Poeschla E, Hoffmann C, Leipzig J, Shinn P, Ecker JR, Bushman F (2005) A role for LEDGF/p75 in targeting HIV DNA integration. Nat Med 11:1287–1289

Coleman J, Eaton S, Merkel G, Skalka AM, Laue T (1999) Characterization of the self association of Avian sarcoma virus integrase by analytical ultracentrifugation. J Biol Chem 274(46):32842–32846

Crowe BL, Larue RC, Yuan C, Hess S, Kvaratskhelia M, Foster MP (2016) Structure of the Brd4 ET domain bound to a C-terminal motif from gamma-retroviral integrases reveals a conserved mechanism of interaction. Proc Natl Acad Sci U S A 113(8):2086–2091

Dar MJ, Monel B, Krishnan L, Shun MC, Di Nunzio F, Helland DE, Engelman A (2009) Biochemical and virological analysis of the 18-residue C-terminal tail of HIV-1 integrase. Retrovirology 6:94

Davies DR, Goryshin IY, Reznikoff WS, Rayment I (2000) Three-dimensional structure of the Tn5 synaptic complex transposition intermediate. Science 289(5476):77–85

De Rijck J, Bartholomeeusen K, Ceulemans H, Debyser Z, Gijsbers R (2010) High-resolution profiling of the LEDGF/p75 chromatin interaction in the ENCODE region. Nucleic Acids Res 38(18):6135–6147

De Rijck J, de Kogel C, Demeulemeester J, Vets S, El Ashkar S, Malani N, Bushman FD, Landuyt B, Husson SJ, Busschots K, Gijsbers R, Debyser Z (2013) The BET family of proteins targets moloney murine leukemia virus integration near transcription start sites. Cell Rep 5(4):886–894

Demeulemeester J, De Rijck J, Gijsbers R, Debyser Z (2015) Retroviral integration: site matters: mechanisms and consequences of retroviral integration site selection. Bioessays 37(11):1202–1214

Desimmie BA, Schrijvers R, Demeulemeester J, Borrenberghs D, Weydert C, Thys W, Vets S, Van Remoortel B, Hofkens J, De Rijck J, Hendrix J, Bannert N, Gijsbers R, Christ F, Debyser Z (2013) LEDGINs inhibit late stage HIV-1 replication by modulating integrase multimerization in the virions. Retrovirology 10:57

Dyda F, Hickman AB, Jenkins TM, Engelman A, Craigie R, Davies DR (1994) Crystal structure of the catalytic domain of HIV-1 integrase: similarity to other polynucleotidyl transferases. Science 266(5193):1981–1986

Eijkelenboom AP, Lutzke RA, Boelens R, Plasterk RH, Kaptein R, Hard K (1995) The DNA-binding domain of HIV-1 integrase has an SH3-like fold. Nat Struct Biol 2(9):807–810

El Ashkar S, De Rijck J, Demeulemeester J, Vets S, Madlala P, Cermakova K, Debyser Z, Gijsbers R (2014) BET-independent MLV-based vectors target away from promoters and regulatory elements. Mol Ther Nucleic Acids 3:e179

Engelman A (1999) In vivo analysis of retroviral integrase structure and function. Adv Virus Res 52:411–426

Engelman A, Cherepanov P (2014) Retroviral integrase structure and DNA recombination mechanism. Microbiol Spectr 2(6):1–22

Engelman AN, Cherepanov P (2017) Retroviral intasomes arising. Curr Opin Struct Biol 47:23–29

Engelman A, Bushman FD, Craigie R (1993) Identification of discrete functional domains of HIV-1 integrase and their organization within an active multimeric complex. EMBO J 12(8):3269–3275

Feng L, Sharma A, Slaughter A, Jena N, Koh Y, Shkriabai N, Larue RC, Patel PA, Mitsuya H, Kessl JJ, Engelman A, Fuchs JR, Kvaratskhelia M (2013) The A128T resistance mutation reveals aberrant protein multimerization as the primary mechanism of action of allosteric HIV-1 integrase inhibitors. J Biol Chem 288(22):15813–15820

Feng L, Larue RC, Slaughter A, Kessl JJ, Kvaratskhelia M (2015) HIV-1 integrase multimerization as a therapeutic target. Curr Top Microbiol Immunol. https://doi.org/10.1007/82_2015_439

Gao K, Butler SL, Bushman F (2001) Human immunodeficiency virus type 1 integrase: arrangement of protein domains in active cDNA complexes. EMBO J 20(13):3565–3576

Grandgenett DP, Vora AC, Schiff RD (1978) A 32,000-dalton nucleic acid-binding protein from avian retravirus cores possesses DNA endonuclease activity. Virology 89(1):119–132

Grandgenett DP, Pandey KK, Bera S, Aihara H (2015) Multifunctional facets of retrovirus integrase. World J Biol Chem 6(3):83–94

Grobler J, McKemma PM, Ly S, Stillmock KA, Bahnck CM, Danovich RM, Dornadula G, Hazuda D, Miller MD (2009) HIV integrase inhibitor dissociation rates correlate with efficacy in vitro. Antivir Ther 14(Supplement 1):A27

Guan R, Aiyer S, Cote ML, Xiao R, Jiang M, Acton TB, Roth MJ, Montelione GT (2017) X-ray crystal structure of the N-terminal region of Moloney murine leukemia virus integrase and its implications for viral DNA recognition. Proteins 85:647–656

Gupta K, Curtis JE, Krueger S, Hwang Y, Cherepanov P, Bushman FD, Van Duyne GD (2012) Solution conformations of prototype foamy virus integrase and its stable synaptic complex with U5 viral DNA. Structure 20:1918–1928

Gupta SS, Maetzig T, Maertens GN, Sharif A, Rothe M, Weidner-Glunde M, Galla M, Schambach A, Cherepanov P, Schulz TF (2013) Bromo- and extraterminal domain chromatin regulators serve as cofactors for murine leukemia virus integration. J Virol 87(23):12721–12736

Gupta K, Turkki V, Sherrill-Mix S, Hwang Y, Eilers G, Taylor L, McDanal C, Wang P, Temelkoff D, Nolte RT, Velthuisen E, Jeffrey J, Van Duyne GD, Bushman FD (2016) Structural basis for inhibitor-induced aggregation of HIV integrase. PLoS Biol 14(12):e1002584

Hare S, Gupta SS, Valkov E, Engelman A, Cherepanov P (2010a) Retroviral intasome assembly and inhibition of DNA strand transfer. Nature 464(7286):232–236

Hare S, Vos AM, Clayton RF, Thuring JW, Cummings MD, Cherepanov P (2010b) Molecular mechanisms of retroviral integrase inhibition and the evolution of viral resistance. Proc Natl Acad Sci U S A 107(46):20057–20062

Hare S, Maertens GN, Cherepanov P (2012) 3'-processing and strand transfer catalysed by retroviral integrase in crystallo. EMBO J 31(13):3020–3028

Hightower KE, Wang R, Deanda F, Johns BA, Weaver K, Shen Y, Tomberlin GH, Carter HL 3rd, Broderick T, Sigethy S, Seki T, Kobayashi M, Underwood MR (2011) Dolutegravir (S/GSK1349572) exhibits significantly slower dissociation than raltegravir and elvitegravir from wild-type and integrase inhibitor-resistant HIV-1 integrase-DNA complexes. Antimicrob Agents Chemother 55(10):4552–4559

Horton R, Mumm SR, Grandgenett DP (1991) Phosphorylation of the avian retrovirus integration protein and proteolytic processing of its carboxyl terminus. J Virol 65(3):1141–1148

Hsieh FK, Kulaeva OI, Patel SS, Dyer PN, Luger K, Reinberg D, Studitsky VM (2013) Histone chaperone FACT action during transcription through chromatin by RNA polymerase II. Proc Natl Acad Sci U S A 110(19):7654–7659

Jones KS, Coleman J, Merkel GW, Laue TM, Skalka AM (1992) Retroviral integrase functions as a multimer and can turn over catalytically. J Biol Chem 267(23):16037–16040

Jurado KA, Wang H, Slaughter A, Feng L, Kessl JJ, Koh Y, Wang W, Ballandras-Colas A, Patel PA, Fuchs JR, Kvaratskhelia M, Engelman A (2013) Allosteric integrase inhibitor potency is determined through the inhibition of HIV-1 particle maturation. Proc Natl Acad Sci U S A 110(21):8690–8695

Kessl JJ, Kutluay SB, Townsend D, Rebensburg S, Slaughter A, Larue RC, Shkriabai N, Bakouche N, Fuchs JR, Bieniasz PD, Kvaratskhelia M (2016) HIV-1 integrase binds the viral RNA genome and is essential during virion morphogenesis. Cell 166(5):1257–1268. e1212

Kirk PD, Huvet M, Melamed A, Maertens GN, Bangham CR (2016) Retroviruses integrate into a shared, non-palindromic DNA motif. Nat Microbiol 2:16212

Knaus RJ, Hippenmeyer PJ, Misra TK, Grandgenett DP, Muller UR, Fitch WM (1984) Avian retrovirus pp32 DNA binding protein. Preferential binding to the promoter region of long terminal repeat DNA. Biochemistry 23(2):350–359

Koh Y, Matreyek KA, Engelman A (2011) Differential sensitivities of retroviruses to integrase strand transfer inhibitors. J Virol 85(7):3677–3682

Krishnan L, Li X, Naraharisetty HL, Hare S, Cherepanov P, Engelman A (2010) Structure-based modeling of the functional HIV-1 intasome and its inhibition. Proc Natl Acad Sci U S A 107:15910–15915

Kulkosky J, Jones KS, Katz RA, Mack JP, Skalka AM (1992) Residues critical for retroviral integrative recombination in a region that is highly conserved among retroviral/retrotransposon integrases and bacterial insertion sequence transposases. Mol Cell Biol 12(5):2331–2338

Kvaratskhelia M, Sharma A, Larue RC, Serrao E, Engelman A (2014) Molecular mechanisms of retroviral integration site selection. Nucleic Acids Res 42(16):10209–10225

Lee SP, Xiao J, Knutson JR, Lewis MS, Han MK (1997) Zn2+ promotes the self-association of human immunodeficiency virus type-1 integrase in vitro. Biochemistry 36(1):173–180

Lesbats P, Engelman AN, Cherepanov P (2016) Retroviral DNA integration. Chem Rev. 10.1021/acs.chemrev.6b00125

Li M, Craigie R (2009) Nucleoprotein complex intermediates in HIV-1 integration. Methods 47:237–242

Li M, Mizuuchi M, Burke TR Jr, Craigie R (2006) Retroviral DNA integration: reaction pathway and critical intermediates. EMBO J 25(6):1295–1304

Li X, Krishnan L, Cherepanov P, Engelman A (2011) Structural biology of retroviral DNA integration. Virology 411(2):194–205

Lin YJ, Umehara T, Inoue M, Saito K, Kigawa T, Jang MK, Ozato K, Yokoyama S, Padmanabhan B, Guntert P (2008) Solution structure of the extraterminal domain of the bromodomain-containing protein BRD4. Protein Sci 17(12):2174–2179

Lubkowski J, Yang F, Alexandratos J, Wlodawer A, Zhao H, Burke TR Jr, Neamati N, Pommier Y, Merkel G, Skalka AM (1998) Structure of the catalytic domain of avian sarcoma virus integrase with a bound HIV-1 integrase-targeted inhibitor. Proc Natl Acad Sci U S A 95(9):4831–4836

Lusic M, Siliciano RF (2017) Nuclear landscape of HIV-1 infection and integration. Nat Rev. Microbiol 15(2):69–82

Lutzke RA, Plasterk RH (1998) Structure-based mutational analysis of the C-terminal DNA-binding domain of human immunodeficiency virus type 1 integrase: critical residues for protein oligomerization and DNA binding. J Virol 72(6):4841–4848

Maertens GN, Hare S, Cherepanov P (2010) The mechanism of retroviral integration from X-ray structures of its key intermediates. Nature 468(7321):326–329

Maskell DP, Renault L, Serrao E, Lesbats P, Matadeen R, Hare S, Lindemann D, Engelman AN, Costa A, Cherepanov P (2015) Structural basis for retroviral integration into nucleosomes. Nature 523(7560):366–369

McCord M, Stahl SJ, Mueser TC, Hyde CC, Vora AC, Grandgenett DP (1998) Purification of recombinant Rous sarcoma virus integrase possessing physical and catalytic properties similar to virion-derived integrase. Protein Expr Purif 14(2):167–177

Mesplede T, Osman N, Wares M, Quashie PK, Hassounah S, Anstett K, Han Y, Singhroy DN, Wainberg MA (2014) Addition of E138K to R263K in HIV integrase increases resistance to dolutegravir, but fails to restore activity of the HIV integrase enzyme and viral replication capacity. J Antimicrob Chemother 69(10):2733–2740

Mitchell RS, Beitzel BF, Schroder AR, Shinn P, Chen H, Berry CC, Ecker JR, Bushman FD (2004) Retroviral DNA integration: ASLV, HIV, and MLV show distinct target site preferences. PLoS Biol 2(8):E234

Mohammed KD, Topper MB, Muesing MA (2011) Sequential deletion of the integrase (Gag-Pol) carboxyl-terminus reveals distinct phenotypic classes of defective HIV-1. J Virol 85:4654–4666

Montano SP, Pigli YZ, Rice PA (2012) The mu transpososome structure sheds light on DDE recombinase evolution. Nature 491(7424):413–417

Moreau K, Torne-Celer C, Faure C, Verdier G, Ronfort C (2000) In vivo retroviral integration: fidelity to size of the host DNA duplication might Be reduced when integration occurs near sequences homologous to LTR ends. Virology 278(1):133–136

Moreau K, Faure C, Verdier G, Ronfort C (2002) Analysis of conserved and non-conserved amino acids critical for ALSV (Avian leukemia and sarcoma viruses) integrase functions in vitro. Arch Virol 147(9):1761–1778

Moreau K, Faure C, Violot S, Gouet P, Verdier G, Ronfort C (2004) Mutational analyses of the core domain of Avian Leukemia and Sarcoma Viruses integrase: critical residues for concerted integration and multimerization. Virology 318(2):566–581

Mumm SR, Horton R, Grandgenett DP (1992) v-Src enhances phosphorylation at Ser-282 of the Rous sarcoma virus integrase. J Virol 66(4):1995–1999

Nowotny M (2009) Retroviral integrase superfamily: the structural perspective. EMBO Rep 10(2):144–151

Oliveira DV, Kato A, Nakamura K, Ikura T, Okada M, Kobayashi J, Yanagihara H, Saito Y, Tauchi H, Komatsu K (2014) Histone chaperone FACT regulates homologous recombination by chromatin remodeling through interaction with RNF20. J Cell Sci 127(Pt 4):763–772

Pandey KK, Bera S, Zahm J, Vora A, Stillmock K, Hazuda D, Grandgenett DP (2007) Inhibition of human immunodeficiency virus type-1 concerted integration by strand transfer inhibitors which recognize a transient structural intermediate. J Virol 81:12189–12199

Pandey KK, Bera S, Vora AC, Grandgenett DP (2010) Physical trapping of HIV-1 synaptic complex by different structural classes of integrase strand transfer inhibitors. Biochemistry 49:8376–8387

Pandey KK, Bera S, Grandgenett DP (2011) The HIV-1 Integrase Monomer Induces a Specific Interaction with LTR DNA for Concerted Integration. Biochemistry 50(45):9788–9796

Pandey KK, Bera S, Korolev S, Campbell M, Yin Z, Aihara H, Grandgenett DP (2014) Rous sarcoma virus synaptic complex capable of concerted integration is kinetically trapped by human immunodeficiency virus integrase strand transfer inhibitors. J Biol Chem 289:19648–19658

Pandey KK, Bera S, Shi K, Aihara H, Grandgenett DP (2017) A C-terminal tail region in the rous sarcoma virus integrase provides high plasticity of functional integrase oligomerization during intasome assembly. J Biol Chem 292:5018–5030

Passos DO, Li M, Yang R, Rebensburg SV, Ghirlando R, Jeon Y, Shkriabai N, Kvaratskhelia M, Craigie R, Lyumkis D (2017) Cryo-EM structures and atomic model of the HIV-1 strand transfer complex intasome. Science 355(6320):89–92

Peletskaya E, Andrake M, Gustchina A, Merkel G, Alexandratos J, Zhou D, Bojja RS, Satoh T, Potapov M, Kogon A, Potapov V, Wlodawer A, Skalka AM (2011) Localization of ASV integrase-DNA contacts by site-directed crosslinking and their structural analysis. PLoS One 6(12):e27751

Petit C, Schwartz O, Mammano F (1999) Oligomerization within virions and subcellular localization of human immunodeficiency virus type 1 integrase. J Virol 73(6):5079–5088

Pommier Y, Johnson AA, Marchand C (2005) Integrase inhibitors to treat HIV/AIDS. Nat Rev Drug Discov 4(3):236–248

Pommier Y, Kiselev E, Marchand C (2015) Interfacial inhibitors. Bioorg Med Chem Lett 25(18):3961–3965

Quashie PK, Mesplede T, Han YS, Oliveira M, Singhroy DN, Fujiwara T, Underwood MR, Wainberg MA (2012) Characterization of the R263K mutation in HIV-1 integrase that confers low-level resistance to the second-generation integrase strand transfer inhibitor dolutegravir. J Virol 86(5):2696–2705

Quercioli V, Di Primio C, Casini A, Mulder LC, Vranckx LS, Borrenberghs D, Gijsbers R, Debyser Z, Cereseto A (2016) Comparative analysis of HIV-1 and murine leukemia virus three-dimensional nuclear distributions. J Virol 90(10):5205–5209

Schroder AR, Shinn P, Chen H, Berry C, Ecker JR, Bushman F (2002) HIV-1 integration in the human genome favors active genes and local hotspots. Cell 110(4):521–529

Sharma A, Larue RC, Plumb MR, Malani N, Male F, Slaughter A, Kessl JJ, Shkriabai N, Coward E, Aiyer SS, Green PL, Wu L, Roth MJ, Bushman FD, Kvaratskhelia M (2013) BET proteins promote efficient murine leukemia virus integration at transcription start sites. Proc Natl Acad Sci U S A 110(29):12036–12041

Shi K, Pandey KK, Bera S, Vora AC, Grandgenett DP, Aihara H (2013) A possible role for the asymmetric C-terminal domain dimer of Rous sarcoma virus integrase in viral DNA binding. PLoS One 8(2):e56892

Skalka AM (2014) Retroviral DNA transposition: themes and variations. Microbiol Spectr 2(5):1–22

Studamire B, Goff SP (2010) Interactions of host proteins with the murine leukemia virus integrase. Viruses 2(5):1110–1145

Vora A, Grandgenett DP (2001) DNase protection analysis of retrovirus integrase at the viral DNA ends for full-site integration in vitro. J Virol 75(8):3556–3567

Vora A, Bera S, Grandgenett D (2004) Structural organization of avian retrovirus integrase in assembled intasomes mediating full-site integration. J Biol Chem 279(18):18670–18678

Wei SQ, Mizuuchi K, Craigie R (1997) A large nucleoprotein assembly at the ends of the viral DNA mediates retroviral DNA integration. EMBO J 16(24):7511–7520

Wei SQ, Mizuuchi K, Craigie R (1998) Footprints on the viral DNA ends in moloney murine leukemia virus preintegration complexes reflect a specific association with integrase. Proc Natl Acad Sci U S A 95(18):10535–10540

Winans S, Larue RC, Abraham CM, Shkriabai N, Skopp A, Winkler D, Kvaratskhelia M, Beemon KL (2017) The FACT complex promotes avian leukosis virus DNA integration. J Virol 91:e00082–e00017

Wu X, Li Y, Crise B, Burgess SM (2003) Transcription start regions in the human genome are favored targets for MLV integration. Science 300(5626):1749–1751

Yang ZN, Mueser TC, Bushman FD, Hyde CC (2000) Crystal structure of an active two-domain derivative of Rous sarcoma virus integrase. J Mol Biol 296(2):535–548

Yin H, Lapkouski M, Yang W, Craigie R (2012) Assembly of prototype foamy virus strand transfer complexes on product DNA bypassing catalysis of integration. Protein Sci 21:1849–1857

Yin Z, Shi K, Banerjee S, Pandey KK, Bera S, Grandgenett DP, Aihara H (2016) Crystal structure of the Rous sarcoma virus intasome. Nature 530(7590):362–366

Zhang W, Zeng F, Liu Y, Shao C, Li S, Lv H, Shi Y, Niu L, Teng M, Li X (2015) Crystal Structure of Human SSRP1 Middle Domain Reveals a Role in DNA Binding. Sci Rep 5:18688

Zheng R, Jenkins TM, Craigie R (1996) Zinc folds the N-terminal domain of HIV-1 integrase, promotes multimerization, and enhances catalytic activity. Proc Natl Acad Sci U S A 93(24):13659–13664

Zhou H, Rainey GJ, Wong SK, Coffin JM (2001) Substrate sequence selection by retroviral integrase. J Virol 75(3):1359–1370

Chapter 11
Structure and Function of the Human Respiratory Syncytial Virus M2–1 Protein

Selvaraj Muniyandi, Georgia Pangratiou, Thomas A. Edwards, and John N. Barr

Introduction

RNA viruses can be divided into positive- and negative-sense groups based on the ability of their RNA genomes to act as messenger RNA (mRNA) for the production of proteins. In positive-sense RNA viruses, the genome can be translated directly, while in negative-sense RNA viruses, the input genome must undergo a copying event to produce coding sense mRNAs. The negative-sense RNA viruses (NSV) can be further divided into segmented (SNSV) and non-segmented negative-strand viruses (NSNSV), depending on whether the genome is a single chain of ribonucleotides or whether it is segmented into two or more separate RNA molecules. The group of NSNSVs encompasses pathogens of humans, animals and plants, and notable examples include rabies virus, *Nipah virus*, Ebola virus (EBOV) and human respiratory syncytial virus (HRSV), many of which are without effective preventative or therapeutic options to prevent disease (Palese et al. 1996; Tao and Ye 2010).

HRSV is the leading cause of lower respiratory tract illness in infants, elderly and immunocompromised individuals, causing bronchiolitis and pneumonia. Early and recurrent infections have been linked to the development of asthma in later life, which has a significant economic burden (Bohmwald et al. 2016). It has been estimated that all infants across the globe are infected with HRSV at least once during their first 2 years of life. Mild infections have a recovery time of 1–2 weeks; however, HRSV infection can be serious and often fatal, and estimates for the number of fatalities per year are put between 50,000 and 199,000 (Scheltema et al. 2017; Shi et al. 2017). Presently, no vaccine exists for HRSV, and the only options to reduce

S. Muniyandi · G. Pangratiou · T. A. Edwards · J. N. Barr (✉)
School of Molecular and Cellular Biology, and The Astbury Centre for Structural Molecular Biology, University of Leeds, Leeds, UK
e-mail: S.Muniyandi@leeds.ac.uk; bs12g2p@leeds.ac.uk; t.a.edwards@leeds.ac.uk; j.n.barr@leeds.ac.uk

© Springer Nature Singapore Pte Ltd. 2018
J. R. Harris, D. Bhella (eds.), *Virus Protein and Nucleoprotein Complexes*, Subcellular Biochemistry 88, https://doi.org/10.1007/978-981-10-8456-0_11

Fig. 11.1 Organization of genes in the HRSV negative-strand non-segmented RNA genome. The gene start (GS) and gene end (GE) are shown as black borders on each gene, with a gene overlap between M2 and L genes

HRSV-mediated disease is preventative administration of a humanized mouse monoclonal antibody directed to a neutralizing epitope on the fusion protein (F) preventing host cell entry. This molecule is known as Palivizumab (or Synagis), and is effective, providing short-term protection to infants at high risk (Brady 2014), although the cost of this treatment is a current barrier to its widespread usage (Hu and Robinson 2010). Post-exposure treatment for HRSV is limited to the FDA-approved administration of nebulized ribavirin (a guanine nucleotide analogue). However, this drug exhibits significant toxicity, and its effectiveness has been questioned (Marcelin et al. 2014). Thus, there is an urgent need to develop new antivirals against HRSV.

This chapter describes the structure and function of the HRSV M2–1 protein, which is an essential transcription factor required for the synthesis of HRSV mRNAs. Determination of the high-resolution structure of M2–1, in combination with defining the structural basis of its function, could aid in the design of new antiviral compounds through structure-based drug design.

The Virion

HRSV is classified in the *Pneumoviridae* family of the order *Mononegavirales*. HRSV virions are pleiomorphic, with a cell-derived lipid membrane surrounding an internal ribonucleocapsid (RNP) complex comprising the HRSV RNA genome in association with multiple viral proteins. The genome is approximately 15,200 nt in length and includes 10 genes coding for 11 proteins (Fig.11.1). Of these 11 proteins, 3 are associated with the RNP: the nucleoprotein (N), the RNA-dependent RNA polymerase (L) and the phosphoprotein (P). Three HRSV proteins are associated with the viral envelope, namely, the fusion protein (F), the so-called attachment glycoprotein (G) and the small hydrophobic protein (SH) (Table 11.1). The matrix protein (M) is situated below the viral envelope and plays a pivotal role in virion assembly. NS1 and NS2 proteins are nonstructural, being absent from the virion, and they play critical roles in overcoming the innate immune defences of the infected host cell. The remaining two proteins are encoded by the M2 gene and are translated from overlapping reading frames on a single M2 mRNA transcript. The M2–1 protein was initially characterized as a matrix protein that associated with the viral envelope (Collins and Wertz, 1985) but is currently described as an essential transcription factor (Collins et al. 1995; Collins et al. 1996) and may also provide a role in virion assembly (Li et al. 2008; Kiss et al. 2014). The M2–2 protein has been

Table 11.1 List of proteins encoded by the HRSV genome and their function

Gene position	Protein	Function
1.	Non-structural protein (NS1)	Types I and III IFN antagonists. IFN α/β antagonist-mediating antiviral state, suppressing maturation of dendritic cells and T-lymphocyte response. Inhibits phosphorylation of IFN response element 3 disrupting binding to IFN promoter and decreases STAT2 production through degradation
2.	Non-structural protein (NS2)	Type I interferon antagonist. Causes degradation of STAT2 and interacts with RIG-I to suppress IFN synthesis
3.	Nucleoprotein (N)	Sequesters viral RNA forming a nucleocapsid (NC) (protein-RNA complex) which is helical. Associates with RNA forming the ribonucleocapsid (RNP) complex
4.	Phosphoprotein (P)	L-protein cofactor that interacts with the NC to place L onto the RNA. Also interacts with M2-1
5.	Matrix protein (M)	Drive HRSV assembly and budding; vital for virus particle formation, having positive and hydrophobic domain important for cytoplasmic membrane binding
6.	Short hydrophobic protein (SH)	Forms a pentameric ion channel, and is able to inhibit tumour necrosis factor alpha (TNFα) signalling, perhaps helping HRSV evade the immune system
7.	Glycoprotein (G)	Involved in viral attachment to the host cell
8.	Fusion protein (F)	Required for fusion of host cell via membranes and promotes syncytia
=9.	M2-1	Transcription factor, with other potential roles as discussed in the text
=9.	M2-2	Inhibits viral transcription up-regulating RNA therefore mediates the regulatory switch from transcription to RNA replication
10.	L-protein	RNA-dependent RNA polymerase (RdRp) transcribing/replicating the viral genome

shown to influence the activity of the viral polymerase (Table 11.1) (Bermingham and Collins 1999), and has been shown to be dispensable for virus multiplication, although viruses in which the M2–2 reading frame is not accessible are growth attenuated.

Transcription in HRSV

Common with all NSNSVs, the gene expression programme of HRSV begins with the input negative-sense RNA genome being copied to produce mRNA transcripts in a process known as primary transcription. This activity is mediated by the resident polymerase, brought into the infected cell with the input negative-sense genome. These transcripts are subsequently translated into either structural proteins that are needed for subsequent rounds of RNA synthesis and for assembly into viral progeny, or they are non-structural proteins that modify the host cell environment facilitating virus multiplication. At a later time point, the input genome is replicated to form positive-sense copies of the input genome, known as anti-genomes, and the events that mediate this switch in polymerase activity from a transcriptase to a replicase are not yet resolved. The promoters for both these activities are located within the initial 44 nucleotides of the genome 3′ end, but the start sites for these district activities are different; replication begins at position 1, whereas transcription is initiated alongside position 3 (Tremaglio et al. 2013). Subsequently, the transcribing polymerase moves along the template in the 3′ to 5′ direction and recognizes and responds to gene-start and gene-end signals that flank each of the HRSV genes. These conserved sequences signal the initiation and termination of a single 5′ capped and a 3′ poly (A) tailed mRNA from each of the 10 HRSV genes (Kuo et al. 1996). In contrast, the replicating polymerase ignores these gene-start and gene-end signals, to synthesize a complementary copy of the genome, which is enwrapped in nucleoprotein (N protein) and subsequently recruits a polymerase complex to form a helical RNP assembly.

Studies using truncated HRSV genomes in which RNP complexes are reconstituted in cells by supplying the protein and RNA components in trans have shown that replication requires P- and L-proteins, whereas transcription requires P, L and also M2–1 (Yu et al. 1995; Collins et al. 1995; Collins et al. 1996). In the absence of M2–1, the abundance of full-length transcription products is reduced, suggesting that the L-protein may be poorly processive, especially during the transcription of long genes (Fearns and Collins 1999). Inclusion of M2–1 in the transcription assay increases the abundance of full-length transcripts, and taken together, these observations have led to the proposal that M2–1 is a processivity factor, allowing the RdRp to reach the end of each transcriptional unit (Collins et al. 1995). In addition, the presence of M2–1 in reconstitution assays appears to influence the ability of the polymerase to terminate transcription at the HRSV conserved gene-end signals, leading to increased synthesis of multicistronic mRNAs spanning two or more transcriptional units (Hardy and Wertz 1998). Whether the polymerase processivity and anti-termination activities of M2–1 are related has not been resolved; it is possible that increased polymerase processivity is observed when M2–1 prevents the polymerase from terminating at sequences resembling gene ends, spuriously located within genes (Sutherland et al. 2001).

More recently, a post-transcriptional role of M2–1 has recently been proposed (Rincheval et al. 2017), as well as recent reports that M2–1 is found in association with M and RNPs within the virion, thus implicating a structural role of M2–1 in virion assembly (Kiss et al. 2014).

Attempts to rescue infectious HRSV from cDNAs from which the M2–1 gene has been omitted have been unsuccessful (Collins et al. 1995), leading to the proposal that M2–1 is essential for virus multiplication, a proposition that is consistent with its apparent and critical role in ensuring full-length transcription of all ten HRSV mRNAs. However, the possibility that the requirement for M2–1 coding capacity is for one of the alternative non-transcription functions cannot be eliminated. Regardless, the requirement of M2–1 for the HRSV lifecycle means it represents a promising target for the design of antiviral molecules that disrupt one or more of its proposed functions and thus block HRSV multiplication and thus disease.

Macromolecular Interactions Involved in M2–1-Mediated Anti-termination

M2–1 is soluble as a tetramer and directly interacts with RNA and with P in a competitive manner (Blondot et al. 2012). It is not clear whether this dual interaction with RNA and P is strictly mutually exclusive or can occur simultaneously on different monomers within the tetramer. P is also a tetramer, and residues within its N-terminus (residues 1–28) interact with N, supplying RNA-free N monomers to encapsidate the newly synthesized replication products in the form of RNPs (Renner et al. 2016). This chaperone activity of P maintains N in a soluble form that otherwise has a tendency to aggregate as helical assemblies, possibly through interacting with host cell RNAs. The C-terminal domain of P, particularly the last nine residues, interacts with N-proteins that are components of the viral RNPs, rather than free N monomers (Tran et al. 2007). Just N-terminal, proximal to this N-interaction region of P, lies the L-protein interacting site (residues 216–239, with crucial hydrophobic residues at 216, 223 and 227) (Sourimant et al. 2015).

The aforementioned proteins, N, P, L and M2–1 form the minimal requirement for the in vitro reconstitution of the HRSV transcription machinery. The variable, overlapping and multivalent molecular interactions involved in formation of the transcriptase and replicase complexes, as well as the large size, represents a significant bottleneck in the in vitro stoichiometric reconstitution and structural analysis of this assembly. All of these proteins, partly based on structural and sequence analysis, appear to have rigid domains connected by hinges facilitating domain movements (Tanner et al. 2014; Tawar et al. 2009; Liang et al. 2015). This can be expected to yield a megadalton complex that is inherently flexible, thereby resulting in conformational heterogeneity in the final assembly, a problem traditionally difficult to address via structural biology. Detailed structural understanding of the molecular interactions involved in HRSV anti-termination would reveal details of the intermolecular interfaces that could be used for structure-based inhibitor design of antiviral therapeutics, in addition to expanding our understanding of the process.

Functional and structural homologs of the above-mentioned replicase and transcriptase complex-associated proteins exist in all other NSNSVs, which also share a broadly similar transcriptional programme. However, it is interesting to note that few NSNSVs possess a separate protein product that acts as a transcription factor, with only the various M2–1 proteins expressed by the pneumoviruses and VP30 expressed by the filoviruses, including Ebola virus (EBOV). Like M2–1, VP30 appears to be dispensable for replication but essential for RdRp transcriptional activities, which is dependent on VP30 phosphorylation at specific sites (Biedenkopf et al. 2016; Lier et al. 2017). Presumably, the many other viruses that do not express such a protein either have incorporated protein modules that possess similar functions elsewhere or utilize a transcription mechanism that has no such requirement. In any case, similarities between M2–1 and VP30 at both the structural and functional levels mean that the knowledge gained from understanding M2–1 structure and function will have important benefits for the other viruses within *Pneumoviridae* and *Filoviridae* families.

Crystal Structure of HRSV M2–1 Protein

Overexpression and Purification of M2–1 for Structural Studies

Full-length HRSV (strain A2) M2–1 (194 residues) has a requirement for zinc ions during high-level expression of folded protein. M2–1 crystallized in two different space groups under conditions containing polyethylene glycol and cadmium ions. X-ray diffraction data was collected, and the crystal structure of M2–1 was solved to 2.5 Å using the anomalous signal from the cadmium ions and from the bound zinc atoms (Tanner et al. 2014).

Molecular Architecture of M2–1

Monomeric M2–1 consists of three distinct regions linked by unstructured flexible sequences (Fig. 11.2a). These are the zinc-binding domain (ZBD: residues 7–25), a tetramerization helix (residues 32–49) and the core domain (residues 69–172). Connecting the tetramerization helix and the core domain is a flexible linker (residues 52–67) that is poorly resolved in the crystal structure, and which includes two sites that can be phosphorylated (S58 and S61) with important consequences relating to the function of M2–1 as a transcription factor. The last 20 residues of the C-terminus are not resolved in the electron density as they are unstructured, and many of these residues are dispensable, as determined by the rescue of infectious HRSV with these residues deleted from the M2–1 ORF. Overall, the molecule has nine alpha helices and no beta strands (Fig. 11.2a). Each monomer of M2–1 interacts with other protomers forming a highly stable tetramer.

M2–1 tetramerisation is mediated by the oligomerization helix (residues 32–49, α1) that buries a series of hydrophobic residues (L36, L43, I46 and M50) on one helix face within a four-helix bundle. There is also extensive interaction between the ZBD of one monomer within the NTD and the core of an adjacent monomer, significantly increasing the buried surface area in the tetramer and consequently increasing its predicted stability. In the context of the tetramer, the ZBD lies on the N-terminal face of the molecule, in close proximity to the RNA-binding surface (Fig. 11.2b), which also includes residues R3 and R4 from the extreme N-terminus of M2–1.

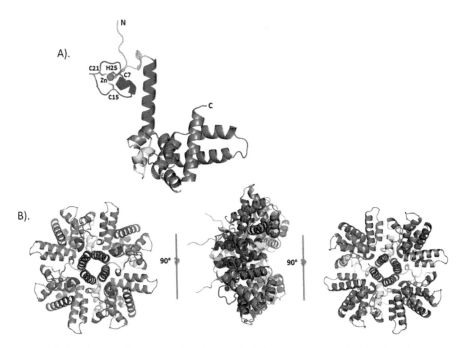

Fig. 11.2 (**a**) The crystal structure of HRSV M2–1. A single protomer is depicted in ribbons representation above. Domains are coloured, from N- to C-terminus: ZBD (*blue*), tetramarization helix (*red*), flexible phosphorylation site (*yellow*) and the core domain (*grey*). Zn^{2+} atom and its coordinating residues are labelled. The green regions represent flexible loops connecting the listed domains in M2–1. (**b**) The tetramer organization in M2–1 shown in three different views

The crystal structure of M2–1 provides a framework for the rational analysis of residues involved in direct M2–1 interactions with RNA, P and M. In addition, the structure will provide a framework on which to explore the mechanisms by which M2–1 performs its various assigned or proposed functions, as well as understand how these multiple functions may be regulated.

RNA Binding by M2–1

Electrostatic potential calculations on M2–1 revealed that its surface is saturated by positively charged residues (Fig. 11.3) that form four prominent tracts on the tetramer. In order to charaterize potential RNA ligands, and thus provide further information regarding the mechanism of M2–1 function, quantitative RNA-binding studies were performed using fluorescence anisotropy (FA) with different RNA sequences. These included polyribonucleotides of A, C, G and U, as well as RNAs representing various positive- and negative-sense HRSV gene-end sequences. These sequences were chosen as all functional studies performed to date reveal M2–1 is

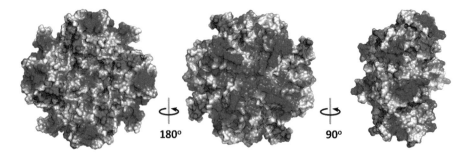

Fig. 11.3 Three views of the electrostatistic potential on HRSV M2–1 surface revealing positively charged surfaces

active using minigenomes in which all HRSV-specific sequences have been eliminated, excepting conserved gene-start and gene-end sequences. The highest binding affinity of all was determined for the polyribonucleotide poly(A) (K_d = 19.1 nM), and comparison of the binding specificities of HRSV gene-end sequences indicated that positive-sense sequences exhibited consistently higher affinities than negative-sense complements, but RNA affinities broadly corresponded with their A content. Variation of up to fivefold were found in M2–1 binding affinity to the different gene-end sequences tested here, with SH gene-end binding the tightest (K_d = 46 nM). These data provide evidence that the function of M2–1 involves binding RNA sequences that are located at the 3′ end of HRSV mRNAs, and this location is consistent with its proposed role in influencing the process of transcription termination.

Mapping M2–1 Residues that Interact with RNA

With M2–1:RNA interactions established through FA and identification of high-affinity RNA ligands, the contribution of specific M2–1 residues towards RNA binding was examined through a variety of M2–1 mutants. These were selected based on the tetramer electrostatics and the previous NMR-based on RNA-binding analysis of the M2–1 globular core (58–177) (Blondot et al. 2012) (Fig. 11.4). Multiple residues that lie within the core resulted in the reduced RNA-binding affinity, including S58D/S61D, K92A, K92D, K150A, R151A, R151D, K150A/R151A and K159A. In addition, residues R3 and R4 that lie within the N-terminal tip of M2–1 were also found to contribute to RNA binding, showing that residues involved in this activity are not restricted only to the core but extended outside.

The S58D/S61D double mutant represents a mimic of phosphorylated M2–1, and these two serine residues are located above the positively charged surface in the 3D structure of M2–1, in a position that is consistent with the observed influence over RNA binding. The reduced affinity of this phosphomimetic mutant for RNA suggests that M2–1 phosphorylation during the infectious cycle may modulate

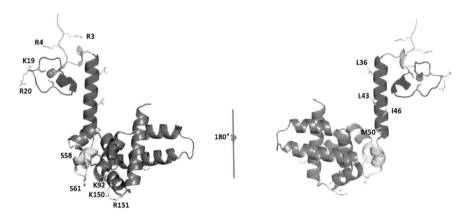

Fig. 11.4 Residues identified, mutated and examined for RNA binding and anti-terminaton activity in M2–1 are lablelled on the ribbons diagram of a single protomer, coloured as for Fig. 11.2. All residues are surface exposed. Hydrophobic residues lining the tetramerization helix (buried in a four-helix bundle in the tetramer) are shown as white sticks

binding affinity to target RNAs, thereby potentially affecting M2–1 function. Comparison between the crystal structures of native M2–1 and the phosphomimetic M2–1 mutant S58D/S61D suggests that phosphorylation does not cause extensive structural rearrangements within the M2–1 tetramer, thus suggesting these functional changes may be due to alterations in surface electrostatic potential and possibly its interaction with an RNA-binding partner.

Identification of M2–1 Residues Involved in Transcription Factor Activity

The aforementioned residues that influence RNA binding were assessed for their contribution towards M2–1 anti-termination activity using a bicistronic HRSV minigenome (Tanner et al. 2014), in which two transcription units were separated by functional gene-start and gene-end sequences, forming a gene junction. This minigenome transcription system was set up so that the expression of luciferase from the downstream gene was dependent on the presence of M2–1, with the implication being that M2–1 was required in order for the polymerase to complete transcription of the long upstream gene and thus gain access to the transcription start site of the downstream luciferase gene. Minigenome analysis found that all previously identified mutants that exhibited a reduction in RNA binding also showed reduced luciferase expression activity *in cellulo*, thus suggesting that the RNA-binding activity of M2–1 was tied to its ability to act as a transcription factor.

Fig. 11.5 Superposition of M2–1 HRSV M2–1 and VP30 of Ebola reveals structural similarity; note the overlay of most helices. The zinc bound to M2–1 is highlighted in green, with the N-termini of both models labelled with blue spheres and the C-termini red spheres

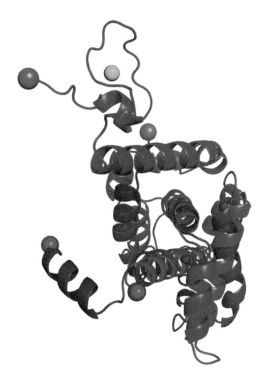

HRSV M2–1 Structure: Similarities and Differences to Other Known Related Structures

The four-helix bundle formed by the oligomerization helix strongly favours the tetrameric organization of M2–1, with the multi-protomer contacts of the ZBD increasing oligomer stability. Comparison of the M2–1 X-ray structure with the NMR structure of the M2–1 core domain (Blondot et al. 2012) (residues 58–177) reveals a similar core fold, suggesting that the core domain can fold independently in the absence of all other M2–1 residues, although residues outside the core contribute to RNA binding as described above (Blondot et al. 2012). Superposition of the full-length M2–1 structure and the EBOV VP30 C-terminal domain shows structural similarities, with helices 3, 6 and 7 aligning well to each other, while helices 4 and 5 share the same orientation (Fig. 11.5). This close structural alignment implies strong evolutionary ties and also raises the possibility that these polymerase accessory proteins may act through common mechanisms. As described above, the EBOV VP30 is one of only a small number of polymerase accessory proteins within the broad group of mononegaviruses, and like M2–1 it possess a ZBD, and its function is modulated by dynamic phosphorylation (Lier et al. 2017). For both M2–1 and VP30, dynamic reversible phosphorylation is required for their respective transcription factor functionality, with VP30 phospho-ablatant mutants being inactive (Biedenkopf et al. 2016).

The Cys3-His ZBD of M2–1 is noncanonical with no structural homologs found by the DALI server (Fig. 11.1a), and similarity to the VP30 ZBD is unknown due to the current lack of a VP30 ZBD structural model. Sequence searches suggest that the M2–1 ZBD is related to the ZBD of Nup475, which was also shown to bind cadmium (Amann et al. 2003).

From the crystal structure of M2–1, it is possible to visualize a track that bound RNA might take across the surface of each monomer, linking residues that are known to influence RNA binding and also encompassing a contiguous region of positive charge. RNA and P have overlapping binding sites on the M2–1 surface and within one protomer; thus, binding is mutually exclusive. M2–1 binds P in a 1:1 stoichiometry at high affinity suggesting that each M2–1 tetramer potentially binds to a P-tetramer. However, the possibility that each monomer within a tetramer might be occupied by either RNA or P ligands cannot be ruled out.

M2–1 from the closely related human metapneumovirus in the *Pneumoviridae* family shares 38% sequence homology with HRSV M2–1 (Leyrat et al. 2014). HMPV M2–1 is also tetrameric with an overall domain organization that is the same as that of HRSV M2–1. The final ten residues in the HMPV M2–1 C-terminus were not resolved in the structure, suggesting it is disordered, as reported for HRSV M2–1. Interestingly, in the crystal structure of HMPV M2–1, one of the monomers has its helical core domain flipped out from the rest of the molecule by 60 Å when compared to symmetrically arranged protomers in the HRSV M2–1 subunit. This structural plasticity in HMPV M2–1 was further explored by molecular dynamics simulations and small-angle X-ray scattering measurements, which were consistent with the possibility that HMPV M2–1 can adopt both open and closed conformations in solution, with dynamic exchange between the two states, with the oligomerization domain as the most rigid part of the structure. This flexibility and domain swinging in HMPV M2–1 was postulated to increase the surface area of the molecule and allow sampling of a larger volume in solution that helps to trap RNA fragments, first proposed through a 'fly-casting' mechanism (Shoemaker et al. 2000).

HMPV M2–1 was also crystallized in the presence of adenosine monophosphate (AMP) and a five nucleotide long DNA fragment. Although these two molecules are non-native binding partners of M2–1, their location on the M2–1 surface corresponded to the RNA-binding site determined for HRSV M2–1 from NMR and mutagenesis studies. A model describing the HPMV M2–1 RNA-binding site proposes that the RNA may bind along an extended and contiguous tract of positive surface charge, extending between the N-terminal and C-terminal faces of each monomer. The length of this surface is consistent with binding a oligoribonucleotide of 13 nucleotides in length, which coincides with the length of the HRSV gene-end consensus sequence.

A Role for M2–1 in Virion Assembly?

The RNP within the HRSV virion comprises the minimal essential components required for transcription, namely, RNA, N, P, L and M2–1. This RNP is enclosed within the lipid bilayer with a matrix protein (M) in between them, and this assembly appears to be mediated by M2–1:M interactions. In a study involving super-resolution microscopy of filamentous HRSV particles isolated from infected cells, the M2–1 and M-proteins were localized within the enveloped virion with M2–1 being closer to the genomic RNA than M (Kiss et al. 2014). Thus, M-protein, by simultaneously interacting with two chemically different entities (M2–1 and the phospholipid bilayer), appears to act as a molecular scaffold. This overall plan of virion architecture is maintained among other members of the order *Mononegavirales*, such as those classified in the *Rhabdoviridae* and *Filoviridae* families, where the corresponding matrix enwraps the RNP. However, the resulting RNP and virion morphologies are less consistent; in the case of the rhabdoviruses, virions are bullet shaped, and the RNP core is found as a highly regular and rigid helical assembly, surrounded by a matrix protein sheath. In the case of filoviruses (e.g. EBOV), the RNPs are helical but exhibit a more extended morphology that is consistent with the filamentous shape of EBOV virions, which suggests RNP flexibility.

HRSV M appears able to adopt multiple different oligomeric states in solution, although it was crystallized as a monomer (Money et al. 2009). M has N- and C-terminal domains connected by a 13-residue region and is globally similar to EBOV VP40 in fold but different in topology. M has a distinct negatively charged lobe at the N-terminus and a positively charged area of 600 Å^2 extending from the NTD to CTD. The N-terminus of the M-protein directly interacts with M2–1, and these two regions are proposed to be responsible for simultaneously interacting with M2–1 and the lipid bilayer, consistent with its role as a virion scaffold. This arrangement would presumably require M2–1 to form a two-dimensional layer below M and also in contact with the RNP. From the crystal structure of HRSV M2–1, it is currently unclear how the tetramer can arrange into a planar lattice along with M, due to its tendency to multimerize as a highly stable tetramer. Whether the alternative conformation of the HMPV M2–1 protein could adopt such as layer, with one core domain per tetramer being extended, is an interesting possibility.

Other M2–1 Functions

Following entry into cells, many NSNSVs are reported to induce the formation of dense cytoplasmic structures that contain viral components, and in HRSV-infected cells, these sites are known as inclusion bodies (IB) (Norrby et al. 1970). Functionally analogous structures have been identified in many other NSNSVs including rhabdoviruses and filoviruses (Lahaye et al. 2009; Dolnik et al. 2015; Baskerville et al.

1985), and they are often used as histological markers for confirming viral infections. Using a combination of state-of-the-art fluorescence microscopy, super-resolution microscopy and pulse-chase techniques, HRSV IBs have been shown to represent the sites of active viral RNA synthesis (Rinchavel et al., 2017), where both transcription and replication take place. Rather than being amorphous aggregates, IBs have a complex organization and contain a spherical substructure named as the 'inclusion body associated granule' (IBAG). These are found to be enriched with nascent mRNA and also M2–1 but devoid of other proteins of the replication complex (N, P and L).

IBAG formation is strictly dependent on viral RNA synthesis, and these structures are dynamic, undergoing continuous assembly and disassembly cycles, as inferred from time-lapse fluorescence studies. The IBAGs were not found to be translationally active, suggesting that they may represent a transient compartment where newly synthesized viral mRNAs that are associated with M2–1 are stored prior to translation in the cytosol. While the interaction between M2–1 and mRNAs is consistent with the previous RNA-binding studies described above, which identified highest affinities for poly(A) sequences, the established strong association of M2–1 with nascent viral RNA in IBAGs may reflect a post-transcriptional function of M2–1 that is novel and needs to be explored.

Conclusions and Future Outlook

Solving the X-ray crystal structure of M2–1 represents an important step towards understanding the mechanism behind the complex functions of M2–1. However, complete clarity of M2–1 functions has yet to be achieved, and M2–1 does not operate in isolation from other viral components. Multiple protein-protein and protein-nucleic acid interactions execute the anti-termination process, and high-resolution structures of these other components will greatly aid in revealing the molecular mechanisms behind M2–1 activities. Though homology modelling and other theoretical methods can help to some extent, accurate understanding requires experimentally determined 3D structures. More important than isolated structures are the structures of binary and ternary complexes of the HRSV transcriptase and replicase complexes.

Structure determination of M2–1 in complex with specific RNA sequences and with full-length and/or crucial fragments of P is the next step for structural biologists. High-resolution structures that represent snapshots of the anti-termination process remain a difficult task; however, through the recent technical advancement in cryo-electron microscopy (cryo-EM) including the development of direct detectors, stable specimen stages, high-energy electron guns, phase plates, movie mode imaging and advanced image processing strategies, the field of cryo-EM is witnessing a 'resolution revolution'. A range of different biological specimens that have otherwise previously resisted structure determination like large complexes, membrane proteins, fibrous assemblies, low-abundant protein complexes and difficult-crystallize huge assemblies like viruses are now feasible targets for structural studies. Unfortunately, there still remains difficulty in structure determination of conformationally flexible, heterogeneous protein assemblies of megadalton size. Further insight into the structure and function of M2–1 can be obtained through the application of recent advances in electron microscopy paired with crystal structures of sub-complexes, in combination of a thorough understanding of its potentially diverse functions from observations made *in cellulo*.

Acknowledgements Work on HRSV M2-1 in the Barr and Edwards laboratories is supported in part by funding from the Medical Research Council (Grant: MR/L007290/1).

References

Amann BT, Worthington MT, Berg JM (2003) A Cys3His zinc-binding domain from Nup475/ tristetraprolin: a novel fold with a disklike structure. Biochemistry 42:217–221

Baskerville SP, Fisher-Hoch GH, Neild AB, Dowsett (1985) Ultrastructural pathology of experimental Ebola haemorrhagic fever virus infection. J Pathology 147:199–209

Bermingham A, Collins PL (1999) The M2-2 protein of human respiratory syncytial virus is a regulatory factor involved in the balance between RNA replication and transcription. Proc Natl Acad Sci U S A. 1999 Sep 28 96(20):11259–11264

Biedenkopf N, Schlereth J, Grunweller A et al (2016) RNA binding of Ebola virus VP30 is essential for activating viral transcription. J Virol 90:7481–7496

Blondot ML, Dubosclard V, Fix J et al (2012) Structure and functional analysis of the RNA- and viral phosphoprotein-binding domain of respiratory syncytial virus M2-1 protein. PLoS Pathog 8:e1002734

Bohmwald K, Espinoza JA, Rey-Jurado E et al (2016) Human respiratory syncytial virus: infection and pathology. Semin Respir Crit Care Med 37:522–537

Brady MT (2014) American Academy of Pediatrics Committee on infectious diseases; American Academy of Pediatrics bronchiolitis guidelines committee. Updated guidance for palivizumab prophylaxis among infants and young children at increased risk of hospitalization for respiratory syncytial virus infection. Pediatrics 134(2):415–420

Collins PL, Hill MG, Camargo E et al (1995) Production of infectious human respiratory syncytial virus from cloned cDNA confirms an essential role for the transcription elongation factor from the 5′ proximal open reading frame of the M2 mRNA in gene expression and provides a capability for vaccine development. Proc Natl Acad Sci U S A 92:11563–11567

Collins PL, Hill MG, Cristina J, Grosfeld H (1996) Transcription elongation factor of respiratory syncytial virus, a nonsegmented negative-strand RNA virus. Proc Natl Acad Sci U S A 93(1):81–85

Dolnik O, Stevermann L, Kolesnikova L et al (2015) Marburg virus inclusions: a virus-induced microcompartment and interface to multivesicular bodies and the late endosomal compartment. Eur J Cell Biol 94:323–331

Fearns R, Collins PL (1999) Role of the M2-1 transcription antitermination protein of respiratory syncytial virus in sequential transcription. J Virol 73:5852–5864

Hardy RW, Wertz GW (1998) The product of the respiratory syncytial virus M2 gene ORF1 enhances readthrough of intergenic junctions during viral transcription. J Virol 72:520–526

Hu J, Robinson JL (2010) Treatment of respiratory syncytial virus with palivizumab: a systematic review. World J Pediatr 6:296–300

Kiss G, Holl JM, Williams GM et al (2014) Structural analysis of respiratory syncytial virus reveals the position of M2-1 between the matrix protein and the ribonucleoprotein complex. J Virol 88:7602–7617

Kuo L, Grosfeld H, Cristina J et al (1996) Effects of mutations in the gene-start and gene-end sequence motifs on transcription of monocistronic and dicistronic minigenomes of respiratory syncytial virus. J Virol 70:6892–6901

Lahaye X, Vidy A, Pomier C et al (2009) Functional characterization of Negri bodies (NBs) in rabies virus-infected cells: evidence that NBs are sites of viral transcription and replication. J Virol 83:7948–7958

Leyrat C, Renner M, Harlos K et al (2014) Drastic changes in conformational dynamics of the antiterminator M2-1 regulate transcription efficiency in Pneumovirinae. elife 3:e02674

Li D, Jans DA, Bardin PG, Meanger J, Mills J, Ghildyal R (2008) Association of respiratory syncytial virus M protein with viral nucleocapsids is mediated by the M2-1 protein. J Virol. 2008 Sep;82(17):8863–8870. https://doi.org/1128/JVI.00343-08. Epub 2008 Jun 25

Liang B, Li Z, Jenni S et al (2015) Structure of the L protein of vesicular stomatitis virus from electron Cryomicroscopy. Cell 162:314–327

Lier C, Becker S, Biedenkopf N (2017) Dynamic phosphorylation of Ebola virus VP30 in NP-induced inclusion bodies. Virology 512:39–47

Marcelin JR, Wilson JW, Razonable RR et al (2014) Oral ribavirin therapy for respiratory syncytial virus infections in moderately to severely immunocompromised patients. Transpl Infect Dis 16:242–250

Money VA, Mcphee HK, Mosely JA et al (2009) Surface features of a Mononegavirales matrix protein indicate sites of membrane interaction. Proc Natl Acad Sci U S A 106:4441–4446

Norrby E, Marusyk H, Orvell C (1970) Morphogenesis of respiratory syncytial virus in a green monkey kidney cell line (Vero). J Virol 6:237–242

Palese P, Zheng H, Engelhardt OG et al (1996) Negative-strand RNA viruses: genetic engineering and applications. Proc Natl Acad Sci U S A 93:11354–11358

Renner M, Bertinelli M, Leyrat C et al (2016) Nucleocapsid assembly in pneumoviruses is regulated by conformational switching of the N protein. elife 5:e12627

Rincheval V, Lelek M, Gault E et al (2017) Functional organization of cytoplasmic inclusion bodies in cells infected by respiratory syncytial virus. Nat Commun 8:563

Scheltema NM, Gentile A, Lucion F et al (2017) Global respiratory syncytial virus-associated mortality in young children (RSV GOLD): a retrospective case series. Lancet Glob Health 5:e984–e991

Shi T, DA MA, O'Brian KL et al (2017) Global, regional and national disease burden estimates of acute lower respiratory infections due to respiratory syncytial virus in young children in 2015: a systematic review and modelling study. Lancet. 2017 Sep 2 390(10098):946–958. https://doi.org/10.1016/S0140-6736(17)30938-8

Shoemaker BA, Portman JJ, Wolynes PG (2000) Speeding molecular recognition by using the folding funnel: the fly-casting mechanism. Proc Natl Acad Sci U S A 97:8868–8873

Sourimant J, Rameix-Welti MA, Gaillard AL et al (2015) Fine mapping and characterization of the L-polymerase-binding domain of the respiratory syncytial virus phosphoprotein. J Virol 89:4421–4433

Sutherland KA, Collins PL, Peeples ME. (2001) Synergistic effects of gene-end signal mutations and the M2-1 protein on transcription termination by respiratory syncytial virus. Virology Sep 30; 288(2):295–307

Tanner SJ, Ariza A, Richard CA et al (2014) Crystal structure of the essential transcription antiterminator M2-1 protein of human respiratory syncytial virus and implications of its phosphorylation. Proc Natl Acad Sci U S A 111:1580–1585

Tao YJ, Ye Q (2010) RNA virus replication complexes. PLoS Pathog 6:e1000943

Tawar RG, Duquerroy S, Vonrhein C et al (2009) Crystal structure of a nucleocapsid-like nucleoprotein-RNA complex of respiratory syncytial virus. Science 326:1279–1283

Tran TL, Castagne N, Bhella D et al (2007) The nine C-terminal amino acids of the respiratory syncytial virus protein P are necessary and sufficient for binding to ribonucleoprotein complexes in which six ribonucleotides are contacted per N protein protomer. J Gen Virol 88:196–206

Tremaglio CZ, Noton SL, Deflube LR et al (2013) Respiratory syncytial virus polymerase can initiate transcription from position 3 of the leader promoter. J Virol 87:3196–3207

Yu Q, Hardy RW, Wertz GW (1995) Functional cDNA clones of the human respiratory syncytial (RS) virus N, P, and L proteins support replication of RS virus genomic RNA analogs and define minimal trans-acting requirements for RNA replication. J Virol 69:2412–2419

Chapter 12
Filamentous Bacteriophage Proteins and Assembly

Suzana K. Straus and Htet E. Bo

Filamentous Bacteriophage: Association with Gram-Negative Bacteria

Filamentous bacteriophages are viruses of the *Inovirus* genus, whose function is to associate with specific bacteria. The first phage to be found to be filamentous was fd, a bacteriophage associated with *Escherichia coli*. Work preceding the discovery by Hartmut Hoffmann-Berling at the Max-Planck Institute in Heidelberg had identified isometric phages, i.e., viruses which were roughly spherical in shape, with diameters of about 250 Å (Sinsheimer 1966). These isometric phages attracted interest not only because of their small size but also because some of them had a single-stranded DNA or RNA genome and because they were specific for male (F+ or Hfr) strains of bacteria, which can transfer DNA to other bacteria (Marvin et al. 2014). Hoffmann-Berling and Marvin characterized the general features of filamentous fd phage and its DNA (Marvin and Hoffmann-Berling 1963a, b; Hofmann-Berling et al. 1963). These landmark studies prompted others to examine their phages and identify a number that were filamentous. For instance, Zinder and coworkers isolated f1, a filamentous phage isolated at the same time as the isometric RNA phage f2 (Loeb and Zinder 1961; Zinder et al. 1963; Zinder 1986). Hofschneider found phage M13, which looked very similar to fd (Hofschneider 1963). It turned out that the three phages fd, f1, and M13 have about 98.5% DNA sequence identity and hence can be considered to be identical (Marvin et al. 2014).

Before introducing other filamentous bacteriophages that are associated with Gram-negative bacteria, let us first briefly elaborate on what makes a bacteriophage filamentous. These phages are filaments of dimension 6 nm in diameter and 1–2 μm

S. K. Straus (✉) · H. E. Bo
Department of Chemistry, University of British Columbia, Vancouver, BC, Canada
e-mail: sstraus@chem.ubc.ca

© Springer Nature Singapore Pte Ltd. 2018
J. R. Harris, D. Bhella (eds.), *Virus Protein and Nucleoprotein Complexes*,
Subcellular Biochemistry 88, https://doi.org/10.1007/978-981-10-8456-0_12

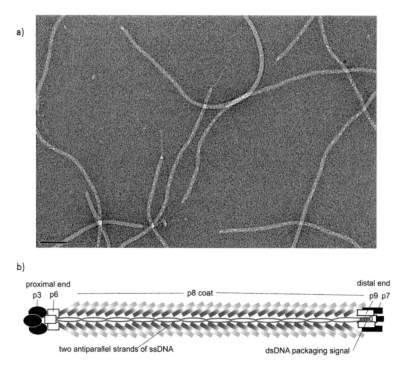

Fig. 12.1 Filamentous bacteriophage structure: (**a**) Transmission electron micrograph of filamentous bacteriophage B5, recorded on a Hitachi H7600 transmission electron microscope (TEM) at the UBC Bioimaging Facility; (**b**) schematic representation of the major coat protein (p8) and minor coat proteins in Ff. The bar in (**a**) represents 50 nm. Figure (**b**) is reproduced from Marvin et al. (2014), with permission

in length, which resemble a semiflexible worm (Fig. 12.1a). The filament is comprised primarily of several thousand copies of identical major coat protein subunits. These coat proteins, which are ca. 50 amino acid residues and α-helical, pack together to encompass the single-stranded circular DNA found in the core of the virion (Fig. 12.1b). The interdigitated packing of the coat protein has been likened to shingles, roof tiles, scales (Marvin et al. 2014), pine cones, or a protea flower. The ends of the virion are capped with minor coat proteins (Fig. 12.1b). These minor coat proteins play a role in infection and/or extrusion of the virion. The encompassed DNA can include foreign DNA, making filamentous bacteriophages useful tools in molecular biology (Anany et al. 2017; Criscuolo et al. 2017) or nanotechnology (Hemminga et al. 2010; Adams et al. 2013). If however the foreign DNA encodes virulence systems, then the resulting filamentous phage can be dangerous as they result in new pathogenic strains of bacteria (Gagic et al. 2016; Shapiro et al. 2016). Bacteriophage replication occurs without killing of the host bacterium. Indeed, filamentous phages differ from other bacteriophages because they are extruded from the host bacteria continuously and the host is not lysed upon progeny extrusion.

Filamentous bacteriophages infect hosts that express sex pili because the sex pili act as receptor sites for phage recognition and infection (Russel 1991). Filamentous phages are classified according to their pilus specificity. The three phages fd, f1, and M13 mentioned above, as well as the less studied filamentous phage ZJ/2 (Bradley 1964), are all specific for *E. coli* bacteria expressing F-pili. These phages are hence often referred to as **Ff** phages, for **F**-specific **f**ilamentous bacteriophage, and have interchangeable gene products (Russel 1991). Other filamentous phages with similar architecture, but different detailed genetics and physiology, include Ike, which infects *E. Coli* expressing N-pili. Ike phages are 50% identical genetically to the Ff phages (Russel 1991). There are also phages that infect *Pseudomonas aeruginosa* via the P-pili, such as Pf1 and Pf3 (Russel 1991; Straus et al. 2011; Marvin et al. 2014). Phage PH75 (Pederson et al. 2001) grows at 70 °C in *Thermus thermophilus*, a thermophilic bacteria and model organism for structural genomics and systems biology.

Filamentous Bacteriophage: Association with Gram-Positive Bacteria

Most filamentous phages are isolated from Gram-negative bacteria, but there is one known instance of a filamentous bacteriophage associated with Gram-positive bacteria. In 2002, Chopin et al. (Chopin et al. 2002) reported the discovery of the B5 filamentous phage. This phage infects the Gram-positive bacteria *Propionibacterium freudenreichii* and has dimensions of 620 nm in length and 12 nm in width (Fig. 12.1a). The B5 phage genome consists of 5806 bases and is organized in a similar way as other filamentous phage genomes. The phage DNA genome has a total G + C content of 64% and has ten open reading frames (Chopin et al. 2002). The constituent proteins for this, as well as all the other filamentous bacteriophages, will be presented in the following section. Interestingly, the gene for the major coat protein of B5 shows homology to the major coat protein gene in PH75 and Pf3 (Chopin et al. 2002).

Constituent Proteins of Filamentous Bacteriophage

Bacteriophage has been used as a tool for a range of applications (Hemminga et al. 2010; Adams et al. 2013; Anany et al. 2017; Criscuolo et al. 2017) because of its simplicity: its genome consists of genes II and V which are needed for DNA synthesis; genes VIII, III, VI, VII, and IX which code for phage structural proteins; and genes I and XI which code for proteins that aid in phage assembly and extrusion. Consequently, phages consist of a very simple makeup of structural proteins forming a protective arrangement around the phage DNA genome. The handful of proteins that make up a phage particle are shown in Fig. 12.1b.

As mentioned above, the filamentous phage is arranged with major coat protein p8 forming a fish scale-like arrangement around the viral genome that consists of a single-stranded, circular DNA (Russel 1991). In Ff phages, there are 2.3 nucleotides per major coat protein, while Pf1 has 1, and Pf3 has 2.4. For B5 phage, this is unknown. Phages can be classified according to the symmetry class, as defined by the packing of their major coat proteins, as identified by their diffraction patterns. Class I phage strains include Ff and Ike phages. Class II phage strains include Pf1 and Pf3 phages. Class I phages have $C_5S_{2.0}$ symmetry (Marvin 1998). Class II phages have $C_1S_{5.4}$ symmetry (Caspar and Makowski 1981).

The minor coat proteins are p3, p6, p7, and p9 (Fig. 12.1b). The distal end of the phage emerges first from the host cell and contains 3–5 copies of p7 and p9. The proximal end of the phage is involved in phage entry and exit and contains five copies each of p3 and p6. Phages lacking minor coat proteins appear as polyphages which are phages 10–20 times the length of a normal phage, are still tethered to the host, and contain more than one complete circular phage genome. Phages lacking p3 and p6 are noninfective, while phages lacking p7 and p9 are still infective. Phages lacking p6 are also unstable (Lopez and Webster 1983). p3 plays a role in not only phage-host cell recognition but also oligomerizes to form a pore large enough for the phage to enter the host cell (Glaser-Wuttke et al. 1989). It was found that in the N-terminal, two thirds of the p3 protein is essential for infectivity (Marvin et al. 2014). The whole phage structure is essentially held together by hydrophobic interactions between the apolar domains in the middle section of p8 and the apolar domains in the minor coat proteins.

The remaining proteins encoded in the phage genome do not form part of the virus particle but are needed for phage assembly. p1, a 348 amino acid long protein, provides the energy needed to drive phage assembly and interacts with both thioredoxin and phage DNA during phage assembly (Russel 1991). Its C-terminus is directed into the periplasm, while the N-terminus is directed into the cytoplasm. The protein p1 is seen to promote the formation of adhesion zones between the inner and outer membrane of Gram-negative host membranes. These adhesion zones are the sites of phage extrusion (Marvin 1998). p11 is a 108 residue protein whose genetic code is within gene I. It has basic residues at the N-terminus, similar to p1, needed to interact with DNA. p4 exists in the outer membrane with its N-terminus oriented toward the periplasm. It forms an oligomer of 10–12 subunits with an internal diameter of 80 Å that acts as a pore for viral extrusion (Papavoine et al. 1998). p4 functions as more than just an exit pore: it aids in phage assembly initiation because in the absence of p4, there is no visible buildup of phage progeny in the host cell (Russel 1991). p2 controls the rate of replicative form (RF) DNA production (Russel 1991). p2 expression is negatively regulated by p5. p5 is 87 amino acids in length and forms a dimer, with its hydrophobic face buried away from the solvent. It binds tightly and cooperatively to phage DNA and protects phage DNA while it is in the host cytoplasm.

Each of these phage proteins plays an important role in assembly, a process that will be reviewed in the following section.

Filamentous Phage Assembly

Although many details of phage assembly are unknown at this juncture (Russel 1991; Russel et al. 1997; Webster 2001; Russel and Model 2006), a number of studies relating in particular to Gram-negative phages have led to a number of models being proposed. By far the most studied phage in this regard is M13, which infects and replicates in *E. coli*.

Entry of M13 into *E. coli* occurs when the phage protein p3 recognizes the F-pilus of *E. coli*. The phage is then brought close to the host cell via retraction of the F-pilus tip (Webster 2001; Russel and Model 2006; Marvin et al. 2014). This hypothesis is known as the "pilus retraction model." Once the phage is close enough to the host cell, infection occurs. p3 oligomerizes and forms a pore large enough for adsorption of the phage particle (Glaser-Wuttke et al. 1989). Glaser-Wuttke et al. (Glaser-Wuttke et al. 1989) have shown that p3 oligomerizes and forms large pores which stay open for seconds.

In addition to p3, adsorption of phage through the host *E. coli* outer membrane, periplasmic space, and inner membrane requires the host proteins tolerant Q (TolQ), TolR, and TolA which are part of the Tol transport system (Karlsson et al. 2003). It is thought that p3 interacts with domain III of TolA via its amino terminus (Webster 1991). This activates the TolQRA complex and leads to phage adsorption. When phage is first adsorbed into the host, p6 is lost from the phage particle, and the phage particle becomes unstable. This instability results in the full disassembly of the phage particle (Russel 1991). The phage coat protein p8 and other minor coat proteins are deposited into the host inner membrane, while the phage DNA enters the host cytoplasm where it is converted into a double-stranded super coiled (RF) DNA via host enzymes (Russel 1991). The RF DNA is the template for not only phage DNA replication but also phage protein synthesis (transcription of RF DNA and translation). All 11 phage proteins are synthesized by transcription of RF DNA via the rolling-circle mechanism and translation of the messenger ribonucleic acid (mRNA) produced (Marvin et al. 2014). The phage structural coat proteins p8, p3, p6, p7, and p9 are deposited into the host inner membrane upon translation. p1, p11, and p4 are proteins required for phage assembly and are deposited in the inner and outer (for p4) membranes of *E. coli* (Fig. 12.2). The phage protein p5 homodimers bind to the newly synthesized phage DNA strand, thus separating the single-stranded phage DNA from the rolling-circle mechanism (Marvin et al. 2014). The DNA strand loops back on itself, and a double-stranded packaging signal is left exposed at the end of the DNA-p5 homodimer complex. Interaction of the double-stranded packaging DNA signal with the p1-thioredoxin complex at the host inner membrane triggers the formation of a pore. The continuous extrusion of the phage particle while it is being assembled requires the replacement of one p5 monomer in the DNA-p5 homodimer complex with two p8 coat proteins. This results in the production of the DNA-p8 complex. The formation of DNA-p8 complex is the pivotal step in phage assembly. The DNA is now protected by the phage coat protein p8. The

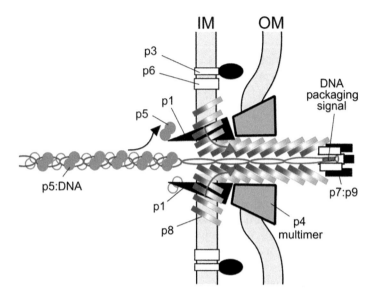

Fig. 12.2 Filamentous bacteriophage assembly process. Assembly is initiated when p1 interacts with the phage DNA packing signal. The major and minor coat proteins assemble around p1. Phage extrudes out of a pore created by p4, while p5 is replaced by p8 during the process. The figure is reproduced from Marvin et al. (2014), with permission

protein p4 oligomerizes in the outer host cell membrane to form the pore for phage extrusion. The assembly process requires aggregation of p8 monomers and their assembly around the DNA complex, as well as changes in the structural properties of p8 in going from the nonpolar environment of the lipid bilayer to an aqueous environment upon extrusion (Opella et al. 2008; Marvin et al. 2014). Addition of the p3 and p6 minor coat proteins to the extruding phage closes the phage particle and stops phage assembly. The phage is then fully released into the medium. Fig. 12.2 shows an illustration of the phage assembly process.

Presumably assembly occurs in an identical fashion of all Gram-negative filamentous bacteriophage. Interestingly, B5 assembly, where extrusion occurs through the single membrane found in Gram-positive bacteria, occurs without p4, which is not encoded in the genome of B5.

In order to understand the assembly process in depth, researchers have carefully studied the structure of p8 in the different environments it experiences as it is being assembled. Indeed, p8 goes from the nonpolar environment of the host lipid membrane where it is not associated with DNA to being associated with the phage DNA as it is shifted upward within the host membrane and eventually extruded out into a polar environment. In the polar environment, the p8 protects the phage DNA and interacts with it. Knowing the structure of p8 in the different environments provides important clues in the phage assembly process and will be discussed below.

Major Coat Protein p8

The major coat proteins of the different phage families are very similar. They are about 50 amino acids in length and consist of an acidic N-terminal end, a hydrophobic central region, and a basic C-terminal end (Marvin et al. 2014). The major coat proteins of the Ff phages differ only by one amino acid: residue 12 in f1 and fd is an aspartic acid, but in M13, it is an asparagine (Marvin et al. 2014) (see Table 1 in (Marvin et al. 2014) or UNIPROT codes given in Fig. 12.3a for amino acid sequences; see Fig. 12.3a for a graphical representation). ZJ/2 differs from M13 in two positions. The major coat protein in B5 phage is not synthesized with a signal sequence, while the major coat protein of Ff phages are synthesized with a 23 amino acid leader signal sequence which is cleaved off upon membrane insertion (Fig. 12.3).

When in the host membrane, p8 is oriented with its N-terminal end directed toward the outer membrane and its C-terminal end directed toward the host cell cytoplasm (Fig. 12.3b). When in the phage, 2700 copies of the major coat protein form a layer around the viral DNA molecule. And because the p8 proteins are held together by hydrophobic interactions in the midsection of p8, the phage is resistant to proteases, salts, detergents, and extreme pH. Phages can only be lysed mechanically or by ether or chloroform (Marvin et al. 2014). In the virion, the acidic N-terminus of p8 is oriented toward the exterior of the phage. The negative charge on the N-terminus is thought to be responsible for phage solubility. The basic C-terminus of p8 points toward the interior of the phage and interacts with the negatively charged DNA via its lysine residues (four in total) (Hunter et al. 1987). There are nonspecific electrostatic interactions independent of DNA sequence between the positively charged lysine residues and the negatively charged DNA (Hunter et al. 1987; Goldbourt et al. 2010; Morag et al. 2015; Abramov et al. 2015).

Structure of p8 in the Membrane

Although fd was discovered first, Hofschneider distributed many samples of M13 (Hofschneider 1963), and hence most structural studies of the major coat protein in a membrane environment involve the p8 subunit from M13. These studies typically involve reconstituting p8 into a membrane mimetic environment best suited to a given technique. As we will examine in this section, the structure of p8 in membranes is diverse.

Solution nuclear magnetic resonance (NMR) is a technique that has been used in the study of the major coat protein structure in membrane mimetics. The coat protein is reconstituted into micelles, with lipids that are as representative as possible of the bacterial host membrane (van de Ven et al. 1993; Williams et al. 1996). For instance, Papavoine et al. (Papavoine et al. 1998) reconstituted the M13 p8 subunit in sodium dodecyl sulfate (SDS) and dodecylphosphocholine (DPC). They found that the M13

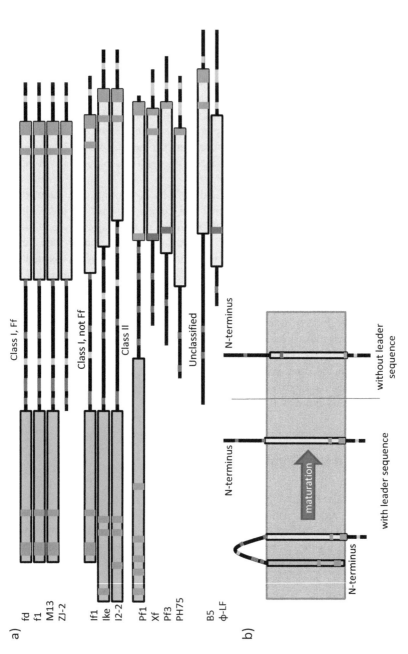

Fig. 12.3 Sequence alignment of p8, as obtained using PRALINE (Bawono and Heringa 2014). The complete amino acid sequences can be obtained by accessing the following UNIPROT codes: P69539 for fd, P69540 for f1, P69541 for M13, P03618 for ZJ-2, P03619 for If1, P03620 for Ike, P15416 for I2–2, P03621 for Pf1, P03622 for Xf, P03622 for Pf3, P82889 for PH75, Q8SCI1 for B5, and P68674 for φ-Lf. Positively charged amino acids are shown in blue, whereas negatively charged residues are shown in orange. The transmembrane region is shown in yellow (predicted using TMHMM v2.0 (Krogh et al. 2001)) and the leader signal sequence in green (as predicted using Phobius (Käll et al. 2004))

protein is composed of three main regions: an amphipathic α-helix (residues 8 to 16), a hinge region (residues 17–24), and a transmembrane hydrophobic helix (residues 25–45). These findings were supported by similar studies (Henry and Sykes 1990, 1992). The amphipathic helix is found on the surface of the micelle at almost a right angle to the transmembrane helix (Papavoine et al. 1998). The residues in the amphipathic helix are arranged around the helix based on their side chains. Amino acids with large side chains such as lysine 8 and phenylalanine 11 are found on the side of helix that is directed away from the solvent, whereas amino acids with smaller side chains such as alanine 9 and serine 13 are found on the side of the helix exposed to solvent. In an actual bacterial membrane, however, the partitioning of residues is based on charge or hydrophobicity (e.g., as for antimicrobial peptides), rather than size. So the distribution of residues found in the solution-state NMR structures may be due more to the artificial environment of the micelle, which is known to induce structural distortions because of its nature (Cross et al. 2011).

Solid-state NMR is another important technique in the study of the major coat protein, in this case from fd. Solid-state NMR involves reconstituting the p8 subunit into lipid bilayers, thereby eliminating the problem of high curvature present in micelles. Solid-state NMR studies have shown that the fd coat protein consists of a 16 Å amphipathic helical segment which lies parallel to the lipid bilayer surface and a 35 Å transmembrane helical segment which lies at a tilt of 16° with respect to the lipid bilayer normal (Opella et al. 2008). The two helices are connected by a short loop made of residues threonine 19 and glutamic acid 20 (Opella et al. 2008). This short segment differs from the substantially longer loop (residues 17–26) found for the same protein in micelles (Opella et al. 2008).

Site-directed spin labeling (SDL) is yet another method used to solve the structure of the major coat protein from M13 by reconstituting p8 in lipid vesicles (Stopar et al. 2006b, a). The shape of the major coat protein obtained from SDL experiments appears to be a continuous helical structure. It consists of a single gently curved helix tilted by 18° with respect to the normal of a unilamellar vesicle membrane and with maximum curvature at residue 20 (Nazarov et al. 2007). Even though SDL is thought to provide the closest representation of the host membrane environment, the helix shows changes in structure when placed in slightly different membrane environments. Stopar et al. (Stopar et al. 2006b) have shown that the first 7 residues in the N-terminus are unstructured when the protein is placed in 22:1 PC and if placed in 14:1 PC, the first 14 residues are unstructured. M13 protein shows a tilt of 33° in 14:1 PC, while it has a tilt of 19° in a 20:1 PC (Stopar et al. 2006a). Spruijt et al. performed SDL experiments on M13 major coat protein, where they engineered M13 mutants with single cysteine substitutions to attach an N-(iodoacetylaminoethyl)-5-naphthylamine-1-sulfonic acid (ADEAS) probe. They performed single cysteine scanning fluorescence microscopy and found that M13 had a banana-like structure, i.e., it was a single slightly curved and tilted molecule with a flexible hinge (Spruijt et al. 2004). The flexible hinge loses its α-helicity when M13 is placed in the thinner 14:1 PC membrane as compared to a 18:1 PC membrane (Spruijt et al. 2004). More recently, Stopar et al. (Stopar et al. 2009) investigated the anchoring of M13 by

6-bromoacetyl-2-dimethylaminonaphthalene (BADAN) labeling. The change in intrinsic fluorescence of tryptophan 26 was analyzed when the M13 protein is placed in 22:1 PC versus 14:1 PC. A blue shift in fluorescence of about 700 cm^{-1} was observed (Stopar et al. 2009). This shift corresponds to the tryptophan going into a more nonpolar environment, suggesting that the tryptophan molecule is "sinking" into the lipid bilayer as the bilayer thickness increases. Residue 46 was also tested for its anchoring ability. BADAN again showed a blue shift in fluorescence when moved into a nonpolar environment. However, when the p8 subunit with the BADAN-labeled cysteine residue 46 was placed in a 22:1 PC versus 14:1 PC, there was a blue shift of only 70 cm^{-1} (Stopar et al. 2009). From these results, it was concluded that the anchoring strength of the C-terminus is five times stronger than that of the N-terminus. The C-terminus consists of two phenylalanine resides and four lysine residues which provide strong anchoring of the protein. Although SDL is carried out in possibly more realistic membrane environments, i.e., vesicles as opposed to closely packed bilayers as found in the solid-state NMR samples (Vos et al. 2009), it also involves the use of large fluorescent moieties that could have an impact on the overall structure.

As we will see below, this diversity in structural models for the p8 subunit is also found for p8 in the virion.

Structure of p8 in the Virion

The structure of p8 in the virion has primarily been determined using two complimentary techniques: X-ray fiber diffraction (Marvin et al. 2014) and solid-state NMR (Opella et al. 2008; Marvin et al. 2014). Since, as mentioned earlier, fd was discovered first, its structure was also investigated first and that by X-ray fiber diffraction. Samples were prepared by aligning concentrated gels of purified phage by flow in capillaries. The X-ray diffraction patterns obtained from this sample showed that the arrangement of the major coat protein is an array of α-helices with the long axes of the helices oriented roughly parallel to the long axis of the fiber (Marvin 1966). The strong intensity at about 10 Å spacing in the equatorial direction, i.e., roughly perpendicular to the fiber axis, is attributed to the distance between close-packed α-helices about that same distance in diameter. Similarly, the strong intensity at about 5.4 Å in the meridional direction, i.e., roughly parallel to the fiber axis, arises because of the 5.4 Å pitch of the α-helix. These features were also confirmed through optical rotatory dispersion measurements (Day 1966). A more detailed account of the fiber diffraction studies on fd can be found in Marvin et al. (Marvin et al. 2014) or more recently in Marvin (Marvin 2017).

Next, fiber diffraction data was obtained for Pf1. It was found that the Pf1 strain yielded better resolved diffraction patterns as compared to fd, making its structural characterization easier. This difference arises from the slight symmetry difference between fd and Pf1, i.e., the distinction between Class I and Class II mentioned earlier. Advances in computational tools as well as the availability of additional data meant that the molecular model for Pf1 was refined and extended to other strains (Marvin et al. 1974, 2014; Marvin and Wachtel 1976). Interestingly, Pf1 undergoes a structural transition at 283 K (Nave et al. 1979). The result is slightly different diffraction patterns from the lower (Pf1[L]) and higher (Pf1[H]) temperature forms of Pf1 (Nave et al. 1979; Welsh et al. 2000). Interconversion between the two forms is completely reversible and is an example of a well-defined transition in a macromolecular assembly (Welsh et al. 2000; Marvin et al. 2014). This structural transition has not been observed for other Class II phages (Fig. 12.3) and may be unique to Pf1 because of its lower DNA/protein ratio (the Pf1 virion is about twice as long and so has about twice as many subunits as other Class II virions).

Soon after the fiber diffraction work was being carried out on Pf1, Opella and co-workers started exploring the use of solid-state NMR to examine the local structure of the individual p8 subunits in fd (Cross and Opella 1980, 1982; Cross et al. 1981, 1983). Since the virion is too large to permit the rapid tumbling needed for solution-state NMR, solid-state NMR methods are required. The study of p8 subunits in the virion is possible by oriented methodology or magic-angle sample (MAS) spinning (Quinn and Polenova 2017). The symmetry and high number of copies of p8 in the virion are extremely advantageous for the application of these methods: data is obtained uniquely for all ca. 50 backbone amide NH groups (with the N in isotopically enriched ^{15}N form), and the concentration of the sample is effectively very high, given that there are over a thousand copies of p8 per virion particle.

The breakthrough in the determination of the p8 structure using solid-state NMR methods came in the early 2000s, with higher magnetic fields, better probes, and better sample preparation techniques, making the study of uniformly ^{15}N–labeled samples possible. Using oriented methods, Zeri et al. (Zeri et al. 2003) proposed that the structure of p8 in fd consists of three α-helical segments, with kinks located between residues 20 and 21 and between residues 38 and 39. Residues 1–6 are unstructured. The kinks in this proposed structure gives p8 a similar shape as its membrane-bound form (Almeida and Opella 1997). Indeed, in this form residues 7–20 are on the surface of the membrane, and residues 21–38 span the membrane bilayer (Opella et al. 2008). Similarly, the structure of p8 in Pf1 was determined using oriented methods and ^{15}N–labeled Pf1 above and below the transition temperature (Thiriot et al. 2004, 2005). Thiriot et al. (Thiriot et al. 2005) suggested that the structure of the individual subunits undergoes no significant change as a function of temperature, confirming previous X-ray fiber diffraction studies indicating that it is the orientation and packing of the subunits, not the structure of the individual subunits, that is changed at the transition temperature. Opella and co-workers, however, suggested that the subunit comprises three distinct α- helix segments, rather than being a single continuous α-helix as in the models based on X-ray fiber diffraction data and as is also the case for fd. Interestingly, the NMR spectra of Pf1,

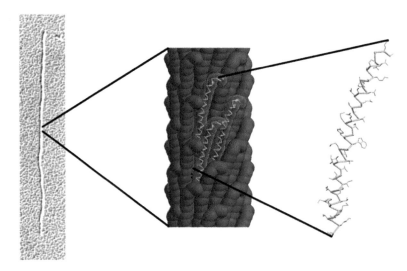

Fig. 12.4 Arrangement of subunits in the fd phage, using the p8 subunit deposited as PDB ID: 2C0X. The virion axis is vertical, and the axial slab in the center corresponds to about 1.4% of the total length of the virion. Each subunit is represented as a red space-filling coil. Three adjacent subunits are shown in atomic detail (yellow lines) within "semitransparent" coils, and a single isolated subunit is displayed at the right. An electron micrograph of a full-length phage, aligned by flow on the electron microscope grid as described by Marvin and Hoffmann-Berling (Marvin and Hoffmann-Berling 1963b), is shown at the left. Reproduced from the cover of Marvin et al. (2006), with permission

in both higher and lower temperature forms, are much better resolved than for fd and are likely due to the fact that it is easier to align Pf1 than fd. More ordered samples give rise to higher-resolution oriented NMR spectra. Finally, MAS experiments were performed on both forms of Pf1 (Goldbourt et al. 2007, 2010; Sergeyev et al. 2017). Using experimentally determined chemical shift values and secondary structure prediction programs, Goldbourt et al. (Goldbourt et al. 2007) presented a model of Pf1 where residues 2–5 are in a non-helical conformation, while residues 6–46 form a continuous helix. This model is more in line with those obtained from X-ray fiber diffraction discussed above (Welsh et al. 2000).

In order to resolve the discrepancies in the p8 models (i.e., X-ray diffraction and MAS continuous helix vs. oriented NMR kinked helix), a joint refinement of the X-ray diffraction and oriented NMR data was carried out (Marvin et al. 2006, 2014; Straus et al. 2008, 2011). For both fd and Pf1, refining the structure against the NMR and the X-ray fiber diffraction information in parallel gave consensus models for the p8 subunit in fd (Marvin et al. 2006) (Fig. 12.4) and in Pf1 (Straus et al. 2011). In both cases, continuous α-helices were found. More extensive details on the refinement process and how the oriented NMR data can be reinterpreted to yield a continuous α-helix are given in Marvin et al. (Marvin et al. 2014). The resulting models, 2C0X.pdb for fd and 2XKM.pdb for Pf1L, not only agree with the fiber diffraction and NMR data but also satisfy additional structural constraints from other methods (e.g., Raman spectroscopy), as reviewed in Marvin et al. (Marvin et al. 2014).

Structure of p8 During Extrusion

Although there is no explicit structural data for the p8 subunit during the process of assembly and extrusion, i.e., in the state shown in Fig. 12.2, a number of models have been proposed over the years on how this process may occur. The models are based on combining structural data of p8 in the membrane and in the virion, as presented in previous sections. To study p8 in the act, as it were, would require the trapping of intermediate forms. Although such forms can be generated by treating the phage with ether or chloroform (Amako and Yasunaka 1977; Griffith et al. 1981; Lopez and Webster 1982; Manning and Griffith 1985; Stopar et al. 1998; Marvin et al. 2014), the application of these intermediates for direct structural studies or reconstitution remains a challenge.

Based on the structural information discussed above, we can consider two extreme cases: one is a model based on the kinked structures of p8 from fd in the membrane (1MZT.pdb) and in the virion (1NH4.pdb); and the second is based on the continuous helix found by SDL (Vos et al. 2009) and 2C0X.pdb, the consensus structure presented earlier (Fig. 12.5). Starting from the kinked structure of p8 in the membrane (Fig. 12.5a, red), one can see that for it to be aligned in the optimal position, i.e., with the amphipathic helix lying on the surface of the lipid bilayer, the transmembrane helix is at a small angle relative to the membrane normal. Indeed, we mentioned earlier that the transmembrane helix lies at a tilt of 16° with respect to the lipid bilayer normal (Opella et al. 2008). If we assume that the structure of p8 in the virion is also present at some point during assembly in the membrane, then a rearrangement between the red helix and the green helix (1NH4.pdb) should occur (Fig. 12.5).

Using recently developed theoretical methods (Lomize et al. 2006a; b; Lomize et al. 2012), it is possible to predict how the green helix would preferentially position and orient itself in the membrane. The methods take into account the hydrophobic, van der Waals, hydrogen bonding, and electrostatic solute-solvent interactions of proteins in lipid bilayers and also account for the preferential solvation of charges and polar groups by water and the effect of hydrophobic mismatch for transmembrane regions (Marvin et al. 2014). Hence, using the server found at http://opm. phar.umich.edu/, the preferential position of the green helix was determined (Fig. 12.5b), as well as that for 2C0X.pdb (blue helix). Both helices have a tilt angle of close to 46–49 ° relative to the membrane normal and a hydrophobic thickness of ca. 25 Å. In *E. coli*, the inner membrane thickness is roughly 40 Å, with 30 Å of that being the hydrophobic region (Briegel et al. 2009) (i.e., comparable to the hydrophobic thickness calculated by the OPM server).

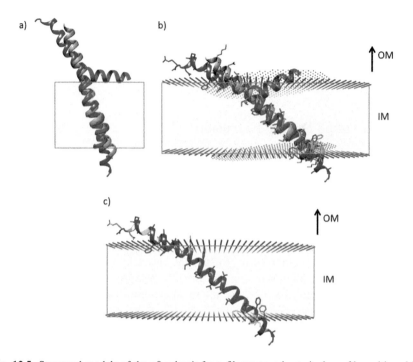

Fig. 12.5 Structural models of the p8 subunit from filamentous bacteriophage fd, positioned in various ways in the membrane (indicated by the gray box): p8 model of the membrane-bound form (1MZT.pdb; red); p8 model of the form in the virion, with kinks (1NH4.pdb; green); and continuous α-helical p8 model of the form in the virion (consensus model, 2C0X.pdb; blue). For all figures, all three helices are aligned so that residues 22–42 overlap. (a) Arrangement that is most optimal for 1MZT.pdb, i.e., with the amphipathic helix lying on the surface of the membrane bilayer. (b) Arrangement based on the output from the OMP server. The blue dots show the boundaries of the calculated apolar membrane core for 2C0X. The green dots show the same boundaries for 1NH4. (c) Arrangement shown for 2C0X alone, with the charged residues shown in cyan and orange (i.e., using the same color code as in Fig. 12.3). IM = inner membrane. OM = outer membrane (not shown)

If one considers an extrusion model based on the kinked helices, then one possibility would be that the amphipathic part of the red helix (Fig. 12.5a) would have to lift itself off the membrane surface in order for the helix to straighten up and insert itself deeper into the membrane and at a larger tilt angle to result in the green helix (Fig. 12.5b). The green helices would then possibly pack together or interact with the DNA to form a DNA-p8 complex. Through one or both of these interactions, the p8 subunits would change their tilt angle from 46° back to the smaller tilt angle found in the virion. Alternatively, these rearrangements could be accompanied by rotation of the amphipathic and transmembrane helices, as suggested by Opella et al. (Opella et al. 2008). Such a "dramatic structural rearrangement" (Opella et al. 2008) between the membrane-bound and virion forms of p8 was also suggested for Pf1.

On the other hand, one can consider a model where only continuous helices are involved. Recall the experimental finding that p8 from M13 forms a continuous α-helix titled at 33° when inserted into lipid bilayers composed of 14:1 PC (Stopar et al. 2006a). Now consider that the *E. coli* inner membrane consists of ca. 75–80% PE and 20% PG (Van Dalen and De Kruijff 2004; Marvin et al. 2014) and lipid acyl chains that are primarily 16:0 or 16:1 (Burnell et al. 1980). In this case, the tilt angle is very likely to be smaller than the 49° predicted by the OPM server, as some of the charged residues would prefer to be closer to the headgroup area (Fig. 12.5c). It is also possible that the tilt angle is smaller than 33°, since Stopar et al. found that this angle becomes smaller with increased membrane thickness or acyl chain length (Stopar et al. 2006a). Regardless of the exact tilt angle, assembly would proceed again through packing together of p8 subunits and/or interaction with the DNA, with only a change in orientation. This model would involve only modest rearrangement of the subunit during extrusion.

Although there are known cases where proteins undergo dramatic structural rearrangements as part of their function (e.g., fusion proteins (Skehel and Wiley 2000)), it seems unlikely that the p8 subunit would do so, as the environment around the single helix does not change if one considers the subunit consisting of the three segments described earlier. In other words, although p8 goes from a nonpolar environment of the host lipid membrane to a polar environment in the virion, locally, the three segments (i.e., N-terminus, central core, and C-terminus) are always experiencing similar interactions (i.e., primarily electrostatic at the termini and hydrophobic in the core region). Also, since it has been demonstrated with Pf1 that the p8 subunit can change its orientation without changing its structure to give rise to the Pf1^L and Pf1^H forms, it seems plausible that a certain orientational flexibility is built into all filamentous bacteriophage major coat proteins.

Conclusions

Assembly is a complex process that involves the movement of multiple components. As illustrated in this chapter, structural information provides key data on two states: of filamentous bacteriophage p8 subunit in the membrane and in the virion. This insight is akin to a "before and after" picture. Much remains to be determined about the steps in between. Future structural studies of, for example, p8 aggregates/ intermediates of phage treated with ether or chloroform or of the minor coat proteins in lipid bilayers could help fill in the gaps. Regardless, filamentous bacteriophage will continue to be an important subject of study and to provide important clues on subcellular processes.

Acknowledgments We would like to thank Dr. Walter Scott for carefully reading through the manuscript and providing constructive criticism. We would also like to thank Dr. Don A. Marvin for a wonderful collaboration and for teaching us all we know about filamentous bacteriophage. This chapter is a "lighter" version of our more in-depth review (Marvin et al. 2014), which we invite you to dive into for further details. Finally, funding from the Natural Sciences and Engineering Research Council of Canada is acknowledged.

References

Abramov G, Morag O, Goldbourt A (2015) Magic-angle spinning NMR of intact bacteriophages: insights into the capsid, DNA and their interface. J Magn Reson 253:80–90. https://doi.org/10.1016/j.jmr.2015.01.011

Adams MC, Belcher A, Keck WM, Grossman JC (2013) Highly-conductive cathode for lithium-ion battery using M13 phage - SWCNT complex

Almeida FCL, Opella SJ (1997) Fd coat protein structure in membrane environments: structural dynamics of the loop between the hydrophobic trans-membrane helix and the amphipathic in-plane helix. J Mol Biol 270:481–495. https://doi.org/10.1006/jmbi.1997.1114

Amako K, Yasunaka K (1977) Ether induced morphological alteration of Pf-1 filamentous phage. Nature 267:862–863

Anany H, Chou Y, Cucic S et al (2017) From bits and pieces to whole phage to Nanomachines: pathogen detection using bacteriophages. Annu Rev Food Sci Technol 8:305–329. https://doi.org/10.1146/annurev-food-041715-033235

Bawono P, Heringa J (2014) PRALINE: a versatile multiple sequence alignment toolkit. Methods Mol Biol 1079:245–262

Bradley DE (1964) The structure of some bacteriophages associated with male strains of Escherichia Coli. J Gen Microbiol 35:471–482. https://doi.org/10.1099/00221287-35-3-471

Briegel A, Ortega DR, Tocheva EI et al (2009) Universal architecture of bacterial chemoreceptor arrays. Proc Natl Acad Sci 106:17181–17186. https://doi.org/10.1073/pnas.0905181106

Burnell E, Alphen L, Van VA, De KB (1980) 31P nuclear magnetic resonance and freeze-fracture electron microscopy studies on escherichia coli. Biochim Biophys Acta 597:492–501

Caspar DL, Makowski L (1981) The symmetries of filamentous phage particles. J Mol Biol 145:611–617

Chopin MC, Rouault A, Dusko Ehrlich S, Gautier M (2002) Filamentous phage active on the gram-positive bacterium Propionibacterium freudenreichii. J Bacteriol 184:2030–2033. https://doi.org/10.1128/JB.184.7.2030-2033.2002

Criscuolo E, Spadini S, Lamanna J et al (2017) Bacteriophages and their immunological applications against infectious threats. J Immunol Res 2017:1–13. https://doi.org/10.1155/2017/3780697

Cross TA, Opella SJ (1980) Structural properties of fd coat protein in sodium dodecyl sulfate micelles. Biochem Biophys Res Commun 92:478–484

Cross TA, Opella SJ (1982) Protein dynamics by solid-state nuclear magnetic resonance spectroscopy. Peptide backbone of the coat protein in fd bacteriophage. J Mol Biol 159:543–549

Cross TA, Gall CM, Opella SJ (1981) NMR studies of filamentous bacteriophage assembly. Prog Clin Biol Res 64:457–465

Cross TA, Tsang P, Opella SJ (1983) Comparison of protein and deoxyribonucleic acid backbone structures in fd and Pf1 bacteriophages. Biochemistry 22:721–726

Cross TA, Sharma M, Yi M, Zhou H-X (2011) Influence of solubilizing environments on membrane protein structures. Trends Biochem Sci 36:117–125. https://doi.org/10.1007/s11103-011-9767-z.Plastid

Day LA (1966) Protein conformation in fd bacteriophage as investigated by optical rotatory dispersion. J Mol Biol 15:395–398

Gagic D, Ciric M, Wen WX et al (2016) Exploring the secretomes of microbes and microbial communities using filamentous phage display. Front Microbiol 7:1–19. https://doi.org/10.3389/fmicb.2016.00429

Glaser-Wuttke G, Keppner J, Rasched I (1989) Pore-forming properties of the adsorption protein of filamentous phage fd. Biochim Biophys Acta 985:239–247

Goldbourt A, Gross BJ, Day LA, McDermott AE (2007) Filamentous phage studied by magic-angle spinning NMR: resonance assignment and secondary structure of the coat protein in Pf1. J Am Chem Soc 129:2338–2344. https://doi.org/10.1021/ja066928u

Goldbourt A, Day LA, McDermott AE (2010) Intersubunit hydrophobic interactions in Pf1 filamentous phage. J Biol Chem 285:37051–37059. https://doi.org/10.1074/jbc.M110.119339

Griffith J, Manning M, Dunn K (1981) Filamentous bacteriophage contract into hollow spherical particles upon exposure to a chloroform-water interface. Cell 23:747–753

Hemminga MA, Vos WL, Nazarov PV et al (2010) Viruses: incredible nanomachines. New advances with filamentous phages. Eur Biophys J 39:541–550. https://doi.org/10.1007/s00249-009-0523-0

Henry GD, Sykes BD (1990) Detergent-solubilized M13 coat protein exists as an asymmetric dimer. J Mol Biol 212:11–14. https://doi.org/10.1016/0022-2836(90)90299-2

Henry GD, Sykes BD (1992) Assignment of amide 1H and 15N NMR resonances in detergent-solubilized M13 coat protein: a model for the coat protein dimer. Biochemistry 31:5284–5297

Hofmann-Berling H, Marvin DA, Duerwald H (1963) A filamentous DNA phage (fd) and a spherical RNA phage (fr), host-specific for the male strain of E. Coli. 1. Preparation and chemical properties of fd and fr. Zeitschrift fur Naturforschung Tl B, Chemie, Biochem Biophys Biol und verwandte Gebiete 18:876–883

Hofschneider P (1963) Untersuchungen über "kleine" E. coli K12 Bacteriophagen. 1 und 2 Mitteilung. Z Naturforsch 18b:203–205

Hunter GJ, Rowitch DH, Perham RN (1987) Interactions between DNA and coat protein in the structure and assembly of filamentous bacteriophage fd. Nature 327:252–254. https://doi.org/10.1038/327252a0

Käll L, Krogh A, Sonnhammer EL (2004) A combined transmembrane topology and signal peptide prediction method. J Mol Biol 338:1027–1036. https://doi.org/10.1016/j.jmb.2004.03.016

Karlsson F, Borrebaeck CA, Nilsson N (2003) The mechanism of bacterial infection by filamentous phages involves molecular interactions between TolA and phage protein 3 domains the mechanism of bacterial infection by filamentous phages involves molecular interactions between TolA and phage protein 3. J Bacteriol 185:2628–2634. https://doi.org/10.1128/JB.185.8.2628

Krogh A, Larsson B, von Heijne G, Sonnhammer EL (2001) Predicting transmembrane protein topology with a hidden markov model: application to complete genomes11. Edited by F. Cohen. J Mol Biol 305:567–580. https://doi.org/10.1006/jmbi.2000.4315

Loeb T, Zinder ND (1961) A Bacteriophage containing RNA. Proc Natl Acad Sci U S A 47:282–289. https://doi.org/10.1073/pnas.47.3.282

Lomize AL, Pogozheva ID, Lomize MA, Mosberg HI (2006a) Positioning of proteins in membranes: a computational approach. Protein Sci 15:1318–1333. https://doi.org/10.1110/ps.062126106

Lomize MA, Lomize AL, Pogozheva ID, Mosberg HI (2006b) OPM: orientations of proteins in membranes database. Bioinformatics 22:623–625. https://doi.org/10.1093/bioinformatics/btk023

Lomize MA, Pogozheva ID, Joo H et al (2012) OPM database and PPM web server: resources for positioning of proteins in membranes. Nucleic Acids Res 40:D370–D376. https://doi.org/10.1093/nar/gkr703

Lopez J, Webster RE (1982) Minor coat protein composition and location of the a protein in bacteriophage f1 spheroids and I-forms. J Virol 42:1099–1107

Lopez J, Webster RE (1983) Morphogenesis of filamentous bacteriophage f1: orientation of extrusion and production of polyphage. Virology 127:177–193

Manning M, Griffith J (1985) Association of M13 I-forms and spheroids with lipid vesicles. Arch Biochem Biophys 236:297–303

Marvin DA (1966) X-ray diffraction and electron microscope studies on the structure of the small filamentous bacteriophage fd. J Mol Biol 15:8–17

Marvin DA (1998) Filamentous phage structure, infection and assembly. Curr Opin Struct Biol 8:150–158

Marvin DA (2017) Fibre diffraction studies of biological macromolecules. Prog Biophys Mol Biol 127:43–87

Marvin DA, Hoffmann-Berling H (1963a) Physical and chemical properties of two new small bacteriophages. Nature 197:517–518. https://doi.org/10.1038/197517b0

Marvin DA, Hoffmann-Berling H (1963b) A fibrous DNA phage (fd) AND a spherical RNA phage (fr) specific for male strains of E. Coli. II Physical characteristics. Zeitschrift fur Naturforschung Tl B, Chemie, Biochem Biophys Biol und verwandte Gebiete 18:884–893

Marvin DA, Wachtel EJ (1976) Structure and assembly of filamentous bacterial viruses. Philos Trans R Soc Lond Ser B Biol Sci 276:81–98

Marvin DA, Wiseman RL, Wachtel EJ (1974) Filamentous bacterial viruses. XI. Molecular architecture of the class II (Pf1, Xf) virion. J Mol Biol 82:121–138

Marvin DA, Welsh LC, Symmons MF et al (2006) Molecular structure of fd (f1, M13) filamentous Bacteriophage refined with respect to X-ray fibre diffraction and solid-state NMR data supports specific models of phage assembly at the bacterial membrane. J Mol Biol 355:294–309. https://doi.org/10.1016/j.jmb.2005.10.048

Marvin DA, Symmons MF, Straus SK (2014) Structure and assembly of filamentous bacteriophages. Prog Biophys Mol Biol 114:80–122. https://doi.org/10.1016/j.pbiomolbio.2014.02.003

Morag O, Sgourakis NG, Baker D, Goldbourt A (2015) The NMR–Rosetta capsid model of M13 reveals a quadrupled hydrophobic packing epitope. Proc Natl Acad Sci 112:971–976. https://doi.org/10.1073/pnas.1415393112

Nave C, Fowler AG, Malsey S et al (1979) Macromolecular structural transitions in Pf1 filamentous bacterial virus. Nature 281:232–234

Nazarov PV, Koehorst RBM, Vos WL et al (2007) FRET study of membrane proteins: determination of the tilt and orientation of the N-terminal domain of M13 major coat protein. Biophys J 92:1296–1305. https://doi.org/10.1529/biophysj.106.095026

Opella SJ, Zeri AC, Park SH (2008) Structure, dynamics, and assembly of filamentous bacteriophages by nuclear magnetic resonance spectroscopy. Annu Rev Phys Chem 59:635–657. https://doi.org/10.1146/annurev.physchem.58.032806.104640

Papavoine CHM, Christiaans BEC, Folmer RHA et al (1998) Solution structure of the M13 major coat protein in detergent micelles: a basis for a model of phage assembly involving specific residues. J Mol Biol 282:401–419. https://doi.org/10.1006/jmbi.1998.1860

Pederson DM, Welsh LC, Marvin DA et al (2001) The protein capsid of filamentous bacteriophage PH75 from Thermus thermophilus. J Mol Biol 309:401–421. https://doi.org/10.1006/jmbi.2001.4685

Quinn CM, Polenova T (2017) Structural biology of supramolecular assemblies by magic-angle spinning NMR spectroscopy. Q Rev Biophys 50:e1. https://doi.org/10.1017/S0033583516000159

Russel M (1991) Filamentous phage assembly. Mol Microbiol 5:1607–1613

Russel M, Model P (2006) Filamentous Phage. In: Calendar R (ed) The bacteriophages, 2nd edn. Oxford University Press, Oxford, UK, pp 146–160

Russel M, Linderoth NA, Šali A (1997) Filamentous phage assembly: variation on a protein export theme. Gene 192:23–32. https://doi.org/10.1016/S0378-1119(96)00801-3

Sergeyev IV, Itin B, Rogawski R et al (2017) Efficient assignment and NMR analysis of an intact virus using sequential side-chain correlations and DNP sensitization. Proc Natl Acad Sci 114:5171–5176. https://doi.org/10.1073/pnas.1701484114

Shapiro JW, Williams ESCP, Turner PE (2016) Evolution of parasitism and mutualism between filamentous phage M13 and *Escherichia coli*. PeerJ 4:e2060. https://doi.org/10.7717/peerj.2060

Sinsheimer RL (1966) φX: Multum in Parvo. In: Cairns J, Stent GS, Watson JD (eds) Phage and the origins of molecular biology. Cold Spring Harbor Laboratory, New York, pp 258–264

Skehel JJ, Wiley DC (2000) Receptor binding and membrane fusion in virus entry: the influenza hemagglutinin. Annu Rev Biochem 69:531–569. https://doi.org/10.1146/annurev.biochem.69.1.531

Spruijt RB, Wolfs CJAM, Hemminga MA (2004) Membrane assembly of M13 major coat protein: evidence for a structural adaptation in the hinge region and a tilted transmembrane domain. Biochemistry 43:13972–13980. https://doi.org/10.1021/bi048437x

Stopar D, Spruijt RB, Wolfs CJ, Hemminga MA (1998) Mimicking initial interactions of bacteriophage M13 coat protein disassembly in model membrane systems. Biochemistry 37:10181–10187. https://doi.org/10.1021/bi9718144

Stopar D, Spruijt RB, Hemminga MA (2006a) Anchoring mechanisms of membrane-associated M13 major coat protein. Chem Phys Lipids 141:83–93. https://doi.org/10.1016/j.chemphyslip.2006.02.023

Stopar D, Strancar J, Spruijt RB, Hemminga MA (2006b) Motional restrictions of membrane proteins: a site-directed spin labeling study. Biophys J 91:3341–3348. https://doi.org/10.1529/biophysj.106.090308

Stopar D, Koehorst RBM, Spruijt RB, Hemminga MA (2009) Asymmetric dipping of bacteriophage M13 coat protein with increasing lipid bilayer thickness. Biochim Biophys Acta Biomembr 1788:2217–2221. https://doi.org/10.1016/j.bbamem.2009.08.008

Straus SK, Scott WRP, Symmons MF, Marvin DA (2008) On the structures of filamentous bacteriophage Ff (fd, f1, M13). Eur Biophys J 37:521. https://doi.org/10.1007/s00249-007-0222-7

Straus SK, Scott WRP, Schwieters CD, Marvin DA (2011) Consensus structure of Pf1 filamentous bacteriophage from X-ray fibre diffraction and solid-state NMR. Eur Biophys J 40:221–234. https://doi.org/10.1007/s00249-010-0640-9

Thiriot DS, Nevzorov AA, Zagyanskiy L et al (2004) Structure of the coat protein in Pf1 Bacteriophage determined by solid-state NMR spectroscopy. J Mol Biol 341:869–879. https://doi.org/10.1016/j.jmb.2004.06.038

Thiriot DS, Nevzorov AA, Opella SJ (2005) Structural basis of the temperature transition of Pf1 bacteriophage. Protein Sci 14:1064–1070. https://doi.org/10.1110/ps.041220305

Van Dalen A, De Kruijff B (2004) The role of lipids in membrane insertion and translocation of bacterial proteins. Biochim Biophys Acta, Mol Cell Res 1694:97–109. https://doi.org/10.1016/j.bbamcr.2004.03.007

van de Ven FJ, van Os JW, Aelen JM et al (1993) Assignment of 1H, 15N, and backbone 13C resonances in detergent-solubilized M13 coat protein via multinuclear multidimensional NMR: a model for the coat protein monomer. Biochemistry 32:8322–8328

Vos WL, Nazarov PV, Koehorst RBM et al (2009) From "I" to "L" and back again: the odyssey of membrane-bound M13 protein. Trends Biochem Sci 34:249–255. https://doi.org/10.1016/j.tibs.2009.01.007

Webster RE (1991) The tol gene products and the import of macromolecules into Escherichia Coli. Mol Microbiol 5:1005–1011

Webster RE (2001) Filamentous phage biology. In: Barbas CF III, Burton DR, Scott JK, Silverman GJ (eds) Phage display : a laboratory manual. Cold Spring Harbor Laboratory Press, New York, pp 1.1–1.37

Welsh LC, Symmons MF, Marvin DA (2000) The molecular structure and structural transition of the α-helical capsid in filamentous bacteriophage Pf1. Acta Crystallogr Sect D Biol Crystallogr 56:137–150. https://doi.org/10.1107/S0907444999015334

Williams KA, Farrow NA, Deber CM, Kay LE (1996) Structure and dynamics of bacteriophage IKe major coat protein in MPG micelles by solution NMR. Biochemistry 35:5145–5157. https://doi.org/10.1021/bi952897w

Zeri AC, Mesleh MF, Nevzorov AA, Opella SJ (2003) Structure of the coat protein in fd filamentous bacteriophage particles determined by solid-state NMR spectroscopy. Proc Natl Acad Sci U S A 100:6458–6463. https://doi.org/10.1073/pnas.1132059100

Zinder ND (1986) Single-stranded DNA-containing bacteriophages. BioEssays 5:84–87. https://doi.org/10.1002/bies.950050209

Zinder ND, Valentine RC, Roger M, Stoeckenius W (1963) f1, a rod-shaped male-specific bacteriophage that contains DNA. Virology 20:638–640. https://doi.org/10.1016/0042-6822(63)90290-3

Chapter 13
Protein-RNA Interactions
in the Single-Stranded RNA Bacteriophages

Jānis Rūmnieks and Kaspars Tārs

Introduction to ssRNA Phages

The bacteriophages with single-stranded RNA (ssRNA) genomes are among the simplest and smallest of the known viruses. Due to their simplicity, these phages have for long been used as models to study fundamental problems in molecular biology, such as translational control mechanisms, protein-RNA interactions, RNA replication, virus evolution, structure, and assembly. In 1976, the ssRNA phage MS2 was the first life form for which the complete genome sequence was determined (Fiers et al. 1976). The ssRNA phages have also been the source for many diverse applications, including vaccine development, imaging tools, and ecological and virus inactivation studies (see Pumpens et al. 2016 for a review).

All of the known ssRNA phages belong to the *Leviviridae* family and have small, approximately 3500 to 4200 nucleotide long genomes that encode just a few proteins (Fig. 13.1). Three of the proteins – the maturation, coat, and replicase - are conserved among all ssRNA phages. Many of the studied phages also have a short open reading frame (ORF) that codes for a lysis protein, which often overlaps with other genes and shows a surprising variation in its location within the genome (Klovins et al. 2002; Kazaks et al. 2011; Rumnieks and Tars 2012). Distantly related lysis proteins lack any sensible sequence identity and have presumably arisen several times independently from each other. All of the lysis polypeptides have one or two predicted transmembrane helices and are thought to cause cell lysis by forming ion-permeable pores in the periplasmic membrane (Goessens et al. 1988), which leads to depolarization of the membrane and subsequent activation of auto-lysins that degrade the cell wall. A subgroup of the ssRNA phages that are assigned

J. Rūmnieks · K. Tārs (✉)
Biomedical Research and Study Center, Riga, Latvia
e-mail: j.rumnieks@biomed.lu.lv; kaspars@biomed.lu.lv

© Springer Nature Singapore Pte Ltd. 2018
J. R. Harris, D. Bhella (eds.), *Virus Protein and Nucleoprotein Complexes*,
Subcellular Biochemistry 88, https://doi.org/10.1007/978-981-10-8456-0_13

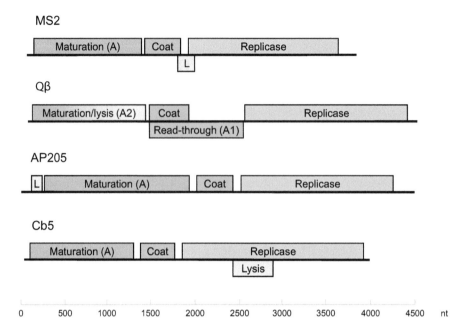

Fig. 13.1 Genome organization of the single-stranded RNA bacteriophages. The genes are represented as boxes; L, lysis gene. In bacteriophage Qβ, the maturation protein mediates cell lysis and A1 is a C-terminally extended variant of the coat protein generated by ribosomal read-through of the coat gene

to a separate *Allolevivirus* genus does not have a dedicated lysis protein; instead, the cell lysis is accomplished by the maturation protein which blocks an enzyme in the peptidoglycan biosynthesis pathway (Karnik and Billeter 1983; Winter and Gold 1983; Bernhardt et al. 2001). Another distinctive feature of the alloleviviruses is the presence of the so-called A1 protein in the capsid, which is an elongated version of the coat protein produced by a translational read-through mechanism (Weiner and Weber 1971). The exact function of the A1 protein is unknown, but it is important for the infectivity of the particles (Hofstetter et al. 1974). The only other recognized genus in the family, *Levivirus*, includes several phages where the lysis ORF is in a position overlapping the coat and replicase genes. However, many other ssRNA phages currently remain unassigned to any genus, and recent metagenomic studies have unraveled a vast array of new ssRNA phage sequences, in many cases very distantly related to the currently described ones (Krishnamurthy et al. 2016; Shi et al. 2016). Therefore, it should be noted that as the true ssRNA phage diversity in nature begins to be realized, a future reclassification within the *Leviviridae* family appears imminent, with a possible dissolution of the currently recognized *Levivirus* and *Allolevivirus* genera.

Structurally, the *Leviviridae* virions are composed of a single RNA molecule packaged inside a protein shell that is approximately 28 nm in diameter and consists of 178 coat protein molecules and a single copy of the maturation protein (Fig. 13.2a).

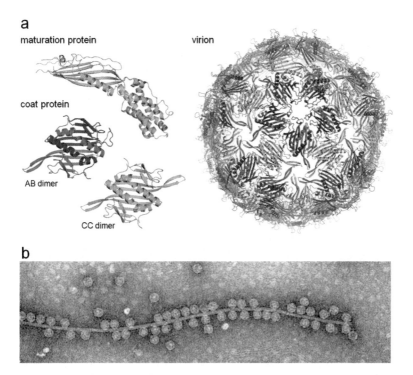

Fig. 13.2 Structure of an ssRNA bacteriophage particle. (**a**) Protein components of the virion. The coat protein dimers exist in two quasi-equivalent conformations in the particle, denoted AB (blue/red) and CC (green). An assembled particle consists of a single copy of the maturation protein, 60 coat protein dimers in the AB conformation and 29 in the CC conformation. The maturation protein replaces a single coat protein CC dimer in the otherwise icosahedrally symmetrical particle. (**b**) An electron micrograph of MS2 bacteriophage particles bound to an F pilus. Figs. 13.2a, 13.4, 13.5, 13.6, and 13.7 were prepared using Pymol version 1.8

The coat protein forms very stable dimers; therefore the capsid is more precisely described as composed of 89 coat protein dimers, and the maturation protein replaces a position otherwise occupied by a single coat protein dimer (Dent et al. 2013; Koning et al. 2016). The maturation protein serves as the attachment protein for the phage and mediates the adsorption of the virion to bacterial pili (Fig. 13.2b), which the ssRNA phages use as receptors for infecting the cell. The pili used by different ssRNA phages are rather distinct, ranging from the F-plasmid-encoded conjugative pili that the *E.coli* phages MS2 and Qβ employ (Crawford and Gesteland 1964), to various genome-encoded pili used by *Pseudomonas* phage PP7 (Bradley 1966), *Acinetobacter* phage AP205 (Klovins et al. 2002), or *Caulobacter* phage Cb5 (Schmidt and Stanier 1965). After adsorption, the maturation protein leaves the capsid together with RNA and guides its entry into host cell via a poorly understood mechanism. A complex of only the maturation protein and the genomic RNA is infectious to the cell (Shiba and Miyake 1975), and the coat protein does not have another role than protecting the genome before the infection takes place.

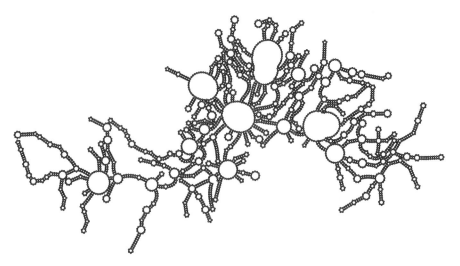

Fig. 13.3 Secondary structure of an ssRNA phage genome. The minimum free energy structure of the MS2 genomic RNA as predicted by the RNAfold software (Zuker and Stiegler 1981). The image was preparedusing RNAfdl (Hecker et al. 2013)

The RNA genome is an equally important structural component of the ssRNA phages. While the notion that the genome of the ssRNA phages is a single-stranded RNA is correct, this is only true in the sense that each virus particle indeed contains a single RNA strand. However, as much as 75% of the RNA bases are involved in short- and long-distance base-pairing interactions within the genome (Skripkin et al. 1990) (Fig. 13.3), which renders most of the genome double stranded and results in a complex three-dimensional structure. The maturation, coat, and replicase proteins are all RNA-binding proteins that recognize specific RNA structures in the genome, and different protein-RNA interactions are of essential importance during the ssRNA phage life cycle.

Replicase-RNA Interactions

As cells do not contain an enzyme capable of synthesizing long RNA molecules from an RNA template, all RNA viruses have to supply their own RNA-dependent RNA polymerase (RdRp) for the purposes of replicating their genome. All ssRNA phages likewise encode a 60–65 kDa polypeptide with the enzymatic RdRp activity; however, the protein alone is not capable of replicating the genome. The phage-encoded protein, often referred to as the "β subunit," recruits three more proteins from the bacterial cell, the ribosomal protein S1 (Wahba et al. 1974), and translation elongation factors EF-Ts and EF-Tu (Blumenthal et al. 1972), that together assemble into the replicase holoenzyme complex. The normal function of EF-Tu in the cell is to bind amino-acyl tRNAs and deliver them to ribosomes, while EF-Ts acts as a guanine nucleotide exchange factor for EF-Tu. The ribosomal protein S1 is a translation initiation factor that consists of six consecutive OB (oligonucleotide/oligosaccharide-binding) domains, of which the two N-terminal domains bind to the small ribosomal subunit, while the rest of the protein interacts with mRNA or autoregulates its own synthesis (Boni et al. 2000). The S1 protein differs from the other subunits in that it is not required for the structural integrity of the replicase complex and a "core replicase" enzyme consisting of only the β subunit, EF-Tu and EF-Ts, is enzymatically active.

The discovery of several host-derived subunits in ssRNA phage replicases seemed rather surprising at first, but since then it has become clear that the ssRNA phages are hardly unique in this respect. The idea of borrowing and repurposing host RNA-binding proteins appears to be fairly popular also among many eukaryotic RNA viruses, which often recruit proteins from the host's translation machinery to assemble a fully functional RdRp complex. Interestingly, several plant and animal viruses use the translational elongation factor eEF1A, a eukaryotic counterpart of EF-Tu, although not exactly in the same way as the ssRNA phages (see Li et al. 2013 for a review). Still, the function of the host-derived proteins in replication is probably the best understood in the ssRNA phage replicases.

Most of what is known about ssRNA phage RdRps, and their RNA replication in general, comes from studies of the enzyme from bacteriophage Qβ. The structure of the Qβ core replicase resembles a boat where the catalytic β subunit is located at one end and EF-Ts and EF-Tu at the other, with the active center facing the inner cavity of the structure (Kidmose et al. 2010; Takeshita and Tomita 2010) (Fig. 13.4). The β subunit has an architecture similar to other RdRps with the right-handed palm, thumb, and finger domains. EF-Tu participates in RNA binding during the elongation stage and forms part of the template exit channel, while the main function of EF-Ts appears to be the stabilization of the other subunits in an active conformation. The S1 protein binds to the opposite side of the β subunit with the same two N-terminal OB domains that are used for binding to the ribosome (Takeshita et al. 2014). The S1 protein is required to recognize and initiate replication of the genomic RNA strand (Kamen et al. 1972), and recently, the two protein-bound N-terminal S1 domains have been also proposed to function as a termination factor for an efficient release of product and template strands in a single-stranded form (Vasilyev et al. 2013).

The Qβ replicase has a remarkable processivity and can generate up to 10^{10} copies of some in vitro selected templates in 10 min (Chetverina and Chetverin 1993). At the same time, the enzyme is strongly selective in which RNAs are replicated well, and the natural template, the Qβ genome, is highly adapted to be efficiently replicated by the Qβ replicase. As the first requirement, the template needs to be single stranded, and the enzyme cannot initiate synthesis on double-stranded RNA (Weissmann et al. 1967). In the phage genome, the large proportion of RNA secondary structures ensures that the (+) and (−) strands do not anneal during replication and remain separate. To initiate RNA synthesis, the Qβ replicase does not require a primer but instead relies on a trinucleotide sequence CCA at the very 3′-terminus of the template (Chetverin and Spirin 1995). Intriguingly, tRNAs also have a CCA-3′

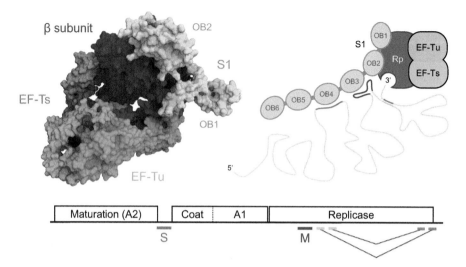

Fig. 13.4 RNA binding of the Qβ replicase. Left, the three-dimensional structure of the replicase complex. The phage-encoded catalytic β subunit (red) recruits three proteins, elongation factors EF-Tu (blue) and EF-Ts (green), and ribosomal protein S1 (yellow) from the host that together assemble into a holoenzyme complex. Right, a model for the genomic RNA recognition by the Qβ replicase. The S1 protein recognizes two internal sites in the Qβ genome (bottom), the S site (green) and the M site (violet). Two long-distance base-pairing interactions (orange/yellow) bridge the 3'-untranslated region of the genome to a nucleotide stretch nearby the M site, which constrains the genome in a particular conformation. Consequently, binding of the replicase-bound S1 protein to the genome positions its 3'-terminus in the active center of the β subunit

sequence which is recognized by EF-Tu, and it was for once thought that the phage has hijacked this ability to recognize its own genome. However, high-resolution structures of the Qβ replicase complex captured in the initiation stage have shown that the 3'-terminus is recognized solely by the β subunit where the 3'-terminal adenosine is kept in position by multiple contacts with the protein and by stacking interactions with the penultimate C and its complementary GTP (Takeshita and Tomita 2012). This way, the 3'-adenosine serves as a stable platform to initiate the replication, which begins at the penultimate cytidine, and not the adenosine itself. Upon termination, the Qβ replicase adds a non-templated adenosine to the newly synthesized strand (Weber and Weissmann 1970). To ensure exponential amplification, the original template thus needs to begin with GG so that the complementary strand ends with CCA and is able to guide the synthesis of another (+) strand.

A high degree of secondary structure and a sequence 5'-GG...CCA-3', however, are not sufficient for an RNA molecule to serve as a good template for the Qβ replicase, and the recognition of the phage genome is considerably more complex. The Qβ genome has two internal sites, the S site and the M site, which are recognized by the replicase holoenzyme (Meyer et al. 1981). The S site is an approximately 100-nucleotide-long uridine-rich stretch preceding the coat protein gene that is recognized by the S1 protein (Miranda et al. 1997). The S site is dispensable for replication but serves a role in coordinating the replication and translation of the genome. The coat protein gene does not have a canonical Shine-Dalgarno sequence to initiate translation and requires the ribosome-bound S1 protein to recruit the ribosome. This creates a situation where the S1 proteins from both the ribosome and the replicase complex compete for the same site in the genome. In the folded phage genome, the only translation initiation site that is available to ribosomes is that of the coat gene, while those of the other genes are buried in secondary structure and inaccessible (Van Duin and Tsareva 2006). Therefore, binding of the replicase complex to the S site prevents ribosomes from translating the genome and grants it exclusive rights for of the (+) strand, which is of particular importance early in infection when it is much more beneficial to actively replicate the genome than translate a few existing copies. In addition, since the replicase has to constantly compete for the (+) strand whereas the complementary strands are always available for copying, the initiation rate on (−) strands is higher, which results in a favorable ten fold excess of the (+) strands (Chetverin and Spirin 1995). The M site is an about 100-nucleotide-long branched stem-loop structure (Schuppli et al. 1998) within the replicase-coding sequence that is part of a bigger RNA structural domain called RD1. The M site is also recognized by the S1 protein (Miranda, Schuppli et al. 1997), but in contrast to the S site, its removal results in a drastic loss of template activity (Schuppli et al. 1998). The M site and the 3′-terminus are more than a thousand nucleotides apart from each other but are brought in close vicinity by long-distance base-pairing that involves the 3′-untranslated region of the genome and a nucleotide stretch adjacent to RD1 (Klovins et al. 1998; Klovins and van Duin 1999). The S site is also apparently close to the M site in the folded genome, as both sites can be bound simultaneously by the S1 protein; the M site is bound primarily by the third OB domain (Takeshita et al. 2014), while the S site is bound most likely by the adjacent C-terminal OB domains. Thus, while the recognition of the genome by the replicase complex is arguably the most extensive of the known protein-RNA interactions in the ssRNA phages, the phage achieves this by a clever recruitment of the cellular S1 protein to exploit its RNA-binding capabilities. Binding of the replicase-constituent S1 protein to the M site apparently positions the complex in such a way that the 3′-terminus of the genome in brought into the active center of the β subunit which allows the RNA synthesis to be initiated (Fig. 13.4).

Another *E.coli* protein, "the host factor for Qβ" (hfq), has been identified that further enhances the replication of Qβ RNA (Franze de Fernandez et al. 1968; Franze de Fernandez et al. 1972). Hfq is an abundant RNA-binding protein with several functions in the cell and directly binds to Qβ genome, where it presumably further increases the availability of the 3′-terminus to the replicase (Van Duin and Tsareva 2006). However, in contrast to the other bacterial proteins that the phage makes use of, the host factor is not essential for Qβ replication, and the phage can quickly adapt to grow without hfq by accumulating a few mutations in the genome (Schuppli et al. 1997; Schuppli et al. 2000).

The ssRNA phage replicases are among the most error-prone polymerases known, resulting in highly divergent sequences. Although much of what is learned from the Qβ replicase likely applies to other ssRNA phages, the differences in genome and protein structure surely have an impact. This appears to hold true for even reasonably closely related phages, such as Qβ and another *E.coli* phage MS2. Although the MS2 genome contains both the 5'-CC and CCA-3' sequences and the S and M sites, it is not recognized as a template by the Qβ replicase, and vice versa, the MS2 enzyme does not copy Qβ RNA (Haruna and Spiegelman 1965). The discrimination is likely caused by differences in the complex three-dimensional structure of the two genomes, which are expected to only become more significant in increasingly distant ssRNA phages. Despite recent advances in the structural studies of the ssRNA phages, the molecular details of how the replicase binds to the whole genome are currently still unknown and await further investigations.

While the Qβ replicase is highly adapted to recognize and replicate the phage genome, there do exist other RNA molecules that the enzyme is capable of replicating. In Qβ-infected cells, a variety of shorter RNA molecules can be detected that had been derived from the Qβ genome, in some cases by recombination with cellular RNAs (Munishkin et al. 1988; Munishkin et al. 1991; Moody et al. 1994; Avota et al. 1998). These shorter RNAs do not seem to have any biological function and are considered as mere by-products of the replicase activity. As shorter RNAs take less time to replicate, they gradually outcompete the longer phage genome, which in the limited time of infection apparently does not cause issues for the phage but are nevertheless interesting to study in vitro. In the most famous experiment, the Qβ replicase was initially allowed to replicate the Qβ genome in vitro, the reaction products transferred to another tube with fresh Qβ replicase and nucleotides but no template, and the transfer was then repeated many times over. After 74 generations, an RNA molecule just 218 nucleotides, dubbed the "Spiegelman's monster," had emerged and was being replicated much better than the original phage genome (Kacian et al. 1972). Since then, many other artificial RNAs replicable by the Qβ replicase have been described (Van Duin and Tsareva 2006). Like the phage genome, these RNAs have the expected 5'-GG...CCA-3' sequence and significant amounts of secondary structure, and apparently some kind of tertiary structure that has selected them as better templates than others. Some in vitro experiments had also suggested that the Qβ replicase can generate RNA spontaneously without any template, given only a mixture of nucleotides (Sumper and Luce 1975; Biebricher et al. 1986; Biebricher and Luce 1993). However, later experiments have arrived at a general conclusion that the sometimes-observed template-free de novo RNA synthesis is an artifact caused by minute amounts of contaminating RNA, present in enzyme preparations, buffer solutions, labware, or even laboratory air (Chetverin et al. 1991).

Coat Protein: RNA Interactions

Repression of the Replicase Gene

The obvious function for the ssRNA phage coat protein is the formation of a protein shell that protects the genome during the extracellular stage of the phage life cycle. Yet another role for the coat protein, at least in a subgroup of the ssRNA phages, is to regulate the translation of the replicase gene. The replicase is a characteristic early gene product that is required at the beginning of the infection, while later, when large amounts of phage RNA have been generated, it becomes beneficial to cease replication and switch to packaging of the genomes in new virus particles. In phage infected-cells, an observation can be made that as the amount of the synthesized coat protein grows, the synthesis of the replicase enzyme correspondingly diminishes. Behind the regulatory mechanism is the specific binding of the coat protein to an RNA hairpin structure at the very beginning of the replicase gene, usually referred to as the "translational operator" or "translational repressor" (Gralla et al. 1974). The hairpin contains the initiation codon of the replicase gene, which upon binding to the coat protein becomes masked form ribosomes, which in turn downregulates the translation of the replicase gene (Weber 1976).

The coat protein RNA operator interaction in the ssRNA phage MS2 has been extensively studied genetically, biochemically, and structurally and is one of the best understood protein-RNA interactions to date. The coat protein consists of an N-terminal β-hairpin, a five-stranded β-sheet, and two C-terminal α-helices. In a coat protein dimer, the two monomers form a continuous ten-stranded β-sheet that in the assembled particle lines the interior of the capsid and forms the RNA-binding surface of the protein (Valegård et al. 1990). The MS2 operator is a 19-nucleotide long RNA hairpin, composed of a seven-base pair stem with a single unpaired adenosine, and a four-nucleotide long loop. While the coat protein dimer is itself symmetric, the RNA operator binds asymmetrically across the RNA-binding surface (Fig. 13.5a), and each of the coat protein monomers interacts differently with the RNA. In the MS2 operator, four nucleotides, A-10, A-7, U-5, and A-4 (the base numbering is relative to the adenosine in the replicase initiation codon), contribute to the specific binding, but the crucial determinants for the interaction are the unpaired A-10 adenosine in the hairpin stem and the A-4 in the loop, which dock into two adenine-recognition pockets in the coat protein dimer (Valegård et al. 1994). Both pockets are identical, each formed by one of the coat protein monomers, but the interaction with the adenine bases is different in each pocket (Fig. 13.5b, c). The protein-RNA interaction is further stabilized by continuous aromatic stacking that via the A-7 and U-5 bases in the loop extends from the RNA stem to a tyrosine side chain in the coat protein. The sugar-phosphate backbone also makes extensive sequence-nonspecific interactions with the protein.

A lot of effort has been put toward characterizing many different MS2 operator variants to determine the exact contribution of the RNA bases for the binding interaction. For example, substitution of the U-5 base in the hairpin loop with a cytidine increases the coat protein-RNA affinity about 50-fold (Lowary and Uhlenbeck 1987). A crystal structure of the "C-variant" complex revealed that while the coat protein-operator interactions are essentially identical to the wild-type, the C-5 forms an additional intramolecular hydrogen bond in the RNA that stabilizes the operator structure and is apparently responsible for the tighter binding (Valegård et al. 1997). Several other structures with substitutions at the −5 position have been determined, with a general conclusion that the existence of the base stack itself is much more important than the identity of the bases (Grahn et al. 2001). In one case, a substitution

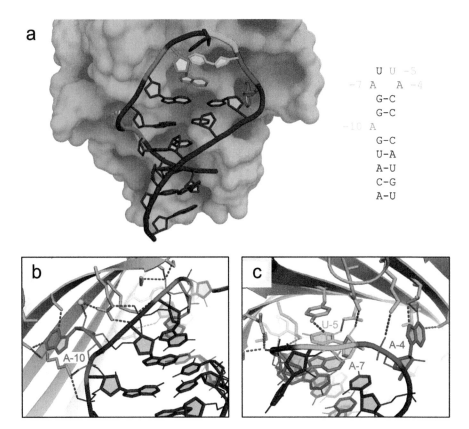

Fig. 13.5 RNA binding of the MS2 coat protein. (**a**) The overall RNA-binding mode. The coat protein dimer (light green/light orange as of the two monomers) binds to a hairpin structure (black) in the phage genome with high affinity. Four nucleotides (colored) are involved in the specific interaction. (**b**, **c**) A close-up view of the coat protein-RNA complex. Interactions around the unpaired adenosine in the hairpin stem (**b**) and the hairpin loop (**c**) are shown in the same colors as in (**a**)

of the −5 uracil with pyridin-4-one led to dramatic conformational rearrangements in the loop that caused the unnatural base to face away from the protein, but the stacking interaction with the tyrosine was still preserved by the neighboring U-6 base (Grahn et al. 2000). Substitutions at the −10 and −7 positions have likewise been tested, with a consensus that the loop is more important for binding than the unpaired base in the stem, but regardless the lower affinity, the coat protein is able to accommodate a wide variety of hairpin variants with only minor structural adjustments (Helgstrand et al. 2002).

The coat protein-RNA interactions have also been studied for several other ssRNA phages, albeit in much less detail compared to MS2. Bacteriophage PRR1 has an operator hairpin similar to MS2, except that it has a five- instead of a four-nucleotide loop. A crystal structure of the PRR1 coat protein-operator complex showed that the interaction, unsurprisingly, is almost identical to that of MS2

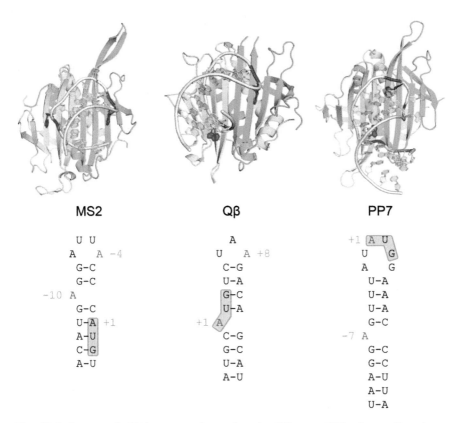

Fig. 13.6 Coat protein-RNA operator interactions in different ssRNA phages. Top, three-dimensional structures of phage MS2, Qβ and PP7 coat protein-operator complexes. The protein is colored in green/yellow as of the two monomers, and the bound operator is shown in light gray. A bulged adenosine that is present in all of the operator stems is indicated in blue, and another adenosine in the hairpin loop that is important for the interaction in red. The corresponding operator sequences are presented below the structures. The numbering of the bases is relative to the initiation codon of the replicase gene (green box)

(Persson et al. 2013). The structure of the coat protein-RNA complex from bacteriophage Qβ turned out to be more interesting (Rumnieks and Tars 2014). The Qβ operator is rather different from that of MS2 with a three-nucleotide loop and the bulged adenosine at a different location relative to the loop (Fig. 13.6). Despite the differences, the overall RNA-binding mode of the MS2 and Qβ coat proteins is similar, and the Qβ adenine-binding pockets are almost identical to those of MS2. In the hairpin loop, the A + 8 in Qβ and A-4 in MS2 operators make virtually identical interactions with one of the pockets, but the other adenine-binding pocket, which in MS2 is occupied by the bulged A-10, is empty in Qβ. Instead, the unpaired A + 1 in the Qβ operator makes a stacking interaction with a tyrosine residue, a mechanism that has not been observed in other ssRNA phages to date. The A + 8 adenine in the hairpin loop is the only sequence-specific requirement for the coat protein-RNA interaction in Qβ, while the identity of the other nucleotides in the loop, as well as the presence of the unpaired base in the stem, is dispensable for the interaction (Lim et al. 1996). Despite the seemingly lower specificity, the Qβ coat protein is still able to discriminate in favor of its cognate operator and binds it with a comparable strength to the other studied ssRNA phages. For strong binding to the Qβ coat protein, the RNA hairpin requires a three-nucleotide loop and an eight-base pair-long stem (Witherell and Uhlenbeck 1989). Compared to MS2, where virtually all of the protein-RNA contacts are located in the region between the bulged adenosine and the loop, in Qβ a significant proportion of the interactions involve the lower part of the stem, which explains the greater length dependence for the RNA. Binding of a three-nucleotide loop with an adenosine in the 3'-most position orients the lower part of the RNA helix at a favorable position for interacting with the distant RNA-binding residues, while a four-nucleotide loop and an unpaired base at an MS2-like position would position the RNA stem differently, resulting in weaker binding.

The replicase operator in bacteriophage PP7 is markedly different from the other studied ssRNA phages, with a six-nucleotide loop and a bulged adenosine four nucleotides prior to the loop (Fig. 13.6). The interaction of the operator with the PP7 coat protein is also very distinct from that of the other phages (Chao et al. 2008). The PP7 interaction relies mostly on sequence-specific interactions between the RNA and the protein (Lim and Peabody 2002) and involves a total of four bases. Similarly to MS2, the unpaired adenosine in the PP7 operator stem and another one in the loop bind to two symmetrical adenine-recognition pockets on the coat protein surface. However, the pockets are unrelated to those of MS2 and are located in a completely different position on the dimer surface. Similar to other phages, also in the PP7 complex, two bases in the hairpin loop continue the base stack from the RNA stem. However, the stack interacts with the protein via a van der Waals interaction with a valine residue, and not via another stacking interaction with an aromatic residue as in the other phages.

The different phages thus show a remarkable variation of how the specificity of the coat protein-RNA interaction is achieved, from several base-specific interactions in PP7 to a recognition mechanism based largely on the RNA backbone orientation in Qβ. Besides the aromatic stacking that extends to the hairpin loop and a functionally conserved binding of a single adenine base in the loop, there appear to be no other common themes in the ssRNA phage coat protein-operator complexes. Still, from the currently available data, two distinct coat protein-RNA-binding modes can be recognized; the first shared by phages MS2, PRR1, and Qβ and the other observed in the PP7 phage. The ssRNA phage coat protein-RNA interaction is a good example for a coevolution of protein and RNA structure, as changes in one of the components have to be complemented with corresponding changes in the other to maintain the binding. While it seems reasonable to assume that the MS2/Qβ and PP7 RNA-binding modes are evolutionary related, they are very distinct, and it is difficult to envision a common ancestor and a step-by-step transition to the two RNA binding modes. It is perhaps worth mentioning that the coat protein-replicase operator interaction is not critical for the phage, as mutants with a nonfunctional operator are still viable and only marginally less fit that the wild type (Peabody 1997; Licis et al. 2000). In addition, despite multiple experimental attempts, there is currently no evidence suggesting that an analogous interaction exists in the more distantly related phages AP205 and Cb5. Therefore, it appears that there might not be a very high pressure for the phage to conserve the interaction, and possibly, the coat protein-mediated replicase repression exists in only a subgroup of the ssRNA phages. It also cannot be excluded that the interaction has arisen more than once in different phage lineages, which might be case for the MS2/Qβ and PP7 RNA recognition mechanisms.

Interactions with the Genome in Virus Particle

The main function of the coat protein, the formation of a protein shell around the genome, also involves RNA binding. While the replicase operator hairpin is apparently the highest-affinity binding site for the coat protein, the protein is able to bind many different RNA stem-loops with lower affinity, which has been extensively characterized both biochemically and structurally. It was therefore obvious to suspect that besides the replicase operator, a number of other genomic RNA structures bind to the coat protein inside the capsid, which a protein-RNA cross-linking study using MS2 virions confirmed experimentally. The study found more than 50 potential coat protein-binding sites in the genome, most of which were clearly predicted to form a hairpin structure (Rolfsson et al. 2016). In a subsequent technological breakthrough, a medium-resolution asymmetric cryo-EM reconstruction of the MS2 bacteriophage allowed for the first time to directly visualize the genome inside the virion (Koning et al. 2016). The structure confirmed that the genome adopts a unique three-dimensional structure in the virus particles and allowed individual interactions between parts of the genome and the virion proteins to be identified. In total, 44 RNA hairpins and 33 double-stranded RNA regions were resolved that were in contact with the coat protein, while only 9 dimers did not have a nearby RNA density. Later, a higher resolution 3D reconstruction of the MS2 virion followed that identified more than 50 RNA hairpins in contact with coat protein, most of which contacted the dimers at the loop region (Dai et al. 2017). Fifteen of the coat protein-interacting stem-loops could be modeled at atomic resolution and turned out to be rather different in sequence and structure, directly demonstrating the flexibility of the coat protein in biding different RNA structures. Most of the coat protein-binding hairpins in the genomic RNA turned out to be asymmetrically distributed and predominantly located in the vicinity of the maturation protein, including the high-affinity replicase operator and two adjacent RNA stem-loops. The multiple interactions between the coat protein and hairpin structures in the genome thus lead to a model where they serve as packaging signals that together with the maturation protein help to recognize the genome and form a nucleation center for virion assembly, discussed in more detail in the next section.

Practical Applications

Besides a purely scientific interest, the ssRNA phage coat proteins and their RNA-binding properties have found a number of different applications, a significant proportion of which make use of ssRNA phage virus-like particles (VLPs). When a cloned coat protein gene is expressed in bacteria, the protein, without the need for the maturation protein or the genome, assembles into shells morphologically very similar to phage capsids (Kastelein et al. 1983; Kozlovskaya et al. 1986; Peabody 1990; Kozlovska et al. 1993). The resulting VLPs package significant amounts of cellular RNA in a largely nonspecific manner, probably via binding to different stem-loops in bacterial ribosomal and messenger RNAs (Pickett and Peabody 1993). A major area where the ssRNA phage VLPs are being explored is their use as antigen carriers in vaccine development (see Pumpens et al. 2016 and Jennings and Bachmann 2008 for reviews), and the RNA contained in the particles may activate toll-like receptors TLR3 and TLR7 that result in an enhanced immune response. The VLPs can also be obtained in vitro by mixing purified coat protein dimers and any heterologous RNA (Hohn 1969), which allows to package specific RNA molecules into the particles. The capsid formation can also be triggered using DNA instead of RNA, which allows to package short sequences of interest such as CpG-containing oligonucleotides into the VLPs to raise a TLR9-enhanced immune response (Bachmann et al. 2003).

A somewhat related application for ssRNA phage VLPs involves the generation of peptide display libraries using a modified version of the MS2 coat protein (Peabody et al. 2008). The method makes use of a surface-exposed region of the MS2 coat protein called the AB loop, which can tolerate short amino acid insertions without compromising its ability to assemble into VLPs. In a specifically designed vector, a randomized oligonucleotide library is ligated in-frame the AB loop, resulting in many bacterial clones each producing VLPs with a different peptide exposed on their surface. Crucially, upon assembly, the VLPs always package some coat protein mRNA into the particles, which is abundant in the cell due to the vector-driven overexpression. After affinity selection, the RNA contents of the target-bound VLPs are extracted, coat protein mRNAs amplified using reverse transcription PCR, and the corresponding peptide sequences recovered using DNA sequencing.

The high-affinity coat protein-operator hairpin interaction has been further employed to produce "armored" RNAs. Various diagnostic assays and other applications often require specific RNA sequences as controls, but due to the ubiquitous presence of ribonucleases in the environment, it is notoriously hard to avoid RNA degradation when working with naked RNA. In the armored RNA technology, the RNA of interest is engineered to contain an MS2 operator hairpin which is then produced in bacteria together with the MS2 coat protein (Pasloske et al. 1998). Due to the high specificity of the interaction, the VLPs that are assembled contain a high proportion of the operator-tagged RNA molecules inside the particles. Once the particles are assembled, the RNA is sealed from the surrounding environment, and the VLPs can then be easily purified using standard protein purification methods without worrying about RNA degradation. The MS2 VLPs are very stable and can be stored for prolonged periods of time without special precautions. The armored RNA technology has been commercialized by Asuragen and is reviewed in detail in Mikel et al. 2015.

The specific coat protein RNA binding is also used as a research tool in molecular biology to identify or track RNA-protein interactions (see Jazurek et al. 2016 for a review). The MS2-BioTRAP method is used for identifying RNA-binding proteins (Bardwell and Wickens 1990; Tsai et al. 2011). The RNA of interest is tagged with tandem repeats of the MS2 replicase hairpin and co-expressed with a modified MS2 coat protein harboring an HB-tag which gets biotynilated in vivo. As a result, the RNA gets decorated with biotynilated MS2 coat protein dimers and can be captured from a cell extract using streptavidin-coated beads. Proteins bound to the RNA of interest can then be stripped off and identified using mass spectrometry or other suitable technique. A related approach can be used to track RNA molecules of interest in living cells. The RNAs are likewise tagged with MS2 operator stem-loops, but the MS2 coat protein is fused with a fluorescent tag such as the green fluorescent protein. The RNAs can then be imaged in confocal fluorescent microscopy to follow the tagged RNAs throughout the cell (Bertrand et al. 1998). The distinct specificities of the MS2 and PP7 coat proteins also allow tracking of two different RNA molecules simultaneously using different fluorescent tags.

Maturation Protein-RNA Interactions

The maturation protein, sometimes referred to as the "A" or "A2" protein in different ssRNA phages, is the least understood of the phage proteins. The maturation protein binds to the genomic RNA and gets incorporated into the capsid along with it, where it later serves as an attachment protein that mediates the binding of the virion to bacterial pili and genome ejection and entry into the host cell. All of the known ssRNA phage maturation proteins are insoluble in an isolated form which has greatly hampered their studies, and for many decades, molecular details explaining how the maturation protein accomplishes any of its different functions had remained unknown. However, a recent high-resolution crystal structure of the maturation protein from bacteriophage Qβ (Rumnieks and Tars 2017) and medium- to high-resolution asymmetric cryo-EM reconstructions of whole MS2 (Dai et al. 2017) and Qβ (Gorzelnik et al. 2016) virions are finally starting to provide some answers about how the ssRNA phage maturation proteins look like and function.

The structural studies have revealed that the ssRNA phage maturation proteins have a rather peculiar highly elongated and bent shape and incorporate into the virion by taking place of a single coat protein dimer in the otherwise symmetrical protein shell (Fig. 13.2a). The maturation protein has a roughly globular α-helical part that faces the capsid interior and an elongated, relatively flat β-part that interacts with the coat protein and points away from the particle at a shallow angle. In phages MS2 and Qβ, both the coat and maturation proteins have approximately 20%

sequence identity, but while the coat protein structure in the two phages is very similar, the same does not hold true for the maturation proteins. Among the two proteins, only the core four-helix bundle of the α-helical region is clearly conserved, while the differences in other parts of the proteins are often too large for a reliable structural alignment. The structure of more distantly ssRNA phage maturation proteins is presumably even more distinct, although all of them probably have an inward-facing α-helical part and a surface-exposed β-region, as suggested by secondary structure predictions.

Like the other ssRNA phage proteins, also the maturation protein is a specific RNA-binding protein, and the high-resolution cryo-EM structure of the MS2 virion provides the first detailed look for any ssRNA phage at how the maturation protein binds to the RNA (Dai et al. 2017). The interaction between the MS2 maturation protein and the genome is rather extensive and involves two distinct RNA-binding surfaces on the protein and four regions in the genome (Fig. 13.7a–c). The first RNA-binding surface is located toward the distal part of the α-helical region and makes contact with two double-helical hairpin stems in the replicase coding region. The interaction is sequence-nonspecific and involves electrostatic interactions between the sugar-phosphate backbone and positively charged residues. The other RNA-binding surface is located around the central part of the protein and is composed of a portion of the central β-sheet and an adjacent part of the α-helical region. It binds two hairpins in the 3′-untranslated region of the genome, the first of which is very short and poorly defined in the structure, while the other is the very 3′-terminal hairpin where the interaction is clearly sequence-specific. Interestingly, binding of the 3′-terminal hairpin is accomplished cooperatively between the maturation protein and two adjacent coat protein dimers (Fig. 13.7d), and a small part of the RNA hairpin is directly exposed to the exterior of the virion (Fig. 13.7e).

The MS2 virion structure also provides important clues about the possible assembly pathway of the virus particle. The loop of the 3′-terminal helix contains a sequence CUGCUU that is fairly conserved among different RNA phages. In the MS2 virion, this sequence forms the stretch that is bound cooperatively by the maturation protein and two adjacent coat protein dimers and is partly exposed on the virion surface. In the Qβ genome, the nucleotide stretch in the 3′-terminal hairpin loop has been shown to form a pseudoknot with a sequence within the replicase coding region, the disruption of which abolishes replication (Klovins and van Duin 1999). The likely explanation for this is that the pseudoknot together with another nearby long-distance interaction positions the M site in an orientation that allows the relpicase complex to bind to the genome and initiate replication. It is not clear whether an equivalent pseudoknot is formed in the MS2 genome, but binding of the maturation protein to the 3′-terminal hairpin clearly renders the nearby 3′-terminus of the genome inaccessible to the replicase and prevents its replication. The cryo-EM structures also revealed that a higher proportion of the stem-loops that bind to the coat protein are located close to the maturation protein compared to the opposite side of the particle. In particular, three adjacent RNA stem-loops were resolved in

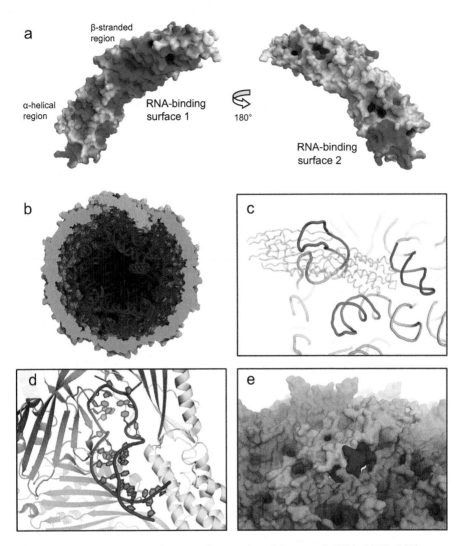

Fig. 13.7 Interactions between the maturation protein and the genomic RNA. (**a**) The MS2 maturation protein has a banana-like shape and consists of an α-helical and a β-stranded region. On the electrostatic surface of the protein, two distinct positively charged areas are present that form the two RNA-binding regions of the protein. (**b**) A cut-away view of the MS2 virion. The maturation protein (light orange in all panels) is partially exposed on the virion surface and makes contact both with the coat protein (gray) and the genome (red). (**c**) Parts of the MS2 genome in contact with the maturation protein. The protein binds to four distinct RNA stem-loops in the genome (colored). Of these, only binding to the 3′-terminal hairpin (red) is sequence-specific. (**d**) A close-up view of the interaction with the 3′-terminal hairpin. The hairpin (red) is bound cooperatively by the maturation protein and two neighboring coat protein dimers (green). In the crevice between the maturation and coat proteins, the RNA is partly exposed onto the virion surface (**e**)

the MS2 genome that bind to coat protein dimers nearby the maturation protein, among those the high-affinity replicase operator. Together, these observations suggest that binding of the maturation protein and two coat protein dimers to the 3′-terminal hairpin likely forms the nucleation center that marks the genome for packaging in new virus particles, and nearby high-affinity stem-loops then recruit more coat protein dimers that result in a rapid formation of a protein shell around the genome.

Together with the high-resolution crystal structure of the Qβ maturation protein, the currently available medium-resolution cryo-EM structure of the Qβ virion (Gorzelnik et al. 2016) allows to explore some of the maturation protein-RNA interaction also in this phage. Like MS2, also the Qβ maturation protein has two distinctive RNA-binding regions located similarly as in the MS2 protein (Rumnieks and Tars 2017). The α-helical region likewise binds several double-helical features in the genome, and the central RNA-binding surface interacts with a single long RNA hairpin, but due to the limited resolution, it is not possible to tell if it this is also the 3′-terminal one like in MS2. The long hairpin approaches the Qβ maturation protein from a different direction, and while it also appears to make contacts with a nearby coat protein dimer, these are not nearly as extensive as in MS2. In Qβ, also no part of the genome becomes exposed to the outer surface of the particle, as the corresponding gap in the virion is plugged by the N-terminal part of the maturation protein that is folded differently than in MS2. Thus, while the non-atomic resolution of the Qβ virion map certainly leaves a room for interpretation errors, the RNA binding of the Qβ maturation appears to be sufficiently different from that of MS2, again suggesting that there might be little conservation and the RNA binding is likely even more distinctive for increasingly further related phages.

Concluding Remarks

It has become increasingly clear that the genetic material of an organism, or a virus, cannot be considered merely as the storage medium for encoding its proteins, and rarely it is more evident than in the case of the ssRNA bacteriophages. Here, the genome is the most central figure that via its complex and dynamic three-dimensional shape orchestrates the phage and host proteins to get replicated and propagated. The story of the phage life cycle is, for the most part, a story of RNA structure and different protein-RNA interactions.

The recent asymmetric cryo-EM reconstructions of ssRNA phage virions have been a major breakthrough that has allowed for the first time to directly visualize the genome inside their particles, reveling its complex and folded three-dimensional structure. We are now closer than ever to a truly molecular-level understanding of the small RNA phages, but there is certainly still much substance for further studies. Among other things, there are some blank spots left of how the ssRNA phage replicases recognize the genomic RNA, and virtually nothing is known about the molecular mechanism of how the maturation protein guides the genome into the host cell. Furthermore, due to the advances in modern sequencing technologies, the true ssRNA phage diversity in nature is just being discovered, which stretches far beyond the few model phages on which almost all of our current understanding of these viruses is built. Surely, many more protein-RNA interactions await their discovery and exploration.

References

Avota E, Berzins V, Grens E, Vishnevsky Y, Luce R, Biebricher CK (1998) The natural 6 S RNA found in Qbeta-infected cells is derived from host and phage RNA. J Mol Biol 276(1):7–17

Bachmann MF, Storni T, Maurer P, Tissot A, Schwarz K, Meijerink E, Lipowsky G, Pumpens P, Cielens I and Renhofa R (2003) Packaging of immunostimulatory cpg into virus-like particles: method of preparation and use. E. P. Office. WO03024481 (A2)

Bardwell VJ, Wickens M (1990) Purification of RNA and RNA-protein complexes by an R17 coat protein affinity method. Nucleic Acids Res 18(22):6587–6594

Bernhardt TG, Wang IN, Struck DK, Young R (2001) A protein antibiotic in the phage Qbeta virion: diversity in lysis targets. Science 292(5525):2326–2329

Bertrand E, Chartrand P, Schaefer M, Shenoy SM, Singer RH, Long RM (1998) Localization of ASH1 mRNA particles in living yeast. Mol Cell 2(4):437–445

Biebricher CK, Luce R (1993) Sequence analysis of RNA species synthesized by Qbeta replicase without template. Biochemistry 32(18):4848–4854

Biebricher CK, Eigen M, Luce R (1986) Template-free RNA synthesis by Qb replicase. Nature 321:89–91

Blumenthal T, Landers TA, Weber K (1972) Bacteriophage Qb replicase contains the protein biosynthesis elongation factors EF Tu and EF Ts. Proc Natl Acad Sci U S A 69(5):1313–1317

Boni IV, Artamonova VS, Dreyfus M (2000) The last RNA-binding repeat of the Escherichia coli ribosomal protein S1 is specifically involved in autogenous control. J Bacteriol 182(20):5872–5879

Bradley DE (1966) The structure and infective process of a Pseudomonas Aeruginosa bacteriophage containing ribonucleic acid. J Gen Microbiol 45:83–96

Chao JA, Patskovsky Y, Almo SC, Singer RH (2008) Structural basis for the coevolution of a viral RNA-protein complex. Nat Struct Mol Biol 15(1):103–105

Chetverin AB, Spirin AS (1995) Replicable RNA vectors: prospects for cell-free gene amplification, expression, and cloning. Prog Nucleic Acid Res Mol Biol 51:225–270

Chetverin AB, Chetverina HV, Munishkin AV (1991) On the nature of spontaneous RNA synthesis by Qbeta replicase. J Mol Biol 222(1):3–9

Chetverina HV, Chetverin AB (1993) Cloning of RNA molecules in vitro. Nucleic Acids Res 21(10):2349–2353

Crawford EM, Gesteland RF (1964) The adsorption of bacteriophage R17. Virology 22:165–167

Dai X, Li Z, Lai M, Shu S, Du Y, Zhou ZH, Sun R (2017) In situ structures of the genome and genome-delivery apparatus in a single-stranded RNA virus. Nature 541(7635):112–116

Dent KC, Thompson R, Barker AM, Hiscox JA, Barr JN, Stockley PG, Ranson NA (2013) The asymmetric structure of an icosahedral virus bound to its receptor suggests a mechanism for genome release. Structure 21(7):1225–1234

Fiers W, Contreras R, Duerinck F, Haegeman G, Iserentant D, Merregaert J, Min Jou W, Molemans F, Raeymaekers A, Vandenberghe A, Volckaert G, Ysebaert M (1976) Complete nucleotide sequence of bacteriophage MS2 RNA: primary and secondary structure of the replicase gene. Nature 260:500–507

Franze de Fernandez MT, Eoyang L, August JT (1968) Factor fraction required for the synthesis of bacteriophage Qbeta-RNA. Nature 219(5154):588–590

Franze de Fernandez MT, Hayward WS, August JT (1972) Bacterial proteins required for replication of phage Qb ribonucleic acid. Pruification and properties of host factor I, a ribonucleic acid-binding protein. J Biol Chem 247(3):824–831

Goessens WHF, Driessen AJM, Wilschut J, van Duin J (1988) A synthetic peptide corresponding to the C-terminal 25 residues of phage MS2 coded lysis protein dissipates the protonmotive force in Escherichia coli membrane vesicles by generating hydrophilic pores. EMBO J 7:867–873

Gorzelnik KV, Cui Z, Reed CA, Jakana J, Young R, Zhang J (2016) Asymmetric cryo-EM structure of the canonical Allolevivirus Qbeta reveals a single maturation protein and the genomic ssRNA in situ. Proc Natl Acad Sci U S A 113(41):11519–11524

Grahn E, Stonehouse NJ, Adams CJ, Fridborg K, Beigelman L, Matulic-Adamic J, Warriner SL, Stockley PG, Liljas L (2000) Deletion of a single hydrogen bonding atom from the MS2 RNA operator leads to dramatic rearrangements at the RNA-coat protein interface. Nucleic Acids Res 28:4611–4616

Grahn E, Moss T, Helgstrand C, Fridborg K, Sundaram M, Tars K, Lago H, Stonehouse NJ, Davis DR, Stockley PG, Liljas L (2001) Structural basis of pyrimidine specificity in the MS2 RNA hairpin-coat-protein complex. RNA 7(11):1616–1627

Gralla J, Steitz JA, Crothers DM (1974) Direct physical evidence for secondary structure in an isolated fragment of R17 bacteriophage mRNA. Nature 248:204–208

Haruna I, Spiegelman S (1965) Specific template requirments of RNA replicases. Proc Natl Acad Sci U S A 54(2):579–587

Hecker N, Wiegels T, Torda AE (2013) RNA secondary structure diagrams for very large molecules: RNAfdl. Bioinformatics 29(22):2941–2942

Helgstrand C, Grahn E, Moss T, Stonehouse NJ, Tars K, Stockley PG, Liljas L (2002) Investigating the structural basis of purine specificity in the structures of MS2 coat protein RNA translational operator hairpins. Nucleic Acids Res 30(12):2678–2685

Hofstetter H, Monstein H, Weissmann C (1974) The readthrough protein A1 is essential for the formation of viable Qb particles. Biochim Biophys Acta 374:238–251

Hohn T (1969) Role of RNA in the assembly process of bacteriophage fr. J Mol Biol 43:191–200

Jazurek M, Ciesiolka A, Starega-Roslan J, Bilinska K, Krzyzosiak WJ (2016) Identifying proteins that bind to specific RNAs - focus on simple repeat expansion diseases. Nucleic Acids Res 44(19):9050–9070

Jennings GT, Bachmann MF (2008) The coming of age of virus-like particle vaccines. Biol Chem 389(5):521–536

Kacian DL, Mills DR, Kramer FR, Spiegelman S (1972) A replicating RNA molecule suitable for a detailed analysis of extracellular evolution and replication. Proc Natl Acad Sci U S A 69(10):3038–3042

Kamen R, Kondo M, Romer W, Weissmann C (1972) Reconstitution of Qb replicase lacking subunit with protein-synthesis-interference factor i. Eur J Biochem 31(1):44–51

Karnik S, Billeter M (1983) The lysis function of RNA bacteriophage Qb is mediated by the maturation (A2) protein. EMBO J 2:1521–1526

Kastelein RA, Berkhout B, Overbeek GP, van Duin J (1983) Effect of the sequences upstream from the ribosome-binding site on the yield of protein from the cloned gene for phage MS2 coat protein. Gene 23:245–254

Kazaks A, Voronkova T, Rumnieks J, Dishlers A, Tars K (2011) Genome structure of caulobacter phage phiCb5. J Virol 85(9):4628–4631

Kidmose RT, Vasiliev NN, Chetverin AB, Andersen GR, Knudsen CR (2010) Structure of the Qbeta replicase, an RNA-dependent RNA polymerase consisting of viral and host proteins. Proc Natl Acad Sci U S A 107(24):10884–10889

Klovins J, van Duin J (1999) A long-range pseudoknot in Qbeta RNA is essential for replication. J Mol Biol 294(4):875–884

Klovins J, Berzins V, van Duin J (1998) A long-range interaction in Qbeta RNA that bridges the thousand nucleotides between the M-site and the 3′ end is required for replication. RNA 4(8):948–957

Klovins J, Overbeek GP, van den Worm SH, Ackermann HW, van Duin J (2002) Nucleotide sequence of a ssRNA phage from Acinetobacter: kinship to coliphages. J Gen Virol 83(Pt 6):1523–1533

Koning RI, Gomez-Blanco J, Akopjana I, Vargas J, Kazaks A, Tars K, Carazo JM, Koster AJ (2016) Asymmetric cryo-EM reconstruction of phage MS2 reveals genome structure in situ. Nat Commun 7:12524

Kozlovska TM, Cielens I, Dreilinna D, Dislers A, Baumanis V, Ose V, Pumpens P (1993) Recombinant RNA phage Qbeta capsid particles synthesized and self-assembled in Escherichia coli. Gene 137(1):133–137

Kozlovskaya TM, Pumpen PP, Dreilina DE, Tsimanis AJ, Ose VP, Tsibinogin VV, Gren EJ (1986) Formation of capsid-like structures as a result of expression of coat protein gene of RNA phage fr. Dokl Akad Nauk SSSR 287:452–455

Krishnamurthy SR, Janowski AB, Zhao G, Barouch D, Wang D (2016) Hyperexpansion of RNA bacteriophage diversity. PLoS Biol 14(3):e1002409

Li D, Wei T, Abbott CM, Harrich D (2013) The unexpected roles of eukaryotic translation elongation factors in RNA virus replication and pathogenesis. Microbiol Mol Biol Rev 77(2):253–266

Licis N, Balklava Z, van Duin J (2000) Forced retroevolution of an RNA bacteriophage. Virology 271:298–306

Lim F, Peabody DS (2002) RNA recognition site of PP7 coat protein. Nucleic Acids Res 30(19):4138–4144

Lim F, Spingola M, Peabody DS (1996) The RNA-binding site of bacteriophage Qb coat protein. J Biol Chem 271:31839–31845

Lowary PT, Uhlenbeck OC (1987) An RNA mutation that increases the affinity of an RNA-protein interaction. Nucleic Acids Res 15:10483–10493

Meyer F, Weber H, Weissmann C (1981) Interaction of Qb replicase with Qb RNA. J Mol Biol 153:631–660

Mikel P, Vasickova P, Kralik P (2015) Methods for preparation of MS2 phage-like particles and their utilization as process control viruses in RT-PCR and qRT-PCR detection of RNA viruses from food matrices and clinical specimens. Food Environ Virol 7:96

Miranda G, Schuppli D, Barrera I, Hausherr C, Sogo JM, Weber H (1997) Recognition of bacteriophage Qbeta plus strand RNA as a template by Qbeta replicase: role of RNA interactions mediated by ribosomal proteins S1 and host factor. J Mol Biol 267(5):1089–1103

Moody MD, Burg JL, DiFrancesco R, Lovern D, Stanick W, Lin-Goerke J, Mahdavi K, Wu Y, Farrell MP (1994) Evolution of host cell RNA into efficient template RNA by Qbeta replicase: the origin of RNA in untemplated reactions. Biochemistry 33(46):13836–13847

Munishkin AV, Voronin LA, Chetverin AB (1988) An in vivo recombinant RNA capable of autocatalytic synthesis by Qbeta replicase. Nature 333(6172):473–475

Munishkin AV, Voronin LA, Ugarov VI, Bondareva LA, Chetverina HV, Chetverin AB (1991) Efficient templates for Q beta replicase are formed by recombination from heterologous sequences. J Mol Biol 221(2):463–472

Pasloske BL, Walkerpeach CR, Obermoeller RD, Winkler M, DuBois DB (1998) Armored RNA technology for production of ribonuclease-resistant viral RNA controls and standards. J Clin Microbiol 36:3590–3594

Peabody DS (1990) Translational repression by bacteriophage MS2 coat protein expressed from a plasmid. A system for genetic analysis of a protein-RNA interaction. J Biol Chem 265:5684–5689

Peabody DS (1997) Role of the coat protein-RNA interaction in the life cycle of bacteriophage MS2. Mol Gen Genet 254:358–364

Peabody DS, Manifold-Wheeler B, Medford A, Jordan SK, do Carmo Caldeira J, Chackerian B (2008) Immunogenic display of diverse peptides on virus-like particles of RNA phage MS2. J Mol Biol 380(1):252–263

Persson M, Tars K, Liljas L (2013) PRR1 coat protein binding to its RNA translational operator. Acta Crystallogr D Biol Crystallogr 69(Pt 3):367–372

Pickett GG, Peabody DS (1993) Encapsidation of heterologous RNAs by bacteriophage MS2 coat protein. Nucleic Acids Res 21:4621–4626

Pumpens P, Renhofa R, Dishlers A, Kozlovska T, Ose V, Pushko P, Tars K, Grens E, Bachmann MF (2016) The true story and advantages of RNA phage capsids as Nanotools. Intervirology 59(2):74–110

Rolfsson O, Middleton S, Manfield IW, White SJ, Fan B, Vaughan R, Ranson NA, Dykeman E, Twarock R, Ford J, Kao CC, Stockley PG (2016) Direct Evidence for Packaging Signal-Mediated Assembly of Bacteriophage MS2. J Mol Biol 428(2 Pt B):431–448

Rumnieks J, Tars K (2012) Diversity of pili-specific bacteriophages: genome sequence of IncM plasmid-dependent RNA phage M. BMC Microbiol 12:277

Rumnieks J, Tars K (2014) Crystal structure of the bacteriophage qbeta coat protein in complex with the RNA operator of the replicase gene. J Mol Biol 426(5):1039–1049

Rumnieks J, Tars K (2017) Crystal structure of the maturation protein from bacteriophage Qbeta. J Mol Biol 429(5):688–696

Schmidt JM, Stanier RY (1965) Isolation and characterization of bacteriophages active against stalked bacteria. J Gen Microbiol 39:95–107

Schuppli D, Miranda G, Tsui HC, Winkler ME, Sogo JM, Weber H (1997) Altered 3′-terminal RNA structure in phage Qbeta adapted to host factor-less Escherichia coli. Proc Natl Acad Sci U S A 94(19):10239–10242

Schuppli D, Miranda G, Qiu S, Weber H (1998) A branched stem-loop structure in the M-site of bacteriophage Qbeta RNA is important for template recognition by Qbeta replicase holoenzyme. J Mol Biol 283(3):585–593

Schuppli D, Georgijevic J, Weber H (2000) Synergism of mutations in bacteriophage Qbeta RNA affecting host factor dependence of Qbeta replicase. J Mol Biol 295(2):149–154

Shi M, Lin XD, Tian JH, Chen LJ, Chen X, Li CX, Qin XC, Li J, Cao JP, Eden JS, Buchmann J, Wang W, Xu J, Holmes EC, Zhang YZ (2016) Redefining the invertebrate RNA virosphere. Nature 540:539

Shiba T, Miyake T (1975) New type of infectious complex of E. coli RNA phage. Nature 254:157–158

Skripkin EA, Adhin MR, de Smit MH, van Duin J (1990) Secondary structure of bacteriophage MS2. Conservation and biological significance. J Mol Biol 211:447–463

Sumper M, Luce R (1975) Evidence for de novo production of self-replicating and environmentally adapted RNA structures by bacteriophage Qbeta replicase. Proc Natl Acad Sci U S A 72(1):162–166

Takeshita D, Tomita K (2010) Assembly of Q{beta} viral RNA polymerase with host translational elongation factors EF-Tu and -Ts. Proc Natl Acad Sci U S A 107(36):15733–15738

Takeshita D, Tomita K (2012) Molecular basis for RNA polymerization by Qbeta replicase. Nat Struct Mol Biol 19(2):229–237

Takeshita D, Yamashita S, Tomita K (2014) Molecular insights into replication initiation by Qbeta replicase using ribosomal protein S1. Nucleic Acids Res 42(16):10809–10822

Tsai BP, Wang X, Huang L, Waterman ML (2011) Quantitative profiling of in vivo-assembled RNA-protein complexes using a novel integrated proteomic approach. Mol Cell Proteomics 10(4):M110 007385

Valegård K, Liljas L, Fridborg K, Unge T (1990) The three-dimensional structure of the bacterial virus MS2. Nature 345:36–41

Valegård K, Murray JB, Stockley PG, Stonehouse NJ, Liljas L (1994) Crystal structure of an RNA bacteriophage coat protein-operator complex. Nature 371:623–626

Valegård K, Murray JB, Stonehouse NJ, van den Worm S, Stockley PG, Liljas L (1997) The three-dimensional structures of two complexes between recombinant MS2 capsids and RNA operator fragments reveal sequence-specific protein-RNA interactions. J Mol Biol 270:724–738

Van Duin J, Tsareva N (2006) Single-stranded RNA phages. In: Calendar R (ed) The bacteriophages. Oxford University Press, New York, pp 175–196

Vasilyev NN, Kutlubaeva ZS, Ugarov VI, Chetverina HV, Chetverin AB (2013) Ribosomal protein S1 functions as a termination factor in RNA synthesis by Qbeta phage replicase. Nat Commun 4:1781

Wahba AJ, Miller MJ, Niveleau A, Landers TA, Carmichael GG, Weber K, Hawley DA, Slobin LI (1974) Subunit I of Qbeta replicase and 30 S ribosomal protein S1 of Escherichia coli. Evidence for the identity of the two proteins. J Biol Chem 249(10):3314–3316

Weber H (1976) The binding site for coat protein on bacteriophage Qb RNA. Biochim Biophys Acta 418:175–183

Weber H, Weissmann C (1970) The 3′-termini of bacteriophage Qbeta plus and minus strands. J Mol Biol 51(2):215–224

Weiner AM, Weber K (1971) Natural read-through at the UGA termination signal of Qbeta coat protein cistron. Nat New Biol 234:206–209

Weissmann C, Feix G, Slor H, Pollet R (1967) Replication of viral RNA. XIV. Single-stranded minus strands as template for the synthesis of viral plus strands in vitro. Proc Natl Acad Sci U S A 57(6):1870–1877

Winter RB, Gold L (1983) Overproduction of bacteriophage Qb maturation (A2) protein leads to cell lysis. Cell 33:877–885

Witherell GW, Uhlenbeck OC (1989) Specific RNA binding by Qb coat protein. Biochemistry 28:71–76

Zuker M, Stiegler P (1981) Optimal computer folding of large RNA sequences using thermodynamics and auxiliary information. Nucleic Acids Res 9(1):133–148

Chapter 14
The Bacteriophage Head-to-Tail Interface

Paulo Tavares

Background

The vast majority of viruses that infect bacteria (phages or bacteriophages) are formed of an icosahedral capsid (or head) and of a tail (Ackermann 2012). This structural organization of bacteriophage particles (or virions) was retained throughout evolution as a highly successful solution for viral infection of bacteria. The function of the capsid is to protect the viral double-stranded DNA (dsDNA) genome. The tail is the device that allows selective recognition of the host surface and delivery of the viral genome from the capsid to the cytoplasm interior. Tailed phages (order *Caudovirales*) can be divided in three families according to their tail morphology: short tail (*Podoviridae*), long non-contractile tail (*Siphoviridae*), and long contractile tail (*Myoviridae*).

The virions of tailed phages species are complex nucleoprotein complexes. They are structurally related and follow a similar assembly pathway to build particles of homogeneous size and shape amounting for several dozens of megadaltons (Fig. 14.1A) (Casjens and Hendrix 1988; Krupovic and Bamford 2011; Abrescia et al. 2012). An icosahedral procapsid (or prohead) is assembled first featuring a specialized vertex characterized by the presence of the *portal protein* (Fig. 14.1). Procapsid assembly likely initiates at this vertex. This ensures the asymmetric incorporation of the portal in the icosahedral lattice (Bazinet and King 1985; Dröge et al. 2000; Newcomb et al. 2005; Fu and Prevelige 2009; Motwani et al. 2017 and references therein). The dodecameric portal structure is organized around a central channel through which dsDNA enters and exits the viral capsid. The portal vertex provides the docking point for the *viral terminase-phage DNA complex* leading to

P. Tavares (✉)
Department of Virology, Institute for Integrative Biology of the Cell (I2BC), CEA, CNRS, Univ Paris-Sud, Université Paris-Saclay, Gif-sur-Yvette, France
e-mail: paulo.tavares@i2bc.paris-saclay.fr

© Springer Nature Singapore Pte Ltd. 2018
J. R. Harris, D. Bhella (eds.), *Virus Protein and Nucleoprotein Complexes*,
Subcellular Biochemistry 88, https://doi.org/10.1007/978-981-10-8456-0_14

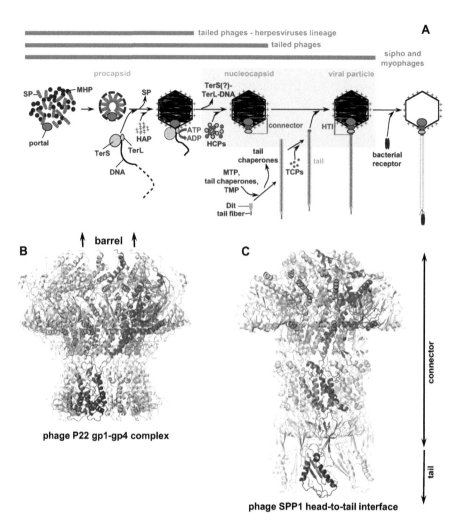

Fig. 14.1 Virion assembly pathway and the head-to-tail interface structure. (**A**) Assembly pathway of a siphovirus and DNA ejection. The different proteins engaged in assembly are identified by their common name or abbreviations (MHP, major head (or capsid) protein; SP, scaffolding protein; TerS, terminase small subunit; TerL, terminase large subunit; HAP, head auxiliary protein(s); HCP, head completion protein(s); Dit, distal tail protein; MTP, major tail protein; TMP, tape measure protein; TCP, tail completion protein(s)). Note that the tail tip represented is the adsorption apparatus found in phages like λ or SPP1, but this part of the virion exhibits great diversity in morphology and complexity among siphoviruses (see text). The connector and the head-to-tail interface (HTI), or neck, are labeled in the figure. Their components are identified in the phage particle cartoon by a color code that is used in the following figures to reflect structural homology: portal protein in blue, SPP1 gp15-like adaptors in magenta, SPP1 gp16 - λ gpFII-like proteins in green, and λ gpU - SPP1 gp17-like tail-head joining proteins (THJPs) in dark red. The first three groups of proteins are connector components, while THJPs are tail completion proteins.

assembly of the DNA packaging motor. Terminases are normally composed of two types of subunits, the small subunits TerS that bind to the viral DNA and the large subunits (TerL) that have endonuclease and ATPase activities (Rao and Feiss 2015). The packaging motor translocates the dsDNA to the procapsid interior energized by the TerL ATPase activity. Expansion of the procapsid to the larger, more angular, capsid form increases its internal space available for DNA packing. At the late stages of genome packaging, the DNA concentration inside the capsid can reach ~500 mg/mL (Chemla and Smith 2012). The resulting tight packing of DNA helices in the capsid confined space generates strong forces that exert pressures above 6 MPa on the capsid structure (Smith et al. 2001) and resist DNA entry. Forces stronger than 100 pN were estimated to be applied by the motor to translocate DNA into the capsid at the late stages of DNA packaging (Chemla and Smith 2012). This energy-driven strategy to package DNA optimizes the amount of genetic information transported by the viral particle. Furthermore, the forces generated by the tight DNA confinement state likely play a critical role to drive initiation of DNA entry in the host cell (Evilevitch et al. 2003; São-José et al. 2007).

With the exception of phages of the ϕ29 group that package genome unit-length DNA molecules (Morais 2012), termination of viral genome packaging results from an endonucleolytic cleavage of the viral DNA concatemeric substrate by TerL. This cut, which can be either sequence-specific or sequence-independent, defines the size of packaged DNA (Tavares et al. 2012). The terminase-viral DNA complex then departs to bind another procapsid to initiate a new encapsidation cycle. Disassembly of the DNA packaging motor at the end of packaging is coordinated with closure of the portal channel. Retention of the packaged DNA in the capsid can be achieved by a structural change in the portal protein, as reported for phage T3 (Donate et al. 1988 and references therein). However, in most phages, the binding of *head completion proteins* seals the portal. The structure formed by the portal protein and these proteins is named *connector*[1] (Fig. 14.1; Lurz et al. 2001; Orlova et al. 2003).

The connector provides an interface for attachment of tail components that yield the mature infectious particles with short tails, the podoviruses (Casjens and Molineux 2012). In contrast, siphoviruses and myoviruses build long tails in an independent assembly pathway. Their assembly starts by building of an *adsorption apparatus* for recognition of the host cell surface receptors. This apparatus provides

---

Fig. 14.1 (continued) The bacterial receptor that triggers DNA release from the virion is depicted in brown. The bars on top of the figure represent the common assembly steps of herpesviruses, tailed phages, and phages with long tails (see Sect. "Principles of Assembly and the Limited Diversity of Phage Head-to-Tail Interfaces"). (**B**) Crystallographic structure of phage P22 connector sub-complex portal gp1-adaptor gp4 (Olia et al. 2011). Two subunits of each oligomer are colored according to the code in (**A**) to highlight the alternate pattern of gp1 and gp4 subunits along the structure height. The long helical barrel of gp1 (Olia et al. 2011) is not shown. (**C**) CryoEM structure of phage SPP1 head-to-tail interface formed by the connector proteins gp6 (portal), gp15 (adaptor), gp16 (stopper), and the tail completion protein gp17 (THJP) (Chaban et al. 2015). Two subunits of each protein are highlighted with the exception of gp17 for which a single subunit is colored

---

[1] Note that some authors use the term connector as synonymous to portal protein.

a platform for assembly of the *tail tube* that is built by helical polymerization of the major tail protein around a tape measure template. The size of the tape measure protein determines the tail tube length (Katsura 1987). The tube is surrounded by a contractile sheath in myoviruses. Long tails are tapered by *tail completion proteins* that build the interface for attachment to the capsid connector, the final assembly step that yields the infectious virion. The structure formed by the connector and the tail completion proteins is named *neck* or *head-to-tail interface* (Fig. 14.1). Here we review its components, structural organization, assembly, and the mechanisms how the interface reversible gates the viral DNA in infectious particles. Reviews on related topics are available for phage structure (Veesler and Cambillau 2011; Aksyuk and Rossmann 2011; Fokine and Rossmann 2014), portal proteins (Cuervo and Carrascosa 2011), phage DNA packaging (Rao and Feiss 2015), connector assembly and function (Tavares et al. 2012), and tail assembly of podoviruses (Casjens and Molineux 2012), siphoviruses (Davidson et al. 2012), and myoviruses (Leiman and Shneider 2012).

Structure of the Head-to-Tail Interface Components

Portal Protein

Portal proteins are cyclical dodecamers (Fig. 14.1B,C) localized at a single fivefold vertex of the capsid from tailed phages and herpesviruses. Their subunits have very different amino acid sequences and show large variation in molecular mass (from 36 kDa in phage φ29 to 83 kDa in phage P22 (Cuervo and Carrascosa 2011)). However, they all share a common core structure, typified by the "minimal" fold of gp10 from phage φ29, and assemble to a turbine-like dodecamer found in phage particles, indicating a common phylogenetic origin. Atomic models were reported for the portal proteins of podovirus φ29 (Simpson et al. 2000; Guasch et al. 2002), siphovirus SPP1 (Lebedev et al. 2007), podovirus P22 (Olia et al. 2011; Lokareddy et al. 2017), myovirus T4 (Sun et al. 2015), siphovirus G20c (Cressiot et al. 2017), and for a phage element encoded by the *Corynebacterium diphtheriae* genome (Tavares et al. 2012) (Fig. 14.2A).

Portal structures can be subdivided into the clip, stem, wings, and crown regions (Fig. 14.2A). The clip has an α/β fold that intertwines the portal subunits stabilizing the overall portal oligomer structure. This region is exposed to the procapsid and capsid exterior, providing the interface for interaction with the terminase in the DNA packaging motor and for the subsequent binding of head completion proteins (Simpson et al. 2000; Lhuillier et al. 2009; Olia et al. 2011; Chaban et al. 2015; Sun et al. 2015). The clip is connected to the wing and crown domains by the stem helices that cross the capsid lattice, as shown for phage P22 (Chen et al. 2011). While the clip and stem are highly conserved among portals, the wings and crown exhibit considerable variation in size and complexity (Fig. 14.2A). The wings are individualized lobular domains that protrude outward from the stem conferring the portal its

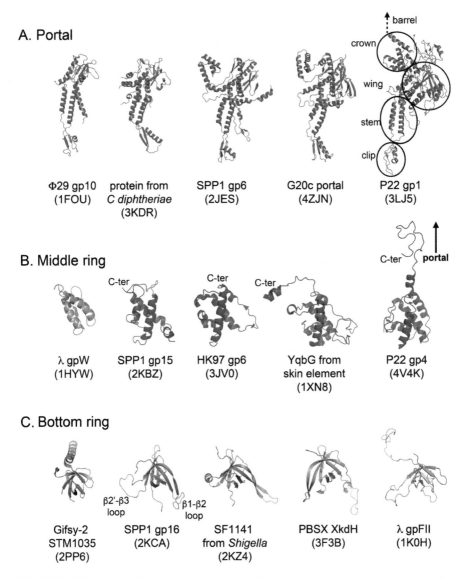

Fig. 14.2 Gallery of single subunits or non-assembled monomers of connectors (see text for details). The connectors of siphoviruses and myoviruses are formed by three stacked rings of different homo-dodecamers as illustrated in Fig. 14.1C: (**A**) portal, (**B**) λ gpW stopper or SPP1 gp15-like adaptor, and (**C**) λ gpFII-SPP1 gp16-like protein. The portal protein domain organization is shown in (**A**) for podophage P22 gp1. The orientation of the phage P22 gp4 adaptor relative to the portal is shown (see Fig. 14.1B). The PDB access codes of the structures are shown within brackets

turbine appearance with a defined handedness. The crown α-helical structure formed by the portal carboxyl terminus is small in phage ϕ29, more elaborated in phages SPP1, T4, and G20c, while an imposing ~200-Å-long α-helical barrel is found in the portal of phage P22 and its relatives. These differences might reflect the development of specialized functions related to the crown close contact with the packaged DNA. In case of phage SPP1, the crown was found to move downward in a mutant affecting the sensor system that measures the amount of DNA packaged inside the capsid (Orlova et al. 1999). The long crown barrel protruding to nearly the capsid centre of phage P22 was proposed to play roles organizing the packaged DNA in the capsid and as sensor of DNA headfilling (Olia et al. 2011). Furthermore, mutational studies showed that shortening of the P22 barrel leads to properly assembled virions that are noninfectious, revealing a function during initiation of phage infection (Tang et al. 2011).

Head Completion Proteins: Adaptors and Stoppers

Head completion proteins bind sequentially to the portal dodecamer after termination of genome encapsidation to prevent release of the tightly packed DNA from the viral capsid. They are distinguished as *adaptors* when they extend the portal channel or *stoppers* if they reversibly close it. This functional classification is based on the phenotype of phage particle assembly intermediates that accumulate in cells infected with mutants defective in the production of a specific head completion protein. Characterization of such intermediates allowed defining the sequential order of adaptors and stoppers binding to the portal system as exemplified, for example, in studies with phage SPP1 (Lurz et al. 2001; Orlova et al. 2003). Cryo-electron microscopy (cryoEM) reconstructions of virions or of their subassemblies provided a glimpse of the head completion proteins positioning in the head-to-tail interface in several phages (reviewed in Johnson and Chiu 2007; Veesler and Cambillau 2011; Casjens and Molineux 2012). High-resolution structures, obtained either by NMR or X-ray crystallography, identified four different families of head completion proteins so far.

The first family includes siphovirus SPP1 gp15, siphovirus HK97 gp6, podovirus P22 gp4, and a protein encoded by the skin phage element of the *Bacillus subtilis* genome (magenta in Fig. 14.2B). The first three proteins are adaptors that bind directly to the portal protein (Fig. 14.1B, C). They are formed by an α-helical bundle from which emerge mobile loops (Lhuillier et al. 2009) and have a flexible carboxyl terminus as found for HK97 gp6 (Cardarelli et al. 2010a) and P22 gp4 (Olia et al. 2011). These flexible regions are stabilized during connector assembly.

The second family features siphovirus λ gpW. GpW is a connector component that likely binds directly to the portal protein (Gaussier et al. 2006; Cardarelli et al. 2010b) and acts as a stopper to prevent leakage of packaged DNA from the λ capsid (Perucchetti et al. 1988). The monomer has two α-helices separated by a β-hairpin

connector

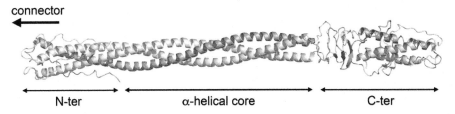

Fig. 14.3 Structure of podovirus P22 gp26, a needle-like stopper (PDB 2POH). Gp26 binds to the gp10 adaptor of the P22 connector through its amino terminus, as labeled. One subunit of the trimer is highlighted in yellow

and an unstructured carboxyl terminus that is essential for function (orange in the left of Fig. 14.2B) (Maxwell et al. 2000, 2001; Sborgi et al. 2011).

The third family includes proteins that are conserved among a large number of siphoviruses and myoviruses. Structures were determined for λ gpFII (Maxwell et al. 2002), SPP1 gp16 (Lhuillier et al. 2009), PBSX XkdH, prophage Gifsy-2 STM1035, and SF1141 encoded by a phage element of the *Shigella flexneri* 2a genome (Fig. 14.2C). They have a common β-barrel fold and several highly mobile loops that participate in assembly reactions. Studies on SPP1 gp15 and λ gpFII showed that they are positioned similarly in the connector to terminate its assembly and to provide an interface for tail attachment. GpFII binds to the gpW stopper to build the connector region of interaction with the λ tail (Casjens et al. 1972; Cardarelli et al. 2010b). SPP1 gp16 has the additional function of acting as a stopper after binding to the gp15 adaptor protein before participating in the connector-tail joining reaction (Lurz et al. 2001; Orlova et al. 2003) (see also Sect. "Assembly and Function of the λ gpFII - SPP1 gp16 Family of Head Completion Proteins").

The fourth family is represented by podovirus P22 gp26 that has a needle like structure very different from the other proteins described previously in this section. Gp26 is a homo-trimeric 240 Å-long fiber whose two coiled coils are interrupted by a triple β-helix (Fig. 14.3). Although the composition and length of phage needles vary significantly within this large family of proteins, possibly reflecting diversity for interaction with different hosts, the amino terminus is highly conserved. This domain was proposed to adopt an extended conformation to plug the connector during assembly of the mature virion (Bhardwaj et al. 2007; Olia et al. 2007b) and to change conformation to a trimer of hairpins when it departs from the viral particle for phage DNA release at the beginning of infection (Bhardwaj et al. 2016). Note that in phage P22 the adaptor gp4 (see above) is separated from the gp26 stopper by a second adaptor, gp10, which extends the portal channel (Strauss and King 1984; Lander et al. 2009). The high-resolution structure of gp10 is not known. P22 gp26-like proteins are conserved in a subgroup of podoviruses where the needle plays the dual role of stopper and piercing device of the cell envelope (Bhardwaj et al. 2009). Puncturing of the bacterium and opening of the portal channel for delivery of the phage genome to the cell cytoplasm are thus intimately linked to the gp26 activity in these viruses.

Tail Completion Proteins

Tail completion proteins bind to the end of the long helical tail tube that is distal from the adsorption apparatus in siphoviruses (Fig. 14.1A) and myoviruses. They build the interface for tail joining to the capsid connector and can play a role to ensure termination of helical polymerization of the tail tube (and sheath in case of myoviruses). There is only one well-studied tail completion protein family. It is typified in siphoviruses by λ gpU (Katsura and Tsugita 1977; Pell et al. 2009b) and SPP1 gp17 (Auzat et al. 2014). A large number of homologous proteins were identified in siphoviruses and myoviruses suggesting that they are an essential component of long tails (Pell et al. 2009b; Lopes et al. 2014). They were initially named tail terminator proteins (Edmonds et al. 2007; Pell et al. 2009b; and references therein) and more recently tail-to-head-joining proteins (THJPs) (Auzat et al. 2014) (see also Sect. "Assembly of the Long Tails Interface for Binding to the Connector"). In myoviruses of the T4 group, they are structurally related to two tail completion proteins. The first, gp3 in T4, closes the tail tube. The second, gp15 in T4, binds to gp3 and to the tail sheath, forming the tail end for attachment to the T4 capsid connector (Zhao et al. 2003; Fokine et al. 2013).

The NMR structures of monomers of λ gpU (Edmonds et al. 2007), SPP1 gp17 (Chagot et al. 2012), and NP_888769.1, encoded by a phage element from *Bordetella bronchiseptica*, show the common three-dimensional fold of THJP monomers (Fig. 14.4A). These assembly-naïve polypeptides have a multi-stranded β-sheet associated with two main α-helices connected by mobile loops. The structure of THJP hexamers, which is the association state found in phage particles, was determined by X-ray crystallography for λ gpU (Pell et al. 2009b), T4 gp15 (Fokine et al. 2013), STM4215 encoded by a putative myo-prophage element of *Salmonella typhimurium*, and the antigen B encoded by a phage element from *Listeria monocytogenes* (Fig. 14.4B). The λ gpU hexamer is a nutlike structure with a central channel delimited by β-sheets and α-helices that define its periphery. The structure is stabilized by hydrophobic and polar contacts including formation of an intersubunit antiparallel β-sheet. Comparison between the λ gpU monomer and its hexamer state showed that gpU maintains its fold during association undergoing localized changes in the putative region of gpU interaction with the tail tube (Pell et al. 2009b). The myovirus T4 gp15 is a larger protein than its siphovirus counterparts. It has a similar overall structure, but large insertions and extension of the carboxyl terminus add structural elements on the outside of the hexamer (Fokine et al. 2013).

Mechanisms of Head-to-Tail Interface Assembly

Closure of the portal vertex is a late step in the sequential program of protein-protein and protein-DNA interactions during assembly of the viral nucleocapsid. Its coordination with termination of DNA packaging and dissociation of the

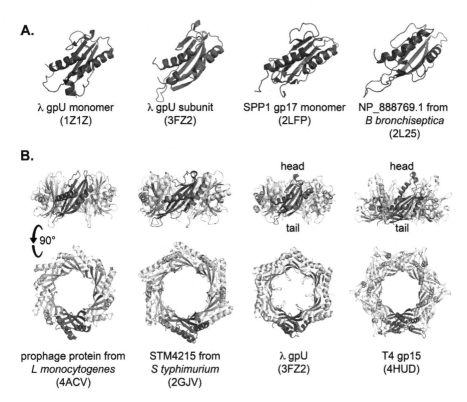

Fig. 14.4 Gallery of tail-to-head joining protein (THJP) structures showing their structural homology. (**A**) NMR structures of assembly-naïve monomers. The subunit extracted from a gpU hexamer (see **B**) is also presented for structural comparison with the non-assembled gpU monomer. (**B**) Structures of THJP hexamers determined by X-ray crystallography. The orientation of the hexamer relative to the phage tail and head is displayed when known. One subunit is highlighted in dark red. The PDB access codes of the structures are shown within brackets

terminase-DNA complex from the portal protein is critical to prevent loss of the packaged DNA. The concomitant interaction of head completion proteins with the portal vertex must be precisely controlled to prevent that it occurs before DNA packaging, which would lead to closed capsids without DNA, and to ensure efficient connector assembly afterward. This illustrates a general problem found in sequential assembly of complex macromolecular machineries. Phage particle building blocks are synthesized rather synchronously in the infected cell. Therefore, the correct order of interactions during tailed phage particle assembly is mainly ensured by the conformational repertoire of viral proteins and their complexes (Casjens and Hendrix 1988).

Coordination of DNA Packaging Termination with Portal Channel Closure

Coordination of the endonucleolytic cleavage, which terminates packaging, with departure of the terminase-DNA complex and closure of the portal channel, is the less understood step in connector assembly. DNA packaging termination can be uncoupled from binding of head completion proteins. In this case, DNA exits leaving the phage expanded capsid empty (Orlova et al. 2003). Structural rearrangements in the clip of the portal protein that precede or result from terminase departure were proposed to retain phage T3 DNA (Donate et al. 1988) and to build, or uncover, the binding site(s) for the first head completion protein. A single amino acid substitution in the clip from the phage SPP1 portal (Fig. 14.5B) was shown to impair binding to the gp15 adaptor without affecting termination of DNA packaging (Isidro et al. 2004a; Chaban et al. 2015). This part of the portal exposed to the capsid exterior features also several mutations that block DNA packaging (Isidro et al. 2004b; Lebedev et al. 2007; Oliveira et al. 2006, 2010), suggesting that the SPP1 terminase and gp15 interact sequentially with the same region of the portal.

Binding of SPP1 gp15-Like Head Completion Proteins to the Portal Protein

The mechanisms ensuring binding of the first head completion protein to the portal after termination of packaging were studied in phages P22, SPP1, and HK97 illustrating different ways to reach a similar goal. The adaptor proteins of the three phages have a similar fold (Fig. 14.2B).

In phage P22, gp4 (adaptor), gp10 (adaptor), and gp26 (stopper) bind sequentially to the portal vertex (Strauss and King 1984; Olia et al. 2007a). Gp4 is a monomer in solution at millimolar concentrations that oligomerizes upon binding to the gp1 portal dodecamer (Olia et al. 2006). The structure of the gp1 and gp4 stacked dodecamers was determined by X-ray crystallography (Olia et al. 2011). The α-helical core of the gp4 subunit sits between two gp1 subunits interacting through helix α4 with the outer surface of the portal, while its carboxyl terminal tail extends to bind between two other adjacent portal subunits (Figs. 14.1B and 14.2B). The interaction of each gp4 chain with four portal subunits to assemble the adaptor ring suggests that the portal cyclical oligomer drives gp4 oligomerization unraveling why isolated gp4 does not form oligomers. However, the gp1-gp4 interaction shall not occur at earlier stages in assembly. This is probably avoided because gp1 is a monomer when free in the infected cell (Moore and Prevelige Jr 2001) and binding sites of gp1 for gp4 are masked in the procapsid state, becoming accessible when the terminase-DNA complex departs from the portal vertex (Tang et al. 2011).

Structures of the SPP1 gp15 adaptor monomer and of its dodecameric ring in the head-to-tail interface were determined by NMR (Fig. 14.2B) (Lhuillier et al. 2009)

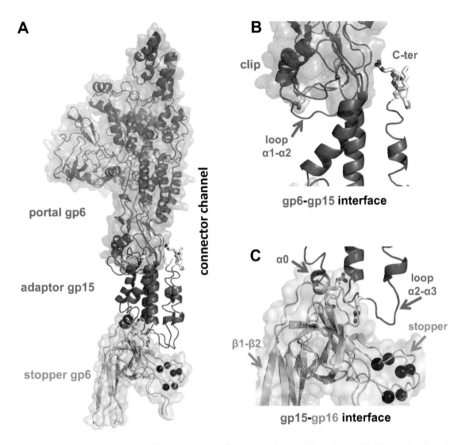

Fig. 14.5 The bacteriophage SPP1 connector in the head-to-tail interface. (**A**) Two subunits of gp6, gp15, and gp16 are shown (see also Fig. 14.1C). (**B**) Gp6-gp15 interface showing gp15 loop α1-α2 and the carboxyl terminus of gp15 that interact with the portal clip. A mutation in residue E294 of gp6, whose side chain is displayed as sticks in the clip, blocks the interaction. Only one subunit of gp15 is displayed for clarity. (**C**) The gp15-gp16 interface showing gp15 helix α0 that binds at the gp16 periphery and gp15 loop α2-α3 that sits between two gp16 subunits. Residues in the gp16 stopper that establish intermolecular disulphide bridges when mutagenized to cysteines are presented as spheres (see text; Lhuillier et al. 2009; Chaban et al. 2015). The gp16 β1-β2 sheet and its connecting loop are involved in the connector-tail joining reaction (see text; Chaban et al. 2015). Figure inspired from Chaban et al. (2015)

and cryoEM (Chaban et al. 2015) (Figs. 14.1C and 14.5), respectively. The main structural change observed during assembly is a rotation of helices α2 and α3 that acquire an orientation roughly parallel to the portal central channel (Chaban et al. 2015). SPP1 gp15 lacks the equivalent to P22 gp4 long helix α4 and its extended carboxyl terminus. Nevertheless, a stretch of basic residues in the gp15 carboxyl terminus is necessary for binding to the portal. In contrast to gp4, the SPP1 adaptor also interacts with the portal via loop α1-α2 that is sandwiched between loops and an α-helix of portal adjacent subunits (Chaban et al. 2015) (Fig. 14.5B). Such interaction differences might explain why gp15 monomers do not bind to isolated SPP1 portal protein oligomers.

In contrast to P22 gp4 and SPP1 gp15, the isolated HK97 gp6 adaptor is found in solution as a mix of cyclical 13-mers and monomers at micromolar concentrations (Cardarelli et al. 2010a). The 13-mer crystal structure provided information how the α-helical subunits pack together to oligomerize and highlighted their subunits, similarity to SPP1 gp15 (Cardarelli et al. 2010a) (Fig. 14.2B). These oligomers are, however, unlikely precursors to match the portal protein dodecameric state for assembling the adaptor ring of the connector. During infection, the weak ribosome binding site of HK97 gene 6 leads to a very low production of gp6 that remains most probably monomeric before binding to the portal vertex after DNA packaging (Cardarelli et al. 2011). Gp6 overproduction is highly detrimental for HK97 multiplication. Such effect is compensated by single mutations that weaken gp6 intersubunit interactions, impairing its auto-oligomerization. In turn, higher amounts of these gp6 oligomerization mutant proteins were necessary for successful HK97 infection (Cardarelli et al. 2011). This elegant work illustrates how gene dosage can control macromolecular assembly and how single amino acid substitutions can strongly impact assembly reactions.

Assembly and Function of the λ gpFII-SPP1 gp16 Family of Head Completion Proteins

Termination of connector assembly in phages with long tails was characterized in siphoviruses SPP1 and λ. The SPP1 gp15 adaptor region distal from the portal binds to gp16. One interaction occurs in the periphery of the connector where the gp15 outer helix α0 contacts the outer top region of a gp16 subunit. Another interaction is mediated by gp15 loop α2-α3 that sits at an internal position between gp16 subunits (Chaban et al. 2015) (Fig. 14.5C). The gp15-gp16 interaction is disrupted by a single amino acid substitution in the gp15 loop α2-α3. In contrast, a double mutant in gp15 α0 allows for gp16 stable binding, but the assembled connector does not retain packaged DNA showing that the gp16 stopper function is impaired (Chaban et al. 2015). This observation led to the model that the gp15-gp16 interaction at the connector periphery acts allosterically, positioning the gp16 subunits β-sheet core for correct assembly of the stopper (or plug) that closes the portal central channel (Chaban et al. 2015). The stopper structure is built by the highly mobile loop β2'-β3 of gp16 monomers (Fig. 14.2C) which folds when gp16 oligomerizes during formation of the connector. Modeling and cysteine cross-linking suggested that loop β2'-β3 folding into a β-strand would lead to formation of a parallel intersubunit β-sheet built by the β-strands of the 12 gp16 subunits, plugging reversibly the portal channel (Lhuillier et al. 2009). The interaction of gp15 α0 with the outer region of gp16 was thus hypothesized to have a long-distance effect across the gp16 structure to enable assembly of the intersubunit stopper at the center of the oligomer.

Docking of the gp16 monomer NMR structure in cryoEM maps of the SPP1 connector (Lhuillier et al. 2009) and of the head-to-tail interface (Chaban et al. 2015) showed that the core β-barrel structure of gp16 is maintained throughout the SPP1 assembly pathway. Major structural changes were observed in the β2'-β3 stopper loop during connector assembly, as described above, and in the gp16 periphery upon binding to the sixfold symmetric tail. In the latter interaction, β-sheet β1/β2 and the long flexible loop between its two strands (Fig. 14.2C), which define the gp16 external surface, move down and inward to embrace the SPP1 tail completion protein gp17 (Fig. 14.1C). They undergo a concomitant shift from their 12-fold symmetry in the connector region to a 6-fold symmetry to match the tail geometry (Chaban et al. 2015).

SPP1 gp16 is structurally homologous to phage λ gpFII. The surfaces of gpFII interaction with gpW in the λ connector and with the tail were identified by structure-driven mutagenesis (Cardarelli et al. 2010b). They are compatible with a positioning of gpFII in the λ head-to-tail interface similar to gp16 in the SPP1 viral particle. However, in phage λ gpW, which has a fold different from the SPP1 gp15 adaptor (Fig. 14.2B) (Sect. "Head Completion Proteins: Adaptors and Stoppers"), retains DNA in capsids lacking gpFII. The gpFII function could thus be limited to bind gpW and provide the connector interface for tail attachment (Casjens et al. 1972). This interface is identified by single amino acid substitutions in different regions of gpFII. Tail attachment is also blocked by deletion of the loop equivalent to the SPP1 gp16 stopper loop, showing that in gpFII it acts in the head-to-tail joining reaction. Proteins of this family might thus have specialized to two different functions by structural evolution. The gpFII fold features a long amino terminal extension necessary for binding to gpW and an internal loop for tail attachment (Cardarelli et al. 2010b). The gp16 fold, which lacks the amino terminus extension, binds to gp15-like partners and closes the connector channel (Lhuillier et al. 2009; Chaban et al. 2015).

Assembly of the Long Tail Interface for Binding to the Connector

In phages with long tails, the connector of the DNA-filled capsid binds to the tail end distal from the adsorption apparatus to form the neck. The tail interface that participates in this reaction is created by tail completion proteins that taper the helical tail structure. Proteins with a λ gpU-like fold (Sect. "Tail Completion Proteins") form a nutlike hexamer at the tail end which is essential for tail-to-head joining (Pell et al. 2009b; Fokine et al. 2013; Auzat et al. 2014). Proteins related to λ gpZ were reported to assist formation of a functional head-to-tail interface (Sect. "DNA Positioning in the Head-to-Tail Interface") (Davidson et al. 2012; Lopes et al. 2014), but their precise role in assembly and structure remains undocumented.

GpU-like proteins were named tail terminator proteins because their absence in λ infections leads to accumulation of abnormally long tails (polytails), revealing that gpU is necessary to halt helical polymerization of the λ gpV major tail protein (Pell et al. 2009b and references therein). However, mutants lacking the related tail completion protein in myoviruses P2 (gpR⁻; Lengyel et al. 1974), SPO1 (gp8⁻; Parker and Eiserling 1983), Mu (gpK⁻; Grundy and Howe 1985), and T4 (gp3⁻; Vianelli et al. 2000) assemble mostly correctly sized tails and only a small population of polytails. SPP1 infection with a gp17 defective mutant leads also to assembly of tails of normal length (Auzat et al. 2014). These proteins were thus renamed tail-to-head joining proteins (THJPs) according to what appears to be their most conserved function (Auzat et al. 2014).

λ gpU remains monomeric in solution at millimolar concentrations suggesting this is its state in vivo. Interaction with the λ tail tube promotes its association, most likely to a hexamer, at the tail end. Multimerization of purified gpU was induced in vitro by 20 mM $MgCl_2$ or by crystallization, mimicking this assembly step. Comparison between the λ gpU monomer and hexamer states showed that association correlates mostly with changes in the surface interacting with the phage tail where folding events extend secondary structure elements (Fig. 14.4A). Motions occur also in this region of the subunits that form an intersubunit antiparallel β-sheet lining the hexamer internal channel (Pell et al. 2009b). Mutational studies defined the surfaces of gpU that interact with the tail tube and with the capsid connector (Fig. 14.4B; Pell et al. 2009b). The SPP1 gp17 ring was directly observed at the tail end that binds to the phage connector (Auzat et al. 2014), and its hexamer organization was established by a cryoEM study of the SPP1 head-to-tail interface (Chaban et al. 2015). The myovirus T4 gp15 is larger than its siphoviruses counterparts. It has a similar α/β core, but large insertions and an additional carboxyl terminus build extra structural elements on the outside of the hexamer (Fokine et al. 2013). These periphery regions taper the tail sheath, which is a unique structure of myoviruses, while the gp15 core stacks on top of gp3 in the internal tail tube (Fokine et al. 2013).

Interestingly, the THJPs, core fold is conserved in proteins forming different regions of the long tail tube that serves as a conduit for DNA ejection in siphoviruses and myoviruses (Cardarelli et al. 2010b; Veesler and Cambillau 2011). An original protein module thus appears to have evolved to assemble long tail tube devices by specializing to build the tail adsorption apparatus hub Dit protein, the helical tail tube building block, and the THJP. This fold is also found in tubes of bacterial type VI secretion systems that deliver virulence factors between competing bacteria in the environment or to eukaryotic cells in pathogenic settings (Leiman et al. 2009; Pell et al. 2009a). Such common structural solution to build devices for delivery of macromolecules to cells adds to the growing evidence blurring the borders between viral and cellular machineries.

DNA Positioning in the Head-to-Tail Interface

A critical requirement during assembly of the head-to-tail interface is that phage DNA is positioned ready for exit when infection initiates. It was early recognized that the last end of the dsDNA linear molecule packaged in phage capsids remains associated to the neck ensuring its polar release from viral particles (Thomas 1974; Chattoraj and Inman 1974; Saigo 1975; Tavares et al. 1996; Loessner et al. 2000). This DNA end exits first from virions when the phage SPP1 genome is ejected *in vitro* (Tavares et al. 1996). Loss of the DNA-neck association in phage λ mutants was shown to correlate with a strong reduction in phage infectivity in spite of the normal morphology of phage particles (Thomas et al. 1978). Such phenotype likely results from a defect in DNA delivery to the host bacterium. One of the mutants lacks gpZ which has a large number of homologous proteins in phages with long tails (Davidson et al. 2012; Lopes et al. 2014). A similar phenotype was observed for phage Mu mutants defective in gpG that assemble complete phage particles whose majority is noninfectious (Grundy and Howe 1985). The gpZ-related proteins gpS from phage P2 and Tap from phage TP901–1 were reported to be necessary for joining tails to capsids (Lengyel et al. 1974; Stockdale et al. 2015). These observations indicate that gpZ-like proteins assist assembly of the functional head-to-tail interface with phage DNA correctly positioned for delivery to the host by a yet unknown mechanism.

Phage λ DNA was reported to penetrate the tail tube of complete virions after the head-to-tail joining reaction (Thomas 1974). This implies that binding of the λ tail to DNA-filled capsids leads to opening of the connector channel, previously closed by gpW and extended by gpFII (see Sect. "Head Completion Proteins: Adaptors and Stoppers" and "Assembly and Function of the λ gpFII - SPP1 gp16 Family of Head Completion Proteins"; Perucchetti et al. 1988), allowing a phage DNA end to enter the tail tube of the mature virion. Such situation differs from phage SPP1 where a 19–29 nm segment of the last packaged DNA end is stably attached to the head-to-tail interface (Tavares et al. 1996). This length fits accurately to the height of the connector channel closed by gp16 (Orlova et al. 2003). DNA confined in the SPP1 channel makes contacts mainly with the portal protein and with a ring of lysines in the stopper that occupy the center of the channel (Chaban et al. 2015). In spite of the differences between λ and SPP1, both phages have their DNA poised for efficient traffic through the phage tail to enter the host bacterium.

DNA positioning in the connectors of podoviruses was observed in asymmetric reconstructions of complete phage particles of T7 (Agirrezabala et al. 2005), φ29 (Xiang et al. 2006; Tang et al. 2008), ε15 (Jiang et al. 2006), P22 (Lander et al. 2006; Chang et al. 2006; Tang et al. 2011), and P-SSP7 (Liu et al. 2010). Although fine details of protein-DNA interactions were hampered by the nanometer resolution, the structures reveal how a phage DNA end is positioned in the connector for efficient ejection. DNA was also found in the head-to-tail interface of myovirus T4 (Leiman et al. 2004).

Genome Egress Through the Head-to-Tail Interface

Genome egress from viral particles is triggered by specific interaction of the phage adsorption apparatus with receptors at the bacterial surface (Vinga et al. 2006). The tail adsorption apparatus varies in complexity from single tail fibers to intricate assemblies of tail spikes and complex baseplates (Casjens and Molineux 2012; Davidson et al. 2012; Leiman and Shneider 2012; Spinelli et al. 2014; and references therein). Their irreversible binding to the host bacterium commits the phage particle to infection. In podoviruses, which have short tails, the receptor binding proteins (RBPs) are physically close or directly attached to the connector. RBPs binding to the cell surface receptors shall thus be tightly coordinated with localized perforation of the bacterial envelope and with the subsequent opening of the connector for genome delivery to the cytoplasm. This might engage the same protein to play sequential roles in the process as the P22 gp26 needlelike stopper (see Sect. "Head Completion Proteins: Adaptors and Stoppers"). Gp26 is the putative P22 tail device that punctures the cell envelope and also the genome gatekeeper that dissociates from the connector at the stage that DNA is delivered to the cytoplasm. The latter step was proposed to be driven by structural changes in the gp26 binding interface to the connector, leading to gp26 release from the phage particle in the conformation found in solution (Fig. 14.3; Bhardwaj et al. 2016). Glimpses of the complete viral particle machinery in the process of infection were obtained by cryo-electron tomography (cryoET) of podoviruses P-SSP7 (Liu et al. 2010), ε15 (Chang et al. 2010), and T7 (Hu et al. 2013) interacting with their Gram-negative hosts. Recently, the initiation of phage φ29 infection of the Gram-positive bacterium *Bacillus subtilis* was also reported (Farley et al. 2017). These studies delivered information on different steps of the interaction of phage particles with the bacterial envelope. The process is initiated by virions searching and binding to receptors, followed by cell envelope penetration, and culminates with naked dsDNA delivery to the host cytoplasm. In spite of the resolution limit at 3–4 nm, such snapshots offer insights on structural changes in the phage particles that accompany those steps (Liu et al. 2010; Chang et al. 2010; Hu et al. 2013; Farley et al. 2017). They also reveal that the phage short tail appears prolonged by a DNA delivery tube crossing the bacterial envelope to build a continuous channel spanning from the phage capsid, through the connector, to the interior of the host cytoplasm (Chang et al. 2010; Hu et al. 2013).

In phages with long tails, binding of the tail adsorption apparatus to bacterial receptors generates a signal to trigger DNA release from the viral particle. Interaction of myoviruses with the cell surface causes contraction of the shaft that surrounds the tail tube driving penetration of the tube to contact the bacterial membrane (Hu et al. 2015; Nováček et al. 2016). This step can be uncoupled from phage DNA ejection (Leiman et al. 2004; Nováček et al. 2016 and references therein). Signaling for genome egress from the viral capsid is thus conceivably transmitted from the adsorption apparatus through the tail tube structure and/or the internal tape measure protein to cause DNA ejection both in siphoviruses and myoviruses. Comparison between the phage tail structure before and after DNA ejection in siphovirus SPP1

(Plisson et al. 2007) and in myovirus ф812 (Nováček et al. 2016) revealed rear-rangements in the tube structure. This led to the proposal that a signal is propagated as a domino-like cascade along the helical tail tube by a sequential conformational change in the major tail protein subunits (Plisson et al. 2007). Recent sub-nanometer structures of bacteriophage T5 tails showed no detectable differences in the tail tube before and after incubation with the bacterial receptor FhuA, but the internal tape measure protein was released upon challenge with FhuA (Arnaud et al. 2017). Departure of the tape measure protein, which spans the complete length of the tail tube lumen, was thus proposed to be the signal that triggers phage DNA release (Arnaud et al. 2017).

The mechanism of neck opening for viral genome egress depends on the position of the phage DNA end in the head-to-tail interface (Sect. "DNA Positioning in the Head-to-Tail Interface"). In case of phage λ, its DNA appears to partially occupy the tail tube in mature virions (Thomas 1974). Release of the tape measure protein promoted by interaction of the tail adsorption apparatus with bacterial receptors opens the way for DNA ejection to the host cell. A different situation was found for phage SPP1. Disulfide bridges engineered in the putative intersubunit β-sheet gp16 stopper which seals the connector channel (see Sect. "Assembly and Function of the λ gpFII - SPP1 gp16 Family of Head Completion Proteins") impair DNA ejection from SPP1. The blockage can be reversed by reduction of the disulfide bonds (Lhuillier et al. 2009; Chaban et al. 2015). Viral DNA is thus locked at the level of the connector gp16 stopper in the mature virion. During infection, the signal that is initiated by SPP1 binding to its receptor YueB (São-José et al. 2006) and transmitted through the tail tube to the connector leads to unzipping of the gp16 intersubunit β-sheet, opening the way for genome ejection. Interestingly, the phage gp16 stopper was found closed after DNA left the tail revealing that it opens only transiently for genome release (Chaban et al. 2015). This is a possible mechanism to avoid loss of cellular components when a continuous hydrophilic channel is established between the bacterial cytoplasm and the phage particle interior (Chaban et al. 2015).

Principles of Assembly and the Limited Diversity of Phage Head-to-Tail Interfaces

The study of bacteriophage head-to-tail assembly and function provided important insights on the molecular principles how large macromolecular structures are built by sequential addition of building blocks. A frequent strategy is the transition from disordered to ordered regions of proteins during the assembly reaction. Numerous non-assembled building blocks of the head-to-tail interface and of long tails have large disordered regions that can account for more than 20% of their amino acid sequence (Maxwell et al. 2001, 2002; Edmonds et al. 2007; Lhuillier et al. 2009; Pell et al. 2009a; Cardarelli et al. 2010b). Some of these regions fold upon binding to their interaction partners in assembly intermediates and might also create the interaction sites for the following protein in the assembly pathway. Frequently, binding sites are created at intersubunit interfaces ensuring that oligomerization of one protein in the macromolecular complex is required for interaction with the downstream partner in the pathway. The alternate pattern of subunits along the height of the resulting structure is illustrated by the portal-adaptor complex of phage P22 (Fig. 14.1B) (Olia et al. 2011) and the head-to-tail interface of phage SPP1 (Fig. 14.1C) (Chaban et al. 2015).

Oligomerization of components of the head-to-tail interface can occur in the absence of other phage proteins. That is the case for most portal proteins. However, portal cyclical oligomers from different phages (Dube et al. 1993; Olia et al. 2011; Sun et al. 2015) and herpes simplex virus-1 (HSV-1) (Trus et al. 2004) can be found in different association states (11-, 13-, or 14-mers) other than the functional 12-mer ring present in phage particles. Since the number of subunits in cyclical oligomers is defined by the angle between subunits (van Heel et al. 1996), it is likely that the interaction partners of the portal during procapsid assembly impose the correct inter-subunit curvature to achieve a homogeneous population of 12-mers. The portal dodecamer in phage capsids can also serve as template to direct the oligomerization of its adjacent adaptor ring as in case of HK97 gp6. Gp6 was found to have the intrinsic capacity to assemble cyclical 13-mers but is most likely a 12-mer in the HK97 connector (see Sect. "Binding of SPP1 gp15-like Head Completion Proteins to the Portal Protein") (Cardarelli et al. 2010a, 2011). The THJP λ gpU wild-type protein assembles pentamers during crystallization, while the single mutation D74A leads to formation of hexamers, the state found in the hexameric phage λ tail. Collectively, these observations suggest that most phage particle building blocks do not have all the structural information necessary to reach efficiently their assembly state. It is the interaction with their partners that ensures fidelity and orderly assembly reactions to build megadalton infectious particles with remarkable precision.

Studies of virions assembly pathways and of their protein structures reveal that ancient viral-like particles rooted lineages of viruses that diverged by coevolution with distinct hosts from the three domains of life (Krupovic and Bamford 2011; Abrescia et al. 2012). This can be rationalized by the complexity of virus particle assembly. Once such complex genome containers and delivery devices were established, their structural design was maintained and adapted throughout the co-evolution of viruses and their hosts. There is extensive evidence that the tailed phages and herpesvirus lineage shares a similar pathway to assemble the procapsid with a portal and to package the viral dsDNA (Fig. 14.1A). Divergence appears to have occurred at the step of portal closure. The protein necessary to retain packaged DNA in the capsid of herpesviruses (UL25 in HSV-1) has a structure different from phage connector adaptors or stoppers (Bowman et al. 2006). The mechanism of closure may also be more complex than adding a plug protein to the portal channel. Changes in the complete capsid surface were observed which might be necessary for packaged DNA retention, for binding of the tegument that surrounds the nucleocapsid, and to engage egress from the cell nucleus (Trus et al. 2007). The divergence point between podoviruses and phages with long tails is more difficult to establish. While there is evidence that phages of the P22 group use the gp4 adaptor protein homologous to SPP1 gp15 to extend their connector channel, the structure of P22 gp10 that prolongs further the channel is not known. It is also not clear if phage φ29 gp11 that binds to the portal after DNA packaging is related to proteins from the head-to-tail interface of known structure. Podoviruses have possibly a more diverse head-to-tail interface organization than phages with long tails (Casjens and Molineux 2012; Lopes et al. 2014). The latter viruses appear to conserve an arrangement of three proteins forming the capsid connector, although some structural (Sect. "Head Completion Proteins: Adaptors and Stoppers"; Fig. 14.2B) and functional differences (see "Binding of SPP1 gp15-like Head Completion Proteins to the Portal Protein" and "Assembly and Function of the λ gpFII - SPP1 gp16 Family of Head Completion Proteins") occur. Furthermore, one or two proteins with a THJP fold taper the tail of siphoviruses and T4-like myoviruses, respectively (see "Assembly of the Long Tails Interface for Binding to the Connector").

Studies on the head-to-tail interface structure and function were carried out with a few model phages. However, bioinformatics data mining showed that the components of their interfaces have a very large number of homologous proteins in phages and prophages. This is particularly the case for the portals, SPP1 gp15 - HK97 gp6 - P22 gp4, λ gpFII - SPP1 gp16, and λ gpU - SPP1 gp17 - T4 gp15 families of proteins (Pell et al. 2009b; Lhuillier et al. 2009; Cardarelli et al. 2010a; b; Lopes et al. 2014). The genes coding these proteins are found clustered in phage and prophage genomes (the gene coding λ gpW-like proteins, sandwiched between the TerL and portal genes, is an exception (Cardarelli et al. 2010b)). Lopes et al. (2014) developed freely available software (http://biodev.extra.cea.fr//virfam/) using amino acid sequence homology and genome context of the head-to-tail proteins coding genes for their identification in genomes. The server was tested against more than 600 tailed phages genomes in the NCBI database assigning more than 90% of them to four types of head-to-tail interfaces: φ29-like (podoviruses), P22-like (podoviruses), SPP1/λ-like (siphoviruses and myoviruses), and T4-like (myoviruses). These results indicate that tailed phages use a limited number of solutions to connect their genome container to the tail delivery device. Nevertheless, phages mark also their differences to achieve the same functional goal, like in case of the λ gpW/gpFII and SPP1 gp15/gp16 connectors (see "Binding of SPP1 gp15-like Head Completion Proteins to the Portal Protein" and "Assembly and Function of the λ gpFII - SPP1 gp16 Family of Head Completion Proteins"), illustrating their immense creativity.

Acknowledgments Marie-Christine Vaney (Institut Pasteur, Paris) is acknowledged for insightful discussions on structure analysis and for invaluable help to prepare manuscript figures.

I thank present and past members of our team and collaborators for their insightful research that established present knowledge on the SPP1 head-to-tail interface.

Work in our laboratory on SPP1 phage assembly is supported by institutional funding from CNRS.

References

Abrescia NG, Bamford DH, Grimes JM, Stuart DI (2012) Structure unifies the viral universe. Annu Rev Biochem 81:795–822

Ackermann HW (2012) Bacteriophage electron microscopy. Adv Virus Res 82:1–32

Agirrezabala X, Martín-Benito J, Castón JR, Miranda R, Valpuesta JM, Carrascosa JL (2005) Maturation of phage T7 involves structural modification of both shell and inner core components. EMBO J 24:3820–3829

Aksyuk AA, Rossmann MG (2011) Bacteriophage assembly. Virus 3:172–203

Arnaud CA, Effantin G, Vivès C, Engilberge S, Bacia M, Boulanger P, Girard E, Schoehn G, Breyton C (2017) Bacteriophage T5 tail tube structure suggests a trigger mechanism for Siphoviridae DNA ejection. Nat Commun 8:1953

Auzat I, Petitpas I, Lurz R, Weise F, Tavares P (2014) A touch of glue to complete bacteriophage assembly: the tail-to-head joining protein (THJP) family. Mol Microbiol 91:1164–1178

Bazinet C, King J (1985) The DNA translocating vertex of dsDNA bacteriophage. Annu Rev Microbiol 39:109–129

Bhardwaj A, Olia AS, Walker-Kopp N, Cingolani G (2007) Domain organization and polarity of tail needle GP26 in the portal vertex structure of bacteriophage P22. J Mol Biol 371:374–387

Bhardwaj A, Walker-Kopp N, Casjens SR, Cingolani G (2009) An evolutionarily conserved family of virion tail needles related to bacteriophage P22 gp26: correlation between structural stability and length of the alpha-helical trimeric coiled coil. J Mol Biol 391:227–245

Bhardwaj A, Sankhala RS, Olia AS, Brooke D, Casjens SR, Taylor DJ, Prevelige PE Jr, Cingolani G (2016) Structural Plasticity of the Protein Plug That Traps Newly Packaged Genomes in Podoviridae Virions. J Biol Chem 291:215–226

Bowman BR, Welschhans RL, Jayaram H, Stow ND, Preston VG, Quiocho FA (2006) Structural characterization of the UL25 DNA-packaging protein from herpes simplex virus type 1. J Virol 80:2309–2317

Cardarelli L, Lam R, Tuite A, Baker LA, Sadowski PD, Radford DR, Rubinstein JL, Battaile KP, Chirgadze N, Maxwell KL, Davidson AR (2010a) The crystal structure of bacteriophage HK97 gp6: defining a large family of head-tail connector proteins. J Mol Biol 395:754–368

Cardarelli L, Pell LG, Neudecker P, Pirani N, Liu A, Baker LA, Rubinstein JL, Maxwell KL, Davidson AR (2010b) Phages have adapted the same protein fold to fulfill multiple functions in virion assembly. Proc Natl Acad Sci U S A 107:14384–14389

Cardarelli L, Maxwell KL, Davidson AR (2011) Assembly mechanism is the key determinant of the dosage sensitivity of a phage structural protein. Proc Natl Acad Sci U S A 108:10168–10173

Casjens S, Hendrix R (1988) Control mechanisms in dsDNA bacteriophage assembly. In: Calendar R (ed) The bacteriophages, vol 1. Plenum Press, New York

Casjens S, Molineux I (2012) Short noncontractile tail machines: adsorption and DNA delivery by podoviruses. Adv Exp Med Biol 726:143–179

Casjens S, Horn T, Kaiser AD (1972) Head assembly steps controlled by genes F and W in bacteriophage lambda. J Mol Biol 64:551–563

Chaban Y, Lurz R, Brasilès S, Cornilleau C, Karreman M, Zinn-Justin S, Tavares P, Orlova EV (2015) Structural rearrangements in the phage head-to-tail Interface during assembly and infection. Proc Natl Acad Sci U S A 112:7009–7014

Chagot B, Auzat I, Gallopin M, Petitpas I, Gilquin B, Tavares P, Zinn-Justin S (2012) Solution structure of gp17 from the Siphoviridae bacteriophage SPP1: insights into its role in virion assembly. Proteins: Struct Funct Bioinf 80:319–326

Chang J, Weigele P, King J, Chiu W, Jiang W (2006) Cryo-EM asymmetric reconstruction of bacteriophage P22 reveals organization of its DNA packaging and infecting machinery. Structure 14:1073–1082

Chang JT, Schmid MF, Haase-Pettingell C, Weigele PR, King JA, Chiu W (2010) Visualizing the structural changes of bacteriophage Epsilon15 and its Salmonella host during infection. J Mol Biol 402:731–740

Chattoraj DK, Inman RB (1974) Location of DNA ends in P2, 186, P4 and lambda bacteriophage heads. J Mol Biol 87:11–22

Chemla YR, Smith DE (2012) Single-molecule studies of viral DNA packaging. Adv Exp Med Biol 726:549–584

Chen DH, Baker ML, Hryc CF, DiMaio F, Jakana J, Wu W, Dougherty M, Haase-Pettingell C, Schmid MF, Jiang W, Baker D, King JA, Chiu W (2011) Structural basis for scaffolding-mediated assembly and maturation of a dsDNA virus. Proc Natl Acad Sci U S A 108:1355–1360

Cressiot B, Greive SJ, Si W, Pascoa TC, Mojtabavi M, Chechik M, Jenkins HT, Lu X, Zhang K, Aksimentiev A, Antson AA, Wanunu M (2017) Porphyrin-Assisted Docking of a Thermophage Portal Protein into Lipid Bilayers: Nanopore Engineering and Characterization. ACS Nano. *in press* 11:11931. https://doi.org/10.1021/acsnano.7b06980

Cuervo A, Carrascosa JL (2011) Viral connectors for DNA encapsulation. Curr Opin Biotechnol 23:529–536

Davidson AR, Cardarelli L, Pell LG, Radford DR, Maxwell KL (2012) Long noncontractile tail machines of bacteriophages. Adv Exp Med Biol 726:115–142

Donate LE, Herranz L, Secilla JP, Carazo JM, Fujisawa H, Carrascosa JL (1988) Bacteriophage T3 connector: three-dimensional structure and comparison with other viral head-tail connecting regions. J Mol Biol 201:91–100

Dröge A, Santos MA, Stiege A, Alonso JC, Lurz R, Trautner TA, Tavares P (2000) Shape and DNA packaging activity of bacteriophage SPP1 procapsid: protein components and interactions during assembly. J Mol Biol 296:117–132

Dube P, Tavares P, Lurz R, van Heel M (1993) Bacteriophage SPP1 portal protein: a DNA pump with 13-fold symmetry. EMBO J 12:1303–1309

Edmonds L, Liu A, Kwan JJ, Avanessy A, Caracoglia M, Yang I, Maxwell KL, Rubenstein J, Davidson AR, Donaldson LW (2007) The NMR structure of the gpU tail-terminator protein from bacteriophage lambda: identification of sites contributing to Mg(II)-mediated oligomerization and biological function. J Mol Biol 365:175–186

Evilevitch A, Lavelle L, Knobler CM, Raspaud E, Gelbart WM (2003) Osmotic pressure inhibition of DNA ejection from phage. Proc Natl Acad Sci U S A 100:9292–9295

Farley MM, Tu J, Kearns DB, Molineux IJ, Liu J (2017) Ultrastructural analysis of bacteriophage Φ29 during infection of Bacillus subtilis. J Struct Biol 197:163–171

Fokine A, Rossmann MG (2014) Molecular architecture of tailed double-stranded DNA phages. Bacteriophage 4:e28281

Fokine A, Zhang Z, Kanamaru S, Bowman VD, Aksyuk AA, Arisaka F, Rao VB, Rossmann MG (2013) The molecular architecture of the bacteriophage T4 neck. J Mol Biol 425:1731–1744

Fu CY, Prevelige PE Jr (2009) In vitro incorporation of the phage Phi29 connector complex. Virology 394:149–153

Gaussier H, Yang Q, Catalano CE (2006) Building a virus from scratch: assembly of an infectious virus using purified components in a rigorously defined biochemical assay system. J Mol Biol 357:1154–1166

Grundy FJ, Howe MM (1985) Morphogenetic structures present in lysates of amber mutants of bacteriophage Mu. Virology 143:485–504

Guasch A, Pous J, Ibarra B, Gomis-Rüth FX, Valpuesta JM, Sousa N, Carrascosa JL, Coll M (2002) Detailed architecture of a DNA translocating machine: the high-resolution structure of the bacteriophages phi29 connector particle. J Mol Biol 315:663–676

van Heel M, Orlova EV, Dube P, Tavares P (1996) Intrinsic versus imposed curvature in cyclical oligomers: the portal protein of bacteriophage SPP1. EMBO J 15:4785–4788

Hu B, Margolin W, Molineux IJ, Liu J (2013) The bacteriophage T7 virion undergoes extensive structural remodeling during infection. Science 339:576–579

Hu B, Margolin W, Molineux IJ, Liu J (2015) Structural remodeling of bacteriophage T4 and host membranes during infection initiation. Proc Natl Acad Sci U S A 112:E4919–E4928

Isidro A, Henriques AO, Tavares P (2004a) The portal protein plays essential roles at different steps of the SPP1 DNA packaging process. Virology 322:253–263

Isidro A, Santos MA, Henriques AO, Tavares P (2004b) The high-resolution functional map of bacteriophage SPP1 portal protein. Mol Microbiol 51:949–962

Jiang W, Chang J, Jakana J, Weigele P, King J, Chiu W (2006) Structure of epsilon15 bacteriophage reveals genome organization and DNA packaging/injection apparatus. Nature 439:612–661

Johnson JE, Chiu W (2007) DNA packaging and delivery machines in tailed bacteriophages. Curr Opin Struct Biol 17:237–243

Katsura I (1987) Determination of bacteriophage lambda tail length by a protein ruler. Nature 327:73–75

Katsura I, Tsugita A (1977) Purification and characterization of the major protein and the terminator protein of the bacteriophage lambda tail. Virology 76:129–145

Krupovic M, Bamford DH (2011) Double-stranded DNA viruses: 20 families and only five different architectural principles for virion assembly. Curr Opin Virol 1:118–124

Lander GC, Tang L, Casjens SR, Gilcrease EB, Prevelige P, Poliakov A, Potter CS, Carragher B, Johnson JE (2006) The structure of an infectious P22 virion shows the signal for headful DNA packaging. Science 312:1791–1795

Lander GC, Khayat R, Li R, Prevelige PE, Potter CS, Carragher B, Johnson JE (2009) The P22 tail machine at subnanometer resolution reveals the architecture of an infection conduit. Structure 17:789–799

Lebedev AA, Krause MH, Isidro AL, Vagin A, Orlova EV, Turner J, Dodson EJ, Tavares P, Antson AA (2007) Structural framework for DNA translocation via the viral portal protein. EMBO J 26:1984–1994

Leiman PG, Shneider MM (2012) Contractile tail machines of bacteriophages. Adv Exp Med Biol 726:93–114

Leiman PG, Chipman PR, Kostyuchenko VA, Mesyanzhinov VV, Rossmann MG (2004) Three-dimensional rearrangement of proteins in the tail of bacteriophage T4 on infection of its host. Cell 118:419–429

Leiman PG, Basler M, Ramagopal UA, Bonanno JB, Sauder JM, Pukatzki S, Burley SK, Almo SC, Mekalanos JJ (2009) Type VI secretion apparatus and phage tail-associated protein complexes share a common evolutionary origin. Proc Natl Acad Sci U S A 106:4154–4159

Lengyel JA, Goldstein RN, Marsh M, Calendar R (1974) Structure of the bacteriophage P2 tail. Virology 62:161–174

Lhuillier S, Gallopin M, Gilquin B, Brasilès S, Lancelot N, Letellier G, Gilles M, Dethan G, Orlova EV, Couprie J, Tavares P, Zinn-Justin S (2009) Structure of bacteriophage SPP1 head-to-tail connection reveals mechanism for viral DNA gating. Proc Natl Acad Sci U S A 106:8507–8512

Liu X, Zhang Q, Murata K, Baker ML, Sullivan MB, Fu C, Dougherty MT, Schmid MF, Osburne MS, Chisholm SW, Chiu W (2010) Structural changes in a marine podovirus associated with release of its genome into Prochlorococcus. Nat Struct Mol Biol 17:830–836

Loessner MJ, Inman RB, Lauer P, Calendar R (2000) Complete nucleotide sequence, molecular analysis and genome structure of bacteriophage A118 of Listeria monocytogenes: implications for phage evolution. Mol Microbiol 35:324–340

Lokareddy RK, Sankhala RS, Roy A, Afonine PV, Motwani T, Teschke CM, Parent KN, Cingolani G (2017) Portal protein functions akin to a DNA-sensor that couples genome-packaging to icosahedral capsid maturation. Nat Commun 8:14310

Lopes A, Tavares P, Petit MA, Guérois R, Zinn-Justin S (2014) Automated classification of tailed bacteriophages according to their neck organization. BMC Genomics 15:1027

Lurz R, Orlova EV, Günther D, Dube P, Dröge A, Weise F, van Heel M, Tavares P (2001) Structural organisation of the head-to-tail interface of a bacterial virus. J Mol Biol 310:1027–1037

Maxwell KL, Davidson AR, Murialdo H, Gold M (2000) Thermodynamic and functional characterization of protein W from bacteriophage lambda. The three C-terminal residues are critical for activity. J Biol Chem 275:18879–18886

Maxwell KL, Yee AA, Booth V, Arrowsmith CH, Gold M, Davidson AR (2001) The solution structure of bacteriophage lambda protein W, a small morphogenetic protein possessing a novel fold. J Mol Biol 308:9–14

Maxwell KL, Yee AA, Arrowsmith CH, Gold M, Davidson AR (2002) The solution structure of the bacteriophage lambda head-tail joining protein, gpFII. J Mol Biol 318:1395–1404

Moore SD, Prevelige PE Jr (2001) Structural transformations accompanying the assembly of bacteriophage P22 portal protein rings in vitro. J Biol Chem 276:6779–6788

Morais MC (2012) The dsDNA packaging motor in bacteriophage φ29. Adv Exp Med Biol 726:511–547

Motwani T, Lokareddy RK, Dunbar CA, Cortines JR, Jarrold MF, Cingolani G, Teschke CM (2017) A viral scaffolding protein triggers portal ring oligomerization and incorporation during procapsid assembly. Sci Adv 3:e1700423

Newcomb WW, Homa FL, Brown JC (2005) Involvement of the portal at an early step in herpes simplex virus capsid assembly. J Virol 79:10540–10546

Nováček J, Šiborová M, Benešík M, Pantůček R, Doškař J, Plevka P (2016) Structure and genome release of Twort-like Myoviridae phage with a double-layered baseplate. Proc Natl Acad Sci U S A 113:9351–9356

Olia AS, Al-Bassam J, Winn-Stapley DA, Joss L, Casjens SR, Cingolani G (2006) Binding-induced stabilization and assembly of the phage P22 tail accessory factor gp4. J Mol Biol 363:558–576

Olia AS, Bhardwaj A, Joss L, Casjens S, Cingolani G (2007a) Role of gene 10 protein in the hierarchical assembly of the bacteriophage P22 portal vertex structure. Biochemistry 46:8776–8784

Olia AS, Casjens S, Cingolani G (2007b) Structure of phage P22 cell envelope-penetrating needle. Nat Struct Mol Biol 14:1221–1226

Olia AS, Prevelige PE Jr, Johnson JE, Cingolani G (2011) Three-dimensional structure of a viral genome-delivery portal vertex. Nat Struct Mol Biol 18:597–603

Oliveira L, Henriques AO, Tavares P (2006) Modulation of the viral ATPase activity by the portal protein correlates with DNA packaging efficiency. J Biol Chem 281:21914–21923

Oliveira L, Cuervo A, Tavares P (2010) Direct interaction of the bacteriophage SPP1 packaging ATPase with the portal protein. J Biol Chem 285:7366–7373

Orlova EV, Dube P, Beckmann E, Zemlin F, Lurz R, Trautner TA, Tavares P, van Heel M (1999) Structure of the 13-fold symmetric portal protein of bacteriophage SPP1. Nature Struct Biol 6:842–846

Orlova EV, Gowen B, Dröge A, Stiege A, Weise F, Lurz R, van Heel M, Tavares P (2003) Structure of a viral DNA gatekeeper at 10 Å resolution by cryo-electron microscopy. EMBO J 22:1255–1262

Parker ML, Eiserling FA (1983) Bacteriophage SPO1 structure and morphogenesis. I. Tail structure and length regulation. J Virol 46:239–249

Pell LG, Kanelis V, Donaldson LW, Howell PL, Davidson AR (2009a) The phage lambda major tail protein structure reveals a common evolution for long-tailed phages and the type VI bacterial secretion system. Proc Natl Acad Sci U S A 106:4160–4165

Pell LG, Liu A, Edmonds L, Donaldson LW, Howell PL, Davidson AR (2009b) The X-ray crystal structure of the phage lambda tail terminator protein reveals the biologically relevant hexameric ring structure and demonstrates a conserved mechanism of tail termination among diverse long-tailed phages. J Mol Biol 389:938–951

Perucchetti R, Parris W, Becker A, Gold M (1988) Late stages in bacteriophage lambda head morphogenesis: in vitro studies on the action of the bacteriophage lambda D-gene and W-gene products. Virology 165:103–114

Plisson C, White HE, Auzat I, Zafarani A, São-José C, Lhuillier S, Tavares P, Orlova EV (2007) Structure of bacteriophage SPP1 tail reveals trigger for DNA ejection. EMBO J 26:2728–3720

Rao VB, Feiss M (2015) Mechanisms of DNA packaging by large double-stranded DNA viruses. Annu Rev Virol 2:351–378

Saigo K (1975) Tail-DNA connection and chromosome structure in bacteriophage T5. Virology 68:154–165

São-José C, Lhuillier S, Lurz R, Melki R, Lepault J, Santos MA, Tavares P (2006) The ectodomain of the viral receptor YueB forms a fiber that triggers DNA ejection of bacteriophage SPP1 DNA. J Biol Chem 281:11464–11470

São-José C, de Frutos M, Raspaud E, Santos MA, Tavares P (2007) Pressure built by DNA packing inside virions: enough to drive DNA ejection in vitro, largely insufficient for delivery into the bacterial cytoplasm. J Mol Biol 374:346–355

Sborgi L, Verma A, Muñoz V, de Alba E (2011) Revisiting the NMR structure of the ultrafast downhill folding protein gpW from bacteriophage λ. PLoS One 6:e26409

Simpson AA, Tao Y, Leiman PG, Badasso MO, He Y, Jardine PJ, Olson NH, Morais MC, Grimes S, Anderson DL, Baker TS, Rossmann MG (2000) Structure of the bacteriophage φ29 DNA packaging motor. Nature 408:745–750

Smith DE, Tans SJ, Smith SB, Grimes S, Anderson DL, Bustamante C (2001) The bacteriophage φ29 portal motor can package DNA against a large internal force. Nature 413:748–752

Spinelli S, Veesler D, Bebeacua C, Cambillau C (2014) Structures and host-adhesion mechanisms of lactococcal siphophages. Front Microbiol 5:3

Stockdale SR, Collins B, Spinelli S, Douillard FP, Mahony J, Cambillau C, van Sinderen D (2015) Structure and Assembly of TP901-1 Virion Unveiled by Mutagenesis. PLoS One 10:e0131676

Strauss H, King J (1984) Steps in the stabilisation of newly packaged DNA during phage P22 morphogenesis. J Mol Biol 172:523–543

Sun L, Zhang X, Gao S, Rao PA, Padilla-Sanchez V, Chen Z, Sun S, Xiang Y, Subramaniam S, Rao VB, Rossmann MG (2015) Cryo-EM structure of the bacteriophage T4 portal protein assembly at near-atomic resolution. Nat Commun 6:7548

Tang J, Olson N, Jardine PJ, Grimes S, Anderson DL, Baker TS (2008) DNA poised for release in bacteriophage phi29. Structure 16:935–943

Tang J, Lander GC, Olia AS, Li R, Casjens S, Prevelige P Jr, Cingolani G, Baker TS, Johnson JE (2011) Peering down the barrel of a bacteriophage portal: the genome packaging and release valve in P22. Structure 19:496–502

Tavares P, Lurz R, Stiege A, Rückert B, Trautner TA (1996) Sequential headful packaging and fate of the cleaved DNA ends in bacteriophage SPP1. J Mol Biol 264:954–967

Tavares P, Zinn-Justin S, Orlova EV (2012) Genome gating in tailed bacteriophage capsids. Adv Exp Med Biol 726:585–600

Thomas JO (1974) Chemical linkage of the tail to the right-hand end of bacteriophage lambda DNA. J Mol Biol 87:1–9

Thomas JO, Sternberg N, Weisberg R (1978) Altered arrangement of the DNA in injection-defective lambda bacteriophage. J Mol Biol 123:149–161

Trus BL, Cheng N, Newcomb WW, Homa FL, Brown JC, Steven AC (2004) Structure and polymorphism of the UL6 portal protein of herpes simplex virus type 1. J Virol 78:12668–12671

Trus BL, Newcomb WW, Cheng N, Cardone G, Marekov L, Homa FL, Brown JC, Steven AC (2007) Allosteric signaling and a nuclear exit strategy: binding of UL25/UL17 heterodimers to DNA-Filled HSV-1 capsids. Mol Cell 26:479–489

Veesler D, Cambillau C (2011) A common evolutionary origin for tailed-bacteriophage functional modules and bacterial machineries. Microbiol Mol Biol Rev 75:423–433

Vianelli A, Wang GR, Gingery M, Duda RL, Eiserling FA, Goldberg EB (2000) Bacteriophage T4 self-assembly: localization of gp3 and its role in determining tail length. J Bacteriol 182:680–688

Vinga I, São-José C, Tavares P, Santos MA (2006) Bacteriophage entry in the host cell. In: Wegrzyn G (ed) Modern bacteriophage biology and biotechnology. Research Signpost, Kerala

Xiang Y, Morais MC, Battisti AJ, Grimes S, Jardine PJ, Anderson DL, Rossmann MG (2006) Structural changes of bacteriophage phi29 upon DNA packaging and release. EMBO J 25:5229–5239

Zhao L, Kanamaru S, Chaidirek C, Arisaka F (2003) P15 and P3, the tail completion proteins of bacteriophage T4, both form hexameric rings. J Bacteriol 185:1693–1700

Chapter 15
Beyond Channel Activity: Protein-Protein Interactions Involving Viroporins

Janet To and Jaume Torres

The Viroporins and Channel Activity Inhibition

The field of viroporin research has its origins in the observation that virus-infected cells show increased membrane permeability (Carrasco 1995). More than two decades later, viroporins have been confirmed in several viral families, e.g., *Orthomyxoviridae* (AM2, PB1-F2, BM2), *Flaviviridae* (p7), *Coronaviridae* (E, 3a, 4a), *Paramyxoviridae* (SH), *Picornaviridae* (2B/2 BC, 3A), *Togaviridae* (6 K), *Retroviridae* (Vpr, Vpu, p13), *Reoviridae* (NSP4 and p10), *Polyomaviridae* (agnoprotein, VP2-VP4), *Papillomaviridae* (E5), or *Rhabdoviridae* (α1). Currently, detailed structural information is limited to only a handful of viroporins (*vide infra*), although these constitute useful templates for the probably hundreds of other unknown viroporins yet to be discovered in reservoir hosts (Anthony et al. 2013).

In most cases, viral attenuation is not only achieved by deletion of the viroporin gene but also simply when their channel activity is suppressed. Indeed, various specific pathogenic roles of viroporin channel activity have been discovered, and attempts have been made to modulate this channel activity, especially that of influenza A virus M2 (IAV M2, or AM2) protein, the first discovered and the best characterized viroporin. In general, however, the road to rational design and discovery of viroporin small-molecule inhibitors has not been successful [see To et al. (2016) for a recent review]. In fact, amantadine and rimantadine are at present the only licensed antiviral drugs that target a viroporin, i.e., IAV M2. However, most circulating strains of IAV are Amtresistant (Deyde et al. 2007; Hayden and De Jong 2011), and neither drug is currently being used in humans.

J. To · J. Torres (✉)
School of Biological Sciences, Nanyang Technological University, Singapore, Singapore
e-mail: janetto@ntu.edu.sg; jtorres@ntu.edu.sg

© Springer Nature Singapore Pte Ltd. 2018
J. R. Harris, D. Bhella (eds.), *Virus Protein and Nucleoprotein Complexes*,
Subcellular Biochemistry 88, https://doi.org/10.1007/978-981-10-8456-0_15

Protein-Protein Interactions (PPIs) Involving Viroporins

Viroporins are involved in many protein-protein interactions (PPIs) that may be also susceptible to therapeutic intervention [see recent reviews (Fischer et al. 2014; Nieva and Carrasco 2015)]. Both intraviral and virus-host interactions, in the form of a myriad of perturbations, can provide important insights into the mechanisms involved in the viral infectious cycle. Also, understanding these PPI networks may aid the design of new antivirals. These strategies depend on both detailed structural and mechanistic information and on the availability of therapeutically relevant targets. In this sense, the initial focused approach to identify viroporin binders is being complemented with genome-wide interactome studies adapted to viral infections using high-throughput technologies, providing a dramatic boost in the search for possible PPIs [see de Chassey et al. (2014) for a recent review].

Useful methods to obtain leads for PPIs involving viroporins include yeast two-hybrid (Y2H) screens, e.g., the genome-wide virus-host PPI screen of HCV was performed almost 10 years ago using a construction of a viral ORFeome and Y2H technology (De Chassey et al. 2008), identifying hundreds of PPIs involving viral and host proteins, 13 of which involving p7 and proteins expressed in the liver. Other Y2H screens have included IAV M2 virus-virus and virus-host interactions (Shapira et al. 2009), or intraviral PPIs in the coronavirus responsible for severe acute respiratory syndrome (SARS-CoV) (von Brunn et al. 2007). However, this method does not measure interactions between proteins in the context of the infected cell, is biased against membrane proteins, and cannot study protein complexes that are weakly or transiently associated. The bias against membrane proteins can be compensated using the split-ubiquitin-based yeast two-hybrid screen, e.g., in a screen to search for binders of the small hydrophobic (SH) protein of the respiratory syncytial virus (RSV) (Li et al. 2015).

More suitable methods detect interactions in the context of the infected cell, using a combination of affinity purification with mass spectrometry, e.g., Wang et al. (2017), although the method has also low sensitivity. Other methods are based on microarrays of deposited purified proteins (Zhu et al. 2001) and protein complementation assay (PCA) (Tarassov et al. 2008). Lastly, an approach that combines on-chip in vitro protein synthesis with an in situ microfluidic affinity assay can detect even weak or transient interactions (Gerber et al. 2009), although host-virus PPIs using this method so far has only been tested for M protein in RSV (Kipper et al. 2015).

From these studies, it is apparent that viroporins, and viral proteins in general, tend to show preference for key host proteins that have a high number of direct interacting partners. Also, interactions may be simultaneous with many cellular proteins, making use of intrinsically disordered protein regions enriched for short linear motifs, e.g., PDZ-binding motifs (Hagai et al. 2014; Meyniel-Schicklin et al. 2012), to compensate for their small proteomes.

The Influenza A Virus Matrix Protein 2 (IAV M2 or AM2)

Influenza viruses belong to the *Orthomyxoviridae* family of segmented, negative-sense, enveloped RNA viruses. The seasonal flu caused by the influenza A virus (IAV) is known for causing pandemics with high mortality rates (Hay et al. 2001; Neumann et al. 2009), although its close relative influenza B accounts for half of the influenza disease in recent years (www.cdc.gov). Generally, influenza virions are spherical in shape ranging from 80 to 120 nm in diameter, although filamentous forms may also occur (Lamb and Choppin 1983). The viral envelope contains three transmembrane proteins, hemagglutinin (HA), neuraminidase (NA), and matrix protein 2 (M2), on the outside and a layer of matrix protein (M1) just underneath the membrane that contains cholesterol-enriched lipid rafts. M1 forms an internal coat that encloses the viral ribonucleoproteins (vRNPs), i.e., the negative-strand viral RNA (vRNA) and nucleoprotein (NP), with small amounts of the nuclear export protein (NEP) and three polymerase (3P) proteins (PA, PB1, and PB2) that form the viral RNA polymerase complex (Fields et al. 2013).

M2 Viroporin

The viroporin M2 in IAV is a homotetrameric channel (Sakaguchi et al. 1997). Each M2 monomer is a 97-amino acid protein comprising an N-terminal ectodomain (24 aa), an α-helical transmembrane domain (TMD, 19 aa), and a highly conserved cytoplasmic tail (CT) domain (54 aa) that is a hotspot for interactions with both viral and host proteins during the IAV life cycle. The latter may therefore constitute an attractive drug target for the development of IAV antivirals. M2 has a pH-activated proton channel activity which is required to complete the uncoating process during virus entry. Upon virus internalization via endocytosis, M2 selectively conducts protons from acidified endosomes into the viral interior. This acidification of virion triggers the dissociation of the M1 protein from the vRNP complex, thereby enabling the transport of vRNPs into the nucleus for replication of viral genetic material (Helenius 1992). For some IAV subtypes, the M2 proton channel raises the pH of the trans-Golgi network (TGN) to protect the viral HA from premature low-pH conformational change during its transport to the cell surface (Takeuchi and Lamb 1994). The channel activity of IAV M2 has been found to be sufficient for the activation of the NLRP3 inflammasome in influenza-infected cells (Ichinohe et al. 2010). Presently, there are more than ten structures of both wild-type and drug-resistant mutant M2 channels in the Protein Data Bank [see review in Gu et al. (2013)].

M1-M2 Interaction

Early works suggested a role for IAV M1-M2 interaction in virus budding and control of virion morphology (filamentous versus spherical). Interaction between M1 with the M2 cytoplasmic tail was first suggested from the analysis of escape mutants (Zebedee and Lamb 1989). Further work revealed a physical interaction between M1 and the M2 cytoplasmic tail at the site of virus budding, to facilitate virus assembly by promoting the recruitment and packaging of viral proteins and viral genome (McCown and Pekosz 2006; Chen et al. 2008). The cytoplasmic tail of M2 contains an amphipathic α-helix (residues 45–62) (Schnell and Chou 2008) that can modulate membrane curvature in a cholesterol-dependent manner. This feature of M2 has been proposed to be implicated in (i) modification of local membrane curvature during virus budding to provide a stabilized scaffold for M1 polymerization and virus filament formation and (ii) alteration of membrane curvature at the neck of budding virions to facilitate membrane scission and virion release (Rossman and Lamb 2011).

Host Interactions

In addition to intraviral interactions, a number of interactions of M2 with host proteins have been described to modulate autophagy, membrane trafficking, host defense, and virus budding. For example, IAV M2 has been reported to arrest autophagy (Gannagé et al. 2009; Beale et al. 2014), a cellular degradation pathway mediated by autophagosomes which delivers cytoplasmic materials to the lysosome that is regulated by autophagy-related genes (Atg). This process involves (i) target engulfment by an isolation crescent membrane (phagophore) to form the autophagosome and (ii) autophagosome-lysosome fusion and degradation of the intra-autophagosomal contents (Fig. 15.1a). IAV subverts this machinery by blocking this fusion, resulting in increased apoptosis of IAV-infected cells. In IAV-infected A549 human lung epithelial cells, M2 coimmunoprecipitates with Atg6/Beclin-1 through interaction with M2 residues 1–60 (Gannagé et al. 2009). Atg6/Beclin1 is part of a complex that regulates autophagosome generation and degradation and is a common target of other viruses for the subversion of autophagy, e.g., herpesviruses and the human immunodeficiency virus (HIV) [reviewed in Münz (2011)].

The C-terminal tail of M2 has also been implicated in the binding to LC3, a protein that normally localizes to autophagosomal membranes (Sou et al. 2006) to recruit autophagy receptors carrying substrates destined for autophagic degradation. These receptors typically contain an LC3-interacting region (LIR), with consensus LIR motif W/FxxI/L/V. In IAV-infected cells, the localization of LC3 changes from the cytoplasm and autophagosomal membranes to the plasma membrane (Beale et al. 2014), a change mediated by a putative LIR motif (residues 91–94, FVSI)

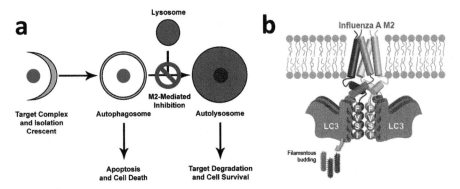

Fig. 15.1 IAV M2 may subvert autophagy by (a) interaction with Atg6/Beclin-1 to block autophagosome maturation, triggering apoptosis [adapted from Rossman and Lamb (2009)] and (b) interaction with LC3 through an M2 LIR motif (FVSI) that hijacks LC3 to the plasma membrane [adapted from Beale et al. (2014)]

present in the cytoplasmic tail of M2 (Fig. 15.1b). Binding of LC3 to the LIR motif of M2 has been confirmed by LUMIER binding assays (Barrios-Rodiles et al. 2005) and GFP pull-down experiments.

The M2-LC3 interaction is also a factor in the budding of IAV, which, depending on the viral strain and host cell type, can produce either spheres or filaments (Bourmakina and García-Sastre 2003), with the latter requiring extensive membrane resources. Cells infected with a filamentous budding IAV strain carrying mutations in the M2 LIR motif that abolished M2-LC3 interactions produced fewer filaments than cells infected with wild-type IAV, suggesting that hijacking of LC3 by M2 may assist in the delivery of LC3-conjugated membranes to the cell surface to facilitate IAV budding (Beale et al. 2014).

The cytoplasmic domain of IAV M2 has been reported to bind caveolin-1 (Cav-1) (Zou et al. 2009), a raft-residing cholesterol-binding protein implicated in the life cycle of viruses that buds from lipid rafts, such as HIV, RSV, and rotavirus. Most Cav-1-associated proteins contain an aromatic-rich caveolin-binding motif (CBM), a consensus sequence of aromatic residues separated by a specific spacing (Couet et al. 1997). The M2-Cav-1 interaction has been confirmed by pull-down and coimmunoprecipitation assays, with the putative CBM in M2 proposed to reside in the cytoplasmic, juxtamembrane region of the M2 tail (Sun et al. 2010). This interaction suggests that Cav-1 may modulate virus budding, possibly through the trafficking of M2 to the plasma membrane.

A yeast two-hybrid screening effort identified the transport protein particle complex 6A (TRAPPC6A) and also its N-internal deleted isoform, TRAPPC6AΔ, as binders to the last six amino acids at the C-terminal end of IAV M2, with highly conserved Leu96 located at the extremity of M2 being indispensable in mediating the interaction (Zhu et al. 2017). TRAPP complexes are multi-subunit tethering

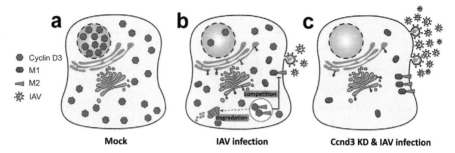

Fig. 15.2 Proposed activity of cyclin D3 in IAV budding. (**a**) In normal cells, cyclin D3 is localized to the nucleus where it regulates cell cycle; (**b**) during IAV infection, cyclin D3 interacts with viral M2 to disrupt M1-M2 binding; (**c**) in the absence of cyclin D3, more infectious progeny virus particles are released [adapted from Fan et al. (2017)]

complexes involved in intracellular membrane trafficking pathways, and TRAPPC6AΔ may be a regulator of M2 transport to the cell surface.

A yeast two-hybrid assay combined with mutagenesis identified the binding of both AM2 and BM2 to the C-terminal domain 1 (CTD1) of Hsp40 (Guan et al. 2010), a molecular chaperone that can regulate the critical PKR signaling pathway against viral infection. Hsp40 associates with the PKR inhibitor, P58[IPK]. Binding studies suggest that M2 also binds to P58[IPK], possibly forming a stable complex with both Hsp40 and P58[IPK] (Guan et al. 2010), which would enhance PKR autophosphorylation and activation to inhibit host protein synthesis.

Results from another yeast two-hybrid screen using the M2 cytoplasmic tail as a bait have identified the human cell cycle regulator cyclin D3 as a binder, and their interaction has been confirmed in infected cells by immunoprecipitation assays (Fan et al. 2017). Using siRNA-mediated knockdown of cyclin D3 expression, the study proposed that cyclin D3 is a negative modulator of IAV infection, competing with M1 for binding to M2 in transfected cells and therefore disrupting the important M1-M2 interaction in the context of an IAV infection. On the other hand, IAV may antagonize cyclin D3 activity by (i) relocating cyclin D3 from the nucleus (Fig. 15.2a) to the cytosol (Fig. 15.2b) to facilitate virus replication by promoting cell cycle arrest and (ii) targeting cyclin D3 for cytosolic proteasomal degradation to prevent its interference with M1-M2 binding (Fig. 15.2b). Cyclin D3 has been also reported to be a target of SARS-CoV viroporin 3a (Yuan et al. 2007)

A recent study has identified M2 as a putative viral antagonist of BST-2 (bone marrow stromal cell antigen 2, also known as tetherin, CD317) (Hu et al. 2017), a protein that may restrict the release of infectious IAV (Mangeat et al. 2012). BST-2 can inhibit the release of a wide range of enveloped viruses by tethering budding virions to the cell surface [reviewed in le Tortorec et al. (2011)], e.g., in HIV-1, in that case antagonized by its viroporin Vpu (Neil et al. 2008). However, the role of BST-2 in limiting IAV release has been disputed (Watanabe et al. 2011), perhaps due to IAV strain-specific susceptibility to BST-2 restriction. This interaction was confirmed by orthogonal assays, and using chimeric and truncated M2, the regions involved in BST-2 downregulation were proposed to be within the M2 extracellular and TMDs.

Another report, using a Y2H screening with M2 cytoplasmic tail as bait, discovered the binding to annexin A6 (AnxA6) (Ma et al. 2012), a Ca^{2+}-regulated membrane-binding protein that controls intracellular cholesterol homeostasis, and regulates membrane fusion and vesicle formation in endocytic and exocytic pathways (Raynal and Pollard 1994). This interaction has been verified by coimmunoprecipitation assays and colocalization studies in infected cells (Ma et al. 2012). Modulation of AnxA6 expression led to corresponding variations in production of infectious IAV, suggesting that AnxA6 is a negative modulator of IAV infection (Ma et al. 2012), as it may impair IAV budding and the release of progeny virus.

A Y2H screen also identified Na^+/K^+ ATPase $\beta1$ subunit (ATP1B1) as binder to the M2 cytoplasmic domain (Mi et al. 2010). SARS-CoV E viroporin and the human papillomavirus E5 viroporin have also been found to bind the host Na^+/K^+ ATPase $\alpha1$ subunit (Nieto-Torres et al. 2011) and vacuolar H^+ ATPase (Andresson et al. 1995), respectively.

PB1-F2 Viroporin

Another protein in IAV, PB1-F2 (Chen et al. 2001), has the hallmarks of a viroporin: it is ~90 residues long, it forms oligomers, and it has been shown to form a nonselective ion channel in planar lipid bilayers and microsomes (Henkel et al. 2010). PB1-F2 is known to induce apoptosis in host immune cells (Chen et al. 2001) via interaction with two mitochondrial proteins, ANT3 and VDAC1, resulting in the loss of mitochondria membrane potential (Varga et al. 2012), although its localization and pro-apoptotic behavior is strain and cell type specific (Varga and Palese 2011). In addition, PB1-F2 from pathogenic strains of IAV can be incorporated into the phagolysosomal compartment to activate the NLRP3 inflammasome, resulting in IL-1β secretion and causing severe pathophysiology (McAuley et al. 2013). Also, binding of PB1-F2 of PR8 to MAVS, a RIG-I-like receptor (RLR) signaling adaptor anchoring to mitochondria, led to antagonism activity on interferon production (Varga et al. 2012). Despite these data, the precise role of PB1-F2 in modulation of IAV-induced immunopathogenesis is still unknown.

IAV Protein Interactome

Many more IAV host-virus PPIs have been detected recently using affinity purification coupled with mass spectrometry in the context of infected cells. For example, the interactome of 11 viral proteins of influenza PR8 IAV and another 3 strains was analyzed (Wang et al. 2017), confirming that M2 protein is one of the major nodes

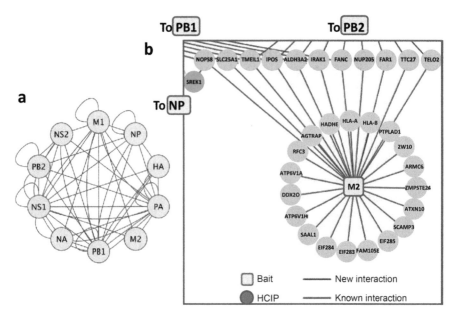

Fig. 15.3 Maps of the IAV intraviral and virus-host protein interactome. (**a**) Result from a Y2H study to identify direct binary contacts among the ten major viral proteins of the PR8 strain (Shapira et al. 2009). This study also detected nine interactions between IAV M2 and host proteins, among them RNA-binding proteins, transcription factors, or proteins involved in signaling pathways, or to detect intraviral PPIs; (**b**) close-up of high-confidence candidate-interacting proteins (HCIPs) associated with multiple IAV strains when M2 was used as bait. Indicated are links to other viral proteins (NP, PB1, and PB2) [adapted from Wang et al. (2017)]

connecting host proteins with roles in immunity and regulation of viral infection. Almost 100 interactions of host with M2 protein were detected, and ~30 were common to at least three of the strains (Fig. 15.3).

The Hepatitis C Virus p7 Protein (HCV-p7)

The hepatitis C virus (HCV) is an enveloped positive-strand RNA virus member of the *Flaviviridae* family (genus *Hepacivirus*) that has chronically infected 170 million people worldwide, causing human liver disease. Hepatitis C is divided into six genotypes, with genotype 1 being the most common and most difficult to treat. Treatment against HCV infection involves drugs targeting both viral and host proteins, e.g., drugs that target NS3/4A protease, the NS5A protein, or the NS5B RNA-dependent RNA polymerase (RdRP). However, the rapid turnover of HCV replication (Neumann et al. 1998) and the error-prone activity of the HCV RNA polymerase lead to a rapid formation of "quasi-species" and therefore resistance to antivirals.

HCV encodes a single polyprotein of ~3000 amino acids that is cleaved by cellular and viral proteases into ten different proteins: three structural proteins (core, E1, and E2) and seven nonstructural proteins (p7, NS2, NS3, NS4A, NS4B, NS5A, and NS5B). The structural proteins Core (C), E1, and E2 are located in the N-terminal region and form the viral particle (Moradpour and Penin 2013), whereas NS3 to NS5B are involved in the replication of the viral genome. p7 and NS2 are dispensable for RNA replication but are critical for virion morphogenesis, which requires both structural and nonstructural proteins (Appel et al. 2008; Steinmann et al. 2007), although the latter are not packaged in viral particles.

HCV replication takes place in double-membrane vesicles (DMVs), while viral assembly sites (AS) have been suggested to be specialized detergent-resistant membranes (DRMs) in the ER or in mitochondria-associated ER membranes rich in cholesterol and sphingolipids (Shanmugam et al. 2015). The core protein concentrates at cytosolic lipid droplets (cLDs) close to the ER-located assembly site and is eventually linked to the vRNA replication site in specialized ER-derived structures.

During HCV assembly, one of the first steps is the interaction of cLD-bound core protein and NS5A (Appel et al. 2008). NS2, probably in complex with p7, interacts with the NS3-4A enzyme, and this retrieves the viral core protein from cLDs into the nascent virus particles [reviewed in Lindenbach and Rice (2013)]. Virus particles transit through the secretory pathway, where they are protected from exposure to low pH by p7, which neutralizes intracellular compartments (Wozniak et al. 2010). More recently, a genetic interaction has been observed between p7 and NS5B proteins, which were found to cooperate to promote virion infectivity by decreasing sphingomyelin content in the virion (Aligeti et al. 2015).

p7 Viroporin The viroporin p7 is produced when E2-p7-NS2 is cleaved by a signal peptidase at the ER (Lin et al. 1994; Mizushima et al. 1994). p7 is a 63-residue-long protein that has two α-helical TMDs and is found mainly at the ER membrane. As mentioned above, p7 is essential for virus particle assembly and release (Steinmann et al. 2007) and for productive HCV propagation in vivo (Sakai et al. 2003), but not necessary for RNA replication. The bovine viral diarrhea virus (BVDV) and the hepacivirus GB virus B (GBV-B), HCV's closest relatives, also have a p7 protein crucial for virus replication.

p7 has channel activity with low cation selectivity (Griffin et al. 2003; Ouyang et al. 2013). p7 has been reported to permeabilize membranes to protons, preventing the acidification of intracellular vesicles (Wozniak et al. 2010), an activity that has been confirmed in vitro using a liposome-based assay (Gan et al. 2014).

The structural model for p7 is that of an α-helical hairpin with two α-helical TMDs kinked in the middle (Cook et al. 2010), or a sequence divided into three helical segments (Ouyang et al. 2013) where the N-terminal half of the polypeptide would face the lumen of the channel and the C-terminal helix, p7(27–63), faces the lipid environment. The channel is formed by either six or seven monomers (Luik et al. 2009; Montserret et al. 2010).

p7-NS2 Interaction

Early genetic analyses suggested that p7 interacts with NS2, a polytopic membrane protein containing three N-terminal TMDs that is essential in the assembly process of the HCV particle (Jirasko et al. 2008). Mutation of residues in one protein to induce the emergence of complementary mutations in the other was used to identify the interaction network of NS2 protein with p7, E1 and E2, and NS3 proteins (Jirasko et al. 2010). Similar conclusions were reached in studies involving chimeric constructs with different genotypes (Pietschmann et al. 2006) which showed that virus release was most efficient when the N-terminal TMD of NS2 was from the same isolate as the core-to-p7 region. In a similar study, adaptive mutations in E1, p7, NS2, and NS3 were detected that were essential for virus assembly and/or release, again suggesting genetic interactions between these proteins (Yi et al. 2007).

Physical interaction between p7 and NS2 was observed during pull-down assays and mutagenesis (Ma et al. 2011), which suggested that p7 may regulate NS2-mediated complexes that are crucial for production of infectious HCV particles. Coimmunoprecipitation and FRET assays also supported a physical interaction between p7 and NS2 (Popescu et al. 2011), suggesting a complex between p7, NS2, and E2 mediated by transmembrane interactions. These interactions were proposed to be required to localize NS proteins and the core-containing cLDs to sites of virus assembly. Overall, these studies demonstrated that NS2, together with p7 protein, plays a central organizing role in HCV particle assembly by bringing together viral structural and nonstructural proteins. Although the exact mechanism linking nucleocapsid assembly with envelope acquisition is unknown, p7 and NS2 have been proposed to play a critical role in the migration of core protein and E1-E2 heterodimers to the virion assembly site (Vieyres et al. 2014).

Another role, in immune evasion, has been identified recently for p7 (Qi et al. 2017). Indeed, HCV acts against the host immune system by downregulating interferon (IFN) production, blocking IFN signaling transduction, and impairing IFN-stimulated gene (ISG) expression. But even when ISGs are expressed, most ISGs have been reported to be ineffective when overexpressed in virus-infected cells due to unknown mechanisms (Schoggins et al. 2011). By constructing a library of mutant HCVs with a 15-nt insertion, p7 was identified as an immune evasion protein that suppresses the antiviral IFN function, forming a complex with the host interferon-inducible protein 6–16 (IFI6–16) that has been verified by coimmunoprecipitation. It was proposed that while IFI6–16 acts to stabilize the mitochondrial membrane potential, p7 counteracts by depolarizing the mitochondrial potential, likely through its ion channel activity. Overall, the findings suggest that p7 antagonizes the antiviral responses of IFN by inhibiting the antiviral function of IFI6–16.

HCV Protein Interactome

A proteome-wide virus-host PPI screen of HCV was performed almost 10 years ago using a construction of a viral ORFeome and Y2H technology (De Chassey et al. 2008). In that study, 314 PPIs were identified involving viral and host proteins, 13 of which involving p7 and proteins expressed in the liver (Fig. 15.4), although these were not confirmed by orthogonal methods. A latter study combined mass spectrometry and functional genomics (Germain et al. 2014), but p7 was not included in the screen.

The interaction network between the ten HCV proteins has been investigated using a flow-cytometry-based FRET assay in living cells. In this study, p7 was found to bind NS2, Core, E1, and E2 (Hagen et al. 2014) (Fig. 15.5a). In 2016, a computational coevolution analysis of HCV attempted to reconstruct the PPI network of the HCV at the residue resolution (Champeimont et al. 2016). Coevolving residues were identified to predict PPIs for further experimental identification of HCV protein complexes (Fig. 15.5b). One of these interactions was p7-NS2 (see Fig. 15.5c).

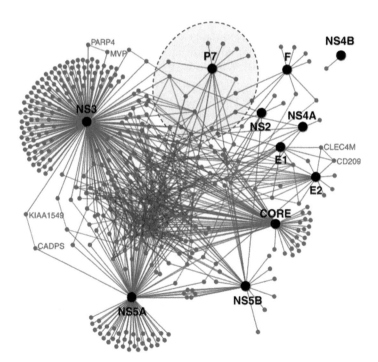

Fig. 15.4 Graphical representation of the HCV virus-host (V-H) human interaction network (*red lines*) with black and red nodes representing viral and human proteins, respectively. Blue lines represent H-H interactions (adapted from De Chassey et al. (2008))

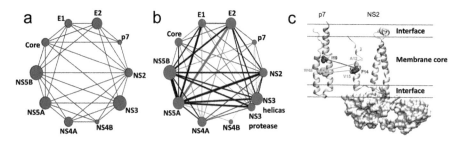

Fig. 15.5 (**a**) Network of reported intraviral HCV PPIs determined experimentally [adapted from Hagen et al. (2014)] compared to (**b**) coevolution links of HCV proteins (Champeimont et al. 2016), where *blue* lines correspond to coevolving links not experimentally reported; (**c**) predicted p7-NS2 interaction between F14 of NS2 and I19 in TM1 of p7 (red line), compared to experimentally reported interactions based on NMR (Cook et al. 2013) between NS2 A12 and V15 and W48 in p7 TM2 (*green* lines) [adapted from Champeimont et al. (2016)]

Coronavirus Viroporins

Coronaviruses (CoV) are vertebrate pathogens which cause human respiratory diseases that typically affect the respiratory tract and gut. CoVs belong to the family *Coronaviridae*, subfamily *Coronavirinae,* and are distributed into four genera α, β, γ, and δ (Enjuanes et al. 2000). While α-CoVs and β-CoVs circulate in mammalian hosts, γ-CoVs and δ-CoVs mainly infect birds. For example, β-CoVs include the murine hepatitis virus (MHV), whereas γ-CoVs include the avian infectious bronchitis virus (IBV). The first coronavirus was isolated in 1937 (IBV), whereas the first human coronavirus (HCoV) was identified in the 1960s. In humans, disease caused by coronaviruses ranges from mild to really severe, e.g., the recent severe acute respiratory syndrome (SARS) and the Middle East respiratory syndrome (MERS).

SARS-CoV appeared in 2002 causing ~10,000 human infections, with a 10% mortality rate (Holmes 2003). MERS coronavirus (MERS-CoV) emerged about 10 years later, and to date (as of July 2017) almost 2040 cases of infection and 712 deaths have been confirmed (http://www.who.int/emergencies/mers-cov/en/), i.e., a mortality of ~35%. Currently, no effective licensed prevention nor treatment exists against coronavirus infection (Lou et al. 2014), although live attenuated vaccines and fusion inhibitors are promising strategies.

CoVs have nonsegmented, exceptionally long genomes (up to 32 kb). One third of the genome hosts the ORFs for structural proteins, i.e., spike (S), envelope (E), membrane (M), and nucleoprotein (N). This part of the genome also encodes other so-called "accessory" proteins, which vary in number and sequence even among CoVs belonging to the same lineage (Enjuanes et al. 2008), e.g., from one in HCoV-NL63 to eight in SARS-CoV. The remaining two thirds of the genome encode nonstructural genes, with open reading frames ORF1a and ORF1b that produce polyproteins pp1a and pp1ab. These are then processed into 16 nonstructural proteins (nsp1 to 16); see Su et al. (2016) for recent general overview of coronaviruses and Forni et al. (2017) for the molecular evolution of HCoVs. In the case of SARS-CoV, the genome is predicted to encode 14 functional open reading frames, leading to the expression of up to 30 structural and nonstructural protein products.

E Viroporin

The E proteins are 76–109 amino acids long with one TMD (Torres et al. 2006; Li et al. 2014a; To et al. 2017), a short lumenal N-terminus and a longer cytoplasmic C-terminal tail (Nieto-Torres et al. 2011) which in SARS E protein tends to form β-structure in isolation, but it is mainly helical in the context of a full length protein (Li et al. 2014a). E proteins are localized particularly in the endoplasmic reticulum-Golgi intermediate compartment (ERGIC), where virus morphogenesis and budding occurs. E protein forms homopentameric channels with poor ion selectivity (Verdia-Baguena et al. 2012). Only the TMD of SARS-CoV E (E-TM) has been characterized in some detail, in lipid membranes (Torres et al. 2006) and in DPC micelles (Pervushin et al. 2009).

SARS-CoV E protein is a virulence factor critical for viral pathogenesis, as SARS-CoV lacking the E gene (rSARS-CoV-ΔE) is attenuated in vivo (DeDiego et al. 2007). Mutations N15A and V25F abolish channel activity in vitro (Torres et al. 2007) and led to attenuation when introduced in a recombinant SARS-CoV (Nieto-Torres et al. 2014). The latter authors showed that channel activity is important for inflammasome activation and elevated production of the pro-inflammatory cytokine IL-1β. SARS-CoV E also regulates host stress response and apoptosis (DeDiego et al. 2011) and improves viral fitness. The importance of the E protein in coronavirus pathogenesis has led to the development of live attenuated vaccines based on E-deleted, E-truncated, or E-mutated virions, e.g., Regla-Nava et al. (2015). In general, E protein plays important roles in coronavirus assembly and morphogenesis, although E protein is not necessary to obtain infectious SARS-CoV (DeDiego et al. 2008).

E-M Interaction

E protein is a known binder of M protein, the most abundant protein component of the virion and the membrane protein responsible for its shape. M protein has three predicted TMDs and a large C-terminal extramembrane domain exposed to the cytoplasm or the interior of the virion. It is this domain that forms contacts with the C-terminal tail of the E protein, although TMD interactions are also likely (Lim and Liu 2001; Hogue and Machamer 2008). These interactions occur at the ERGIC, the budding compartment of the host cell. Since M-M interactions are major drivers of viral envelope formation, these contacts are likely to be important for particle assembly. The E-M interaction has long been reported in infectious bronchitis virus (IBV) and mouse hepatitis virus (MHV) by coimmunoprecipitation in virus-infected or virus-transfected cells (Corse and Machamer 2003; Lim and Liu 2001; Maeda et al. 1999) and was proposed to be crucial for the formation of virus-like particles (VLPs) and virions. In SARS-CoV, coexpression of M and E in a baculovirus expression system was sufficient for the assembly of VLPs (Ho et al. 2004). The deletion of the E gene (ΔE) in the murine coronavirus, the mouse hepatitis virus (MHV), produced revertants where the M gene appeared to have been duplicated, creating new variants of M protein that lacked most of its C-terminal cytoplasmic tail (Kuo and Masters 2010). These results suggested a role for E proteins in "dispersing or de-aggregating" M protein during packaging.

Other CoV Viroporins Although the most studied viroporin in CoVs is the envelope (E) protein, other viroporins have been found in an accessory gene present in all CoVs, between the S and E gene loci. In SARS-CoV (SARS-ORF3a) and in HCoV-229E (229E–ORF4a), these proteins form ion channels (Zhang et al. 2014; Lu et al. 2006), whereas HCoV-NL63-ORF3 has also been proposed to be a viroporin (Zhang et al. 2015). In HCoV-OC43, this genomic segment encodes ORF5, which has been reported to facilitate virion morphogenesis (Zhang et al. 2015). The latter has only a single TMD, in contrast with SARS-CoV ORF3a, HCoV-229E ORF4a, and HCoV-NL63 ORF3.

3a Viroporin SARS-CoV 3a protein is a 274-amino-acid-long protein with three TMDs and forms homotetrameric complexes that have ion channel activity (Lu et al. 2006). 3a protein has a cysteine-rich domain (residues 127–133) responsible for homo- and hetero-dimerization (Lu et al. 2006). In addition to a Yxx domain and a diacidic domain, it has a C-terminal domain (Tan et al. 2006). The C-terminal domain has RNA-binding activity (Sharma et al. 2007). Protein 3a is suggested to play a structural role in the SARS-CoV life cycle, since it interacts with S, E, and M proteins (Tan et al. 2004) and it is incorporated into newly packaged matured SARS-CoV virions (Shen et al. 2005; Ito et al. 2005). Also, it regulates various cellular responses of host cells, e.g., the upregulation of fibrinogen gene expression (Tan et al. 2005), and the increase of IL-8 and NF-B promoter activities (Kanzawa et al. 2006), possibly through its RNA-binding activity (Sharma et al. 2007). SARS-CoV 3a protein induces caspase-dependent apoptosis both in vivo and in vitro (Wong et al. 2005).

Intraviral Interactions

Two yeast-two-hybrid (Y2H) studies have been conducted to study intraviral SARS-CoV protein interactions (von Brunn et al. 2007; Pan et al. 2008), although only a few of these interactions have been verified. Von Brunn et al. (von Brunn et al. 2007) reported interactions of E protein with the nonstructural proteins nsp1, nsp8, and nsp11, as well as with the accessory proteins ORF3b, ORF7b, and ORF9b, whereas ORF3a interacted with M and S. However, not all these interactions were confirmed by coimmunoprecipitation in mammalian cells. Overall, however, only 13% of the intraviral SARS interactions known at that time were identified. This is likely due to the bias of Y2H against membrane proteins, which prevents the transfer of expressed prey and/or bait fusion proteins to the nucleus to activate transcription. Pan et al. (Pan et al. 2008) also reported a genome-wide analysis of intraviral PPIs in SARS-CoV replication, using a mammalian two-hybrid system screen, although only two interactions of E and 3a were found here. In comparison with a similar screen in yeast, native posttranslational modifications and folding should be present, but the two methods share a similar limitation in terms of bias against membrane proteins.

Later, a tandem affinity purification (TAP) study coupled to mass spectrometry (Álvarez et al. 2010), using dual-tagged SARS-CoV E protein in infected cells as bait, identified viral proteins nsp3, S, and M and host proteins dynein heavy chain, fatty acid synthase, aminopeptidase puromycin sensitive, phosphofructokinase platelet, tubulin alpha and beta, actin beta, transmembrane protein 43, and lactate dehydrogenase A as binders. Nsp3 is the largest nonstructural protein of SARS-CoV (1922 amino acids) which is proposed to act as a replication/transcription scaffolding protein (Imbert et al. 2008). Interaction with nsp3 was confirmed by coimmunoprecipitation and was localized to one of the nsp3 seven domains, i.e., the N-terminal acidic domain (nsp3a), that has a ubiquitin-like fold (Serrano et al. 2007). Colocalization of E and nsp3 in the cytoplasm of the infected cell suggested nsp3 may bring E protein into the vicinity of the replication/transcription complex.

The PLpro domain of nsp3 has deubiquitinating activity (Lindner et al. 2005) and might act to protect the viral replication complex from proteasomal degradation via deubiquitination. The authors proposed that E-nsp3 interaction could control ubiquitination of E protein during infection. Interaction between nsp3a and SARS-CoV E was shown to involve residues in the C-terminal domain of the latter (Li et al. 2014a).

Host Interactions

More recently, two Y2H studies searched for host interacting partners using the C-terminal tail of SARS-CoV E as a bait (Teoh et al. 2010; Jimenez-Guardeño et al. 2014). The first of these reported the protein associated with *Caenorhabditis elegans* lin-7 protein 1 (PALS1) as a binder (Teoh et al. 2010). PALS1 is a tight junction-associated protein and part of a complex that maintains epithelial cell polarity (Fig. 15.6). Alterations of lung epithelia integrity were consistent with E protein hijacking PALS1 to the ERGIC/Golgi region. The E-PALS1 interaction is mediated by (i) a Postsynaptic density protein-95/Discs Large/Zonula occludens-1 (PDZ) domain present in PALS1 and (ii) the last four C-terminal residues of E protein which represent a putative type II PDZ-binding motif (PBM) (Harris and Lim 2001).

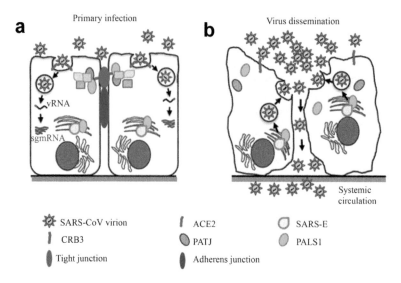

Fig. 15.6 Model of the potential consequences of SARS-CoV infection on polarity and intercellular junctions formed by alveolar epithelial cells. (**a**) The inner surface of human lung alveolae is lined with a monolayer of polarized epithelial cells. Components of CRB and PAR polarity complexes, including PALS1, are shown close to the apical domain. During infection, structural proteins accumulate in the ERGIC compartment, where SARS-CoV E could bind to PALS1 to disrupt its trafficking to the tight junction; (**b**) disruption of the tight junction and virus dissemination [adapted from Teoh et al. (2010)]

PDZ domains are ~80–90 amino acids long and typically bind the C-terminal tails of proteins, although internal binding sites have also been reported [reviewed in Ye and Zhang (2013)]. They are usually found in signaling proteins that alter signaling pathways, with over 250 nonredundant PDZ domains being recognized in the human proteome (Wang et al. 2010).

The C-terminal tail of SARS-CoV E protein, which includes the proposed PBM, forms a random coil secondary structure (Li et al. 2014a) in a variety of environments. However, PBMs usually fold as β-strands (Ye and Zhang 2013), which suggests that a β-structure conformation may be induced by target binding. Another similar Y2H study that used the same bait described a similar PDZ domain-containing binder, the syndecan-binding protein (syntenin) (Jimenez-Guardeño et al. 2014). Syntenin is a scaffolding protein that can initiate a signaling cascade resulting in the phosphorylation and activation of p38 mitogen-activated protein kinase (p38-MAPK), leading to expression of pro-inflammatory cytokines. The authors showed that the proposed C-terminal PBM in SARS-CoV E protein is a determinant of virulence. Since SARS-CoV-infected patients show an exacerbated inflammatory response that leads to epithelial and endothelial damage, edema, and acute respiratory distress syndrome (ARDS), the disruption of this pathway may have therapeutic implications. Overall, an involvement of this PBM in E protein in epithelial integrity and inflammatory responses is likely. Several other viruses, e.g., influenza A virus or human papillomavirus, have been described to enhance pathogenesis through viral proteins containing PBMs [reviewed in Javier and Rice (2011)], which probably constitute a common viral strategy.

CoV E proteins may also interact with, and modulate, host channels to support the virus life cycle. In *Xenopus* oocytes, it has been shown that coexpression of SARS-CoV E with human epithelial sodium transporter (ENaC) reduced amiloride-sensitive current through PKC activation followed by reduction of ENaC surface levels (Ji et al. 2009). A similar direct or indirect inhibitory effect on other endogenous channels was proposed from patch-clamp experiments using SARS-CoV E-transfected cells (Nieto-Torres et al. 2011). For IBV E, interaction with endogenous channels or SNAREs has been suggested to justify the Golgi complex rearrangement in response to IBV E expression (Ruch and Machamer 2011), although this observation may also involve the IBV E channel itself. For example, ion homeostasis at the Golgi could affect Na^+/H^+ exchangers that are critical for maintaining low luminal pH. Interactions of viroporins with Golgi channels or transporters are largely unexplored in the viroporins field, but notable cases have been already reported. For example, oncogenic protein E5 from papillomavirus (Wetherill et al. 2012) is able to bind the 16 K subunit of the lumen-acidifying V-ATPase (Goldstein et al. 1991), preventing assembly of the pump and leading to alkalinization of the Golgi lumen (Schapiro et al. 2000).

The Respiratory Syncytial Virus Small Hydrophobic Protein (RSV-SH)

Human respiratory syncytial virus (hRSV) belongs to the *Paramyxoviridae* family in the pneumovirus genus. This enveloped virus has a negative-sense single-strand RNA genome 15.2 kb long that encodes 10 sub-genomic mRNAs and 11 proteins (Fields et al. 2013). These 11 proteins include three membrane proteins accessible to the surface of the virion: the two that generate most RSV-neutralizing antibodies, fusion (F) and attachment (G), and the small hydrophobic (SH) protein.

RSV affects more than 30 million children below 5 years old and is the leading cause of bronchiolitis and pneumonia in infants and elderly (Dowell et al. 1996). Disease caused by RSV is responsible for 200,000 deaths worldwide which mostly occur in developing countries. hRSV exists as two antigenically distinct subgroups, A and B, both capable of inducing severe lower respiratory tract (LRT) disease in humans (Hall et al. 1990).

Although the virus was isolated more than half a century ago, no effective licensed treatment or vaccine is available for the general population, despite promising RSV vaccine candidates in clinical trials. Palivizumab is a humanized monoclonal antibody (IgG) directed against the F protein that is recommended for infants <2 years old with high risk. However, it is not effective therapeutically and is only moderately effective at preventing infection. Since it costs $4500 per treatment course (Weiner et al. 2011), its use is limited to a small fraction of patients worldwide. The only licensed drug for therapeutic use is a nucleoside analog which has limited efficacy.

SH Viroporin

The SH protein in hRSV is only 64 (subgroup A) or 65 (subgroup B) amino acids long, but its sequence is well conserved, especially the N-terminal extramembrane domain (Tapia et al. 2014). It has a single TM α-helical hydrophobic region, with C- (lumenal or extracellular) and N- terminal (cytoplasmic) extramembrane domains (Collins and Mottet 1993). The N-terminal cytoplasmic domain forms a short α-helix (residues 5–14) (Fig. 15.7a), almost coincident with a "10-residue" conserved sequence between hRSV and MuV SH protein sequences. SH proteins in MuV, PIV5, and JPV have extremely short lumenal domains (nine, two, and ten residues, respectively) compared with their much longer N-terminal cytoplasmic domains, which are likely involved in PPIs. The C-terminal extramembrane domain forms an extended β-hairpin. In bicelles, the α-helix of the TMD extends up to residue His-51 (Li et al. 2014b), resulting in both protonatable residues of SH protein, His-22, and His-51, oriented toward the lumen of the channel.

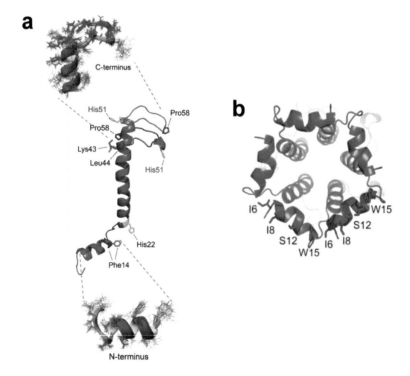

Fig. 15.7 Structural model of SH protein monomer. (**a**) Comparison of models of monomeric SH protein obtained in micelles (*red*) and in bicelles (*blue*), with residues prolonging the TM domain up to His-51 (Li et al. 2014b); (**b**) residues in SH involved in interaction with BAP31; N-terminal cytoplasmic helix of SH protein, with residues perturbed (*red*) after addition of BAP31 cytoplasmic domain to labeled SH protein in detergent micelles (Li et al. 2015)

SH protein forms homo-oligomers (pentamers), and this oligomeric form is responsible for ion channel (IC) activity (Gan et al. 2012; Gan et al. 2008) that has poor ion selectivity. In infected cells, most SH protein accumulates at the membranes of the Golgi complex, but it is also found in the ER or plasma membrane (Rixon et al. 2004). SH has potential glycosylation sites in both the C- and N-terminal domains (Collins et al. 1990). In infected cells, the SH protein of strain A2 accumulates in four different forms (Olmsted and Collins 1989; Collins et al. 1984; Collins and Mottet 1993), but the most abundant is a full-length unglycosylated form. The G protein forms G-F and G-SH complexes, but direct interactions between SH and F have not been observed (Low et al. 2008).

SH and apoptosis. It has been proposed that SH protein blocks apoptosis through inhibition of the TNF-α pathway (Fuentes et al. 2007), but the mechanism of this inhibition is not clear. A similar anti-apoptotic effect of SH protein has been reported for other members of the *Paramyxoviridae* family that encode SH proteins, e.g., mumps virus (MuV) and the parainfluenza virus 5 (PIV5).

Incidentally, an anti-apoptotic effect has also been noted for other similar viral channels (viroporins), e.g., E5 in the human papillomavirus type 16 (HPV-16) (Kabsch et al. 2004), or the envelope (E) protein, a viroporin in the severe acute respiratory syndrome (SARS) virus (DeDiego et al. 2011).

SH and the Inflammasome

SH protein is also involved in inflammasome regulation, but the mechanism involved is not known. Indeed, some authors have proposed that RSV SH has a role in regulation of the NLRP3 inflammasome (Russell et al. 2015). The latter is "primed" after the recognition of viral genomic RNA (vRNA) by pattern recognition receptors (PRRs) and subsequent activation of NF-kB. This priming involves the expression of inflammasome components, e.g., NLRP3 and inactive procaspase-1 (Elliott and Sutterwala 2015). Various virus-induced damage-associated molecular patterns (DAMPs) induce the assembly and activation of the NLRP3 inflammasome. This leads to processing of procaspase-1 into active caspase-1, which in turn cleaves inactive pro-IL-1β into the mature form IL-1β. The latter is a potent pro-inflammatory cytokine crucial in resolving infectious processes.

Various viruses can activate the inflammasome by disrupting ion homeostasis through the expression of viroporins. For example, influenza A virus (IAV) activates NLRP3 as a result of H^+ or ion flux from Golgi mediated by the M2 channel (Ichinohe et al. 2010). The 2B protein in picornaviruses induce NLRP3 cytoplasmic relocalization and inflammasome activation in an intracellular Ca^{2+}-mediated manner (Ito et al. 2012), while a similar mechanism has been proposed for SARS-CoV E (Nieto-Torres et al. 2015). The latter triggered inflammation in the lungs of mice, leading to epithelial cell damage and death (Nieto-Torres et al. 2014), and this was correlated to high levels of mature IL-1β in the airways of infected animals.

Similarly, RSV SH protein has been suggested to activate the NLRP3 inflamma-some through its IC activity and ion leakage from the Golgi (Triantafilou et al. 2013), similar to the mechanism proposed for SARS-CoV E (Nieto-Torres et al. 2015). Another study (Russell et al. 2015) proposed that IL-1β is overproduced when SH is absent from RSV. The study also showed attenuation in mice when infected with RSV ΔSH. That deletion of RSV SH leads to an increase in IL-1β is also supported by studies in bovine RSV (bRSV), where a ΔSH vaccine strain induced higher levels of IL-1β in the lungs of infected cattle (Taylor et al. 2014). Consistent with this, lung macrophages infected with RSV did not lead to increased IL-1β, although other pro-inflammatory cytokines were overexpressed (Ravi et al. 2013). Overall, it has been proposed (Russell et al. 2015) that SH protein blocks IL-1β production, preventing the clearance of infected cells. Indeed, blockade of IL-1β prior to infection increased the viral load, supporting the idea that SH might enable immunomodulation. The interaction between SH and G protein has also been shown previously to have an immunomodulatory role (Polack et al. 2005).

Although in cell culture RSV ΔSH is still viable, grows to similar titer to wild-type RSV, and still forms syncytia, SH-deleted RSV (RSV ΔSH) is significantly attenuated in a variety of hosts (Taylor et al. 2014; Bukreyev et al. 1997; Russell et al. 2015). Indeed, in the last few years, one of the leading RSV LAVs have included, among other modifications, a deletion in the SH gene (Karron et al. 2005). The cause of attenuation is not known, although it may have to do with effective transmission of the virus (Bukreyev et al. 1997).

A transcriptome analysis comparing RSV with and without SH protein could help to decipher the role of SH during infection and the cause of attenuation in vivo and to associate these responses to specific SH domains or features.

Host Interactions

Recently, a membrane-based yeast two-hybrid system (MbY2H) was used to iden-tify a cellular binding partner of hRSV SH protein, the B-cell receptor-associated protein 31 (BAP31) (Li et al. 2015), in a human lung cDNA library. BAP31 is a membrane protein located at the ER that has an essential role in sorting newly syn-thesized membrane proteins. Additionally, BAP31 has a cytoplasmic C-terminus with two coiled coils (Quistgaard et al. 2013), one of them containing a variant of the death effector domain (vDED) flanked by two caspase-8 cleavage sites. This domain is excised upon activation of caspase-8 to produce a fragment p20, known to function as a proapoptotic factor (Breckenridge et al. 2003). This interaction was confirmed using co-transfection, pull-down assay and immunofluorescence colocalization, and also using endogenous BAP31 and was localized to the first N-terminal 44 residues (Li et al. 2015). When ^{15}N–labeled SH protein was titrated with cytoplasmic C-terminal domain of human BAP31, major shifts were observed at residues I6, I8, S12, and W15 (Fig. 15.7b). It can be hypothesized that this inter-action could interfere with the interaction between BAP31 and caspase-8, blocking the cleavage sites and preventing conversion to the pro-apoptotic form of BAP31, i.e., p20, thus delaying apoptosis. Incidentally, the viroporin E5 from the high-risk human papillomavirus HPV-16 and HPV-31 was also found to interact with BAP31, where it is similarly thought to regulate apoptosis in addition to its roles in immunomodulation (Regan and Laimins 2008) (see below).

Intraviral Interactions

The interaction between RSV SH and G proteins has been reported previously in infected cells (Low et al. 2008; Rixon et al. 2005), although its significance is not yet clear. F protein seems to be the main determinant of host cell specificity during virus entry, and both F and G proteins are able to bind heparin sulfate, the putative cell receptor for RSV (Kargel et al. 2001). However, a tri-component complex between SH, G, and F proteins was not observed (Low et al. 2008). Both G and F proteins have one predicted TMD, and interaction with SH protein can be both through the TMDs or extramembrane domains of the latter.

Until now, all studies to determine the role of SH protein in RSV infection have used wild-type RSV and ΔSH RSV (SH gene deleted) and compared the effects of this deletion on various parameters in infected cells, or in animal models. The effects caused by transfection of SH protein in readouts that depend on, for example, inflammasome activation or apoptosis, have also been explored. However, comprehensive datasets that aim at elucidating the contribution of SH to virulence observed in vivo, and a rationale for the attenuation observed when SH is deleted, are lacking.

The Human Immunodeficiency Virus Viral Protein U (HIV-1-Vpu)

The human immunodeficiency virus type 1 (HIV-1) is an enveloped virus that causes AIDS. It has a single-stranded, positive-sense RNA genome of 9.8 kb which encodes for nine genes: the structural genes *gag*, *pol*, and *env*, the regulatory genes *tat* and *rev*, and additional genes *nef*, *vif*, *vpr*, and *vpu* which encode for accessory proteins. One of these accessory proteins, the Vpu (viral protein U) (Cohen et al. 1988), is an 81-residue small-membrane protein consisting of an N-terminal transmembrane helix and a C-terminal cytoplasmic domain which contains two helices linked by a flexible loop region [reviewed in Opella (2015)] (Fig. 15.8a). Vpu can oligomerize to form cation-selective channels in membranes, although the rationale for this channel activity is not well defined.

Vpu has two primary roles during HIV-1 infection: (i) enhancement of virion release (Terwilliger et al. 1989; Strebel et al. 1988) and (ii) degradation of host CD4 receptor (Willey et al. 1992). Absence of Vpu in infected cells correlates with

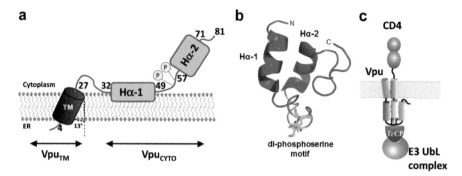

Fig. 15.8 (a) Predicted secondary structure of Vpu showing N-terminal TMD (*blue*) and two α-helices of the cytoplasmic (CYTO) domain (*red*). Phosphorylated S52 and S56 are represented as circles (adapted from Dubé et al. (2010)). (b) Solution NMR structure of Vpu$_{CYTO}$ in DPC micelles (Protein Data Bank code: 2K7Y). Helix 1 (Ile39-Glu48) and helix 2 (Leu64-Arg70) are shown as ribbons. β-TrCP-binding DSGxxS motif is in blue. Side chains of phosphorylated serines are shown as balls and sticks. Vpu$_{CYTO}$ residues showing substantial chemical shift changes upon addition of 1 mM CD4mut are in red (adapted from Singh et al. (2012)); (c) Vpu-mediated degradation of CD4. Binding of Vpu to CD4 is mediated by the cytoplasmic helices and TM helices. Vpu has two conserved phosphoserines which constitute a binding motif for β-TrCP, leading to assembly of an E3 ubiquitin ligase (UbL) complex that results in CD4 ubiquitination and proteasomal degradation (modified from Strebel (2014))

reduction of viral release and intracellular accumulation of HIV-1 viral proteins (Klimkait et al. 1990). The host protein BST-2 is a restriction factor of HIV-1 release, and its activity can be neutralized by Vpu (Neil et al. 2008; Van Damme et al. 2008).

Host Interactions A number of host factors have been reported to bind Vpu, including the immunoreceptors major histocompatibility complex class I (MHC-I) (Kerkau et al. 1997), CD1d (Moll et al. 2010), NK-T-B-antigen (NTB-A) (Shah et al. 2010), poliovirus receptor (PVR) (Matusali et al. 2012) and human leukocyte antigen C (HLA-C) (Apps et al. 2016), C-C chemokine receptor type 7 (CCR7) (Ramirez et al. 2014) and CD62L (Vassena et al. 2015), tetraspanins (Haller et al. 2014; Lambelé et al. 2015), K$^+$ channel TASK-1 (Hsu et al. 2004), metabolic transporter, sodium-coupled neutral amino acid transporter 1 (SNAT1) (Matheson et al. 2015), and the most recently reported intercellular adhesion molecule 1 (ICAM-I) (Sugden et al. 2017). However, the most important binders are CD4 and BST-2.

Vpu-Mediated Degradation of CD4

The host CD4 is a cell surface receptor critical for HIV-1 entry into target cells by endocytosis, but its expression at the cell surface prevents the release of infectious virions from infected cells. To counteract this host defense mechanism, viral Vpu,

Env, and Nef downmodulate CD4 surface expression [see review in Doms and Trono (2000)]. Vpu interacts with newly synthesized CD4 at the ER to mediate CD4 degradation (Willey et al. 1992). Vpu-mediated degradation of CD4 requires physical interaction between the two proteins. In CD4, the interaction domain has been mapped to a specific motif ($L_{414}SEKKT_{419}$) and a membrane-proximal α-helix (Bour et al. 1995). For Vpu, residues in both cytoplasmic α-helices of Vpu experienced chemical shift perturbations upon CD4 binding (Fig. 15.8b) (Singh et al. 2012), although involvement of the TMDs of both proteins has also been suggested (Magadán and Bonifacino 2012; Do et al. 2013). However, the scrambling or replacement of the whole Vpu TMD appears to have no effect on Vpu-mediated degradation of CD4 (Willey et al. 1994; Schubert et al. 1996), whereas a single amino acid substitution at the TMD, W22 L, abolished CD4 degradation but did not disrupt the CD4-Vpu interaction (Magadán and Bonifacino 2012), suggesting that TM interactions between CD4 and Vpu may function beyond the expected role of stabilizing the protein complex for CD4 degradation.

Overall, it has been proposed that Vpu acts as an adapter protein to link CD4 to the host ubiquitin-proteasome machinery for degradation. This binding triggers the recruitment of the host beta-transducin repeat-containing protein (β-TrCP). A conserved di-phosphoserine motif ($D_{51}SGxxS_{56}$) located within the loop region that connects the two cytoplasmic α-helices of Vpu is necessary for this process. Interaction of the phosphoserines in Vpu with the WD boxes of β-TrCP enables the formation of a CD4-Vpu-β-TrCP ternary complex (Margottin et al. 1998), bringing CD4 and other components of the SCF$^{β\text{-}TrCP}$ E3 ubiquitin ligase complex (Skp1 and Cullin1) in close proximity to facilitate the *trans*-ubiquitination of the CD4 cytosolic tail (Binette et al. 2007) on lysine and serine/threonine residues (Magadán et al. 2010) and subsequently its transportation to the cytosol for degradation (Fig. 15.8c).

Vpu-Mediated Antagonism of BST-2

Vpu enhances virus dissemination by antagonizing the host BST-2, a host restriction factor with antiviral capabilities [reviewed in Simon et al. (2015)]. BST-2 is an interferon-inducible type II integral membrane protein located at the budding site of HIV-1. BST-2 has an N-terminal cytoplasmic domain, a TMD, followed by a coiled-coil ectodomain and finally a C-terminal glycosyl-phosphatidylinositol (GPI) membrane anchor. Its unusual topology enables it to tether virions by inserting, preferentially its GPI anchor, into the envelope of assembling virion particles, while itself remains embedded in its host cell membrane (Venkatesh and Bieniasz 2013; Neil et al. 2006).

It has been proposed that Vpu engages BST-2 through interaction between their respective TMDs, with involvement of Vpu's A14, W22, and A18 (Vigan and Neil 2010) and BST-2's I34, L37, and L41 (Kobayashi et al. 2011). An NMR study described an antiparallel interaction between Vpu and BST-2 TMDs in DHPC

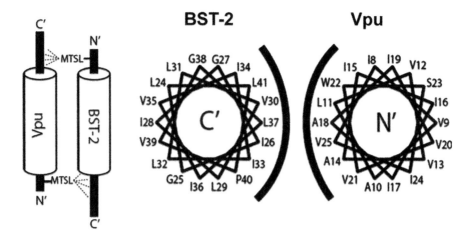

Fig. 15.9 *Left*, the spin label (MTSL) at the N-terminus of Vpu selectively decreased the intensity of C-terminal residues of BST-2, whereas the spin label at the N-terminus of BST-2 selectively decreased the intensity of C-terminal residues of Vpu; *right*, helical wheel diagrams of the BST-2 and Vpu TMDs. The TMDs of BST-2 and Vpu are depicted in their anti parallel orientation (adapted from Skasko et al. (2012))

micelles, in an orientation where A18 of Vpu faces L37 of BST-2 (Skasko et al. 2012) (Fig. 15.9). The conserved residues within the Vpu membrane-proximal cytoplasmic hinge region ($E_{28}YRKIL_{33}$) have also been found to be important for Vpu-mediated BST-2 antagonism (Lukhele and Cohen 2017).

The mechanism of BST-2 neutralization by Vpu appears to be multifaceted and under debate, although the key mechanism appears to be the direct displacement of BST-2 from the virus assembly sites at the plasma membrane (McNatt et al. 2013). Accordingly, a C-terminal Trp residue at the Vpu cytoplasmic tail has been reported to contribute to this displacement of cell surface BST-2 (Lewinski et al. 2015). Other proposed mechanisms have been proposed, e.g., Vpu can disrupt intracellular BST-2 trafficking by sequestering both newly synthesized and recycling BST-2 within intracellular compartments such as the TGN (Dubé et al. 2010; Hauser et al. 2010). This effectively blocks the resupply of BST-2 to the plasma membrane and thereby reduce BST-2 surface density (Dubé et al. 2011). Vpu-mediated BST-2 mis-trafficking has been reported to involve the host clathrin-dependent membrane trafficking pathways which are mediated by clathrin adaptor protein (AP) complexes (Kueck and Neil 2012; Lau et al. 2011). It was proposed that Vpu is able to hijack the clathrin-dependent trafficking machinery via a mimicked canonical acidic di-leucine sorting motif ($E_{59}xxxLV_{64}$) within its second cytoplasmic α-helix to recruit the AP complexes, forming a Vpu-BST-2-AP ternary complex (McNatt et al. 2013; Jia et al. 2014; Kueck et al. 2015). In addition, the conserved di-phosphoserine motif ($D_{51}SGxxS_{56}$) in Vpu may also be required for this recruitment (Kueck et al. 2015).

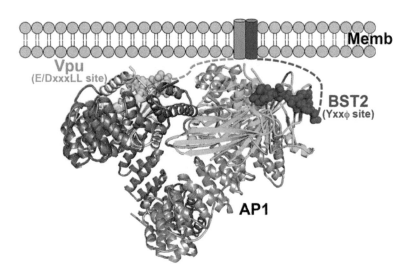

Fig. 15.10 Schematics of Vpu hijacking of AP1 to target BST2. AP1 is colored by subunit ($\beta1$ in gray, γ in *orange*, $\mu1$ in *green*, and $\sigma1$ in *yellow*). Vpu$_{\text{CYTO}}$ (*cyan*) binds to the acidic di-leucine-binding pocket of $\gamma/\sigma1$, and BST-2$_{\text{CYTO}}$ (*magenta*) binds to the tyrosine-binding pocket in $\mu1$. Transmembrane helices are represented by cylinders (adapted from Jia et al. (2014))

A crystal structure of a Vpu$_{\text{CYTO}}$-BST-2$_{\text{CYTO}}$ fusion protein in complex with the AP1 core has been solved (Jia et al. 2014). In this model, the cytoplasmic domains of Vpu and BST-2 do not interact directly. Instead, Vpu seems to act as a chaperone to enhance binding of AP1 to BST-2. Stability of this complex is achieved by pair-wise binary interactions between Vpu and BST-2 TMDs and between Vpu and BST-2 cytoplasmic domains to several parts of AP1 (Fig. 15.10). Thereafter, the Vpu-BST-2 complex is thought to proceed through the clathrin-mediated trafficking pathway for β-TrCP-dependent ubiquitination of BST-2 before subsequent ESCRT-mediated endo-lysosomal degradation.

Recently, it has been reported that Vpu may hijack the function of the host actin cross-linking regulator filamin A (FLNa) during its quest of BST-2 modulation (Dotson et al. 2016). Vpu may also exploit an LC3-associated noncanonical autophagy pathway to restrict BST-2 (Madjo et al. 2016). The refinement of current models, together with the discovery of additional host and/or intraviral factors involved, will ultimately form a complete and accurate picture of Vpu-mediated BST-2 antagonism.

The Polyomavirus JC Agnoprotein

Polyomaviruses are small, non-enveloped DNA viruses with their closed circular genome packaged within an icosahedral viral capsid. Progressive multifocal leuko-encephalopathy (PML), a deadly demyelinating disease of the brain, is attributed to

Fig. 15.11 NMR structure of agnoprotein contains two main α-helical structures (*red*) and two unstructured regions (*yellow*) [adapted from Coric et al. (2017)]

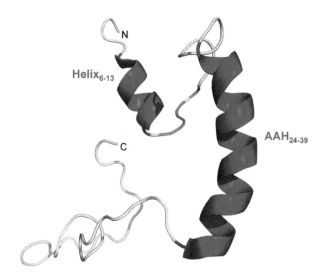

the human polyomavirus JCV (John Cunningham virus). JCV encodes for a small and highly basic protein called agnoprotein which has important regulatory roles in the JCV replication cycle. Besides JCV, the agnoprotein can also be found in other polyomaviruses including BK virus (BKV) and simian virus 40 (SV40) (Sariyer et al. 2011). The 71-residue agnoprotein has a central hydrophobic region which has been reported to form an amphipathic α-helix (Lys23-Phe39) (Coric et al. 2014). This helix is characteristically rich in Leu/Ile/Phe, which is required for protein stability and oligomerization (Saribas et al. 2013). A recently solved NMR structure in organic solvent has revealed a second α-helix, albeit minor, spanning residues Leu6-Lys13 (Fig. 15.11) (Coric et al. 2017).

Agnoprotein demonstrates several key features which are commonly shared among viroporins (Suzuki et al. 2010a; Suzuki et al. 2013; Suzuki et al. 2010b), e.g., it is a membrane protein that associates into homo-oligomers that increased membrane permeability leading to influx of extracellular Ca^{2+} and enhancement of virus release. In addition, agnoprotein-deleted mutants have defective virion release and viral propagation.

Intraviral and Host Interactions The JCV agnoprotein has been shown to interact with a number of viral proteins: large T-antigen (LT-ag) (Safak et al. 2001), small t-antigen (St-ag) (Sariyer et al. 2008), HIV-1 Tat (Kaniowska et al. 2006), and capsid protein VP1 (Suzuki et al. 2012). It also interacts with cellular proteins, including the Y-box-binding factor 1 (YB-1) (Safak et al. 2002), tumor suppressor p53 (Darbinyan et al. 2002), tubulin (Endo et al. 2003), DNA damage repair protein Ku70 (Darbinyan et al. 2004), fasciculation and elongation protein zeta 1 (FEZ1) (Suzuki et al. 2005), heterochromatin protein 1 alpha (HP-1α) (Okada et al. 2005), protein phosphatase 2A (PP2A) (Sariyer et al. 2008), and the adaptor protein complex 3 (AP3) δ subunit (Suzuki et al. 2013).

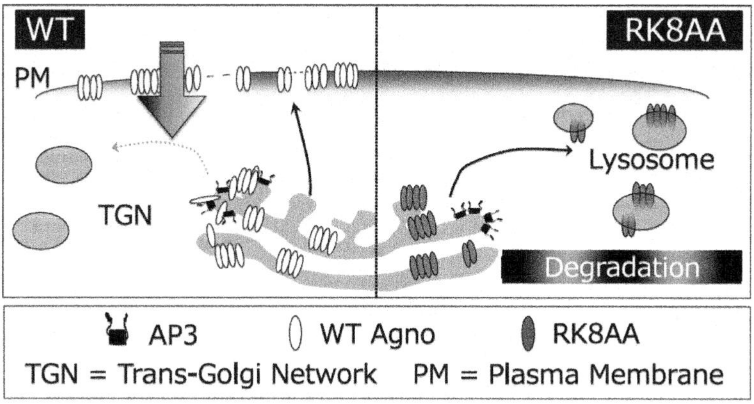

Fig. 15.12 Involvement of AP-3 in membrane permeabilization and virion release involving WT (*left*) and mutant RK8AA (*right*) agnoprotein. Both WT and RK8AA form homo-oligomers as integral membrane proteins in cytoplasmic organelles. WT disrupts AP3-mediated vesicular trafficking, is translocated to plasma membrane, and functions as a viroporin, resulting in the promotion of virion release. In contrast, the RK8AA mutant fails to bind to AP3D and does not disrupt AP3-mediated vesicular trafficking and is transported to lysosomes and degraded. RK8AA agnoprotein cannot promote virion release and is defective in viral propagation [adapted from Suzuki et al. (2013)]

One of the interesting host factors targeted by the JCV agnoprotein is the AP3 (Suzuki et al. 2013). Interaction of agnoprotein with the δ subunit of AP3 (AP3D) appears to hijack the AP3-mediated intracellular vesicular trafficking to prevent the targeted lysosomal degradation of agnoprotein. This phenomenon is reminiscent of the special features of viroporins such as M2 and Vpu, which also manipulate host trafficking pathways. Agnoprotein is then allowed to be translocated to the plasma membrane to act as a viroporin and promote virion release (Suzuki et al. 2013). Alanine substitutions of two basic residues (Arg8 and Lys9) in the N-terminus of agnoprotein disrupt its viroporin activity (Suzuki et al. 2010a) and also disrupt binding to AP3D, ensuing its transport to the lysosomes and subsequent lysosomal degradation (Fig. 15.12) (Suzuki et al. 2013). These basic residues are part of an ordered helical structure (Coric et al. 2017) and may constitute an important regulatory and/ or interaction motif.

Rotavirus NSP4

Rotaviruses are members of the *Reoviridae* family of non-enveloped viruses which consist of segmented, double-stranded RNA genomes surrounded by multiple concentric protein capsids (Coombs 2006). Rotaviruses are a leading cause of severe viral gastroenteritis and dehydrating diarrhea in infants and young children, resulting in a high global child mortality rate of 215,000 in 2013 (in children <5 years old)

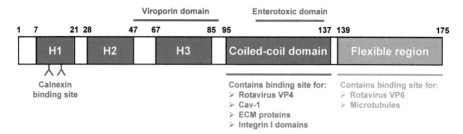

Fig. 15.13 Schematic representation of rotavirus NSP4 structural domains. H1, H2, and H3, hydrophobic domains 1, 2, and 3, respectively. NSP4 viroporin domain (residues 47–90) and enterotoxic domain (residues 114–135) are indicated. Putative binding sites of NSP4 interaction partners are also indicated

(Tate et al. 2016). The rotavirus nonstructural protein 4 (NSP4) is a 175-amino-acid transmembrane ER glycoprotein and the first virus-encoded enterotoxin to be discovered (Ball et al. 1996). Besides its primary ER localization, NSP4 can also be secreted as a soluble enterotoxin (Bugarčić and Taylor 2006) or colocalize with the autophagy protein LC3 in cap-like structures that associate with viroplasms (Berkova et al. 2006).

NSP4 consists of three hydrophobic domains (H1, H2, and H3) followed by a long cytoplasmic region containing a coiled-coil domain (CCD) and a flexible tail region (Estes and Greenberg 2013) (Fig. 15.13). Notably, its distinctive functional domains include (i) an enterotoxic domain (ETD, residues 114–135), which can function as a diarrhea-inducing enterotoxin similar to the full-length protein in young mice (Ball et al. 1996), and (ii) a recently discovered viroporin domain (VPD, residues 47–90) which is made of a penta-lysine domain and the amphipathic helix H3 (Hyser et al. 2010) and exhibits cation-selective channel activity in artificial lipid bilayers (Pham et al. 2017).

An alteration in cellular calcium homeostasis is critical for rotavirus replication and cytopathogenesis, and this has been correlated with the NSP4 viroporin [see review in Hyser and Estes (2015)]. Earlier studies reported that NSP4 colocalize with the autophagosome marker LC3 in "cap-like structures" associated with viroplasms (Berkova et al. 2006), sparking interest of whether the host autophagy machinery is manipulated during rotavirus infection. Indeed, NSP4 appears to orchestrate a series of events which ultimately lead to host autophagy. NSP4 viroporin activity at the ER can activate the ER calcium sensor stromal interaction molecule 1 (STIM1), which triggers an activation of store-operated calcium entry (SOCE), which in turn facilitates Ca^{2+} influx through the plasma membrane (Hyser et al. 2013). This increase in intracellular Ca^{2+} activates the Ca^{2+}/calmodulin-dependent protein kinase kinase-β (CaMKK-β) to initiate autophagy (Anderson et al. 1999; Crawford et al. 2012). The autophagy membrane trafficking pathway is then hijacked by the virus to transport viral proteins from the ER to viroplasms for assembly of infectious virus (Crawford et al. 2012).

Intraviral and Host Interactions

Even before the discovery of its viroporin activity, NSP4 has been described to perform multiple functions through interacting with a number of viral and host factors, and the cytoplasmic region of NSP4 is an interaction hotspot. For instance, the NSP4 cytoplasmic CCD has been reported to be the binding site for the rotavirus spike protein VP4 (NSP4 aa112–148) (Au et al. 1993), host extracellular matrix (ECM) proteins laminin-β3 and fibronectin (NSP4 aa87–145) (Boshuizen et al. 2004), caveolin-1 (Cav-1) (NSP4 aa114–135) (Parr et al. 2006; Ball et al. 2013), and the integrin I domains (NSP4 aa114–130) (Seo et al. 2008). In addition, the flexible region of the NSP4 cytoplasmic tail interacts with VP6 to serve as an intracellular receptor for the viral double-layered particles (DLPs) (NSP4 aa161–175) (Au et al. 1989; Taylor et al. 1996) and can also bind microtubules (NSP4 aa120–175) (Xu et al. 2000). NSP4 can also bind the host calnexin via the two N-linked high-mannose oligosaccharide residues within the NSP4 H1 domain (Mirazimi et al. 1998). The putative binding sites for these interactions are summarized (Fig. 15.13).

While mechanistic and structural information on membrane insertion and oligomerization of the full-length NSP4 is still lacking, a topology model of NSP4 as a three-pass transmembrane protein has been proposed (Fig. 15.14). In addition, crystal structures have revealed that the NSP4 CCD from two different rotavirus strains can form a tetramer and a pentamer, respectively (Bowman et al. 2000; Chacko et al. 2011) and that the tetrameric NSP4 CCD, but not the pentameric form, can bind Ca^{2+} at its core. Later studies clarified that the oligomeric status of NSP4 CCD can be regulated by pH and Ca^{2+}; at neutral pH it forms a tetramer that binds Ca^{2+},

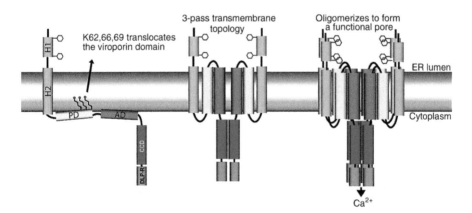

Fig. 15.14 Model of NSP4 as a three-pass transmembrane protein. Left: initial ER membrane insertion of NSP4 mediated by uncleaved signal sequence in H2 domain. Lysine residues interact with membrane phospholipid, promoting insertion of the viroporin domain (PD + AD) as an anti-parallel α-helical hairpin. Center: insertion of the viroporin domain generates a three-pass transmembrane topology. Right: NSP4 oligomerization around the amphipathic α-helix generates an aqueous pore for the passage of ER Ca^{2+} [adapted from Hyser et al. (2010)]

but at low pH it forms a pentamer that does not bind Ca^{2+} (Sastri et al. 2014). While the NSP4 CCD appears to be an interaction hotspot, it remains to be revealed how environmental cues may influence its binding conformations and affinity with inter-action partners. For instance, the CCD may act as a cytoplasmic pH/Ca^{2+} sensor that alters the NSP4 oligomeric state and conformation to regulate its binding to a cer-tain interaction partner, or Ca^{2+} may even act as a cofactor for binding.

Human Papillomavirus E5

Human papillomaviruses (HPVs) are double-stranded DNA viruses which are small, non-enveloped, and epitheliotropic. They are known to infect mucosal and cutaneous epithelia of the anogenital tract and the hands/feet regions. High-risk HPVs, mainly HPV-16 and HPV-18, are responsible for 70% of cervical cancers and precancerous cervical lesions. The HPV-16 E5 (16E5), a small hydrophobic oncoprotein, is a recent addition to the viroporin family as it exhibits ion channel activity in vitro (Wetherill et al. 2012). 16E5 is an 83-residue protein with three putative TMDs and an N-terminal luminal, C-terminal cytoplasmic topology (Krawczyk et al. 2010). 16E5 monomers oligomerize as homodimers or hexamers (Kell et al. 1994; Gieswein et al. 2003; Wetherill et al. 2012). 16E5 has roles in cel-lular transformation, mitogenic signaling, immune evasion, intracellular protein trafficking, and apoptosis [reviewed in Müller et al. (2015)].

Host Interactions

It has been reported that 16E5 forms a stable complex with the epidermal growth factor receptor (EGFR) in co-transfected cells (Hwang et al. 1995) and with the 16 K subunit of the vacuolar H^+-ATPase (V-ATPase) (Conrad et al. 1993), although the 16E5 binding site with the latter is under debate (Adam et al. 2000; Rodríguez et al. 2000). The C-terminal domain of 16E5 has been reported to bind the nuclear transport receptor karyopherin β3 (KNβ3) (Krawczyk et al. 2008) and the $Ca^{2+}/$ phospholipid-/actin-binding protein calpactin I (Krawczyk et al. 2011). Other reported interaction targets of 16E5 include the gap junction protein connexin 43 (Oelze et al. 1995; Tomakidi et al. 2000), growth factor receptor ErbB4 (Chen et al. 2007), zinc transporter ZnT-1 (Lazarczyk et al. 2008), transmembrane channel-like proteins EVER1 and EVER2 (Lazarczyk et al. 2008), the putative ER ion channel A4 (Kotnik Halavaty et al. 2014), and the Golgi-resident transmembrane protein YIPF4 (Müller et al. 2015).

Interactions between 16E5 and host proteins have been implicated in the modu-lation of host defense. For instance, 16E5 can help HPV escape from immunesur-veillance by downregulating expression of antigen-presenters at the host cell surface. 16E5 binds and retains MHC-I in the ER and Golgi to prevent its trafficking

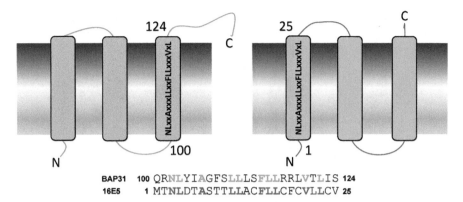

Fig. 15.15 E5 and BAP31 topology. Top, topology of E5 and Bap31, with TMDs indicated; TM3 in BAP31 and TM1 in 16E5 share a similar motif [adapted from Cortese et al. (2010)]. The alignment of 16E5 TM1 and Bap31 TM3 showing the ten-residue identity which may be involved in interactions among 16E5, MHC-I and/or other associated proteins. Identical residues are in bold

to the cell surface, in an interaction involving two leucine pairs in the first TMD (TM1) of 16E5 and the heavy chain of MHC-I (Ashrafi et al. 2005; Cortese et al. 2010; Ashrafi et al. 2006). The TM1 of 16E5 may also bind and cripple the function of the MHC-I chaperone, Bap31 (Ladasky et al. 2006; Regan and Laimins 2008; Cortese et al. 2010). Intriguingly, a motif consisting of ten identical residues between the 16E5 TM1 and Bap31 TM3 has been discovered (Fig. 15.15) and could represent a case of molecular piracy used by 16E5 to displace Bap31 from MHC-I. In addition, 16E5-mediated ER retention of MHC-I may also involve an interaction with the ER chaperone calnexin, since surface downregulation of MHC-I is not observed in calnexin-deficient cells (Gruener et al. 2007). In the same study, a triprotein complex of 16E5, MHC-I, and calnexin could be obtained based on a coimmunoprecipitation assay. The 16E5-calnexin interaction also reduced CD1d surface levels by retaining it in the ER and subsequently redirecting it for proteasomal degradation (Miura et al. 2010).

BPV E5: A Case Study

E5 also plays an important role in cell tumorigenic transformation. One associated cellular target of the bovine papillomavirus BPV-1 E5 is the platelet-derived growth factor β receptor (PDGFβR), a transmembrane receptor tyrosine kinase (Fig. 15.16, left). Under normal circumstances, the ligand PDGF binds to the extracellular domain of its receptor to induce receptor dimerization, leading to the autophosphorylation of tyrosine residues at the receptor intracellular domain, activation of their intrinsic tyrosine kinase activity, and subsequent signal transduction resulting in mitogenesis (Fig. 15.16, middle). While the BPV-1 E5 is not a natural ligand of

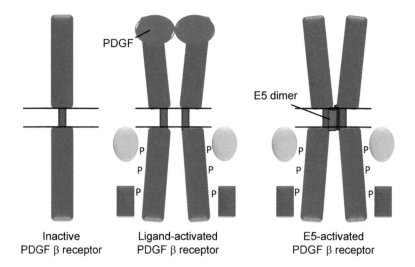

Fig. 15.16 Model for E5-PDGFβR interaction. *Left*, monomer of inactive PDGFβR; middle, PDGFβR activation by PDGF binding in the extracellular domain, leading to receptor dimerization, tyrosine phosphorylation in the intracellular domain, and recruitment of cellular signaling substrates (*green* and *purple*); right, PDGFβR activation by binding of a BPV-1 E5 dimer to the TMD of the receptor. Horizontal lines represent cell membrane, with the cytoplasm beneath [adapted from DiMaio and Petti (2013)]

PDGFβR, it can constitutively activate its receptor tyrosine kinase activity (Fig. 15.16, right) (Drummond-Barbosa et al. 1995) and is therefore an oncoprotein that can lead to host cell transformation.

Coimmunoprecipitation experiments have shown that the E5-PDGFβR interaction is stable and is important in E5-induced cell transformation (Petti and DiMaio 1992; Goldstein et al. 1992). Deletion of the extracellular ligand-binding domain of PDGFβR did not prevent its cooperation with E5, demonstrating that E5 activation of PDGFβR is ligand independent (Drummond-Barbosa et al. 1995). Also, mutagenesis and chimeric studies of PDGFβR have mapped the binding region to the TMD of the receptor (Cohen et al. 1993; Nappi et al. 2002). The interaction between E5 and PDGFβR is also highly specific, since cooperation was not observed between E5 and PDGFαR, a closely related receptor tyrosine kinase (Goldstein et al. 1994). Molecular dynamics experiments have proposed a model of the E5 dimer, where the Gln17 of monomer 1 and the Asp33 of monomer 2 are on the same face of the dimer that interacts with a molecule of PDGFβR (Surti et al. 1998). NMR studies of E5 peptides in detergent micelles also favor E5 dimerization for complex formation with PDGFβR through its TMD (King et al. 2011).

Concluding Remarks

The interplay between the host cell responses and viral offensive mechanisms yields the final outcome of an infection. For efficient viral replication, the virus must be resourceful in harnessing or disabling the host cellular machinery. In recent years, the role of viroporins as essential players in viral pathogenesis has been established. However, in addition to disrupting cellular ion homeostasis by their channel activity, viroporins also interact with host factors and coordinate with other viral proteins in structural roles. The structural features of these complexes remain poorly understood. Advances in this field will provide useful insights in the design of new antivirals.

References

Adam JL, Briggs MW, McCance DJ (2000) A mutagenic analysis of the E5 protein of human papillomavirus type 16 reveals that E5 binding to the vacuolar H+-ATPase is not sufficient for biological activity, using mammalian and yeast expression systems. Virology 272(2):315–325. https://doi.org/10.1006/viro.2000.0376

Aligeti M, Roder A, Horner SM (2015) Cooperation between the hepatitis C virus p7 and NS5B proteins enhances Virion infectivity. J Virol 89(22):11523–11533. https://doi.org/10.1128/JVI.01185-15

Álvarez E, DeDiego ML, Nieto-Torres JL, Jiménez-Guardeño JM, Marcos-Villar L, Enjuanes L (2010) The envelope protein of severe acute respiratory syndrome coronavirus interacts with the non-structural protein 3 and is ubiquitinated. Virology 402(2):281–291. https://doi.org/10.1016/j.virol.2010.03.015

Anderson KA, Means RL, Huang QH, Kemp BE, Goldstein EG, Selbert MA, Edelman AM, Fremeau RT, Means AR (1999) Components of a calmodulin-dependent protein kinase cascade: molecular cloning, functional characterization and cellular localization of Ca2+/calmodulin-dependent protein kinase kinase β. J Biol Chem 273(48):31880–31889. https://doi.org/10.1074/jbc.273.48.31880

Andresson T, Sparkowski J, Goldstein DJ, Schlegel R (1995) Vacuolar H+-ATPase mutants transform cells and define a binding site for the papillomavirus E5 oncoprotein. J Biol Chem 270(12):6830–6837. https://doi.org/10.1074/jbc.270.12.6830

Anthony SJ, Epstein JH, Murray KA, Navarrete-Macias I, Zambrana-Torrelio CM, Solovyov A, Ojeda-Flores R, Arrigo NC, Islam A, Khan SA, Hosseini P, Bogich TL, Olival KJ, Sanchez-Leon MD, Karesh WB, Goldstein T, Luby SP, Morse SS, Mazet JAK, Daszak P, Lipkin WI (2013) A strategy to estimate unknown viral diversity in mammals. MBio 4(5). ARTN e00598-13. https://doi.org/10.1128/mBio.00598-13

Appel N, Zayas M, Miller S, Krijnse-Locker J, Schaller T, Friebe P, Kallis S, Engel U, Bartenschlager R (2008) Essential role of domain III of nonstructural protein 5A for hepatitis C virus infectious particle assembly. PLoS Path 4(3). https://doi.org/10.1371/journal.ppat.1000035

Apps R, Del Prete GQ, Chatterjee P, Lara A, Brumme ZL, Brockman MA, Neil S, Pickering S, Schneider DK, Piechocka-Trocha A, Walker BD, Thomas R, Shaw GM, Hahn BH, Keele BF, Lifson JD, Carrington M (2016) HIV-1 Vpu mediates HLA-C Downregulation. Cell Host and Microbe 19(5):686–695. https://doi.org/10.1016/j.chom.2016.04.005

Ashrafi GH, Haghshenas M, Marchetti B, Campo MS (2006) E5 protein of human papillomavirus 16 downregulates HLA class I and interacts with the heavy chain via its first hydrophobic domain. Int J Cancer 119(9):2105–2112. https://doi.org/10.1002/ijc.22089

Ashrafi GH, Haghshenas MR, Marchetti B, O'Brien PM, Campo MS (2005) E5 protein of human papillomavirus type 16 selectively downregulates surface HLA class I. Int J Cancer 113(2):276–283. https://doi.org/10.1002/ijc.20558

Au KS, Chan WK, Burns JW, Estes MK (1989) Receptor activity of rotavirus nonstructural glycoprotein NS28. J Virol 63(11):4553–4562

Au KS, Mattion NM, Estes MK (1993) A subviral particle binding domain on the rotavirus nonstructural glycoprotein NS28. Virology 194(2):665–673. https://doi.org/10.1006/viro.1993.1306

Ball JM, Schroeder ME, Williams CV, Schroeder F, Parr RD (2013) Mutational analysis of the rotavirus NSP4 enterotoxic domain that binds to caveolin-1. Virol J 10. https://doi.org/10.1186/1743-422X-10-336

Ball JM, Tian P, Zeng CQY, Morris AP, Estes MK (1996) Age-dependent diarrhea induced by a rotaviral nonstructural glycoprotein. Science 272(5258):101–104

Barrios-Rodiles M, Brown KR, Ozdamar B, Bose R, Liu Z, Donovan RS, Shinjo F, Liu Y, Dembowy J, Taylor IW, Luga V, Przulj N, Robinson M, Suzuki H, Hayashizaki Y, Jurisica I, Wrana JL (2005) High-throughput mapping of a dynamic signaling network in mammalian cells. Science 307(5715):1621–1625. https://doi.org/10.1126/science.1105776

Beale R, Wise H, Stuart A, Ravenhill BJ, Digard P, Randow F (2014) A LC3-interacting motif in the influenza a virus M2 protein is required to subvert autophagy and maintain virion stability. Cell Host and Microbe 15(2):239–247. https://doi.org/10.1016/j.chom.2014.01.006

Berkova Z, Crawford SE, Trugnan G, Yoshimori T, Morris AP, Estes MK (2006) Rotavirus NSP4 induces a novel vesicular compartment regulated by calcium and associated with viroplasms. J Virol 80(12):6061–6071. https://doi.org/10.1128/JVI.02167-05

Binette J, Dubé M, Mercier J, Halawani D, Latterich M, Cohen EA (2007) Requirements for the selective degradation of CD4 receptor molecules by the human immunodeficiency virus type 1 Vpu protein in the endoplasmic reticulum. Retrovirology 4. https://doi.org/10.1186/1742-4690-4-75

Boshuizen JA, Rossen JWA, Sitaram CK, Kimenai FFP, Simons-Oosterhuis Y, Laffeber C, Büller HA, Einerhand AWC (2004) Rotavirus enterotoxin NSP4 binds to the extracellular matrix proteins laminin-β3 and fibronectin. J Virol 78(18):10045–10053. https://doi.org/10.1128/JVI.78.18.10045-10053.2004

Bour S, Schubert U, Strebel K (1995) The human immunodeficiency virus type 1 Vpu protein specifically binds to the cytoplasmic domain of CD4: implications for the mechanism of degradation. J Virol 69(3):1510–1520

Bourmakina SV, García-Sastre A (2003) Reverse genetics studies on the filamentous morphology of influenza a virus. J Gen Virol 84(3):517–527. https://doi.org/10.1099/vir.0.18803-0

Bowman GD, Nodelman IM, Levy O, Lin SL, Tian P, Zamb TJ, Udem SA, Venkataraghavan B, Schutt CE (2000) Crystal structure of the oligomerization domain of NSP4 from rotavirus reveals a core metal-binding site. J Mol Biol 304(5):861–871. https://doi.org/10.1006/jmbi.2000.4250

Breckenridge DG, Stojanovic M, Marcellus RC, Shore GC (2003) Caspase cleavage product of BAP31 induces mitochondrial fission through endoplasmic reticulum calcium signals, enhancing cytochrome c release to the cytosol. J Cell Biol 160(7):1115–1127. https://doi.org/10.1083/jcb.200212059

Bugarčić A, Taylor JA (2006) Rotavirus nonstructural glycoprotein NSP4 is secreted from the apical surfaces of polarized epithelial cells. J Virol 80(24):12343–12349. https://doi.org/10.1128/JVI.01378-06

Bukreyev A, Whitehead SS, Murphy BR, Collins PL (1997) Recombinant respiratory syncytial virus from which the entire SH gene has been deleted grows efficiently in cell culture and exhibits site-specific attenuation in the respiratory tract of the mouse. J Virol 71(12):8973–8982

Carrasco L (1995) Modification of membrane permeability by animal viruses. Adv Virus Res 45:61–112

Chacko AR, Arifullah M, Sastri NP, Jeyakanthan J, Ueno G, Sekar K, Read RJ, Dodson EJ, Rao DC, Suguna K (2011) Novel pentameric structure of the diarrhea-inducing region of the rotavirus enterotoxigenic protein NSP4. J Virol 85(23):12721–12732. https://doi.org/10.1128/JVI.00349-11

Champeimont R, Laine E, Hu SW, Penin F, Carbone A (2016) Coevolution analysis of hepatitis C virus genome to identify the structural and functional dependency network of viral proteins. Sci Rep 6. https://doi.org/10.1038/srep26401

Chen BJ, Leser GP, Jackson D, Lamb RA (2008) The influenza virus M2 protein cytoplasmic tail interacts with the M1 protein and influences virus assembly at the site of virus budding. J Virol 82(20):10059–10070

Chen SL, Lin ST, Tsai TC, Hsiao WC, Tsao YP (2007) ErbB4 (JM-b/CYT-1)-induced expression and phosphorylation of c-Jun is abrogated by human papillomavirus type 16 E5 protein. Oncogene 26(1):42–53. https://doi.org/10.1038/sj.onc.1209768

Chen W, Calvo PA, Malide D, Gibbs J, Schubert U, Bacik I, Basta S, O'Neill R, Schickli J, Palese P, Henklein P, Bennink JR, Yewdell JW (2001) A novel influenza a virus mitochondrial protein that induces cell death. Nat Med 7(12):1306–1312. https://doi.org/10.1038/nm1201-1306

Cohen BD, Goldstein DJ, Rutledge L, Vass WC, Lowy DR, Schlegel R, Schiller JT (1993) Transformation-specific interaction of the bovine papillomavirus E5 oncoprotein with the platelet-derived growth factor receptor transmembrane domain and the epidermal growth factor receptor cytoplasmic domain. J Virol 67(9):5303–5311

Cohen EA, Terwilliger EF, Sodroski JG, Haseltine WA (1988) Identification of a protein encoded by the vpu gene of HIV-1. Nature 334(6182):532–534

Collins PL, Huang YT, Wertz GW (1984) Identification of a tenth mRNA of respiratory syncytial virus and assignment of polypeptides to the 10 viral genes. J Virol 49(2):572–578

Collins PL, Mottet G (1993) Membrane orientation and oligomerization of the small hydrophobic protein of human respiratory syncytial virus. J Gen Virol 74:1445–1450

Collins PL, Olmsted RA, Johnson PR (1990) The small hydrophobic protein of human respiratory syncytial virus: comparison between antigenic subgroups a and B. J Gen Virol 71(Pt 7):1571–1576

Conrad M, Bubb VJ, Schlegel R (1993) The human papillomavirus type 6 and 16 E5 proteins are membrane-associated proteins which associate with the 16-kilodalton pore-forming protein. J Virol 67(10):6170–6178

Cook GA, Dawson LA, Tian Y, Opella SJ (2013) Three-dimensional structure and interaction studies of hepatitis C virus p7 in 1,2-dihexanoyl- sn -glycero-3-phosphocholine by solution nuclear magnetic resonance. Biochemistry 52(31):5295–5303. https://doi.org/10.1021/bi4006623

Cook GA, Zhang H, Park SH, Wang Y, Opella SJ (2010) Comparative NMR studies demonstrate profound differences between two viroporins: p7 of HCV and Vpu of HIV-1. Biochim Biophys Acta 1808 (2):554–560. https://doi.org/S0005-2736(10)00286-5 [pii] https://doi.org/10.1016/j.bbamem.2010.08.005

Coombs KM (2006) Reovirus structure and morphogenesis. Current Topics in Microbiology and Immunology, vol 309

Coric P, Saribas AS, Abou-Gharbia M, Childers W, Condra JH, White MK, Safak M, Bouaziz S (2017) Nuclear magnetic resonance structure of the human Polyoma JC virus Agnoprotein. J Cell Biochem. https://doi.org/10.1002/jcb.25977

Coric P, Saribas AS, Abou-Gharbia M, Childers W, White MK, Bouaziz S, Safak M (2014) Nuclear magnetic resonance structure revealed that the human polyomavirus JC virus agnoprotein contains an α-helix encompassing the Leu/Ile/Phe-rich domain. J Virol 88(12):6556–6575. https://doi.org/10.1128/JVI.00146-14

Corse E, Machamer CE (2003) The cytoplasmic tails of infectious bronchitis virus E and M proteins mediate their interaction. Virology 312(1):25–34

Cortese MS, Ashrafi GH, Campo MS (2010) All 4 di-leucine motifs in the first hydrophobic domain of the E5 oncoprotein of human papillomavirus type 16 are essential for surface MHC class I downregulation activity and E5 endomembrane localization. Int J Cancer 126(7):1675–1682. https://doi.org/10.1002/ijc.25004

Couet J, Li S, Okamoto T, Ikezu T, Lisanti MP (1997) Identification of peptide and protein ligands for the caveolin- scaffolding domain. Implications for the interaction of caveolin with caveolae-associated proteins. J Biol Chem 272(10):6525–6533. https://doi.org/10.1074/jbc.272.10.6525

Crawford SE, Hyser JM, Utama B, Estes MK (2012) Autophagy hijacked through viroporin-activated calcium/calmodulin-dependent kinase kinase-β signaling is required for rotavirus replication. Proc Natl Acad Sci U S A 109(50):E3405–E3413. https://doi.org/10.1073/pnas.1216539109

Darbinyan A, Darbinian N, Safak M, Radhakrishnan S, Giordano A, Khalili K (2002) Evidence for dysregulation of cell cycle by human polyomavirus, JCV, late auxiliary protein. Oncogene 21(36):5574–5581. https://doi.org/10.1038/sj.onc.1205744

Darbinyan A, Siddiqui KM, Slonina D, Darbinian N, Amini S, White MK, Khalili K (2004) Role of JC virus agnoprotein in DNA repair. J Virol 78(16):8593–8600. https://doi.org/10.1128/JVI.78.16.8593-8600.2004

de Chassey B, Meyniel-Schicklin L, Vonderscher J, André P, Lotteau V (2014) Virus-host interactomics: new insights and opportunities for antiviral drug discovery. Genome Med 6(11). https://doi.org/10.1186/s13073-014-0115-1

De Chassey B, Navratil V, Tafforeau L, Hiet MS, Aublin-Gex A, Agaugué S, Meiffren G, Pradezynski F, Faria BF, Chantier T, Le Breton M, Pellet J, Davoust N, Mangeot PE, Chaboud A, Penin F, Jacob Y, Vidalain PO, Vidal M, André P, Rabourdin-Combe C, Lotteau V (2008) Hepatitis C virus infection protein network. Mol Syst Biol 4. https://doi.org/10.1038/msb.2008.66

DeDiego ML, Álvarez E, Almazán F, Rejas MT, Lamirande E, Roberts A, Shieh WJ, Zaki SR, Subbarao K, Enjuanes L (2007) A severe acute respiratory syndrome coronavirus that lacks the E gene is attenuated in vitro and in vivo. J Virol 81(4):1701–1713. https://doi.org/10.1128/JVI.01467-06

DeDiego ML, Nieto-Torres JL, Jimenez-Guardeno JM, Regla-Nava JA, Alvarez E, Oliveros JC, Zhao J, Fett C, Perlman S, Enjuanes L (2011) Severe acute respiratory syndrome coronavirus envelope protein regulates cell stress response and apoptosis. PLoS Path 7(10):e1002315. https://doi.org/10.1371/journal.ppat.1002315

DeDiego ML, Pewe L, Alvarez E, Rejas MT, Perlman S, Enjuanes L (2008) Pathogenicity of severe acute respiratory coronavirus deletion mutants in hACE-2 transgenic mice. Virology 376(2):379–389. https://doi.org/10.1016/j.virol.2008.03.005

Deyde VM, Xu X, Bright RA, Shaw M, Smith CB, Zhang Y, Shu Y, Gubareva LV, Cox NJ, Klimov AI (2007) Surveillance of resistance to adamantanes among influenza a(H3N2) and a(H1N1) viruses isolated worldwide. J Infect Dis 196(2):249–257

DiMaio D, Petti LM (2013) The E5 proteins. Virology 445(1–2):99–114. https://doi.org/10.1016/j.virol.2013.05.006

Do HQ, Wittlich M, Glück JM, Möckel L, Willbold D, Koenig BW, Heise H (2013) Full-length Vpu and human CD4(372-433) in phospholipid bilayers as seen by magic angle spinning NMR. Biol Chem 394(11):1453–1463. https://doi.org/10.1515/hsz-2013-0194

Doms RW, Trono D (2000) The plasma membrane as a combat zone in the HIV battlefield. Genes Dev 14(21):2677–2688. https://doi.org/10.1101/gad.833300

Dotson D, Woodruff EA, Villalta F, Dong X (2016) Filamin a is involved in HIV-1 Vpu-mediated evasion of host restriction by modulating tetherin expression. J Biol Chem 291(8):4236–4246. https://doi.org/10.1074/jbc.M115.708123

Dowell SF, Anderson LJ, Gary HE Jr, Erdman DD, Plouffe JF, File TM Jr, Marston BJ, Breiman RF (1996) Respiratory syncytial virus is an important cause of community-acquired lower respiratory infection among hospitalized adults. J Infect Dis 174(3):456–462

Drummond-Barbosa D, Vaillancourt RR, Kazlauskas A, Dimaio D (1995) Ligand-independent activation of the platelet-derived growth factor β receptor: requirements for bovine papillomavirus E5-induced mitogenic signaling. Mol Cell Biol 15(5):2570–2581

Dubé M, Bego MG, Paquay C, Cohen TA (2010) Modulation of HIV-1-host interaction: role of the Vpu accessory protein. Retrovirology 7. https://doi.org/10.1186/1742-4690-7-114

Dubé M, Paquay C, Roy BB, Bego MG, Mercier J, Cohen EA (2011) HIV-1 Vpu antagonizes BST-2 by interfering mainly with the trafficking of newly synthesized BST-2 to the cell surface. Traffic 12(12):1714–1729. https://doi.org/10.1111/j.1600-0854.2011.01277.x

Elliott EI, Sutterwala FS (2015) Initiation and perpetuation of NLRP3 inflammasome activation and assembly. Immunol Rev 265(1):35–52. https://doi.org/10.1111/imr.12286

Endo S, Okada Y, Orba Y, Nishihara H, Tanaka S, Nagashima K, Sawa H (2003) JC virus agnoprotein colocalizes with tubulin. J Neurovirol 9(SUPPL. 1):10–14

Enjuanes L, Brian D, Cavanagh D, Holmes K, Lai MMC, Laude H, Masters P, Rottier P, Siddell SG, Spaan WJM, Taguchi F, Talbot P (2000) Coronaviridae. In: van MHV R, Fauquet CM, Bishop DHL et al (eds) Virus taxonomy. Classification and nomenclature of viruses. Academic Press, San Diego, pp 835–849

Enjuanes L, Gorbalenya AE, de Groot RJ, Cowley JA, Ziebuhr J, Snijder EJ (2008) The Nidovirales. Encyclopedia of Virology:419–430

Estes MK, Greenberg HB (2013) Rotaviruses. Fields Virology:1347–1401

Fan Y, Mok CKP, Chan MCW, Zhang Y, Nal B, Kien F, Bruzzone R, Sanyal S (2017) Cell cycle-independent role of Cyclin D3 in host restriction of influenza virus infection. J Biol Chem 292(12):5070–5088. https://doi.org/10.1074/jbc.M117.776112

Fields BN, Knipe DM, Howley PM (2013) Fields virology. Wolters Kluwer Health/Lippincott Williams & Wilkins, Philadelphia

Fischer WB, Li LH, Mahato DR, Wang YT, Chen CP (2014) Viral channel proteins in intracellular protein-protein communication: Vpu of HIV-1, E5 of HPV16 and p7 of HCV. Biochim Biophys Acta 1838(4):1113–1121. https://doi.org/10.1016/j.bbamem.2013.08.017

Forni D, Cagliani R, Clerici M, Sironi M (2017) Molecular evolution of human coronavirus genomes. Trends Microbiol 25(1):35–48. https://doi.org/10.1016/j.tim.2016.09.001

Fuentes S, Tran KC, Luthra P, Teng MN, He B (2007) Function of the respiratory syncytial virus small hydrophobic protein. J Virol 81(15):8361–8366

Gan SW, Ng L, Xin L, Gong X, Torres J (2008) Structure and ion channel activity of the human respiratory syncytial virus (hRSV) small hydrophobic protein transmembrane domain. Protein Sci 17:813–820

Gan SW, Surya W, Vararattanavech A, Torres J (2014) Two different conformations in hepatitis C virus p7 protein account for proton transport and dye release. PLoS One 9(1):e78494. https://doi.org/10.1371/journal.pone.0078494

Gan SW, Tan E, Lin X, Yu D, Wang J, Tan GM-Y, Vararattanavech A, Yeo CY, Soon CH, Soong TW, Pervushin K, Torres J (2012) The small hydrophobic protein of the human respiratory syncytial virus forms pentameric ion channels. J Biol Chem 287(29):24671–24689

Gannagé M, Dormann D, Albrecht R, Dengjel J, Torossi T, Rämer PC, Lee M, Strowig T, Arrey F, Conenello G, Pypaert M, Andersen J, García-Sastre A, Münz C (2009) Matrix protein 2 of influenza a virus blocks Autophagosome fusion with lysosomes. Cell Host and Microbe 6(4):367–380. https://doi.org/10.1016/j.chom.2009.09.005

Gerber D, Maerkl SJ, Quake SR (2009) An in vitro microfluidic approach to generating protein-interaction networks. Nat Methods 6(1):71–74. https://doi.org/10.1038/nmeth.1289

Germain MA, Chatel-Chaix L, Gagné B, Bonneil E, Thibault P, Pradezynski F, De Chassey B, Meyniel-Schicklin L, Lotteau V, Baril M, Lamarre D (2014) Elucidating novel hepatitis C virus-host interactions using combined mass spectrometry and functional genomics approaches. Mol Cell Proteomics 13(1):184–203. https://doi.org/10.1074/mcp.M113.033803

Gieswein CE, Sharom FJ, Wildeman AG (2003) Oligomerization of the E5 protein of human papillomavirus type 16 occurs through multiple hydrophobic regions. Virology 313(2):415–426. https://doi.org/10.1016/S0042-6822(03)00296-4

Goldstein DJ, Andresson T, Sparkowski JJ, Schlegel R (1992) The BPV-1 E5 protein, the 16 kDa membrane pore-forming protein and the PDGF receptor exist in a complex that is dependent on hydrophobic transmembrane interactions. EMBO J 11(13):4851–4859

Goldstein DJ, Fmbow ME, Andresson T, McLean P, Smith K, Bubb V, Schlegel R (1991) Bovine papillomavirus E5 oncoprotein binds to the 16K component of vacuolar H+-ATPases. Nature 352(6333):347–349

Goldstein DJ, Li W, Wang LM, Heidaran MA, Aaronson S, Shinn R, Schlegel R, Pierce JH (1994) The bovine papillomavirus type 1 E5 transforming protein specifically binds and activates the β-type receptor for the platelet-derived growth factor but not other related tyrosine kinase-containing receptors to induce cellular transformation. J Virol 68(7):4432–4441

Griffin SDC, Beales LP, Clarke DS, Worsfold O, Evans SD, Jaeger J, Harris MPG, Rowlands DJ, Klenk HD (2003) The p7 protein of hepatitis C virus forms an ion channel that is blocked by the antiviral drug, amantadine. FEBS Lett 535(1–3):34–38

Gruener M, Bravo IG, Momburg F, Alonso A, Tomakidi P (2007) The E5 protein of the human papillomavirus type 16 down-regulates HLA-I surface expression in calnexin-expressing but not in calnexin-deficient cells. Virol J 4. https://doi.org/10.1186/1743-422X-4-116

Gu RX, Liu LA, Wei DQ (2013) Structural and energetic analysis of drug inhibition of the influenza a M2 proton channel. Trends Pharmacol Sci 34(10):571–580. https://doi.org/10.1016/j.tips.2013.08.003

Guan Z, Liu D, Mi S, Zhang J, Ye Q, Wang M, Gao GF, Yan J (2010) Interaction of Hsp40 with influenza virus M2 protein: implications for PKR signaling pathway. Protein and Cell 1(10):944–955. https://doi.org/10.1007/s13238-010-0115-x

Hagai T, Azia A, Babu MM, Andino R (2014) Use of host-like peptide motifs in viral proteins is a prevalent strategy in host-virus interactions. Cell Rep 7:1729–1739

Hagen N, Bayers K, Rošch K, Schindler M (2014) The intraviral protein interaction network of hepatitis C virus. Mol Cell Proteomics 13(7):1676–1689. https://doi.org/10.1074/mcp.M113.036301

Hall CB, Walsh EE, Schnabel KC, Long CE, Mcconnochie KM, Hildreth SW, Anderson LJ (1990) Occurrence of group-a and group-B of respiratory syncytial virus over 15 years - associated epidemiologic and clinical characteristics in hospitalized and ambulatory children. J Infect Dis 162(6):1283–1290

Haller C, Müller B, Fritz JV, Lamas-Murua M, Stolp B, Pujol FM, Keppler OT, Fackler OT (2014) HIV-1 Nef and Vpu are functionally redundant broad-spectrum modulators of cell surface receptors, including tetraspanins. J Virol 88(24):14241–14257. https://doi.org/10.1128/JVI.02333-14

Harris BZ, Lim WA (2001) Mechanism and role of PDZ domains in signaling complex assembly. J Cell Sci 114(18):3219–3231

Hauser H, Lopez LA, Yang SJ, Oldenburg JE, Exline CM, Guatelli JC, Cannon PM (2010) HIV-1 Vpu and HIV-2 Env counteract BST-2/tetherin by sequestration in a perinuclear compartment. Retrovirology 7. https://doi.org/10.1186/1742-4690-7-51

Hay AJ, Gregory V, Douglas AR, Yi PL (2001) The evolution of human influenza viruses. Philosophical Transactions of the Royal Society B: Biological Sciences 356(1416):1861–1870. https://doi.org/10.1098/rstb.2001.0999

Hayden FG, De Jong MD (2011) Emerging influenza antiviral resistance threats. J Infect Dis 203(1):6–10. https://doi.org/10.1093/infdis/jiq012

Helenius A (1992) Unpacking the incoming influenza virus. Cell 69(4):577–578. https://doi.org/10.1016/0092-8674(92)90219-3

Henkel M, Mitzner D, Henklein P, Meyer-Almes FJ, Moroni A, DiFrancesco ML, Henkes LM, Kreim M, Kast SM, Schubert U, Thiel G (2010) Proapoptotic influenza a virus protein PB1-F2 forms a nonselective ion channel. PLoS One 5(6). https://doi.org/10.1371/journal.pone.0011112

Ho Y, Lin PH, Liu CYY, Lee SP, Chao YC (2004) Assembly of human severe acute respiratory syndrome coronavirus-like particles. Biochem Biophys Res Commun 318(4):833–838. https://doi.org/10.1016/j.bbrc.2004.04.111

Hogue BG, Machamer CE (2008) Coronavirus structural proteins and virus assembly. In: Perlman S, Gallagher T, Snijder EJ (eds) Nidoviruses, Washington DC, pp 179–200

Holmes KV (2003) SARS coronavirus: a new challenge for prevention and therapy. J Clin Invest 111(11):1605–1609. https://doi.org/10.1172/JCI18819

Hsu K, Seharaseyon J, Dong P, Bour S, Marbán E (2004) Mutual functional destruction of HIV-1 Vpu and host TASK-1 channel. Mol Cell 14(2):259–267. https://doi.org/10.1016/S1097-2765(04)00183-2

Hu S, Yin L, Mei S, Li J, Xu F, Sun H, Liu X, Cen S, Liang C, Li A, Guo F (2017) BST-2 restricts IAV release and is countered by the viral M2 protein. Biochem J 474(5):715–730. https://doi.org/10.1042/BCJ20160861

Hwang ES, Nottoli T, Dimaio D (1995) The HPV16 E5 protein: expression, detection, and stable complex formation with transmembrane proteins in COS cells. Virology 211(1):227–233. https://doi.org/10.1006/viro.1995.1395

Hyser JM, Collinson-Pautz MR, Utama B, Estes MK (2010) Rotavirus disrupts calcium homeostasis by NSP4 viroporin activity. MBio 1(5). https://doi.org/10.1128/mBio.00265-10

Hyser JM, Estes MK (2015) Pathophysiological consequences of calcium-conducting Viroporins. Annual Review of Virology 2:473–496. https://doi.org/10.1146/annurev-virology-100114-054846

Hyser JM, Utama B, Crawford SE, Broughman JR, Estes MK (2013) Activation of the endoplasmic reticulum calcium sensor STIM1 and store-operated calcium entry by rotavirus requires NSP4 viroporin activity. J Virol 87(24):13579–13588. https://doi.org/10.1128/JVI.02629-13

Ichinohe T, Pang IK, Iwasaki A (2010) Influenza virus activates inflammasomes via its intracellular M2 ion channel. Nat Immunol 11(5):404–410. https://doi.org/10.1038/ni.1861

Imbert I, Snijder EJ, Dimitrova M, Guillemot JC, Lécine P, Canard B (2008) The SARS-coronavirus PLnc domain of nsp3 as a replication/transcription scaffolding protein. Virus Res 133(2):136–148. https://doi.org/10.1016/j.virusres.2007.11.017

Ito M, Yanagi Y, Ichinohe T (2012) Encephalomyocarditis virus viroporin 2B activates NLRP3 inflammasome. PLoS Path 8(8):e1002857. https://doi.org/10.1371/journal.ppat.1002857

Ito N, Mossel EC, Narayanan K, Popov VL, Huang C, Inoue T, Peters CJ, Makino S (2005) Severe acute respiratory syndrome coronavirus 3a protein is a viral structural protein. J Virol 79(5):3182–3186. https://doi.org/10.1128/JVI.79.5.3182-3186.2005

Javier RT, Rice AP (2011) Emerging theme: cellular PDZ proteins as common targets of pathogenic viruses. J Virol 85(22):11544–11556. https://doi.org/10.1128/JVI.05410-11

Ji HL, Song W, Gao Z, Su XF, Nie HG, Jiang Y, Peng JB, He YX, Liao Y, Zhou YJ, Tousson A, Matalon S (2009) SARS-CoV proteins decrease levels and activity of human ENaC via activation of distinct PKC isoforms. Am J Physiol Lung Cell Mol Physiol 296(3):L372–L383. https://doi.org/10.1152/ajplung.90437.2008

Jia X, Weber E, Tokarev A, Lewinski M, Rizk M, Suarez M, Guatelli J, Xiong Y (2014) Structural basis of HIV-1 Vpu-mediated BST2 antagonism via hijacking of the clathrin adaptor protein complex 1. elife 2014(3). https://doi.org/10.7554/eLife.02362

Jimenez-Guardeño JM, Nieto-Torres JL, DeDiego ML, Regla-Nava JA, Fernandez-Delgado R, Castaño-Rodriguez C, Enjuanes L (2014) The PDZ-binding motif of severe acute respiratory syndrome coronavirus envelope protein is a determinant of viral pathogenesis. PLoS Path 10(8). https://doi.org/10.1371/journal.ppat.1004320

Jirasko V, Montserret R, Appel N, Janvier A, Eustachi L, Brohm C, Steinmann E, Pietschmann T, Penin F, Bartenschlagerd R (2008) Structural and functional characterization of nonstructural protein 2 for its role in hepatitis C virus assembly. J Biol Chem 283(42):28546–28562. https://doi.org/10.1074/jbc.M803981200

Jirasko V, Montserret R, Lee JY, Gouttenoire J, Moradpour D, Penin F, Bartenschlager R (2010) Structural and functional studies of nonstructural protein 2 of the hepatitis C virus reveal its key role as organizer of virion assembly. PLoS Path 6(12). https://doi.org/10.1371/journal.ppat.1001233

Kabsch K, Mossadegh N, Kohl A, Komposch G, Schenkel J, Alonso A, Tomakidi P (2004) The HPV-16 E5 protein inhibits TRAIL- and FasL-mediated apoptosis in human keratinocyte raft cultures. Intervirology 47(1):48–56. https://doi.org/10.1159/000076642

Kaniowska D, Kaminski R, Amini S, Radhakrishnan S, Rappaport J, Johnson E, Khalili K, Del Valle L, Darbinyan A (2006) Cross-interaction between JC virus agnoprotein and human immunodeficiency virus type 1 (HIV-1) tat modulates transcription of the HIV-1 long terminal repeat in glial cells. J Virol 80(18):9288–9299. https://doi.org/10.1128/JVI.02138-05

Kanzawa N, Nishigaki K, Hayashi T, Ishii Y, Furukawa S, Niiro A, Yasui F, Kohara M, Morita K, Matsushima K, Le MQ, Masuda T, Kannagi M (2006) Augmentation of chemokine production by severe acute respiratory syndrome coronavirus 3a/X1 and 7a/X4 proteins through NF-κB activation. FEBS Lett 580(30):6807–6812. https://doi.org/10.1016/j.febslet.2006.11.046

Kargel A, Schmidt U, Buchholz UJ (2001) Recombinant bovine respiratory syncytial virus with deletions of the G or SH genes: G and F proteins binds heparin. J Gen Virol 82(3):631–640

Karron RA, Wright PF, Belshe RB, Thumar B, Casey R, Newman F, Polack FP, Randolph VB, Deatly A, Hackell J, Gruber W, Murphy BR, Collins PL (2005) Identification of a recombinant live attenuated respiratory syncytial virus vaccine candidate that is highly attenuated in infants. J Infect Dis 191(7):1093–1104. https://doi.org/10.1086/427813

Kell B, Jewers RJ, Cason J, Pakarian F, Kaye JN, Best JM (1994) Detection of E5 oncoprotein in human papillomavirus type 16-positive cervical scrapes using antibodies raised to synthetic peptides. J Gen Virol 75(9):2451–2456

Kerkau T, Bacik I, Bennink JR, Yewdell JW, Hünig T, Schimpl A, Schubert U (1997) The human immunodeficiency virus type 1 (HIV-1)vpu protein interferes with an early step in the biosynthesis of major histocompatibility complex (MHC) class I molecules. J Exp Med 185(7):1295–1305. https://doi.org/10.1084/jem.185.7.1295

King G, Oates J, Patel D, Van Den Berg HA, Dixon AM (2011) Towards a structural understanding of the smallest known oncoprotein: investigation of the bovine papillomavirus E5 protein using solution-state NMR. Biochim Biophys Acta Biomembr 1808(6):1493–1501. https://doi.org/10.1016/j.bbamem.2010.11.004

Kipper S, Hamad S, Caly L, Avrahami D, Bacharach E, Jans DA, Gerber D, Bajorek M (2015) New host factors important for respiratory syncytial virus replication revealed by a novel microfluidics screen for interactors of matrix protein. Mol Cell Proteomics

Klimkait T, Strebel K, Hoggan MD, Martin MA, Orenstein JM (1990) The human immunodeficiency virus type 1-specific protein vpu is required for efficient virus maturation and release. J Virol 64(2):621–629

Kobayashi T, Ode H, Yoshida T, Sato K, Gee P, Yamamoto SP, Ebina H, Strebel K, Sato H, Koyanagi Y (2011) Identification of amino acids in the human tetherin transmembrane domain responsible for HIV-1 Vpu interaction and susceptibility. J Virol 85(2):932–945. https://doi.org/10.1128/JVI.01668-10

Kotnik Halavaty K, Regan J, Mehta K, Laimins L (2014) Human papillomavirus E5 oncoproteins bind the A4 endoplasmic reticulum protein to regulate proliferative ability upon differentiation. Virology 452-453:223–230. https://doi.org/10.1016/j.virol.2014.01.013

Krawczyk E, Hanover JA, Schlegel R, Suprynowicz FA (2008) Karyopherin β3: a new cellular target for the HPV-16 E5 oncoprotein. Biochem Biophys Res Commun 371(4):684–688. https://doi.org/10.1016/j.bbrc.2008.04.122

Krawczyk E, Suprynowicz FA, Hebert JD, Kamonjoh CM, Schlegel R (2011) The human papillomavirus type 16 E5 oncoprotein translocates calpactin I to the perinuclear region. J Virol 85(21):10968–10975. https://doi.org/10.1128/JVI.00706-11

Krawczyk E, Suprynowicz FA, Sudarshan SR, Schlegel R (2010) Membrane orientation of the human papillomavirus type 16 E5 oncoprotein. J Virol 84(4):1696–1703. https://doi.org/10.1128/JVI.01968-09

Kueck T, Foster TL, Weinelt J, Sumner JC, Pickering S, Neil SJD (2015) Serine phosphorylation of HIV-1 Vpu and its binding to Tetherin regulates interaction with Clathrin adaptors. PLoS Path 11(8). https://doi.org/10.1371/journal.ppat.1005141

Kueck T, Neil SJD (2012) A cytoplasmic tail determinant in HIV-1 vpu mediates targeting of tetherin for endosomal degradation and counteracts interferon-induced restriction. PLoS Path 8(3). https://doi.org/10.1371/journal.ppat.1002609

Kuo L, Masters PS (2010) Evolved variants of the membrane protein can partially replace the envelope protein in murine coronavirus assembly. J Virol 84(24):12872–12885. https://doi.org/10.1128/JVI.01850-10

Ladasky JJ, Boyle S, Seth M, Li H, Pentcheva T, Abe F, Steinberg SJ, Edidin M (2006) Bap31 enhances the endoplasmic reticulum export and quality control of human class I MHC molecules. J Immunol 177(9):6172–6181

Lamb RA, Choppin PW (1983) The gene structure and replication of influenza virus. Annu Rev Biochem 52:467–506

Lambelé M, Koppensteiner H, Symeonides M, Roy NH, Chan J, Schindler M, Thali M (2015) Vpu is the main determinant for tetraspanin downregulation in HIV-1-infected cells. J Virol 89(6):3247–3255. https://doi.org/10.1128/JVI.03719-14

Lau D, Kwan W, Guatelli J (2011) Role of the endocytic pathway in the counteraction of BST-2 by human lentiviral pathogens. J Virol 85(19):9834–9846. https://doi.org/10.1128/JVI.02633-10

Lazarczyk M, Pons C, Mendoza JA, Cassonnet P, Jacob Y, Favre M (2008) Regulation of cellular zinc balance as a potential mechanism of EVER-mediated protection against pathogenesis by cutaneous oncogenic human papillomaviruses. J Exp Med 205(1):35–42. https://doi.org/10.1084/jem.20071311

le Tortorec A, Willey S, Neil SJD (2011) Antiviral inhibition of enveloped virus release by Tetherin/BST-2: action and counteraction. Virus 3(5):520–540. https://doi.org/10.3390/v3050520

Lewinski MK, Jafari M, Zhang H, Opella SJ, Guatelli J (2015) Membrane anchoring by a C-terminal tryptophan enables HIV-1 Vpu to displace bone marrow stromal antigen 2 (BST2) from sites of viral assembly. J Biol Chem 290(17):10919–10933. https://doi.org/10.1074/jbc.M114.630095

Li Y, Jain N, Limpanawat S, To J, Quistgaard EM, Nordlund P, Thanabalu T, Torres J (2015) Interaction between human BAP31 and respiratory syncytial virus small hydrophobic (SH) protein. Virology 482:105–110. https://doi.org/10.1016/j.virol.2015.03.034

Li Y, Surya W, Claudine S, Torres J (2014a) Structure of a conserved Golgi complex-targeting signal in coronavirus envelope proteins. J Biol Chem 289(18):12535–12549. https://doi.org/10.1074/jbc.M114.560094

Li Y, To J, Verdia-Baguena C, Dossena S, Surya W, Huang M, Paulmichl M, Liu DX, Aguilella VM, Torresa J (2014b) Inhibition of the human respiratory syncytial virus small hydrophobic protein and structural variations in a Bicelle environment. J Virol 89(20):11899–11914. https://doi.org/10.1128/Jvi.00839-14

Lim KP, Liu DX (2001) The missing link in coronavirus assembly. Retention of the avian coronavirus infectious bronchitis virus envelope protein in the pre-Golgi compartments and physical interaction between the envelope and membrane proteins. J Biol Chem 276(20):17515–17523

Lin C, Lindenbach BD, Prágai BM, McCourt DW, Rice CM (1994) Processing in the hepatitis C virus E2-NS2 region: identification of p7 and two distinct E2-specific products with different C termini. J Virol 68(8):5063–5073

Lindenbach BD, Rice CM (2013) The ins and outs of hepatitis C virus entry and assembly. Nat Rev Microbiol 11(10):688–700. https://doi.org/10.1038/nrmicro3098

Lindner HA, Fotouhi-Ardakani N, Lytvyn V, Lachance P, Sulea T, Ménard R (2005) The papain-like protease from the severe acute respiratory syndrome coronavirus is a deubiquitinating enzyme. J Virol 79(24):15199–15208. https://doi.org/10.1128/JVI.79.24.15199-15208.2005

Lou Z, Sun Y, Rao Z (2014) Current progress in antiviral strategies. Trends Pharmacol Sci 35(2):86–102. https://doi.org/10.1016/j.tips.2013.11.006

Low KW, Tan T, Ng K, Tan BH, Sugrue RJ (2008) The RSV F and G glycoproteins interact to form a complex on the surface of infected cells. Biochem Biophys Res Commun 366(2):308–313. https://doi.org/10.1016/j.bbrc.2007.11.042

Lu W, Zheng BJ, Xu K, Schwarz W, Du LY, Wong CKL, Chen JD, Duan SM, Deubel V, Sun B (2006) Severe acute respiratory syndrome-associated coronavirus 3a protein forms an ion channel and modulates virus release. Proc Natl Acad Sci U S A 103(33):12540–12545. https://doi.org/10.1073/pnas.0605402103

Luik P, Chew C, Aittoniemi J, Chang J, Wentworth P, Jr., Dwek RA, Biggin PC, Venien-Bryan C, Zitzmann N (2009) The 3-dimensional structure of a hepatitis C virus p7 ion channel by electron microscopy. Proc Natl Acad Sci U S A 106 (31):12712–12716. https://doi.org/0905966106 [pii] doi:https://doi.org/10.1073/pnas.0905966106

Lukhele S, Cohen A (2017) Conserved residues within the HIV-1 Vpu transmembrane-proximal hinge region modulate BST2 binding and antagonism. Retrovirology 14(1). https://doi.org/10.1186/s12977-017-0345-6

Ma H, Kien F, Manière M, Zhang Y, Lagarde N, Tse KS, Poon LLM, Nal B (2012) Human annexin A6 interacts with influenza a virus protein M2 and negatively modulates infection. J Virol 86(3):1789–1801. https://doi.org/10.1128/JVI.06003-11

Ma Y, Anantpadma M, Timpe JM, Shanmugam S, Singh SM, Lemon SM, Yi M (2011) Hepatitis C virus NS2 protein serves as a scaffold for virus assembly by interacting with both structural and nonstructural proteins. J Virol 85(1):86–97. https://doi.org/10.1128/JVI.01070-10

Madjo U, Leymarie O, Frémont S, Kuster A, Nehlich M, Gallois-Montbrun S, Janvier K, Berlioz-Torrent C (2016) LC3C contributes to Vpu-mediated antagonism of BST2/Tetherin restriction on HIV-1 release through a non-canonical autophagy pathway. Cell Rep 17(9):2221–2233. https://doi.org/10.1016/j.celrep.2016.10.045

Maeda J, Maeda A, Makino S (1999) Release of coronavirus E protein in membrane vesicles from virus- infected cells and E protein-expressing cells. Virology 263(2):265–272. https://doi.org/10.1006/viro.1999.9955

Magadán JG, Bonifacino JS (2012) Transmembrane domain determinants of CD4 downregulation by HIV-1 Vpu. J Virol 86(2):757–772. https://doi.org/10.1128/jvi.05933-11

Magadán JG, Pérez-Victoria FJ, Sougrat R, Ye Y, Strebel K, Bonifacino JS (2010) Multilayered mechanism of CD4 downregulation by HIV-1 vpu involving distinct ER retention and ERAD targeting steps. PLoS Path 6(4):1–18

Mangeat B, Cavagliotti L, Lehmann M, Gers-Huber G, Kaur I, Thomas Y, Kaiser L, Piguet V (2012) Influenza virus partially counteracts restriction imposed by tetherin/BST-2. J Biol Chem 287(26):22015–22029. https://doi.org/10.1074/jbc.M111.319996

Margottin F, Bour SP, Durand H, Selig L, Benichou S, Richard V, Thomas D, Strebel K, Benarous R (1998) A novel human WD protein, h-βTrCP, that interacts with HIV-1 Vpu connects CD4 to the ER degradation pathway through an F-box motif. Mol Cell 1(4):565–574

Matheson NJ, Sumner J, Wals K, Rapiteanu R, Weekes MP, Vigan R, Weinelt J, Schindler M, Antrobus R, Costa ASH, Frezza C, Clish CB, Neil SJD, Lehner PJ (2015) Cell surface proteomic map of HIV infection reveals antagonism of amino acid metabolism by Vpu and Nef. Cell Host and Microbe 18(4):409–423. https://doi.org/10.1016/j.chom.2015.09.003

Matusali G, Potestá M, Santoni A, Cerboni C, Doria M (2012) The human immunodeficiency virus type 1 Nef and Vpu proteins downregulate the natural killer cell-activating ligand PVR. J Virol 86(8):4496–4504. https://doi.org/10.1128/JVI.05788-11

McAuley JL, Tate MD, MacKenzie-Kludas CJ, Pinar A, Zeng W, Stutz A, Latz E, Brown LE, Mansell A (2013) Activation of the NLRP3 Inflammasome by IAV virulence protein PB1-F2 contributes to severe pathophysiology and disease. PLoS Path 9(5). https://doi.org/10.1371/journal.ppat.1003392

McCown MF, Pekosz A (2006) Distinct domains of the influenza a virus M2 protein cytoplasmic tail mediate binding to the M1 protein and facilitate infectious virus production. J Virol 80(16):8178–8189. https://doi.org/10.1128/JVI.00627-06

McNatt MW, Zang T, Bieniasz PD (2013) Vpu binds directly to Tetherin and displaces it from nascent Virions. PLoS Path 9(4). https://doi.org/10.1371/journal.ppat.1003299

Meyniel-Schicklin L, De Chassey B, André P, Lotteau V (2012) Viruses and interactomes in translation. Mol Cell Proteomics 11(7). https://doi.org/10.1074/mcp.M111.014738

Mi SF, Li Y, Yan JH, Gao GF (2010) Na+/K+-ATPase β1 subunit interacts with M2 proteins of influenza a and B viruses and affects the virus replication. Sci China Life Sci 53(9):1098–1105. https://doi.org/10.1007/s11427-010-4048-7

Mirazimi A, Nilsson M, Svensson L (1998) The molecular chaperone calnexin interacts with the NSP4 enterotoxin of rotavirus in vivo and in vitro. J Virol 72(11):8705–8709

Miura S, Kawana K, Schust DJ, Fujii T, Yokoyama T, Iwasawa Y, Nagamatsu T, Adachi K, Tomio A, Tomio K, Kojima S, Yasugi T, Kozuma S, Taketani Y (2010) CD1d, a sentinel molecule bridging innate and adaptive immunity, is downregulated by the human papillomavirus (HPV) E5 protein: a possible mechanism for immune evasion by HPV. J Virol 84(22):11614–11623. https://doi.org/10.1128/JVI.01053-10

Mizushima H, Hijikata M, Asabe SI, Hirota M, Kimura K, Shimotohno K (1994) Two hepatitis C virus glycoprotein E2 products with different C termini. J Virol 68(10):6215–6222

Moll M, Andersson SK, Smed-Sörensen A, Sandberg JK (2010) Inhibition of lipid antigen presentation in dendritic cells by HIV-1 Vpu interference with CD1d recycling from endosomal compartments. Blood 116(11):1876–1884. https://doi.org/10.1182/blood-2009-09-243667

Montserret R, Saint N, Vanbelle C, Salvay AG, Simorre JP, Ebel C, Sapay N, Renisio JG, Bockmann A, Steinmann E, Pietschmann T, Dubuisson J, Chipot C, Penin F (2010) NMR structure and ion channel activity of the p7 protein from hepatitis C virus. J Biol Chem 285(41):31446–31461

Moradpour D, Penin F (2013) Hepatitis C virus proteins: from structure to function. In: Hepatitis C virus: from molecular virology to antiviral therapy 369, pp 113–142. https://doi.org/10.1007/978-3-642-27340-7_5

Müller M, Prescott EL, Wasson CW, MacDonald A (2015) Human papillomavirus E5 oncoprotein: function and potential target for antiviral therapeutics. Futur Virol 10(1):27–39. https://doi.org/10.2217/fvl.14.99

Münz C (2011) Beclin-1 targeting for viral immune escape. Virus 3(7):1166–1178. https://doi.org/10.3390/v3071166

Nappi VM, Schaefer JA, Petti LM (2002) Molecular examination of the transmembrane requirements of the platelet-derived growth factor β receptor for a productive interaction with the bovine papillomavirus E5 oncoprotein. J Biol Chem 277(49):47149–47159. https://doi.org/10.1074/jbc.M209582200

Neil SJ, Eastman SW, Jouvenet N, Bieniasz PD (2006) HIV-1 Vpu promotes release and prevents endocytosis of nascent retrovirus particles from the plasma membrane. PLoS Path 2(5). https://doi.org/10.1371/journal.ppat.0020039

Neil SJD, Zang T, Bieniasz PD (2008) Tetherin inhibits retrovirus release and is antagonized by HIV-1 Vpu. Nature 451(7177):425–430. https://doi.org/10.1038/nature06553

Neumann AU, Lam NP, Dahari H, Gretch DR, Wiley TE, Layden TJ, Perelson AS (1998) Hepatitis C viral dynamics in vivo and the antiviral efficacy of interferon-alpha therapy. Science 282(5386):103–107. https://doi.org/10.1126/science.282.5386.103

Neumann G, Noda T, Kawaoka Y (2009) Emergence and pandemic potential of swine-origin H1N1 influenza virus. Nature 459(7249):931–939

Nieto-Torres JL, DeDiego ML, Álvarez E, Jiménez-Guardeño JM, Regla-Nava JA, Llorente M, Kremer L, Shuo S, Enjuanes L (2011) Subcellular location and topology of severe acute respiratory syndrome coronavirus envelope protein. Virology 415(2):69–82. https://doi.org/10.1016/j.virol.2011.03.029

Nieto-Torres JL, Dediego ML, Verdia-Baguena C, Jimenez-Guardeno JM, Regla-Nava JA, Fernandez-Delgado R, Castano-Rodriguez C, Alcaraz A, Torres J, Aguilella VM, Enjuanes L (2014) Severe acute respiratory syndrome coronavirus envelope protein ion channel activity promotes virus fitness and pathogenesis. PLoS Path 10(5):e1004077. https://doi.org/10.1371/journal.ppat.1004077

Nieto-Torres JL, Verdia-Báguena C, Jimenez-Guardeño JM, Regla-Nava JA, Castaño-Rodriguez C, Fernandez-Delgado R, Torres J, Aguilella VM, Enjuanes L (2015) Severe acute respiratory syndrome coronavirus E protein transports calcium ions and activates the NLRP3 inflammasome. In press, Virology

Nieva JL, Carrasco L (2015) Viroporins: structures and functions beyond cell membrane permeabilization. Virus 7(10):5169–5171. https://doi.org/10.3390/v7102866

Oelze I, Kartenbeck J, Crusius K, Alonso A (1995) Human papillomavirus type 16 E5 protein affects cell-cell communication in an epithelial cell line. J Virol 69(7):4489–4494

Okada Y, Suzuki T, Sunden Y, Orba Y, Kose S, Imamoto N, Takahashi H, Tanaka S, Hall WW, Nagashima K, Sawa H (2005) Dissociation of heterochromatin protein 1 from lamin B receptor induced by human polyomavirus agnoprotein: role in nuclear egress of viral particles. EMBO Rep 6(5):452–457. https://doi.org/10.1038/sj.embor.7400406

Olmsted RA, Collins PL (1989) The 1A protein of respiratory syncytial virus is an integral membrane protein present as multiple, structurally distinct species. J Virol 63(5):2019–2029

Opella SJ (2015) Relating structure and function of viral membrane-spanning miniproteins. Curr Opin Virol 12:121–125. https://doi.org/10.1016/j.coviro.2015.05.006

Ouyang B, Xie S, Berardi MJ, Zhao X, Dev J, Yu W, Sun B, Chou JJ (2013) Unusual architecture of the p7 channel from hepatitis C virus. Nature 498(7455):521–525

Pan J, Peng X, Gao Y, Li Z, Lu X, Chen Y, Ishaq M, Liu D, DeDiego ML, Enjuanes L, Guo D (2008) Genome-wide analysis of protein-protein interactions and involvement of viral proteins in SARS-CoV replication. PLoS One 3(10). https://doi.org/10.1371/journal.pone.0003299

Parr RD, Storey SM, Mitchell DM, McIntosh AL, Zhou M, Mir KD, Ball JM (2006) The rotavirus enterotoxin NSP4 directly interacts with the caveolar structural protein caveolin-1. J Virol 80(6):2842–2854. https://doi.org/10.1128/JVI.80.6.2842-2854.2006

Pervushin K, Tan E, Parthasarathy K, Lin X, Jiang FL, Yu D, Vararattanavech A, Tuck WS, Ding XL, Torres J (2009) Structure and inhibition of the SARS coronavirus envelope protein ion channel. PLoS Path 5(7)

Petti L, DiMaio D (1992) Stable association between the bovine papillomavirus E5 transforming protein and activated platelet-derived growth factor receptor in transformed mouse cells. Proc Natl Acad Sci U S A 89(15):6736–6740. https://doi.org/10.1073/pnas.89.15.6736

Pham T, Perry JL, Dosey TL, Delcour AH, Hyser JM (2017) The rotavirus NSP4 Viroporin domain is a calcium-conducting Ion Channel. Sci Rep 7. https://doi.org/10.1038/srep43487

Pietschmann T, Kaul A, Koutsoudakis G, Shavinskaya A, Kallis S, Steinmann E, Abid K, Negro F, Dreux M, Cosset FL, Bartenschlager R (2006) Construction and characterization of infectious intragenotypic and intergenotypic hepatitis C virus chimeras. Proceedings of the Natural Academy of Sciences USA 103(19):7408–7413

Polack FP, Irusta PM, Hoffman SJ, Schiatti MP, Melendi GA, Delgado MF, Laham FR, Thumar B, Hendry RM, Melero JA, Karron RA, Collins PL, Kleeberger SR (2005) The cysteine-rich region of respiratory syncytial virus attachment protein inhibits innate immunity elicited by the virus and endotoxin. Proc Natl Acad Sci U S A 102(25):8996–9001. https://doi.org/10.1073/pnas.0409478102

Popescu CI, Callens N, Trinel D, Roingeard P, Moradpour D, Descamps V, Duverlie G, Penin F, Héliot L, Rouillé Y, Dubuisson J (2011) NS2 protein of hepatitis C virus interacts with structural and non-structural proteins towards virus assembly. PLoS Path 7(2). https://doi.org/10.1371/journal.ppat.1001278

Qi H, Chu V, Wu NC, Chen Z, Truong S, Brar G, Su SY, Du Y, Arumugaswami V, Olson CA, Chen SH, Lin CY, Wu TT, Sun R (2017) Systematic identification of anti-interferon function on hepatitis C virus genome reveals p7 as an immune evasion protein. Proc Natl Acad Sci U S A 114(8):2018–2023. https://doi.org/10.1073/pnas.1614623114

Quistgaard EM, Low C, Moberg P, Guettou F, Maddi K, Nordlund P (2013) Structural and biophysical characterization of the cytoplasmic domains of human BAP29 and BAP31. PLoS One 8(8):e71111. https://doi.org/10.1371/journal.pone.0071111

Ramirez P, Famiglietti M, Sowrirajan B, DePaula-Silva A, Rodesch C, Barker E, Bosque A, Planelles V (2014) Downmodulation of CCR7 by HIV-1 Vpu results in impaired migration and chemotactic Signaling within CD4+ T cells. Cell Rep 7(6):2019–2030. https://doi.org/10.1016/j.celrep.2014.05.015

Ravi LI, Li L, Sutejo R, Chen H, Wong PS, Tan BH, Sugrue RJ (2013) A systems-based approach to analyse the host response in murine lung macrophages challenged with respiratory syncytial virus. BMC Genomics 14:190. https://doi.org/10.1186/1471-2164-14-190

Raynal P, Pollard HB (1994) Annexins: the problem of assessing the biological role for a gene family of multifunctional calcium- and phospholipid-binding proteins. BBA - Reviews on Biomembranes 1197(1):63–93. https://doi.org/10.1016/0304-4157(94)90019-1

Regan JA, Laimins LA (2008) Bap31 is a novel target of the human papillomavirus E5 protein. J Virol 82(20):10042–10051. https://doi.org/10.1128/JVI.01240-08

Regla-Nava JA, Nieto-Torres JL, Jimenez-Guardeño JM, Fernandez-Delgado R, Fett C, Castaño-Rodríguez C, Perlman S, Enjuanes L, De Diego ML (2015) Severe acute respiratory syndrome coronaviruses with mutations in the E protein are attenuated and promising vaccine candidates. J Virol 89(7):3870–3887. https://doi.org/10.1128/JVI.03566-14

Rixon HW, Brown G, Aitken J, McDonald T, Graham S, Sugrue RJ (2004) The small hydrophobic (SH) protein accumulates within lipid-raft structures of the Golgi complex during respiratory syncytial virus infection. J Gen Virol 85(Pt 5):1153–1165

Rixon HW, Brown G, Murray JT, Sugrue RJ (2005) The respiratory syncytial virus small hydrophobic protein is phosphorylated via a mitogen-activated protein kinase p38-dependent tyrosine kinase activity during virus infection. J Gen Virol 86(Pt 2):375–384

Rodríguez MI, Finbow ME, Alonso A (2000) Binding of human papillomavirus 16 E5 to the 16 kDa subunit c (proteolipid) of the vacuolar H+-ATPase can be dissociated from the E5-mediated epidermal growth factor receptor overactivation. Oncogene 19(33):3727–3732

Rossman JS, Lamb RA (2009) Autophagy, apoptosis, and the influenza virus M2 protein. Cell Host and Microbe 6(4):299–300. https://doi.org/10.1016/j.chom.2009.09.009

Rossman JS, Lamb RA (2011) Influenza virus assembly and budding. Virology 411(2):229–236. https://doi.org/10.1016/j.virol.2010.12.003

Ruch TR, Machamer CE (2011) The hydrophobic domain of infectious bronchitis virus e protein alters the host secretory pathway and is important for release of infectious virus. J Virol 85(2):675–685. https://doi.org/10.1128/JVI.01570-10

Russell RF, McDonald JU, Ivanova M, Zhong Z, Bukreyev A, Tregoning JS (2015) Partial attenuation of respiratory syncytial virus with a deletion of a small hydrophobic gene is associated with elevated interleukin-1beta responses. J Virol 89(17):8974–8981. https://doi.org/10.1128/JVI.01070-15

Safak M, Barrucco R, Darbinyan A, Okada Y, Nagashima K, Khalili K (2001) Interaction of JC virus Agno protein with T antigen modulates transcription and replication of the viral genome in glial cells. J Virol 75(3):1476–1486. https://doi.org/10.1128/JVI.75.3.1476-1486.2001

Safak M, Sadowska B, Barrucco R, Khalili K (2002) Functional interaction between JC virus late regulatory agnoprotein and cellular Y-box binding transcription factor, YB-1. J Virol 76(8):3828–3838. https://doi.org/10.1128/JVI.76.8.3828-3838.2002

Sakaguchi T, Tu QA, Pinto LH, Lamb RA (1997) The active oligomeric state of the minimalistic influenza virus M-2 ion channel is a tetramer. Proc Natl Acad Sci U S A 94(10):5000–5005

Sakai A, Claire MS, Faulk K, Govindarajan S, Emerson SU, Purcell RH, Bukh J (2003) The p7 polypeptide of hepatitis C virus is critical for infectivity and contains functionally important genotype-specific sequences. Proc Natl Acad Sci U S A 100(20):11646–11651

Saribas AS, Abou-Gharbia M, Childers W, Sariyer IK, White MK, Safak M (2013) Essential roles of Leu/Ile/Phe-rich domain of JC virus agnoprotein in dimer/oligomer formation, protein stability and splicing of viral transcripts. Virology 443(1):161–176. https://doi.org/10.1016/j.virol.2013.05.003

Sariyer IK, Khalili K, Safak M (2008) Dephosphorylation of JC virus agnoprotein by protein phosphatase 2A: inhibition by small t antigen. Virology 375(2):464–479. https://doi.org/10.1016/j.virol.2008.02.020

Sariyer IK, Saribas AS, White MK, Safak M (2011) Infection by agnoprotein-negative mutants of polyomavirus JC and SV40 results in the release of virions that are mostly deficient in DNA content. Virol J 8. https://doi.org/10.1186/1743-422x-8-255

Sastri NP, Viskovska M, Hyser JM, Tanner MR, Horton LB, Sankaran B, Venkataram Prasad BV, Estes MK (2014) Structural plasticity of the coiled-coil domain of rotavirus NSP4. J Virol 88(23):13602–13612. https://doi.org/10.1128/JVI.02227-14

Schapiro F, Sparkowski J, Adduci A, Suprynowicz F, Schlegel R, Grinstein S (2000) Golgi alkalinization by the papillomavirus E5 oncoprotein. J Cell Biol 148(2):305–315. https://doi.org/10.1083/jcb.148.2.305

Schnell JR, Chou JJ (2008) Structure and mechanism of the M2 proton channel of influenza a virus. Nature 451(7178):591–595

Schoggins JW, Wilson SJ, Panis M, Murphy MY, Jones CT, Bieniasz P, Rice CM (2011) A diverse range of gene products are effectors of the type i interferon antiviral response. Nature 472(7344):481–485. https://doi.org/10.1038/nature09907

Schubert U, Ferrer-Montiel AV, Oblatt-Montal M, Henklein P, Strebel K, Montal M (1996) Identification of an ion channel activity of the Vpu transmembrane domain and its involvement in the regulation of virus release from HIV-1-infected cells. FEBS Lett 398(1):12–18

Seo NS, Zeng CQY, Hyser JM, Utama B, Crawford SE, Kim KJ, Höök M, Estes MK (2008) Integrins $\alpha1\beta1$ and $\alpha2\beta1$ are receptors for the rotavirus enterotoxin. Proc Natl Acad Sci U S A 105(26):8811–8818. https://doi.org/10.1073/pnas.0803934105

Serrano P, Johnson MA, Almeida MS, Horst R, Herrmann T, Joseph JS, Neuman BW, Subramanian V, Saikatendu KS, Buchmeier MJ, Stevens RC, Kuhn P, Wüthrich K (2007) Nuclear magnetic resonance structure of the N-terminal domain of nonstructural protein 3 from the severe acute respiratory syndrome coronavirus. J Virol 81(21):12049–12060. https://doi.org/10.1128/JVI.00969-07

Shah AH, Sowrirajan B, Davis ZB, Ward JP, Campbell EM, Planelles V, Barker E (2010) Degranulation of natural killer cells following interaction with HIV-1-infected cells is hindered by downmodulation of NTB-A by Vpu. Cell Host and Microbe 8(5):397–409. https://doi.org/10.1016/j.chom.2010.10.008

Shanmugam S, Saravanabalaji D, Yi M (2015) Detergent-resistant membrane association of NS2 and E2 during hepatitis C virus replication. J Virol 89(8):4562–4574. https://doi.org/10.1128/JVI.00123-15

Shapira SD, Gat-Viks I, Shum BOV, Dricot A, de Grace MM, Wu L, Gupta PB, Hao T, Silver SJ, Root DE, Hill DE, Regev A, Hacohen N (2009) A physical and regulatory map of host-influenza interactions reveals pathways in H1N1 infection. Cell 139(7):1255–1267. https://doi.org/10.1016/j.cell.2009.12.018

Sharma K, Surjit M, Satija N, Liu B, Chow VTK, Lai SK (2007) The 3a accessory protein of SARS coronavirus specifically interacts with the 5'UTR of its genomic RNA, using a unique 75 amino acid interaction domain. Biochemistry 46(22):6488–6499. https://doi.org/10.1021/bi062057p

Shen S, Lin PS, Chao YC, Zhang A, Yang X, Lim SG, Hong W, Tan YJ (2005) The severe acute respiratory syndrome coronavirus 3a is a novel structural protein. Biochem Biophys Res Commun 330(1):286–292. https://doi.org/10.1016/j.bbrc.2005.02.153

Simon V, Bloch N, Landau NR (2015) Intrinsic host restrictions to HIV-1 and mechanisms of viral escape. Nat Immunol 16(6):546–553. https://doi.org/10.1038/ni.3156

Singh SK, Möckel L, Thiagarajan-Rosenkranz P, Wittlich M, Willbold D, Koenig BW (2012) Mapping the interaction between the cytoplasmic domains of HIV-1 viral protein U and human CD4 with NMR spectroscopy. FEBS J 279(19):3705–3714. https://doi.org/10.1111/j.1742-4658.2012.08732.x

Skasko M, Wang Y, Tian Y, Tokarev A, Munguia J, Ruiz A, Stephens EB, Opella SJ, Guatelli O (2012) HIV-1 Vpu protein antagonizes innate restriction factor BST-2 via lipid-embedded helix-helix interactions. J Biol Chem 287(1):58–67. https://doi.org/10.1074/jbc.M111.296772

Sou YS, Tanida I, Komatsu M, Ueno T, Kominami E (2006) Phosphatidylserine in addition to phosphatidylethanolamine is an in vitro target of the mammalian Atg8 modifiers, LC3, GABARAP, and GATE-16. J Biol Chem 281(6):3017–3024. https://doi.org/10.1074/jbc.M505888200

Steinmann E, Penin F, Kallis S, Patel AH, Bartenschlager R, Pietschmann T (2007) Hepatitis C virus p7 protein is crucial for assembly and release of infectious virions. PLoS Path 3(7):0962–0971

Strebel K (2014) HIV-1 Vpu - an ion channel in search of a job. Biochim Biophys Acta 1838(4):1074–1081. https://doi.org/10.1016/j.bbamem.2013.06.029

Strebel K, Klimkait T, Martin MA (1988) A novel gene of HIV-1, vpu, and its 16-kilodalton product. Science 241(4870):1221–1223

Su S, Wong G, Shi W, Liu J, Lai ACK, Zhou J, Liu W, Bi Y, Gao GF (2016) Epidemiology, genetic recombination, and pathogenesis of coronaviruses. Trends Microbiol 24(6):490–502. https://doi.org/10.1016/j.tim.2016.03.003

Sugden SM, Pham TNQ, Cohen ÉA (2017) HIV-1 Vpu downmodulates ICAM-1 expression, resulting in decreased killing of infected CD4+ T cells by NK cells. J Virol 91(8). https://doi.org/10.1128/JVI.02442-16

Sun L, Hemgård GV, Susanto SA, Wirth M (2010) Caveolin-1 influences human influenza A virus (H1N1) multiplication in cell culture. Virol J 7. https://doi.org/10.1186/1743-422X-7-108

Surti T, Klein O, Aschheim K, DiMaio D, Smith SO (1998) Structural models of the bovine papillomavirus E5 protein. Proteins Struct Funct Genet 33(4):601–612. https://doi.org/10.1002/(SICI)1097-0134(19981201)33:4<601::AID-PROT12>3.0.CO;2-I

Suzuki T, Okada Y, Sembat S, Orba Y, Yamanouchii S, Endo S, Tanaka S, Fujita T, Kuroda S, Nagashima K, Sawa H (2005) Identification of FEZ1 as a protein that interacts with JC virus agnoprotein and microtubules: role of agnoprotein-induced dissociation of FEZ1 from microtubules in viral propagation. J Biol Chem 280(26):24948–24956. https://doi.org/10.1074/jbc.M411499200

Suzuki T, Orba Y, Makino Y, Okada Y, Sunden Y, Hasegawa H, Hall WW, Sawa H (2013) Viroporin activity of the JC polyomavirus is regulated by interactions with the adaptor protein complex 3. Proc Natl Acad Sci U S A 110(46):18668–18673. https://doi.org/10.1073/pnas.1311457110

Suzuki T, Orba Y, Malcino Y, Okada Y, Sunden Y, Kimura T, Hasegawa H, Sata T, Hall WW, Sawa H (2010a) Disruption of intracellular vesicular trafficking by Agnoprotein is essential for Viroporin activity and JC virus replication. J Neurovirol 16:84–84

Suzuki T, Orba Y, Okada Y, Sunden Y, Kimura T, Tanaka S, Nagashima K, Hall WW, Sawa H (2010b) The human polyoma JC virus agnoprotein acts as a viroporin. PLoS Path 6(3):e1000801. https://doi.org/10.1371/journal.ppat.1000801

Suzuki T, Semba S, Sunden Y, Orba Y, Kobayashi S, Nagashima K, Kimura T, Hasegawa H, Sawa H (2012) Role of JC virus agnoprotein in virion formation. Microbiol Immunol 56(9):639–646. https://doi.org/10.1111/j.1348-0421.2012.00484.x

Takeuchi K, Lamb RA (1994) Influenza virus M2 protein ion channel activity stabilizes the native form of fowl plague virus hemagglutinin during intracellular transport. J Virol 68(2):911–919

Tan YJ, Lim SG, Hong W (2006) Understanding the accessory viral proteins unique to the severe acute respiratory syndrome (SARS) coronavirus. Antivir Res 72(2):78–88. https://doi.org/10.1016/j.antiviral.2006.05.010

Tan YJ, Teng E, Shen S, Tan THP, Goh PY, Fielding BC, Ooi EE, Tan HC, Lim SG, Hong W (2004) A novel severe acute respiratory syndrome coronavirus protein, U274, is transported to the cell surface and undergoes endocytosis. J Virol 78(13):6723–6734. https://doi.org/10.1128/JVI.78.13.6723-6734.2004

Tan YJ, Tham PY, Chan DZL, Chou CF, Shen S, Fielding BC, Tan THP, Lim SG, Hong W (2005) The severe acute respiratory syndrome coronavirus 3a protein up-regulates expression of fibrinogen in lung epithelial cells. J Virol 79(15):10083–10087. https://doi.org/10.1128/JVI.79.15.10083-10087.2005

Tapia LI, Shaw CA, Aideyan LO, Jewell AM, Dawson BC, Haq TR, Piedra PA (2014) Gene sequence variability of the three surface proteins of human respiratory syncytial virus (HRSV) in Texas. PLoS One 9(3). https://doi.org/10.1371/journal.pone.0090786

Tarassov K, Messier V, Landry CR, Radinovic S, Serna Molina MM, Shames I, Malitskaya Y, Vogel J, Bussey H, Michnick SW (2008) An in vivo map of the yeast protein interactome. Science 320(5882):1465–1470. https://doi.org/10.1126/science.1153878

Tate JE, Burton AH, Boschi-Pinto C, Parashar UD, Agocs M, Serhan F, De Oliveira L, Mwenda JM, Mihigo R, Ranjan Wijesinghe P, Abeysinghe N, Fox K, Paladin F (2016) Global, regional, and National Estimates of rotavirus mortality in children <5 years of age, 2000-2013. Clin Infect Dis 62:S96–S105. https://doi.org/10.1093/cid/civ1013

Taylor G, Wyld S, Valarcher JF, Guzman E, Thom M, Widdison S, Buchholz UJ (2014) Recombinant bovine respiratory syncytial virus with deletion of the SH gene induces increased apoptosis and pro-inflammatory cytokines in vitro, and is attenuated and induces protective immunity in calves. J Gen Virol 95:1244–1254. https://doi.org/10.1099/Vir.0.064931-0

Taylor JA, O'Brien JA, Yeager M (1996) The cytoplasmic tail of NSP4, the endoplasmic reticulum-localized non-structural glycoprotein of rotavirus, contains distinct virus binding and coiled coil domains. EMBO J 15(17):4469–4476

Teoh KT, Siu YL, Chan WL, Schluter MA, Liu CJ, Peiris JS, Bruzzone R, Margolis B, Nal B (2010) The SARS coronavirus E protein interacts with PALS1 and alters tight junction formation and epithelial morphogenesis. Mol Biol Cell 21(22):3838–3852. https://doi.org/10.1091/mbc.E10-04-0338

Terwilliger EF, Cohen EA, Lu Y, Sodroski JG, Haseltine WA (1989) Functional role of human immunodeficiency virus type 1 vpu. Proc Natl Acad Sci U S A 86(13):5163–5167

To J, Surya W, Fung TS, Li Y, Verdià-Bàguena C, Queralt-Martin M, Aguilella VM, Liu DX, Torres J (2017) Channel-inactivating mutations and their revertant mutants in the envelope protein of infectious bronchitis virus. J Virol 91(5). https://doi.org/10.1128/JVI.02158-16

To J, Surya W, Torres J (2016) Targeting the channel activity of Viroporins. Adv Protein Chem Struct Biol 104:307–355. https://doi.org/10.1016/bs.apcsb.2015.12.003

Tomakidi P, Cheng H, Kohl A, Komposch G, Alonso A (2000) Connexin 43 expression is down-regulated in raft cultures of human keratinocytes expressing the human papillomavirus type 16 E5 protein. Cell Tissue Res 301(2):323–327

Torres J, Maheswari U, Parthasarathy K, Ng LF, Liu DX, Gong XD (2007) Conductance and amantadine binding of a pore formed by a lysine-flanked transmembrane domain of SARS coronavirus envelope protein. Protein Sci 16(9):2065–2071. https://doi.org/10.1110/ps.062730007

Torres J, Parthasarathy K, Lin X, Saravanan R, Kukol A, Ding XL (2006) Model of a putative pore: the pentameric α-helical bundle of SARS coronavirus E protein in lipid bilayers. Biophys J 91(3):938–947

Triantafilou K, Kar S, Vakakis E, Kotecha S, Triantafilou M (2013) Human respiratory syncytial virus viroporin SH: a viral recognition pathway used by the host to signal inflammasome activation. Immunology 140:87–88

Van Damme N, Goff D, Katsura C, Jorgenson RL, Mitchell R, Johnson MC, Stephens EB, Guatelli J (2008) The interferon-induced protein BST-2 restricts HIV-1 release and is Downregulated from the cell surface by the viral Vpu protein. Cell Host and Microbe 3(4):245–252. https://doi.org/10.1016/j.chom.2008.03.001

Varga ZT, Grant A, Manicassamy B, Palese P (2012) Influenza virus protein pb1-f2 inhibits the induction of type I interferon by binding to mavs and decreasing mitochondrial membrane potential. J Virol 86(16):8359–8366. https://doi.org/10.1128/JVI.01122-12

Varga ZT, Palese P (2011) The influenza a virus protein PB1-F2: killing two birds with one stone? Virulence 2(6):542–546

Vassena L, Giuliani E, Koppensteiner H, Bolduan S, Schindler M, Doria M (2015) HIV-1 Nef and Vpu interfere with L-selectin (CD62L) cell surface expression to inhibit adhesion and signaling in infected CD4⁺ T lymphocytes. J Virol 89(10):5687–5700. https://doi.org/10.1128/JVI.00611-15

Venkatesh S, Bieniasz PD (2013) Mechanism of HIV-1 Virion entrapment by Tetherin. PLoS Path 9(7). https://doi.org/10.1371/journal.ppat.1003483

Verdia-Baguena C, Nieto-Torres JL, Alcaraz A, DeDiego ML, Torres J, Aguilella VM, Enjuanes L (2012) Coronavirus E protein forms ion channels with functionally and structurally-involved membrane lipids. Virology 432(2):485–494. https://doi.org/10.1016/j.virol.2012.07.005

Vieyres G, Dubuisson J, Pietschmann T (2014) Incorporation of hepatitis C virus E1 and E2 glycoproteins: the keystones on a peculiar virion. Virus 6(3):1149–1187. https://doi.org/10.3390/v6031149

Vigan R, Neil SJD (2010) Determinants of tetherin antagonism in the transmembrane domain of the human immunodeficiency virus type 1 Vpu protein. J Virol 84(24):12958–12970. https://doi.org/10.1128/JVI.01699-10

von Brunn A, Teepe C, Simpson JC, Pepperkok R, Friedel CC, Zimmer R, Roberts R, Baric R, Haas J (2007) Analysis of intraviral protein-protein interactions of the SARS coronavirus ORFeome. PLoS One 2(5). https://doi.org/10.1371/journal.pone.0000459

Wang CK, Pan L, Chen J, Zhang M (2010) Extensions of PDZ domains as important structural and functional elements. Protein and Cell 1(8):737–751. https://doi.org/10.1007/s13238-010-0099-6

Wang L, Fu B, Li W, Patil G, Liu L, Dorf ME, Li S (2017) Comparative influenza protein inter-actomes identify the role of plakophilin 2 in virus restriction. Nat Commun 8. https://doi.org/10.1038/ncomms13876

Watanabe R, Leser GP, Lamb RA (2011) Influenza virus is not restricted by tetherin whereas influ-enza VLP production is restricted by tetherin. Virology 417(1):50–56. https://doi.org/10.1016/j.virol.2011.05.006

Weiner LB, Polak MJ, Masaquel A, Mahadevia PJ (2011) Cost-effectiveness of respiratory syn-cytial virus prophylaxis with Palivizumab among preterm infants covered by Medicaid in the United States. Value Health 14(3):A118–A118

Wetherill LF, Holmes KK, Verow M, Müller M, Howell G, Harris M, Fishwick C, Stonehouse N, Foster R, Blair GE, Griffin S, Macdonald A (2012) High-risk human papillomavirus E5 oncoprotein displays channel-forming activity sensitive to small-molecule inhibitors. J Virol 86(9):5341–5351. https://doi.org/10.1128/JVI.06243-11

Willey RL, Buckler-White A, Strebel K (1994) Sequences present in the cytoplasmic domain of CD4 are necessary and sufficient to confer sensitivity to the human immunodeficiency virus type 1 Vpu protein. J Virol 68(2):1207–1212

Willey RL, Maldarelli F, Martin MA, Strebel K (1992) Human immunodeficiency virus type 1 Vpu protein induces rapid degradation of CD4. J Virol 66(12):7193–7200

Wong SLA, Chen Y, Chak MC, Chan CSM, Chan PKS, Chui YL, Kwok PF, Waye MMY, Tsui SKW, Chan HYE (2005) In vivo functional characterization of the SARS-coronavirus 3a pro-tein in drosophila. Biochem Biophys Res Commun 337(2):720–729. https://doi.org/10.1016/j.bbrc.2005.09.098

Wozniak AL, Griffin S, Rowlands D, Harris M, Yi M, Lemon SM, Weinman SA (2010) Intracellular proton conductance of the hepatitis C virus p7 protein and its contribution to infectious virus production. PLoS Pathog 6(9):e1001087. https://doi.org/10.1371/journal.ppat.1001087

Xu A, Bellamy AR, Taylor JA (2000) Immobilization of the early secretory pathway by a virus glycoprotein that binds to microtubules. EMBO J 19(23):6465–6474

Ye F, Zhang M (2013) Structures and target recognition modes of PDZ domains: recurring themes and emerging pictures. Biochem J 455(1):1–14. https://doi.org/10.1042/BJ20130783

Yi M, Ma Y, Yates J, Lemon SM (2007) Compensatory mutations in E1, p7, NS2, and NS3 enhance yields of cell culture-infectious intergenotypic chirneric hepatitis C virus. J Virol 81(2):629–638

Yuan X, Yao Z, Wu J, Zhou Y, Shan Y, Dong B, Zhao Z, Hua P, Chen J, Cong Y (2007) G1 phase cell cycle arrest induced by SARS-CoV 3a protein via the Cyclin D3/pRb pathway. Am J Respir Cell Mol Biol 37(1):9–19. https://doi.org/10.1165/rcmb.2005-0345RC

Zebedee SL, Lamb RA (1989) Growth restriction of influenza a virus by M2 protein antibody is genetically linked to the M1 protein. Proc Natl Acad Sci U S A 86(3):1061–1065

Zhang R, Wang K, Lv W, Yu W, Xie S, Xu K, Schwarz W, Xiong S, Sun B (2014) The ORF4a protein of human coronavirus 229E functions as a viroporin that regulates viral production. Biochim Biophys Acta Biomembr 1838(4):1088–1095. https://doi.org/10.1016/j.bbamem.2013.07.025

Zhang R, Wang K, Ping X, Yu W, Qian Z, Xiong S, Sun B (2015) The ns12.9 accessory protein of human coronavirus OC43 is a viroporin involved in virion morphogenesis and pathogenesis. J Virol 89(22):11383–11395. https://doi.org/10.1128/JVI.01986-15

Zhu H, Bilgin M, Bangham R, Hall D, Casamayor A, Bertone P, Lan N, Jansen R, Bidlingmaier S, Houfek T, Mitchell T, Miller P, Dean RA, Gerstein M, Snyder M (2001) Global analysis of pro-tein activities using proteome chips. Science 293(5537):2101–2105. https://doi.org/10.1126/science.1062191

Zhu P, Liang L, Shao X, Luo W, Jiang S, Zhao Q, Sun N, Zhao Y, Li J, Wang J, Zhou Y, Zhang J, Wang G, Jiang L, Chen H, Li C (2017) Host cellular protein TRAPPC6AΔ interacts with influ-enza a virus M2 protein and regulates viral propagation by modulating M2 trafficking. J Virol 91(1). https://doi.org/10.1128/JVI.01757-16

Zou P, Wu F, Lu L, Huang JH, Chen YH (2009) The cytoplasmic domain of influenza M2 protein interacts with caveolin-1. Arch Biochem Biophys 486(2):150–154. https://doi.org/10.1016/j.abb.2009.02.001

Chapter 16
Protein Complexes and Virus-Like Particle Technology

Andris Zeltins

Introduction

Proteins are the most multifunctional macromolecules in living organisms. As products of translated genetic information from corresponding coding genes, they play a central role in such relevant functions as catalysis of thousands of biochemical processes, providing molecular architecture for different subcellular components and controlling such crucial processes as cell division and intracellular signaling. Structurally, proteins are built up from approximately 1000 protein domain types, and these are involved in approximately 10,000 types of different protein-protein interactions (Aloy and Russell 2004).

Proteins perform their functions predominantly in a form of different complexes, which are products of specific protein-protein, protein-nucleic acid, protein-lipid, protein-polysaccharide and protein-low-molecular compound interactions. For example, human cells contain approximately 4300 proteins which are involved in 13,900 identified interactions, whereas 630 interaction partners are known for proteins encoded by bacteriophage lambda. However, these numbers may represent only a minor part of all possible protein interactions in living cells (Hao et al. 2016).

Protein complexes are formed by physical interactions between proteins and their binding partners based on electrostatic, hydrophobic, hydrogen bonds and Van der Waals forces. In special cases, the protein complex formation results in covalent bonds, which is the strongest and most stable association between components of the complex.

The formation activity of a complex can be characterized by such biophysical parameter as binding constant, or association constant K_a, which characterizes the

A. Zeltins (✉)
Latvian Biomedical Research and Study Centre, Riga, Latvia
e-mail: anze@biomed.lu.lv

© Springer Nature Singapore Pte Ltd. 2018
J. R. Harris, D. Bhella (eds.), *Virus Protein and Nucleoprotein Complexes*,
Subcellular Biochemistry 88, https://doi.org/10.1007/978-981-10-8456-0_16

equilibrium between formed complexes and free, unbound components. The dissociation constant K_d is the inverse of the association constant. To characterize the complex stability, the apparent dissociation constant K_D is frequently used, which is approximately equal to that critical concentration, when the concentrations of free binding partners and the complex are equivalent (Katen and Zlotnick 2009).

The stability of protein complexes can be highly different. Their dissociation constant K_D values can cover very broad range from femtomolar (fM) to millimolar (mM) concentrations. Protein interactions with their binding partners with the K_D values in the fM to nM range are considered as stable, whereas protein complexes with μM or mM values are regarded as transient. It should be noted that transient interactions are dominating in living cells, because the ability of proteins for reversible binding is a prerequisite of many biochemical and biophysical processes (Liu et al. 2016). Typical examples are enzyme-catalysed reactions, DNA replication, signal transduction cascades and many others. Also viral structural proteins interact mostly transiently with each other as well as with host cell proteins, because assembly/disassembly processes are an essential part of viral life cycle. Viral capsids have to be stable to ensure the protection of the enclosed genome during passage between susceptible hosts; on the other hand, they have to be able to disassemble in infected cells to start a new round of infection.

Since virus discovery more than 100 years ago, the knowledge about the structure and functions of virus-encoded proteins as well as about specific interactions in virus-infected cells is continuously growing. Accordingly, Google Scholar searches using a key phrase "virus-host interaction" result in 26,000 entries by 1987, 141,000 by 1997 and 1,100,000 by 2007; actual number of entries reaches more than 1,700,000 (2017). The studies of last years about virus-host interactomas revealed thousands of complexes which contributed considerably to our understanding of the cell processes during viral infections. It is important to note that these interactions are not universal but can be totally different depending of the virus type (see recent reviews Lum and Cristea 2016; Wang 2015; Korth et al. 2013).The interaction studies are important not only from academic point of view, but also for technology development, providing the knowledge about potential targets for antiviral therapies.

Viruses can be regarded as natural systems of nanometer scale that "exist at the interface of living organisms and nonliving biological machines" (Schwarz et al. 2017). As suggested by Watson and Crick already more than 60 years ago, morphologically viruses represent either rod- or sphere-like structures (Crick and Watson 1956). Structural studies during subsequent decades revealed that viruses are more diverse in composition, size and three-dimensional morphology and are mainly assembled in particles of different icosahedral or helical structures (Fig.16.1). However, some viral structures are not organized symmetrically. Simplest viruses are built up from single or few structural proteins and nucleic acid(s), whereas more complex viruses contain also lipid envelopes and different sugar molecules. Viral genomes are encoded by single- or double-stranded DNA or RNA. More detailed information about the diversity of viral structures can be obtained in recent review

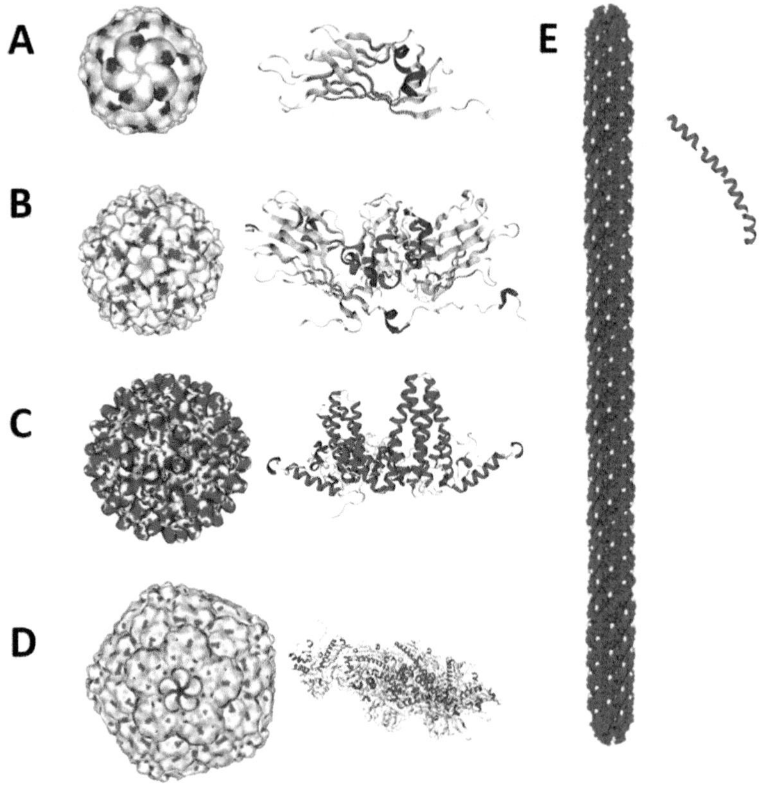

Fig. 16.1 Structures of icosahedral and helical viruses.
Images were created using Protein Data Bank and NGL 3D viewer (Rose and Hildebrand 2015).
On the left side of corresponding panel, the surface model of the virus is shown, on the right – side view of the asymmetric unit of structure. α-helices are shown in red, β-sheets – in yellow.
A satellite tobacco mosaic virus (STMV) structure (T = 1 symmetry, diameter 17 nm). Image of 1A34 (Larson et al. 1998); B cowpea chlorotic mottle virus (CCMV) structure (T = 3 symmetry, diameter 29 nm). Image of 1CWP (Speir et al. 1995); C hepatitis B virus core (HBc) structure (T = 4 symmetry, diameter 35 nm). Image of 3J2V (Yu et al. 2013); D bacteriophage P22 structure (T = 7 symmetry, diameter 50 nm). Image of 5UU5 (Hryc et al. 2017); E bacteriophage fd structure (flexible filaments, 7 nm × 880 nm). Image of 2HI5 (Wang et al. 2006)

articles (Zlotnick et al. 2016; Tars 2016; Solovyev and Makarov 2016; Mannige and Brooks III 2010).

As viruses structurally represent perfect natural designs, their structures and functional properties serve as inspiring generator of ideas for new nanomaterials (Narayanan and Han 2017). Biotechnologists working with virus-derived structures frequently use the term "virus-like particles" (VLPs). The phrase "virus-like bodies" or particles is known since the 1930s of the last century and was used to describe the structures found in exudates of rheumatic patients (Eagles 1939). Today, we define the virus-like particles as multisubunit protein complexes capable to self-assembly, structurally resembling their progenitor native viruses. VLPs can be obtained from heterogeneous, recombinant host expression systems in nearly unlimited amounts. They principally differ from native viruses by the fact that they do not contain original virus nucleic acids and therefore are not infectious. If recombinant technology is applied to VLPs, it allows different manipulations with VLPs, including targeted insertions of functional amino acid stretches in VLP structures at defined locations. These properties of VLPs make them as nearly ideal building blocks for generating of wide variety of new nanomaterials for different applications (Zeltins 2016).

This chapter is a brief overview of assembly mechanisms of icosahedral and helical viruses and VLPs and provides most interesting examples on how continuously growing information about viral and non-viral protein complexes influenced and still continue to influence the development of virus-like particle technology.

Protein Complexes and Interactions in Virus Assembly Processes

Virus life cycle can be divided in the following steps: virus attachment to the cell surface, virus entry in the cells, virus disassembly or uncoating, genome replication and virus-encoded gene expression, assembly of viral particles and, finally, the release of infectious particles (Lum and Cristea 2016). Virus assembly in infected cells is a highly complicated, multistep process, where specific protein interactions with their binding partners ensure the formation of reproducible viral particles in the cellular milieu, which contains thousands of other proteins, nucleic acids, low-molecular compounds and subcellular structures (Perlmutter and Hagan 2015).

As virus-like technology is based on virus-inspired artificial structures, the intrinsic ability of viral structural proteins to self-assemble at controlled conditions is absolutely the central prerequisite for process developments. Therefore, the basic knowledge about the mechanisms of viral particle formation is highly important for designing of industrial applications. This section focuses on the mechanisms – how the viruses form their perfectly ordered structures.

First of all, the ability of viral structural proteins to self-assemble is determined by their amino acid (AA) composition and three-dimensional structure. Viral coat proteins are structurally flexible molecules which frequently contain positively charged domains, including arginine-rich motifs (ARMs) to ensure the interactions with specific nucleic acids, as well as different hydrophobic and charged AA stretches for making contacts between subunits. However, the formation of these contacts is primarily determined by hydrophobic interactions. Electrostatic attraction/repulsion together with van de Waals and hydrogen bonding interactions ensures the necessary directional specificity for further particle formation (Perlmutter and Hagan 2015).

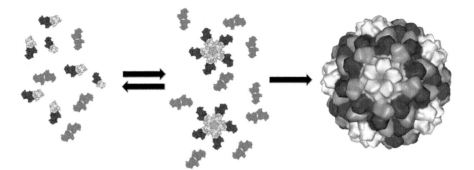

Fig. 16.2 Schematic representation of assembly mechanism for icosahedral virus.
The model is built based on cowpea chlorotic mottle virus (CCMV) structural studies. Virus assembly starts with the formation of different quasi-equivalent conformations of CP dimers (differently colored in the image). Then, after formation of relatively stable pentamers of dimers (POD), the assembly completed by using both POD and free dimers (Johnson et al. 2004). Images were created using Protein Data Bank (1CWP; Speir et al. 1995) and NGL 3D viewer (Rose and Hildebrand 2015)

Structurally, viral coat proteins (CPs) can be built up from α-helices (Fig.16.1), for example, hepatitis B core protein (HBc; Wynne et al. 1999) and tobacco mosaic virus (TMV; Holmes et al. 1975), as well as from β-sheets, which forms characteristic β-barrel structures and is exploited by many plant and animal virus CPs (Zlotnick et al. 2000).

Icosahedral Viruses

For icosahedral viruses, the assembly of viral capsid starts with formation of stable CP-CP dimers and/or oligomers. It is known that some viral CPs are able to form empty particles without encapsidated nucleic acids; these VLPs have been used as model systems to elucidate the mechanisms of capsid formation. As shown in experiments with HBc (Wingfield et al. 1995), MS2 bacteriophage (Lima et al. 2006) and cowpea chlorotic mottle virus (CCMV; Adolph and Butler 1974), the CP dimers serve as a nucleus for subsequent generation of viral structures (Fig. 16.2); CP monomers cannot be isolated at native conditions. Moreover, it is possible to abolish the viral particle formation and obtain stable CP-CP dimers by introducing of just a single mutation in CP sequence, as shown with Sesbania mosaic virus (SeMV; Pappachan et al. 2009).

The dissociation constant K_D values for HBc dimer CPs are in the millimolar range, suggesting weak interactions between individual homodimer blocks. However, a network of these weak interactions finally results in relatively stable viral capsids.

Moreover, interaction network in the assembled particle results in considerably weaker concentration dependence for capsid dissociation than one can expect from assembly experiments. In other words, already formed viral particles are stable at very low concentrations (hysteresis effect; Singh and Zlotnick 2003).

In vitro capsid assembly tests have shown that ionic strength and pH of the solution influence significantly the capsid formation. Apparent K_D values or pseudo-critical concentrations of HBc assembly can vary up to 20 times at different salt concentrations (0.77–14 µM; Ceres and Zlotnick 2002). For plant virus CCMV, both pH and salt concentration are dominating factors influencing the capsid formation (Comas-Garcia et al. 2014), whereas the assembly reaction is not temperature-dependent over the range 5–25 °C (Adolph and Butler 1976; Zlotnick and Mukhopadhyay 2011).

The minimal oligomeric structure for the initiation of capsid formation can be also pentamer of CPs, for example, poliovirus SV40, which is built up of 72 pentamers (Liddington et al. 1991) or other CP oligomers (Prevelige et al. 1993). These oligomeric structures can be further stabilized by permanent or transient disulfide bonds (Kushima et al. 2010; Kobayashi et al. 2013). Additionally, many viruses contain cations, which ensure further stabilization of the particles. For example, five adjacent acidic AA from two CP molecules in plant virus CCMV structure bind Ca^{2+} ions with comparably low affinity (K_D = 1.97 mM; Basu et al. 2003).

Many viruses do not form particles without encapsidated nucleic acids. For some viruses, nucleic acid secondary structures which are located in viral genomes are suggested as nucleation signals of assembly. These signals interact specifically with positively charged segments of CP oligomers and ensure encapsidation of viral genomes. Moreover, binding of nucleic acid elicits the conformational changes in CPs and maximizes the subunit-subunit contacts, which are necessary for particle formation. Latest studies on RNA viruses suggest that viral genomes can contain not only single origin of assembly (OAS) per genome but even several tens of short RNA stem-loop structures. These RNA secondary structures bind corresponding CP molecules with high affinity in the nanomolar concentration range; additionally, higher number of such stemloops results in lower CP concentrations necessary for particle formation (Stockley et al. 2016). It agrees well to the idea that viruses and VLPs preferently encapsidate longer nucleic acids (Cerqueira et al. 2015).

Additionally, RNA secondary structure predictions suggest that viral genomic RNAs can be more tightly packaged than non-viral RNAs present in host cells. The nucleotide sequence-dependent intrinsinc property of viral RNA to form compact structures allows to encapsidate very long nucleic acid(s) into spatially limited volume of isometric viral capsids (Yoffe et al. 2008).

Typically, viral nucleic acid size determines the lengths helical viruses (this aspect will be discussed in next section). However, it influences also the shape of icosahedral VLPs. As shown in model experiments with such plant multipartite RNA virus as CCMV, the CP/RNA complex is determining the VLP formation, and, surprisingly, if the RNA sizes exceed the lengths of specific viral RNA, it can stimulate the assembly up to 4 VLPs sharing the same 12,000-nt-long RNA molecule

(Cadena-Nava et al. 2012). Similar observations have been made also using negatively charged synthetic polymer instead of RNA (Brasch and Cornelissen 2012).

As suggested by Perlmutter and Hagan (2015), RNA can be encapsidated in viral particles according two assembly mechanisms. According to the "nucleation and growth" mechanism, capsid assembly starts with CP oligomer formation which is stabilized by nucleic acid. Then, during assembly growth phase, CP oligomers are reversibly added to the growing capsid until the particle formation is complete. Alternatively, according to "en masse" mechanism, CP subunits randomly bind to RNA and then due to CP-CP interactions rearrange into complete, icosahedral particles. The first mechanism is suggested to be effective at high ionic strength conditions and acidic pH, when CP-CP interactions dominate and their interactions with nucleic acids are weak. The "en masse" capsid assembly can take place at low salt conditions, which involves weak CP-CP interactions.

Some viruses are involving special proteins in assembly process called "scaffolding" proteins, for example, herpesviruses and bacteriophages. The scaffolding proteins are involved in procapsid formation, whereas viral DNA is encapsidated in an ATP-dependent process only after formation of the procapsid (Speir and Johnson 2012). Similar to nucleic acid-mediated stabilization, the scaffolding proteins stabilize the CP-CP interactions, which are important for procapsid formation with a correct, assembly competent structure (Zlotnick et al. 2012).

Helical Viruses

The first virus, which assembly process was studied in details already more than 70 years ago, was a plant virus TMV. Authors of the study (Fraenkel-Conrat and Williams 1955) demonstrated the disassembly and successful reassembly of TMV particles, which were infectious for test plants after these manipulations in vitro.

TMV is a typical morphological example of helical viruses, in which 2130 CP molecules form 300-nm-long, right-handed helical particles with an outer diameter of 18 nm. TMV genomic RNA is encapsidated in the inner channel of 4 nm (Pieters et al. 2016). The three-dimensional structure of TMV was elucidated using fiber diffraction method (Namba and Stubbs 1986).

Purified, nucleic acid-free TMV CPs represent a mixture of oligomers, mostly up to subunit hexamers. The assembly process of TMV starts with formation of doubled discs (Fig.16.3) containing 34 CP molecules, which recognise a special loop sequence in TMV genomic RNA (OAS). When the assembly progresses and the next disc is binding to the first, the OAS loop is pulled through the growing TMV particle, until the whole RNA is encapsidated in the viral coat. Native TMV particles are rigid, rod-shaped particles, which are highly stable at different pH in the range 3–9, as well as up to 90 °C (Butler 1999; Pieters et al. 2016).

This mechanism suggests that the length of helical viruses is directly depending on nucleic acid size. The affinity of TMV discs to specific OAS is comparably high – in vitro experiments with specific ribonucleotides resulted in $K_D = 0.7$ μM

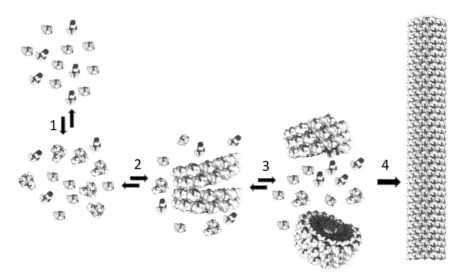

Fig. 16.3 Schematic representation of assembly mechanism for helical virus.
The model is built based on tobacco mosaic virus (TMV) structural studies. Virus assembly starts with the formation of CP oligomers (stage 1). Next, the ring-shaped aggregates, containing 34 CP molecules (20S disc), are formed (stage 2), which may undergo conformational change into a "«lockwasher»" structure (stage 3). Then, the assembly nucleation complex is growing by addition of further disks, until the virus structure is complete (stage 4). Simultaneously, viral RNA is packaged in the central channel of the virus (Koch et al. 2016). Images were created using Protein Data Bank (4UDV; Fromm et al. 2015) and NGL 3D viewer (Rose and Hildebrand 2015)

(Butler 1999), which is a comparable value for nanomolar range CP-nucleic acid dissociation constants for icosahedral viruses, as discussed earlier. However, the virus organization in helical structures demonstrates principal differences to icosahedral assemblies – helical symmetry results in higher surface area, more interactions between viral nucleic acid and CPs and less packaging efficiency. If icosahedral viruses contain up to 30% of nucleic acids, the proportion of nucleic acids in helical viruses constitutes less than 10% of virion mass (Narayanan and Han 2017; Solovyev and Makarov 2016).

Also other viruses can be structurally organized into helical structures, for example, plant viruses from *Carlavirus*, *Closterovirus*, *Potexvirus* and *Potyvirus* groups, as well as filamentous bacteriophages and such mammalian viruses as influenza and rabies. Due to experimental difficulties in obtaining of crystals for X-ray analysis, structural information about helical viruses is known from X-ray fibre diffraction, cryo-electron microscopy, as well as from different biochemical, biophysical and *in silico* modelling experiments. However, in most cases, no high-resolution structural data are available for filamentous viruses.

As reviewed by Solovyev and Makarov (2016), recent structural studies demonstrate significant differences between rod-shaped and filamentous viruses of helical morphology in terms of CP monomeric structures and interactions between subunits, resulting in different flexibility of the particles. Additionally, helical viruses seem to be less stable than previously proposed. This instability can be bound with such stages of virus life cycle as assembly/disassembly and virus transfer between host cells. The differences between few rod-shaped and numerous filamentous viruses suggest that classical, rigid structure of TMV can be regarded rather as an exception among helical viruses.

Artificial Viral Protein Complexes

Basic virus assembly studies provide numerous ideas for different developments of virus-derived technology. This section summarizes the principles and application examples, where structural and assembly/disassembly properties of viral structures are exploited for creation of new nanomaterials.

Assembly Inhibitors

The knowledge about virus assembly mechanisms can suggest targets for preventing viral diseases. As shown for several viruses, different low-molecular compounds can form stable complexes with viral structural proteins, influence their conformation and therefore affect viral assembly processes.

As demonstrated by Zlotnick and coworkers in a detailed study (Stray et al. 2005), heteroaryldihydropyrimidines (HAPs) are efficient inhibitors of infectious HBV production. HAP molecules bind strongly to the HBc dimers (K_D approx. 30 nM) and stimulate the formation of hexameric CP structures. As a result, the assembly process is redirected towards unusual structures (sheets or 120-nm-long tubes instead of 30–42 nm icosahedra), which do not encapsidate specific HBV nucleic acids and, as a result, interrupt the infection. The assembly inhibitors are found also for other viruses, such as enteroviruses (Plevka et al. 2013), HIV (Prevelige 2011) and bacteriophage P22 (Teschke et al. 1993).

As virus infections influence considerably the agriculture, there is a necessity also for new inhibitors of plant viruses. One of such inhibitors, ningnanmycin (NNM) is a cytosine nucleoside-derived pesticide, which can be obtained from soil bacterium *Streptomyces*. NNM is shown to bind an assembly intermediate of TMV at four-layer aggregate disc stage ($K_D = 3.3$ μM), resulting in TMV disassembly back to CP trimers and in loss of infectivity at experimental conditions (Li et al. 2016).

Disassembly and Reassembly

For technology developments, frequently different manipulations with viral structural proteins are necessary. One of most important manipulation is the disassembly and reassembly process, which is used to improve the structural properties or to introduce the chosen functional molecules in the VLP structure. The ideas for construction of new virus-like nanomaterials can be generated from basic assembly studies, which are discussed in previous sections.

Viral structures can be easily disassembled using chaotropic agents (urea, guanidinium salts) or strong detergents in combination with reducing agents. However, such treatments can result in irreversibly denatured viral structural proteins, which are not suitable for subsequent reassembly. Therefore, for most viruses it is necessary to find the conditions allowing to disassemble the viral structure, simultaneously preserving the structure of CPs. As summarized in Table 16.1, different buffer systems are used to disassemble viruses or VLPs, which ensure further reconstruction of viral structure.

Taking in account the diversity of viruses, there is no universal disassembly/reassembly system for all viruses. As viruses are natural metastable protein complexes with an intrinsic ability to disassemble and assemble, the experimental conditions for these processes probably should be close to the environments that viruses encounter during their life cycle in the host cells. As an example, adenoviruses partially disassemble in endosomes, where the conditions are acidic at comparably high salt concentrations (Nemerow and Stewart 2016).

However, detailed mechanisms of virus life cycle are not elucidated yet for most part of viruses. Additionally, different host factors are involved in these processes, therefore, the disassembly/reassembly conditions for technology development of new viral objects have to be found experimentally. For virus and VLP disassembly, buffer systems with strongly acidic or basic pH and/or extremely high ionic strength are typically used; in some cases, denaturing agents and low-salt solutions are necessary (Table 16.1). For metal ion-containing viruses, the dissociation of viral particles in CP oligomers requires the addition of chelating agents.

The reassembly processes are "opposite" in terms of reaction conditions; for example, if VLPs disassemble at high-salt conditions, then the VLP structures can be re-established in low-salt buffer systems; similar effects can be expected by change of pH and presence or absence of metal ions (Table 16.1). As already discussed in previous sections, the addition of specific nucleic acid can stimulate the assembly of corresponding VLPs, especially when particle formation is not achievable without nucleic acids.

Disassembly and subsequent reassembly can considerably improve the quality of target VLPs. Several examples demonstrate the importance of these processes in product development, especially in the field of vaccines. It is known from development studies of human prophylactic vaccine against cervical cancer that during natural human pappiloma virus (HPV) infection, L1 proteins are synthesized in cytoplasm and form L1 pentamers; HPV virions are assembled in nucleus. As such parameters as ionic strength, pH and other factors in heterologous host cells mostly

Table 16.1 Examples of assembly and disassembly conditions for different viruses or VLPs

Virus/VLP	Disassembly conditions	Reassembly conditions	K_D*	Morphology	Reference
Bacteriophages					
MS2	66% acetic acid	50 mM ammonium acetate, 1 mM EDTA, 1 mM DTT, 0.05% Tween 20, pH 7.5 specific nucleic acid	nM range	Icosahedral, T=3	Borodavka et al. (2012)
PP7	6 M urea, 10 mM DTT	Tris, pH 8.5, RNA	n.d.**	Icosahedral, T=3	Caldeira and Peabody (2007)
P22	9 M urea (from inclusion bodies)	20 mM Na phosphate, pH 7 Scaffolding protein	nM range	Icosahedral, T=4	Zlotnick et al. (2012)
Plant viruses					
BMV	50 mM Tris-HCl, pH 7.5	50 mM Tris-HCl, pH 7.2	n.d.	Icosahedral T=3	Choi and Rao (2000)
	0.5 M CaCl$_2$, 1 mM EDTA	50 mM NaCl, 10 mM KCl 5 mM MgCl$_2$, nucleic acids			
CCMV	1.0 M NaCl	0.9 M NaCl, 0.01 M MgCl$_2$	n.d.	Icosahedral, T=3	Lavelle et al. (2007)
	20 mM Tris, 1 mM DTT pH 7.5	0.1 M Na acetate, pH 4.8			
CCMV		0.1 M Na acetate, 0.25 M NaCl	2 µM (pH 4.8) 4 µM (pH 5.0) 10 µM (pH 5.25)	Icosahedral, T=3	Johnson et al. (2005)
CCMV	0.5 M CaCl$_2$	0.9 M KCl,	n.d.	Icosahedral, T=3	Adolph and Butler (1974)
	0.1 M Tris, pH 8	0.1 M Na acetate, pH 5			
CCMV		0.4 M KCl	n.d.	Tubular structures	Adolph and Butler (1974)
		0.1 M Na acetate, pH 5			
CCMV		0.1 M Na acetate, pH 5	n.d.	Multishelled structures	Adolph and Butler (1974)
PVA	3.5 M LiCl, at −20oC	5 mM Tris, pH 7.8	n.d.	>30 nm long filaments	Ksenofontov et al. (2017)
PVX	2 M LiCl, at −20oC	10 mM Tris, pH 7.6	n.d.	30–150-nm-long filaments	Karpova et al. (2006)

(continued)

Table 16.1 (continued)

Virus/VLP	Disassembly conditions	Reassembly conditions	K_D*	Morphology	Reference
STNV	50 mM HEPES, 1 M NaCl 3 mM EDTA, pH 9	50 mM HEPES, 100 mM NaCl, 3 mM MgCl$_2$, 3 mM CaCl$_2$, pH 7.5	n.d.	Icosahedral T=1	Bunka et al. (2011)
TMV	66% acetic acid	75 mM Na phosphate, pH 7.2 specific nucleic acid	0.7 µM***	60-nm-long rods	Lam et al. (2016)
Mammalian viruses					
HBcAg	50 mM Tris pH 9.5 3.5 M urea	50 mM HEPES, pH 7.5 1 mM DTT, variable NaCl	14 µM (0.15 M NaCl) 0.77 µM (0.7 M NaCl)	Icosahedral T=4	Ceres and Zlotnick (2002)
HEV	4.0 M urea (from inclusion bodies)	137 mM NaCl, 2.7 mM KCl, 10 mM Na$_2$HPO$_4$, 2 mM KH$_2$PO$_4$	n.d.	Isometric 23 nm	Li et al. (2005)
HPV	20 mM Tris pH 8.2 0.166 M NaCl, 2 mM DTT	1 M NaCl, pH 6–7	n.d.	Isometric	Mach et al. (2006)
Norwalk virus	10 mM CHES, pH 9.0	50 mM MES pH 6.0 50 mM NaCl	n.d.	Icosahedral, T=3	Tresset et al. (2013)
SV40	20 mM Tris, pH 8.9 50 mM NaCl, 2 mM DTT 5 mM EDTA	0.1 M MOPS, pH 7.2 0.25 M NaCl, nucleic acid	0.6–2.1 µM	Icosahedral, T=1	Kler et al. (2012)

*K_D, apparent = [CP] / [capsid] as index of assembly;
**data not found
***from Butler 1999

are different from that in native host; it can result in VLPs with different immunogenity and antigenity properties. Therefore, the recombinant L1 protein oligomers are purified and assembled in vitro after expression of HPV L1 structural proteins in baculovirus/insect cell system. The VLP assembly process is carried out at high-salt conditions; resulting HPV L1 VLPs structurally are highly similar to native virions, including conformational epitopes, which are necessary for virus neutralization. Additionally, the reassembly process guarantees the necessary batch-to-batch consistency and VLP stability at different conditions, including the vaccine formulations with adjuvants (Deschuyteneer et al. 2010).

Similar purification strategy is used also for other virus-derived recombinant vaccines. The main component of vaccine against hepatitis E (HEV) is truncated structural protein of the virus. It is produced in *E. coli* cells in the form of insoluble inclusion bodies, which are solubilized in 4 M urea, purified by column chromatography and refolded in the presence of 0.5 M ammonium sulphate, using tangential filtration technique. Resulting HEV VLPs are safe and efficacious, as demonstrated in a large clinical trial with more than 100.000 volunteers (Li et al. 2015).

Recently, in vitro assembly is suggested as a part of production scheme also for rotavirus candidate vaccine. In *E. coli*, rotavirus structural proteins do not form VLPs spontaneously. Instead, both proteins VP2 and VP6 form CP oligomers, which can be purified by column chromatography. Mixing of both rotavirus proteins at optimized conditions results in formation of double-layered particles, morphologically similar to authentic rotavirus structures. These VLPs as main vaccine component are able to efficiently protect animals against diarrhoea caused by the rotavirus (Li et al. 2014).

For enveloped viruses, the reassembly processes are more complex. As shown with alphavirus VLPs, these can be produced in vitro by mixing of recombinant CP and single-stranded nucleic acid. To obtain lipid-enveloped VLPs, assembled particles are transfected in mammalian cells, expressing alphavirus glycoproteins. Resulting alphavirus VLPs are released from the cells and contain typical lipid envelope and glycoproteins; engineered alphavirus particles can be used as delivery agents for gene and drug therapies (Cheng et al. 2013). The intrinsic property of structural proteins of enveloped viruses to integrate in phospholipid membranes is exploited in the so-called virosome technology. Several developed virosomes are suggested as prophylactic or therapeutic vaccines, including against influenza, malaria and HIV (for review see Trovato and De Berardinis 2015).

Other application examples where VLP disassembly and reassembly are used for introduction of different functional molecules in the interior of viral particles can be found in a recent review article (Zeltins 2016).

Solutions to Obtain VLPs

The detailed information about virus assembly mechanisms can be highly helpful to find solutions how to achieve the VLP formation in heterologous hosts, especially when different functional molecules have to be introduced in VLPs.

As shown by Hwang et al. (1994), TMV CP oligomers ("discs", Fig. 16.3) can be produced also in *E. coli* cells and assembled in the presence of specific OAS containing RNAs both in vitro and in situ by coexpression of specific RNA in recombinant bacteria. Recent study (Brown et al. 2013) demonstrates that empty TMV nanoparticles without encapsdated RNA can be obtained from *E. coli* cells by expression of genetically modified TMV coat protein gene. In previous research, authors (Lu et al. 1998) identified two AA, which ensure the repulse action between two CP subunits and stimulate disassembly of native TMV virions in plant cells. The substitution of these acidic AA with neutral ones resulted in formation of stable TMV nanoparticles in bacterial cells. The recombinant, empty TMV VLPs, containing different functional molecules, are suggested as core elements for different nanotechnological applications.

It is known from icosahedral virus assembly studies that stable CP dimers serve as virion building blocks (see previous sections). This fact stimulated an idea that covalent CP dimers should be still able to form VLPs. Recombinant tandem CP VLPs are obtained from bacteriophages PP7 (Caldeira and Peabody 2007), MS2 (Chackerian et al. 2011), Qβ (Fiedler et al. 2012), and HBc (Peyret et al. 2015). For these viruses, the N-terminal and C-terminal parts of CPs are localized spatially close to each other, allowing to introduce the covalent linkage between both molecules. In VLPs of tandem CP, both CP proteins are connected with shorter or longer AA linkers. Surprisingly, tandem CPs containing PP7 and Qβ VLPs do not reveal enhanced thermal stability. The main advantage of such tandem VLPs is the versatility allowing to introduce comparably long protein domains in the VLP structure genetically. As a result, only a half of CP subunits in the VLP structure is fused with foreign protein, allowing the self-assembly of tandem VLPs directly in host cells. This approach ensures the exposition of the chosen protein on surface of VLPs, when direct fusion with each CP does not support the VLP assembly due to spatial limitations. Additionally, the tandem CP system demonstrates that intra-dimer interactions are more important for VLP thermal stability and assembly than inter-dimer interactions (Fiedler et al. 2012).

Another interesting example, when artificial VLPs for nanotechnological applications are constructed based on virus assembly mechanism, is the so-called split-core system (Walker et al. 2011). It is based on intrinsic property of HBc structural protein to form four-helix bundles, which are clearly identifiable on the capsid surface as protruding spikes (Fig. 16.1, C). These α-helices containing structures serve as dimerization interfaces between two CP molecules. The insertions of foreign, short peptides in the loop between these α-helices in HBc VLP structure are well known (Pushko et al. 2013), whereas introduction of longer proteins results in loss of VLP structure in most cases, with some exceptions (Kratz et al. 1999). The suggested "split-core" system exploits the HBc CP gene which is separated in two independently expressing parts and codes N-terminal and C-terminal fragments of CP. The strong interactions between mentioned α-helices ensure the VLP formation from split CP fragments directly in *E. coli* cells without additional in vitro assembly steps. The technological advantage of this system is similar to that of tandem CP; it allows to expose up to 300 AA long protein molecules on VLP surface, when corresponding genes are genetically fused to gene fragments coding for N-terminal or C-terminal part of splitted HBc protein. Also this system can be useful for different applications, including for usage in vaccine technologies.

Molecules involved in viral assembly processes can be useful for construction of artificial VLPs with desired properties. To enable the targeted packaging into VLPs, chosen nucleic acids can be fused with packaging signals of corresponding virus. Probably, the best example is the translation operator sequence from bacteriophage MS2, which strongly interacts with coat protein dimers and is shown to stimulate the efficient encapsidation in MS2 VLPs even after fusion with foreign functional RNA. These and other development aspects of bacteriophage-based nanotechnologies are discussed in recent excellent review (Pumpens et al. 2016).

Along with specific nucleic acids, also scaffolding proteins (SP) can be used for construction of VLPs for desired applications. Already mentioned bacteriophage P22 is shown as very efficient packaging system for different recombinant proteins.

The P22 VLPs are assembly products of two proteins, containing 420 copies of CP and up to 300 copies of SP. The SPs form stable complexes with CP molecules in the interior of VLPs through the non-covalent binding. Whereas the C-terminal part of SP is involved in VLP assembly, the N-terminal part can be replaced with different proteins via genetic fusions. Such design results in controlled encapsidation of chosen protein molecules in recombinant bacterial cells during the cultivation; no additional in vitro assembly steps are necessary to encapsidate several hundred enzyme molecules per VLP. Additionally, the P22 encapsidation system allows to obtain soluble, enzymatically active proteins, which otherwise form insoluble inclusion bodies after expression in bacterial cells (Patterson et al. 2012, 2013). Moreover, the system is suitable for packaging of CRISP-Cas9 nuclease and can be potentially developed as cell-specific genome engineering nanoparticle (Qazi et al. 2016).

Viral Protein Domains

For construction of VLPs, typically full-sized viral CPs are used. However, specific fragments of structural proteins can be sufficient to obtain protein complexes, resembling virus structures, as shown by results of several recent studies.

Synthetic, 24 AA long peptide derived from β-annulus domain of tomato bushy stunt virus (TBSV) is appropriate to form virus-like nanocapsules in sizes between 30 and 50 nm. In native TBSV, the β-annulus of CP is involved in the formation of ordered internal structure of the virion. The process of nanocapsule assembly can be initiated at comparably low peptide concentrations (25 μM), indicating on thermodynamic stability of the peptide complex (Matsuura et al. 2010). Interestingly, when Ni^{2+}-chelating ligand nitrilotriacetic acid (Ni-NTA) molecule is added to the N-terminus of the peptide, the critical concentration of capsule formation is reduced even more than 400 times, indicating the stabilizing role of Ni ion complexes with

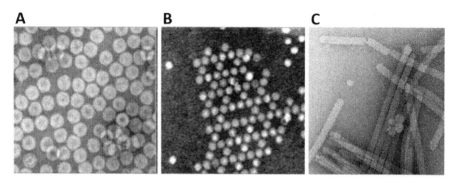

Fig. 16.4 Rice yellow mottle virus (RYMV) VLP structural variants (Resevica, Balke, Ose, Zeltins, unpublished).
A Electronmicropy image of RYMV VLPs obtained from recombinant *E. coli* cells (T = 3 symmetry, 28 nm); B RYMV VLPs, obtained after expression of 5'end truncated CP (69 AA removed from N-terminus) gene in *E. coli* cells (T = 1 symmetry, 19 nm); C tubular structures of RYMV CP, obtained after in vitro disassembly/reassembly in the presence of oligonucleotides (28 × 180–480 nm)

NTA. Latest design allows to encapsidate His-tagged model proteins. Authors suggest that virus-derived nanocapsules with the minimized design still can serve as DNA or protein carriers for vaccines and vaccine adjuvants (Matsuura et al. 2016a). The same group of researchers suggest similar strategy for self-assembling β-annulus peptide from another plant virus (Sesbania mosaic virus, SeMV). In this case, the addition of stabilizing, antiparallel β-sheet forming FKFE sequence at C-terminus of SeMV β-annulus is necessary to self-assemble the 30 nm nanocapsules (Matsuura et al. 2016b).

Interestingly, it is possible to manipulate with spatial structure at the AA level of the CP, as shown in the experiments with plant sobemoviruses. Deletion of first 36 AA from N-terminus of SeMV CP results in unstable VLPs of pseudo T = 2 geometry, whereas CPs missing 65 AA assemble exclusively in stable T = 1 particles (Satheshkumar et al. 2005). We also observed the formation of similar structures from CPs of related sobemovirus RYMV (Fig. 16.4).

Non-viral Protein Complexes in VLP Applications

Virus-like particle laboratory applications are known for more than 30 years. Since that time, more than 100 VLPs from different virus species are obtained, using recombinant expression systems (Zeltins 2013).

Unmodified recombinant VLPs derived from mammalian viruses are excellent candidates for prophylactic and therapeutic vaccines, as discussed in previous sections. However, for many applications the inclusion of different functional molecules, like proteins, peptides, nucleic acids and low-molecular compounds in viral structures, is prerequisite for new technology development. Additionally, such modifications allow to exploit also VLPs obtained from bacterial, plant and insect viruses in medical and veterinary technologies. Already first VLP solutions demonstrate the

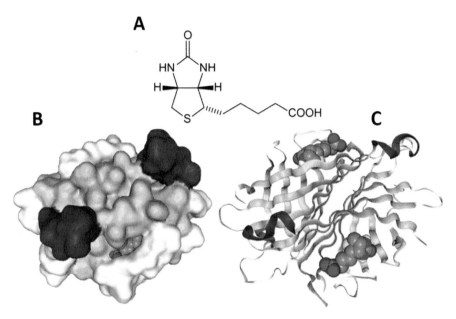

Fig. 16.5 Schematic representation of binding of biotin to streptavidin.
A Chemical structure of biotin; B surface model of streptavidin; C cartoon model.
Bound biotin is shown as grey space-filled molecule. Structure of the biotin-streptavidin dimer shows an antiparallel β-barrel structure of streptavidin with the biotin molecule into the binding pocket. The high affinity of biotin to streptavidin is ensured by a spacious network of hydrogen bonds. Images were created using Protein Data Bank (2IZF; Katz 1997) and NGL 3D viewer (Rose and Hildebrand 2015)

option to include desired functional peptides by adding corresponding coding DNA sequences to viral structural genes via gene synthesis (Haynes et al. 1986). The genetic approach along with chemical coupling technique is still one of the most important solutions in construction of new virus-derived nanomaterials (for review, see Zeltins 2016). As an alternative to genetic and chemical processes, different physical methods are frequently used for VLP construction, including well-characterized protein complexes for introduction of foreign molecules in VLP structures.

One of the best characterized protein complexes is the avidin and biotin pair, which is supposed to be the strongest non-covalent interaction in nature with a K_D of about 10^{-15} M. Avidin is a host defence protein, which is induced during inflammation processes caused by toxic agents and different pathogens, including biotin-auxotrophic bacteria and yeasts (Tuohimaa et al. 1989). The complex is highly stable at physiological conditions and can be dissociated only at strongly denaturing conditions in a mixture of ethanol and acetic acid at 70 °C (Garlick and Giese 1988). Avidin and its bacterial analogue streptavidin are widely used in different biotechnological applications, including sensitive analytical methods (Airenne et al. 1999). The highly stable biotin-streptavidin complexes (Fig. 16.5) are used also in development of different VLP-based technologies, for example, for prophylactic and therapeutic candidate vaccines, for targeted packaging of functional proteins and for other applications. Examples and corresponding literature citations can be found in Table 16.2.

As already discussed, different peptides can be comparably easy introduced in the VLP structure by chemical or genetic methods. Consequently, it allows to exploit well-characterized peptide interactions for construction of VLPs with desired properties.

Possibly, the strong histidine-tag (six or more His molecules) interactions with metal ions ($K_D = 14$ µM; Lata et al. 2005) are the most popular peptide interaction used in biotechnology since the introduction in 1975 (Porath et al. 1975). Similar to other recombinant proteins, sometimes also viral CPs are designed with His-tags to enable the single-step chromatographic purification and facilitate the downstream processing, as shown in Manuel-Cabrera et al. (2016) and Zhang et al. (2002). Additionally, the affinity of hexahistidine tag to different metal ions is shown to be useful for immobilization of gold nanoparticles on surface of VLPs (Wnęk et al. 2013). Moreover, His-tag and Ni-NTA interactions are efficient for peptide immobilization on VLP surface (Koho et al. 2015) and creation of new virus-like structures, using NTA-modified virus-derived peptide and non-viral protein with His-tag (Matsuura et al. 2016a, b).

Different complexes formed between polyanionic and polycationic molecules play an important role in cellular processes, including nucleic acid packaging in viruses and a number of diseases, like Parkinson's disease and cystic fibrosis (Wong and Pollack 2010). Such complexes are used in different applications, for example, for packaging of nucleic acids for gene delivery (Toncheva et al. 1998). As shown in experiments with two types of VLPs, a short octameric Glu peptide, if engineered on exterior of VLPs, can electrostatically bind 8 Arg residue containing model proteins with sufficient affinity. If these peptides contain also Cys residues, the adjacent sulfhydryl groups after oxidation form covalent disulphide bond and immobilize the peptides or even antibodies on VLP surface (Table 16.2).

In some cases, also unusual peptide interactions can be adapted for construction of VLPs with unexpected properties. The elastin-like polypetides (ELP) are insoluble proteins which ensure the elasticity of the skin and blood vessels. Recombinant variants of ELP are soluble at low temperatures and self-assemble if the temperature rises to physiological. ELP-oligopeptide-modified CCMV CPs form dimers at low temperatures and self-assemble into T = 1 VLPs at 35 °C (van Eldijk et al. 2012).

Interesting solution for foreign protein packaging inside of VLPs is shown with the same CCMV using small leucine zipper-like peptides (E-coil and K-coil), which are genetically built in the structures of CP and model protein. The E-coil- and K-coil-containing proteins bind in vitro with a high affinity ($K_D = 70$ nM; Litowski and Hodges 2002). Therefore, the assembly reaction between E-coil-GFP and K-coil-derived CPs in the presence of purified unmodified CPs results in encapsidation of 15 GFP molecules inside of CCMV VLPs (Minten et al. 2009).

Another protein complex-based application which is widely used in construction of new VLPs is based on the ability of *Staphylococcus* protein A (SPA) to strongly interact with different immunoglobulins in a nonantigenic manner. SPA is the cell

Table 16.2 Examples of peptide and protein complexes used for introduction of functional molecules in virus-like particle structures

Complex	K_D	Virus / VLP	Functional molecule(s)	Application	Reference
Biotin/(strept) avidin	fM range	HPV16	*Plasmodium* antigen	Malaria vaccine candidate	Thrane et al. (2015)
		CMV	Biotinylated oligonucleotides, streptavidin	Packaging of proteins	Lu et al. (2012)
		MCV	Streptavidin	Reference in immunoassays	Chen et al. (2012)
		PVX	Streptavidin	Virus functionalization studies	Steinmetz et al. (2010)
		TMV	papillomavirus L2 protein	Model vaccines	Smith et al. (2006)
		BPV	TNF-α	Arthritis therapeutic vaccine candidate	Chackerian et al. (2001)
$(Glu)_8$ / $(Arg)_8$ peptides	n.d.	BPV	Human mucin-1 peptide	Cancer vaccine candidate	Pejawar-Gaddy et al. (2010)
		Polyoma virus	Tumour-specific antibody fragment	Antibody presentation on VLP surface	Stubenrauch et al. (2001)
$(His)_6$ / Ni-NTA	14 μM	TBSV domain with Ni-NTA	$(His)_6$-GFP	Artificial GFP virus-like capsules	Matsuura et al. (2016a, b)
		NoV with $(His)_6$	NTA-modified peptides	Drug delivery model	Koho et al. (2015)
E-coil/ K-coil peptides	70 nM	CCMV with K-coil (KIAALKE)$_3$	GFP with E-coil (EIAALEK)$_3$	Encapsidation of proteins	Minten et al. (2009)
Elastin-like domains	n.d.	CCMV with ELP peptides	ELP peptide $(V_4L_4G_1)_9$	Temperature-dependent VLP assembly	van Eldijk et al. (2012)
Protein A domains/IgG	70–700 nM	SeMV with B-domain	Human mAb's	Intracellular delivery of mAb's	Abraham et al. (2016)
		PVM with Z-domain	Rabbit IgG	Immobilization of IgGs on VLP surface	Kalnciema et al. (2015)
		JCPyV with Z-domain	IgG	Targeted gene delivery model	Deng et al. (2015)
		Qβ with Z-domain	IgG	Production of antibody-binding VLPs	Brown et al. (2009)
		TVCV with Protein A fragment	mAb's	Single-use immunoadsorbent for antibody purification	Werner et al. (2006)

Abbreviations of virus names: BPV, bovine papillomavirus; CCMV, cowpea chlorotic mottle virus; CMV, cucumber mosaic virus; HPV, human papillomavirus; JCPyV, JC polyomavirus; MCV – Merkel cell polyomavirus; NoV, norovirus; PVM, potato virus M; PVX, potato virus X, Qβ, bacteriophage Qβ; SeMV, Sesbania mosaic virus; TBSV, tomato bushy stunt virus; TMV, tobacco mosaic virus; TVCV, turnip vein clearing virus

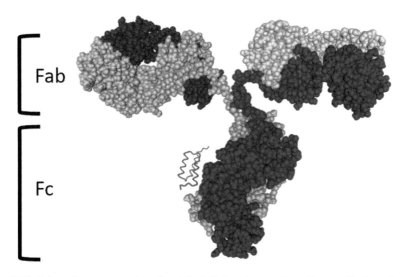

Fig. 16.6 Schematic representation of protein A Z-domain complex with an antibody molecule. The Z-domain binding site is located in the interface between CH2 and CH3 domains in Fc fragment of the antibody (Justiz-Vaillant et al. 2017). IgG chains are coloured as follows: dark blue and braun, constant chains; red and light blue, variable chains. The relative localization of antigen binding fragments (Fab) and crystallisable fragment (Fc) domains is shown. Z-domain is drawn as red molecule next to Fc domain of IgG. An antibody image was created using Protein Data Bank (2IGT; Harris et al. 1997) and NGL 3D viewer (Rose and Hildebrand 2015). Z-domain structural model was generated from AA sequence using I-tasser program (Yang et al. 2015)

wall protein of *Staphylococcus aureus*, which binds the Fc domain of many mammalian immunoglobulins (IgG) with high affinity (Fig. 16.6), especially IgGs of human, rabbit, pig and guinea pig origins. Other immunoglobulins such as IgA, IgE and IgM are reacting with SPA weakly (Brown et al. 1998). As shown in antibody binding experiments with recombinant, artificial fragments of SPA (Z-domains), the dissociation constants K_D vary between 70 and 700 nM, depending on the number of Z-domains in recombinant protein (Madan et al. 2013). If SPA fragments are introduced in VLP structure, VLPs bind up to 90 antibody molecules per viral particle and are suggested for intracellular delivery of therapeutic mAb's (Abraham et al. 2016). Other applications for SPA-decorated VLPs include targeted gene delivery and new materials for purification of mAb's (examples and literature citations are provided in Table 16.2).

Concluding Remarks

Analysis of capsid assembly and examples discussed here suggest several considerations, which are important for virus-like particle technology developments: (1) at low CP concentrations, the capsid formation is inefficient. If VLPs are produced in heterologous expression system, first, the system itself and the conditions for cell cultivation have to be found in order to achieve sufficient concentrations of target CP in the cells to induce the VLP formation; alternatively, viral CPs can be first purified, refolded if necessary and assembled in vitro; (2) increase of CP subunit concentrations or interactions between them leads to efficient VLP formation in vitro as well as in producing cells; for in vivo production, the presence of specific or non-specific nucleic acids in host cells can stimulate the VLP formation; (3) as CP mRNA is the only "virus-related" nucleic acid in recombinant, heterologous cells, it is advisable to use the original coding sequence in the expression system without codon adapting for specific expression host; codon adaptation can change mRNA structure and negatively influence the VLP formation in the cells; (4) overproduction and very strong interactions between viral CPs, as well as suboptimal conditions in cells of heterologous host, can lead to the assembly of VLP intermediates, "half" and malformed capsids, which lack typical icosahedral or helical structures. However, taking into account the virus diversity, these and other nanoparticle construction principles (Zeltins 2016) have to be adapted individually.

The number of characterized viruses is continuously growing, especially due to broad usage of next-generation nucleic acid sequencing technologies, allowing to discover new viral sequences even in environmental samples without isolation of individual viruses. From structural point of view, the newest developments in X-ray crystallography, macromolecular NMR and cryoelectron microscopy allow to rapidly characterize newly discovered viruses and constructed VLPs at resolutions close to atomic level. As one of the basic principles of nanotechnology is the manipulations with materials at atomic and molecular level, the structural and functional knowledge about natural systems gained from basic research is one of the central prerequisites for technology developments. Virus-based recombinant technologies of industrial level are known since 1980s of the last century, when the first human VLP-based vaccine against hepatitis B was introduced to prevent the disease. Since that time two additional VLP-derived vaccines against human papillomaviruses and hepatitis E are licensed, and several tens of human and veterinary VLP vaccines are at different stages of clinical trials. These aspects are enormous high stimulating factors driving the developments of different virus-based technologies also in other areas than vaccinology, such as new gene therapy agents, new drug delivery tools and even new inorganic materials, inspired from perfect viral structures.

The development of viral technologies is strongly influenced also from other areas of research. As discussed in this chapter, many well-characterized non-viral protein complexes are successfully applied to create new viral nanoparticles with desired properties, as well as to simplify the construction and production processes. The simplicity aspect is especially important for subsequent technology transfers and industrial production. It can be expected that also in the future the knowledge originating from new virus studies and from protein complex research, as well as from chemistry and physics will be successfully adapted for construction of new virus-based nanomaterials, will result in industrial technologies, producing new medical products and advanced materials.

Acknowledgements I wish to thank Prof. Dr. P. Pumpens (Riga) and Prof. Dr. M. Bachmann (Bern) for the helpful discussions. Dr. I. Balke, Dr. V. Ose and MSc. G. Resevica are acknowledged for providing of unpublished results. The writing of the review was supported by the Grant No. SP 672/14 provided by Latvian Science Concil.

References

Abraham A, Natraj U, Karande AA, Gulati A, Murthy MR, Murugesan S, Mukunda P, Savithri HS (2016) Intracellular delivery of antibodies by chimeric Sesbania mosaic virus (SeMV) virus like particles. Sci Rep 6:21803

Adolph KW, Butler PJ (1974) Studies on the assembly of a spherical plant virus. I States of aggregation of the isolated protein. J Mol Biol 88:327–341

Adolph KW, Butler PJ (1976) Assembly of a spherical plant virus. Philos Trans R Soc Lond Ser B Biol Sci 276:113–122

Airenne KJ, Marjomäki VS, Kulomaa MS (1999) Recombinant avidin and avidin-fusion proteins. Biomol Eng 16:87–92

Aloy P, Russell RB (2004) Ten thousand interactions for the molecular biologist. Nat Biotechnol 22:1317–1321

Basu G, Allen M, Willits D, Young M, Douglas T (2003) Metal binding to cowpea chlorotic mottle virus using terbium(III) fluorescence. J Biol Inorg Chem 8:721–725

Borodavka A, Tuma R, Stockley PG (2012) Evidence that viral RNAs have evolved for efficient, two-stage packaging. Proc Natl Acad Sci U S A 109:15769–15774

Brasch M, Cornelissen JJLM (2012) Relative size selection of a conjugated polyelectrolyte in virus-like protein structures. Chem Commun 48:1446–1448

Brown NL, Bottomley SP, Scawen MD, Gore MG (1998) A study of the interactions between an IgG-binding domain based on the B domain of staphylococcal protein A and rabbit IgG. Mol Biotechnol 10:9–16

Brown SD, Fiedler JD, Finn MG (2009) Assembly of hybrid bacteriophage Qbeta virus-like particles. Biochemistry 48:11155–11157

Brown AD, Naves L, Wang X, Ghodssi R, Culver JN (2013) Carboxylate-directed in vivo assembly of virus-like nanorods and tubes for the display of functional peptides and residues. Biomacromolecules 14:3123–3129

Bunka DH, Lane SW, Lane CL, Dykeman EC, Ford RJ, Barker AM, Twarock R, Phillips SE, Stockley PG (2011) Degenerate RNA packaging signals in the genome of satellite tobacco necrosis virus: implications for the assembly of a T=1 capsid. J Mol Biol 413:51–65

Butler PJ (1999) Self-assembly of tobacco mosaic virus: the role of an intermediate aggregate in generating both specificity and speed. Philos Trans R Soc Lond Ser B Biol Sci 354:537–550

Cadena-Nava RD, Comas-Garcia M, Garmann RF, Rao AL, Knobler CM, Gelbart WM (2012) Self-assembly of viral capsid protein and RNA molecules of different sizes: requirement for a specific high protein/RNA mass ratio. J Virol 86:3318–3326

Caldeira JC, Peabody DS (2007) Stability and assembly in vitro of bacteriophage PP7 virus-like particles. J Nanobiotechnol 5:10

Ceres P, Zlotnick A (2002) Weak protein-protein interactions are sufficient to drive assembly of hepatitis B virus capsids. Biochemistry 41:11525–11531

Cerqueira C, Pang YY, Day PM, Thompson CD, Buck CB, Lowy DR, Schiller JT (2015) A cell-free assembly system for generating infectious human papillomavirus 16 capsids implicates a size discrimination mechanism for preferential viral genome packaging. J Virol 90:1096–1107

Chackerian B, Lowy DR, Schiller JT (2001) Conjugation of a self-antigen to papillomavirus-like particles allows for efficient induction of protective autoantibodies. J Clin Invest 108:415–423

Chackerian B, Caldeira JC, Peabody J, Peabody DS (2011) Peptide epitope identification by affinity selection on bacteriophage MS2 virus-like particles. J Mol Biol 409:225–237

Chen T, Hedman L, Mattila PS, Jartti L, Jartti T, Ruuskanen O, Söderlund-Venermo M, Hedman K (2012) Biotin IgM antibodies in human blood: a previously unknown factor eliciting false results in biotinylation-based immunoassays. PLoS One 7(8):e42376

Cheng F, Tsvetkova IB, Khuong YL, Moore AW, Arnold RJ, Goicochea NL, Dragnea B, Mukhopadhyay S (2013) The packaging of different cargo into enveloped viral nanoparticles. Mol Pharm 10:51–58

Choi YG, Rao AL (2000) Molecular studies on bromovirus capsid protein: VII. Selective packaging on BMV RNA4 by specific N-terminal arginine residuals. Virology 275:207–217

Comas-Garcia M, Garmann RF, Singaram SW, Ben-Shaul A, Knobler CM, Gelbart WM (2014) Characterization of viral capsid protein self-assembly around short single-stranded RNA. J Phys Chem B 118:7510–7519

Crick FH, Watson JD (1956) Structure of small viruses. Nature 177:473–475

Deng YN, Zeng JY, Su H, Qu QM (2015) Recombinant VLP-Z of: a novel vector for targeting gene delivery. Intervirology 58:363–368

Deschuyteneer M, Elouahabi A, Plainchamp D, Plisnier M, Soete D, Corazza Y, Lockman L, Giannini S, Deschamps M (2010) Molecular and structural characterization of the L1 virus-like particles that are used as vaccine antigens in Cervarix™, the AS04-adjuvanted HPV-16 and -18 cervical cancer vaccine. Hum Vaccin 6:407–419

Eagles GH (1939) A Virus in Rheumatism. Ann Rheum Dis 1:18–26

van Eldijk MB, Wang JC, Minten IJ, Li C, Zlotnick A, Nolte RJ, Cornelissen JJ, van Hest JC (2012) Designing two self-assembly mechanisms into one viral capsid protein. J Am Chem Soc 134:18506–18509

Fiedler JD, Higginson C, Hovlid ML, Kislukhin AA, Castillejos A, Manzenrieder F, Campbell MG, Voss NR, Potter CS, Carragher B, Finn MG (2012) Engineered mutations change the structure and stability of a virus-like particle. Biomacromolecules 13:2339–2348

Fraenkel-Conrat H, Williams RC (1955) Reconstitution of active tobacco mosaic virus from its inactive protein and nucleic acid components. Proc Natl Acad Sci U S A 41:690–698

Fromm SA, Bharat TAM, Jakobi AJ, Hagen WJH, Sachse C (2015) Seeing tobacco mosaic virus through direct electron detectors. J Struct Biol 189:87–97

Garlick RK, Giese RW (1988) Avidin binding of radiolabeled biotin derivatives. J Biol Chem 263:210–215

Hao T, Peng W, Wang Q, Wang B, Sun J (2016) Reconstruction and application of protein-protein interaction network. Int J Mol Sci 17(6):pii:E907

Harris LJ, Larson SB, Hasel KW, McPherson A (1997) Refined structure of an intact IgG2a monoclonal antibody. Biochemistry 36:1581–1597

Haynes JR, Cunningham J, von Seefried A, Lennick M, Garvin RT, Shen SH (1986) Development of a genetically-engineered, candidate polio vaccine employing the selfassembling properties of the tobacco mosaic virus coat protein. Biotechnology 4:637–641

Holmes KC, Stubbs GJ, Mandelkow E, Gallwitz U (1975) Structure of tobacco mosaic virus at 6.7A resolution. Nature 254:192–196

Hryc CF, Chen DH, Afonine PV, Jakana J, Wang Z, Haase-Pettingell C, Jiang W, Adams PD, King JA, Schmid MF, Chiu W (2017) Accurate model annotation of a near-atomic resolution cryo-EM map. Proc Natl Acad Sci U S A 114:3103–3108

Hwang DJ, Roberts IM, Wilson TM (1994) Expression of tobacco mosaic virus coat protein and assembly of pseudovirus particles in Escherichia coli. Proc Natl Acad Sci U S A 91:9067–9071

Johnson JM, Willits DA, Young MJ, Zlotnick A (2004) Interaction with capsid protein alters RNA structure and the pathway for in vitro assembly of cowpea chlorotic mottle virus. J Mol Biol 335:455–464

Johnson JM, Tang J, Nyame Y, Willits D, Young MJ, Zlotnick A (2005) Regulating self-assembly of spherical oligomers. Nano Lett 5:765–770

Justiz-Vaillant A, McFarlane-Anderson N, Smikle M (2017) Bacterial immunoglobulin (Ig)-receptors: past and present perspectives. Am J Microbiol Res 5:44–50

Kalnciema I, Balke I, Skrastina D, Ose V, Zeltins A (2015) Potato virus M-like nanoparticles: construction and characterization. Mol Biotechnol 57:982–992

Karpova OV, Zayakina OV, Arkhipenko MV, Sheval EV, Kiselyova OI, Poljakov VY, Yaminsky IV, Rodionova NP, Atabekov JG (2006) Potato virus X RNA-mediated assembly of single-tailed ternary 'coat protein-RNA-movement protein' complexes. J Gen Virol 87:2731–2740

Katen S, Zlotnick A (2009) The thermodynamics of virus capsid assembly. Methods Enzymol 455:395–417

Katz BA (1997) Binding of biotin to streptavidin stabilizes intersubunit salt bridges between Asp61 and His87 at low pH. J Mol Biol 274:776–800

Kler S, Asor R, Li C, Ginsburg A, Harries D, Oppenheim A, Zlotnick A, Raviv U (2012) RNA encapsidation by SV40-derived nanoparticles follows a rapid two-state mechanism. J Am Chem Soc 134:8823–8830

Kobayashi S, Suzuki T, Igarashi M, Orba Y, Ohtake N, Nagakawa K, Niikura K, Kimura T, Kasamatsu H, Sawa H (2013) Cysteine residues in the major capsid protein, Vp1, of the JC virus are important for protein stability and oligomer formation. PLoS One 8(10):e76668

Koch C, Eber FJ, Azucena C, Förste A, Walheim S, Schimmel T, Bittner AM, Jeske H, Gliemann H, Eiben S, Geiger FC, Wege C (2016) Novel roles for well-known players: from tobacco mosaic virus pests to enzymatically active assemblies. Beilstein J Nanotechnol 7:613–629

Koho T, Ihalainen TO, Stark M, Uusi-Kerttula H, Wieneke R, Rahikainen R, Blazevic V, Marjomäki V, Tampé R, Kulomaa MS, Hytönen VP (2015) His-tagged norovirus-like particles: a versatile platform for cellular delivery and surface display. Eur J Pharm Biopharm 96:22–31

Korth MJ, Tchitchek N, Benecke AG, Katze MG (2013) Systems approaches to influenza-virus host interactions and the pathogenesis of highly virulent and pandemic viruses. Semin Immunol 25:228–239

Kratz PA, Bottcher B, Nassal M (1999) Native display of complete foreign protein domains on the surface of hepatitis B virus capsids. Proc Natl Acad Sci U S A 96:1915–1920

Ksenofontov AL, Dobrov EN, Fedorova NV, Serebryakova MV, Prusov AN, Baratova LA, Paalme V, Järvekülg L, Shtykova EV (2017) Isolated potato virus a coat protein possesses unusual properties and forms different short virus-like particles. J Biomol Struct Dyn 8:1–11

Kushima Y, Wakita T, Hijikata M (2010) A disulfide-bonded dimer of the core protein of hepatitis C virus is important for virus-like particle production. J Virol 84:9118–9127

Lam P, Gulati NM, Stewart PL, Keri RA, Steinmetz NF (2016) Bioengineering of tobacco mosaic virus to create a non-infectious positive control for Ebola diagnostic assays. Sci Rep 6:23803

Larson SB, Day J, Greenwood A, McPherson A (1998) Refined structure of satellite tobacco mosaic virus at 1.8 a resolution. J Mol Biol 277:37–59

Lata S, Reichel A, Brock R, Tampé R, Piehler J (2005) High-affinity adaptors for switchable recognition of histidine-tagged proteins. J Am Chem Soc 127:10205–10215

Lavelle L, Michel JP, Gingery M (2007) The disassembly, reassembly and stability of CCMV protein capsids. J Virol Methods 146:311–316

Li SW, Zhang J, He ZQ, Gu Y, Liu RS, Lin J, Chen YX, Ng MH, Xia NS (2005) Mutational analysis of essential interactions involved in the assembly of hepatitis E virus capsid. J Biol Chem 280:3400–3406

Li T, Lin H, Zhang Y, Li M, Wang D, Che Y, Zhu Y, Li S, Zhang J, Ge S, Zhao Q, Xia N (2014) Improved characteristics and protective efficacy in an animal model of E. coli-derived recombinant double-layered rotavirus virus-like particles. Vaccine 32:1921–1931

Li SW, Zhao Q, Wu T, Chen S, Zhang J, Xia NS (2015) The development of a recombinant hepatitis E vaccine HEV 239. Hum Vaccin Immunother 11:908–914

Li X, Chen Z, Jin L, Hu D, Yang S (2016) New strategies and methods to study interactions between tobacco mosaic virus coat protein and its inhibitors. Int J Mol Sci 17:252

Liddington R, Yan Y, Moulai J, Sahli R, Benjamin T, Harrison S (1991) Structure of simian virus 40 at 3.8-Å resolution. Nature 354:278–284

Lima SM, Vaz AC, Souza TL, Peabody DS, Silva JL, Oliveira AC (2006) Dissecting the role of protein-protein and protein-nucleic acid interactions in MS2 bacteriophage stability. FEBS J 273:1463–1475

Litowski JR, Hodges RS (2002) Designing heterodimeric two-stranded alpha-helical coiled-coils. Effects of hydrophobicity and alpha-helical propensity on protein folding, stability, and specificity. J Biol Chem 277:37272–37279

Liu Z, Gong Z, Dong X, Tang C (2016) Transient protein-protein interactions visualized by solution NMR. Biochim Biophys Acta 1864:115–122

Lu B, Taraporewala F, Stubbs G, Culver JN (1998) Intersubunit interactions allowing a carboxylate mutant coat protein to inhibit tobamovirus disassembly. Virology 244:13–19

Lu X, Thompson JR, Perry KL (2012) Encapsidation of DNA, a protein and a fluorophore into virus-like particles by the capsid protein of cucumber mosaic virus. J Gen Virol 93:1120–1126

Lum KK, Cristea IM (2016) Proteomic approaches to uncovering virus-host protein interactions during the progression of viral infection. Expert Rev Proteomics 13:325–340

Mach H, Volkin DB, Troutman RD, Wang B, Luo Z, Jansen KU, Shi L (2006) Disassembly and reassembly of yeast-derived recombinant human papillomavirus virus-like particles (HPV VLPs). J Pharm Sci 95:2195–2206

Madan B, Chaudhary G, Cramer SM, Chen W (2013) ELP-z and ELP-zz capturing scaffolds for the purification of immunoglobulins by affinity precipitation. J Biotechnol 163:10–16

Mannige RV, Brooks CL III (2010) Periodic table of virus capsids: implications for natural selection and design. PLoS One 5(3):e9423

Manuel-Cabrera CA, Vallejo-Cardona AA, Padilla-Camberos E, Hernández-Gutiérrez R, Herrera-Rodríguez SE, Gutiérrez-Ortega A (2016) Self-assembly of hexahistidine-tagged tobacco etch virus capsid protein into microfilaments that induce IgG2-specific response against a soluble porcine reproductive and respiratory syndrome virus chimeric protein. Virol J 13:196

Matsuura K, Watanabe K, Matsuzaki T, Sakurai K, Kimizuka N (2010) Self-assembled synthetic viral capsids from a 24-mer viral peptide fragment. Angew Chem Int Ed 49:9662–9665

Matsuura K, Nakamura T, Watanabe K, Noguchi T, Minamihata K, Kamiya N, Kimizuka N (2016a) Self-assembly of Ni-NTA-modified β-annulus peptides into artificial viral capsids and encapsulation of His-tagged proteins. Org Biomol Chem 14:7869–7874

Matsuura K, Mizuguchi Y, Kimizuka N (2016b) Peptide nanospheres self-assembled from a modified β-annulus peptide of Sesbania mosaic virus. Biopolymers 106:470–475

Minten IJ, Hendriks LJ, Nolte RJ, Cornelissen JJ (2009) Controlled encapsulation of multiple proteins in virus capsids. J Am Chem Soc 131:17771–17773

Namba K, Stubbs G (1986) Structure of tobacco mosaic virus at 3.6 Å resolution: implications for assembly. Science 231:1401–1406

Narayanan KB, Han SS (2017) Helical plant viral nanoparticles-bioinspired synthesis of nanomaterials and nanostructures. Bioinspir Biomim 12:031001

Nemerow GR, Stewart PL (2016) Insights into adenovirus Uncoating from interactions with Integrins and mediators of host immunity. Viruses 8(12):pii: E337

Pappachan A, Chinnathambi S, Satheshkumar PS, Savithri HS, Murthy MR (2009) A single point mutation disrupts the capsid assembly in Sesbania mosaic virus resulting in a stable isolated dimer. Virology 392:215–221

Patterson DP, Prevelige PE, Douglas T (2012) Nanoreactors by programmed enzyme encapsulation inside the capsid of the bacteriophage P22. ACS Nano 6:5000–5009

Patterson DP, LaFrance B, Douglas T (2013) Rescuing recombinant proteins by sequestration into the P22 VLP. Chem Commun (Camb) 49:10412–10414

Pejawar-Gaddy S, Rajawat Y, Hilioti Z, Xue J, Gaddy DF, Finn OJ, Viscidi RP, Bossis I (2010) Generation of a tumor vaccine candidate based on conjugation of a MUC1 peptide to polyionic papillomavirus virus-like particles. Cancer Immunol Immunother 59:1685–1696

Perlmutter JD, Hagan MF (2015) Mechanisms of virus assembly. Annu Rev Phys Chem 66:217–239

Peyret H, Gehin A, Thuenemann EC, Blond D, El Turabi A, Beales L, Clarke D, Gilbert RJ, Fry EE, Stuart DI, Holmes K, Stonehouse NJ, Whelan M, Rosenberg W, Lomonossoff GP, Rowlands DJ (2015) Tandem fusion of hepatitis B core antigen allows assembly of virus-like particles in bacteria and plants with enhanced capacity to accommodate foreign proteins. PLoS One 10(4):e0120751

Pieters BJ, van Eldijk MB, Nolte RJ, Mecinović J (2016) Natural supramolecular protein assemblies. Chem Soc Rev 45:24–39

Plevka P, Perera R, Yap ML, Cardosa J, Kuhn RJ, Rossmann MG (2013) Structure of human enterovirus 71 in complex with a capsid-binding inhibitor. Proc Natl Acad Sci U S A 110:5463–5467

Porath J, Carlsson J, Olsson I, Belfrage G (1975) Metal chelate affinity chromatography, a new approach to protein fractionation. Nature 258:598–599

Prevelige PE (2011) New approaches for antiviral targeting of HIV assembly. J Mol Biol 410:634–664

Prevelige PE, Thomas D, King J (1993) Nucleation and growth phases in the polymerization of coat and scaffolding subunits into icosahedral Procapsid shells. Biophys J 64:824–835

Pumpens P, Renhofa R, Dishlers A, Kozlovska T, Ose V, Pushko P, Tars K, Grens E, Bachmann MF (2016) The true story and advantages of RNA phage capsids as Nanotools. Intervirology 59:74–110

Pushko P, Pumpens P, Grens E (2013) Development of virus-like particle technology from small highly symmetric to large complex virus-like particle structures. Intervirology 56:141–165

Qazi S, Miettinen HM, Wilkinson RA, McCoy K, Douglas T, Wiedenheft B (2016) Programmed self-assembly of an active P22-Cas9 Nanocarrier system. Mol Pharm 13:1191–1196

Rose AS, Hildebrand PW (2015) NGL viewer: a web application for molecular visualization. Nucl Acids Res 43(W1):W576–W579

Satheshkumar PS, Lokesh GL, Murthy MR, Savithri HS (2005) The role of arginine-rich motif and beta-annulus in the assembly and stability of Sesbania mosaic virus capsids. J Mol Biol 353:447–458

Schwarz B, Uchida M, Douglas T (2017) Biomedical and catalytic opportunities of virus-like particles in nanotechnology. Adv Virus Res 97:1–60

Singh S, Zlotnick A (2003) Observed hysteresis of virus capsid disassembly is implicit in kinetic models of assembly. J Biol Chem 278:18249–18255

Smith ML, Lindbo JA, Dillard-Telm S, Brosio PM, Lasnik AB, McCormick AA, Nguyen LV, Palmer KE (2006) Modified tobacco mosaic virus particles as scaffolds for display of protein antigens for vaccine applications. Virology 348:475–788

Solovyev AG, Makarov VV (2016) Helical capsids of plant viruses: architecture with structural lability. J Gen Virol 97:1739–1754

Speir JA, Johnson JE (2012) Nucleic acid packaging in viruses. Curr Opin Struct Biol 22:65–71

Speir JA, Munshi S, Wang G, Baker TS, Johnson JE (1995) Structures of the native and swollen forms of cowpea chlorotic mottle virus determined by X-ray crystallography and cryo-electron microscopy. Structure 3:63–77

Steinmetz NF, Mertens ME, Taurog RE, Johnson JE, Commandeur U, Fischer R, Manchester M (2010) Potato virus X as a novel platform for potential biomedical applications. Nano Lett 10:305–312

Stockley PG, White SJ, Dykeman E, Manfield I, Rolfsson O, Patel N, Bingham R, Barker A, Wroblewski E, Chandler-Bostock R, Weiß EU, Ranson NA, Tuma R, Twarock R (2016) Bacteriophage MS2 genomic RNA encodes an assembly instruction manual for its capsid. Bacteriophage 6(1):e1157666

Stray SJ, Bourne CR, Punna S, Lewis WG, Finn MG, Zlotnick A (2005) A heteroaryldihydropyrimidine activates and can misdirect hepatitis B virus capsid assembly. Proc Natl Acad Sci U S A 102:8138–8143

Stubenrauch K, Gleiter S, Brinkmann U (2001) Conjugation of an antibody Fv fragment to a virus coat protein: cell-specific targeting of recombinant polyoma-virus-like particles. Biochem J 356:867–873

Tars K (2016) Introduction to capsid architecture. In: Khudyakov YE, Pumpens P (eds) Viral nanotechnology. CRC Press, Boca Raton, pp 3–11

Teschke CM, King J, Prevelige PE Jr (1993) Inhibition of viral capsid assembly by 1,1'-bi(4-anilinonaphthalene-5-sulfonic acid). Biochemistry 32:10658–11065

Thrane S, Janitzek CM, Agerbæk MØ, Ditlev SB, Resende M, Nielsen MA, Theander TG, Salanti A, Sander AF (2015) A novel virus-like particle based vaccine platform displaying the placental malaria antigen VAR2CSA. PLoS One 10(11):e0143071

Toncheva V, Wolfert MA, Dash PR, Oupicky D, Ulbrich K, Seymour LW, Schacht EH (1998) Novel vectors for gene delivery formed by self-assembly of DNA with poly(L-lysine) grafted with ydrophilic polymers. Biochim Biophys Acta 1380:354–368

Tresset G, Decouche V, Bryche JF, Charpilienne A, Le Cœur C, Barbier C, Squires G, Zeghal M, Poncet D, Bressanelli S (2013) Unusual self-assembly properties of norovirus Newbury2 virus-like particles. Arch Biochem Biophys 537:144–152

Trovato M, De Berardinis P (2015) Novel antigen delivery systems. World J Virol 4:156–168

Tuohimaa P, Joensuu T, Isola J, Keinänen R, Kunnas T, Niemelä A, Pekki A, Wallén M, Ylikomi T, Kulomaa M (1989) Development of progestin-specific response in the chicken oviduct. Int J Dev Biol 33:125–134

Walker A, Skamel C, Nassal M (2011) SplitCore: an exceptionally versatile viral nanoparticle for native whole protein display regardless of 3D structure. Sci Rep 1:5

Wang A (2015) Dissecting the molecular network of virus-plant interactions: the complex roles of host factors. Annu Rev Phytopathol 53:45–66

Wang YA, Yu X, Overman S, Tsuboi M, Thomas GJ, Egelman EH (2006) The structure of a filamentous bacteriophage. J Mol Biol 361:209–215

Werner S, Marillonnet S, Hause G, Klimyuk V, Gleba Y (2006) Immunoabsorbent nanoparticles based on a tobamovirus displaying protein a. Proc Natl Acad Sci U S A 103:17678–17683

Wingfield PT, Stahl SJ, Williams RW, Steven AC (1995) Hepatitis core antigen produced in Escherichia Coli: subunit composition, conformational analysis, and in vitro capsid assembly. Biochemistry 34:4919–4932

Wnęk M, Górzny ML, Ward MB, Wälti C, Davies AG, Brydson R, Evans SD, Stockley PG (2013) Fabrication and characterization of gold nano-wires templated on virus-like arrays of tobacco mosaic virus coat proteins. Nanotechnology 24:025605

Wong GC, Pollack L (2010) Electrostatics of strongly charged biological polymers: ion-mediated interactions and self-organization in nucleic acids and proteins. Annu Rev Phys Chem 61:171–189

Wynne SA, Crowther RA, Leslie AG (1999) The crystal structure of the human hepatitis B virus capsid. Mol Cell 3:771–780

Yang J, Yan R, Roy A, Xu D, Poisson J, Zhang Y (2015) The I-TASSER suite: protein structure and function prediction. Nat Methods 12:7–8

Yoffe AM, Prinsen P, Gopal A, Knobler CM, Gelbart WM, Ben-Shaul A (2008) Predicting the sizes of large RNA molecules. Proc Natl Acad Sci U S A 105:16153–16158

Yu X, Jin L, Jih J, Shih C, Zhou ZH (2013) 3.5 angstrom cryoEM structure of hepatitis B virus Core assembled from full-length Core protein. PLoS One 8(9):e69729–e69729

Zeltins A (2013) Construction and characterization of virus-like particles: a review. Mol Biotechnol 53:92–107

Zeltins A (2016) Viral nanoparticles: principles of construction and characterization. In: Khudyakov YE, Pumpens P (eds) Viral nanotechnology. CRC Press, Boca Raton, pp 93–119

Zhang HG, Xie J, Dmitriev I, Kashentseva E, Curiel DT, Hsu HC, Mountz JD (2002) Addition of six-his-tagged peptide to the C terminus of adeno-associated virus VP3 does not affect viral tropism or production. J Virol 76:12023–12031

Zlotnick A, Mukhopadhyay S (2011) Virus assembly, allostery and antivirals. Trends Microbiol 19:14–23

Zlotnick A, Aldrich R, Johnson JM, Ceres P, Young MJ (2000) Mechanism of capsid assembly for an icosahedral plant virus. Virology 277:450–456

Zlotnick A, Suhanovsky MM, Teschke CM (2012) The energetic contributions of scaffolding and coat proteins to the assembly of bacteriophage procapsids. Virology 428:64–69

Zlotnick A, Francis S, Lee LS, Wang JC (2016) Self-assembling virus-like and virus-unlike particles. In: Khudyakov YE, Pumpens P (eds) Viral nanotechnology. CRC Press, Boca Raton, pp 13–26

Chapter 17
The Role of Flaviviral Proteins in the Induction of Innate Immunity

L. Cedillo-Barrón, J. García-Cordero, G. Shrivastava, S. Carrillo-Halfon,
M. León-Juárez, J. Bustos Arriaga, Pc León Valenzuela,
and B. Gutiérrez Castañeda

Introduction

The genus *Flavivirus* contains enveloped, positive, single-stranded (ss) RNA viruses. These viruses can be categorized into two main groups according to the vector of transmission: mosquito-borne and tick-borne viruses (Holbrook 2017). Some of the members of this genus cause important diseases in humans and animals. Among them, dengue is the most prevalent mosquito-borne viral disease worldwide. Additionally, Zika virus (ZIKV) is a well-known flavivirus that has spread rapidly around the world. Other medically relevant flaviviruses include yellow fever virus (YFV), West Nile virus (WNV), Japanese encephalitis virus (JEV), tick-borne encephalitis virus (TBEV), and Kyasanur Forest disease virus (KFDV) (Pardigon 2017).

Diseases caused by members of the *Flavivirus* genus manifest a wide range of clinical forms, from mild illness, such as rash, fever, and joint pain, to more severe symptoms, such as haemorrhagic fever and fatal encephalitis. ZIKV infection has

L. Cedillo-Barrón (✉) · J. García-Cordero · G. Shrivastava · S. Carrillo-Halfon
P. León Valenzuela
Departamento de Biomedicina Molecular, CINVESTAV IPN, México, D.F, Mexico
e-mail: lcedillo@cinvestav.mx; gshrivastava@cinvestav.mx; pvalenzuela@cinvestav.mx

M. León-Juárez
Department of Immunobiochemistry, National Institute of Perinatology, México City, Mexico

J. Bustos Arriaga
Unidad de Biomedicina. Facultad de Estudios Superiores-Iztacala, Universidad Nacional Autonoma de México, Edo. de México, Mexico

B. Gutiérrez Castañeda
Immunology Department UMF Facultad de Estudios Superiores-Iztacala, Universidad Nacional Autonoma de México, Edo. de México, Mexico

© Springer Nature Singapore Pte Ltd. 2018
J. R. Harris, D. Bhella (eds.), *Virus Protein and Nucleoprotein Complexes*,
Subcellular Biochemistry 88, https://doi.org/10.1007/978-981-10-8456-0_17

407

recently been reported to have major symptoms, including encephalitis, microcephaly, acute flaccid paralysis, and Guillain-Barré syndrome, which can be fatal (Diamond and Pierson 2015).

Replication of the flaviviral ssRNA genome occurs in the cytoplasm at a membranous web that is associated with the endoplasmic reticulum (ER). The genesis of these membranous webs is related to viral proteins of infected cells (Pierson and Kielian 2013).

The innate immune response is critical for the early control of all flaviviruses and precedes induction of the adaptive response. The innate antiviral response is initiated in the skin following inoculation with the virus, where diverse target cells are present, e.g. dendritic cells (DCs), natural killer (NK) cells, neutrophils, keratinocytes, and fibroblasts. These cells may control viral replication and dissemination after the viral genome is released into the cytoplasm, where the viral RNA is sensed by retinoic acid-inducible gene I (RIG-I) and melanoma differentiation-associated gene (MDA5) or RNA-dependent protein kinases (PKRs). This triggers the innate immune response through the expression of type I interferon (IFN-I) and proinflammatory cytokines. The IFN-I signalling pathway in infected cells induces more than 300 IFN-dependent genes with antiviral functions that impair the virus cycle (Munoz-Jordan 2010; Hollidge et al. 2011). Thus, the viral RNA is sensed and translated into a polyprotein, which then is then targeted to the ER to be processed by viral and host proteases to yield three structural proteins, i.e. capsid protein (C), prM, and enveloped protein (E), as well as seven nonstructural proteins (NS1, NS2A,NS2B, NS3, NS4A, NS4B, and NS5). The nonstructural proteins in all flaviviruses are involved in viral replication, processing, and virion assembly (Iglesias and Gamarnik 2011) (Fig. 17.1).

In addition to their roles in the important and concerted steps of the viral replicative cycle, flaviviral proteins are also involved in intricate mechanisms to evade host immune responses and efficiently establish infection in the host. Viral proteins manipulate cell-induction stress to promote viral replication. Furthermore, all members of this genus antagonize the host IFN-I response by preventing Janus-activated kinase (JAK)/signal transducer and activator of transcription (STAT) signalling (Castillo Ramirez and Urcuqui-Inchima 2015)

In this chapter, we will review current knowledge of the interactions between the flaviviruses and their individual proteins with components of the host innate immune response.

Biology of Human Flaviviruses

Flaviviruses are small, enveloped positive-sense ssRNA viruses, of approximately 50 nm in diameter (Scherwitzl et al. 2017). Different members of the *Flavivirus* genus share a common mechanism of replication in the host cell. To initiate the cycle of infection, the viral particle attaches to the surface of the cell membrane through glycoprotein E; then, the viral particle interacts with one or several cell

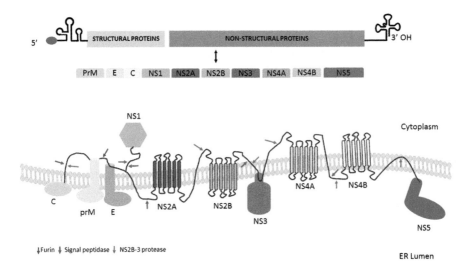

Fig. 17.1 Schematic representation of dengue polyprotein. The genome encodes a single long open reading frame (ORF), flanked by highly structured 5' and 3' untranslated regions (UTRs) which are translated into a single polypeptide. This protein is cleavage by host and viral proteases (the site of cleavage by the proteases is marked with arrows). Proteolysis yields ten proteins, the three structural proteins C, prM, and E and seven nonstructural proteins NS1, NS2A, NS2B, NS3, N S4A, NS4B, and NS5. All these proteins are necessary for the viral replication

surface receptors with high affinity, which triggers internalization of the viral particle by clathrin-mediated endocytosis (Gerold et al. 2017).

The viral particle is composed of an icosahedral capsid covered by both the envelope (E) protein and the M protein, which are anchored to the viral membrane; these two proteins are the major surface flaviviral proteins. The E protein is organized into 180 homodimers associated with the host-derived lipid membrane (Modis et al. 2004; Mukhopadhyay et al. 2005; Wang and Shi 2015). M protein is a small protein that is hidden underneath the E protein layer in all flaviviruses (Maier et al. 2007; Zhang et al. 2013). Furthermore, the nucleocapsid is formed by the C protein; each monomer of the C protein is associated with the viral genome and viral membrane lipids. A unique characteristic of the flaviviral nucleocapsid is the lack of a well-ordered viral assembly process, generating an amorphous capsid covered by the enveloping membrane.

Although the structure that the viral RNA adopts within the virus is unknown, different studies have characterized important regions in the viral genome involved in replication and viral translation. The viral genome size is approximately 11,000 nucleotides, although the size varies depending on the species. A common characteristic of flaviviral genomes is the presence of the highly structured 5' and 3' untranslated regions (UTRs). All flaviviruses possess a short 5'UTR (100 nt), which harbours a type I cap structure (m^{7}GpppAmG) at the 5' ends. The size of the 3'UTR ranges from 340 to 700 nt, and this region lacks a polyadenylated tail (Lindenbach and Rice 2003, Blitvich and Firth 2015). Complementary sequences are present in

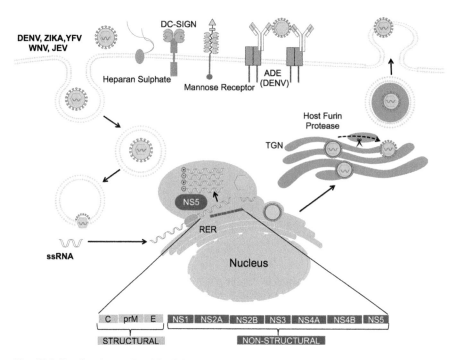

Fig. 17.2 **Replication cycle of flaviviruses**. Schematic representation of the flaviviral life cycle. The virus binds its host cell receptor (DC-SIGN, heparan sulphate, mannose receptor, etc.) and enters the cell through endocytosis. Once in the endosome, the pH changes promote the uncoating of the nucleocapsid and the release of the viral RNA to the cytoplasm. In the cytoplasm, the RNA binds to ribosomes in the endoplasmic reticulum and generates one single polyprotein, which is proteolytically cleaved to yield three structural and seven nonstructural proteins. NS5 polymerase replicates the genome, and the resulting RNA binds to the nucleocapsid, and the assembly takes place in the ER. The immature virion translocates to the Golgi apparatus, and the host furin protease cleaves the pr peptide to generate mature virions, which exit the cell to repeat the cycle

both UTRs, promoting interaction between the 3′ and 5′ termini of the flaviviral genome; these sequences are required for viral cyclization process and are important for viral replication (Gebhard et al. 2011; Ng et al. 2017). Finally, all members of the *Flavivirus* genus encode a long open reading frame (ORF) of 3400 codons, which translates into a single polyprotein. This polyprotein is processed by the viral protease NS2B-NS3 and different cellular proteases (Lobigs et al. 2010; Roby et al. 2015) to yield three structural proteins that comprise the viral particle (E, prM, and C) and seven nonstructural proteins involved in viral replication (NS1, NS2A, NS2B, NS3, NS4A, NS4B, and NS5) (Bartholomeusz and Wright 1993) (Fig. 17.2).

The viral surface glycoprotein E undergoes a conformational rearrangement in a pH-dependent manner that results in exposure of peptide fusion, enabling the fusion of viral and endosomal membranes. Disassembly of the viral capsid is not yet understood (Rey et al. 1995; Pastorino et al. 2010). In the ER, the viral genome is recognized as a mRNA, allowing it to use the translation cellular machinery and generate a single polyprotein. This polyprotein is processed by the viral protease NS2B-NS3 and different host cellular proteases (Lobigs et al. 2010; Roby et al. 2015).

The hallmark of all members of the *Flaviviridae* family is the rearrangement and redistribution of host intracellular membranes, leading to the development of organelle-like structures and creating platforms for viral replication. Viral proteins and cellular factors involved in viral replication are thus concentrated at those platforms (Welsch et al. 2009). Once the viral RNA translation starts, the capsid proteins interact with copies of the genome and lipid droplets, contributing to morphogenesis and viral genome assembly (Samsa et al. 2009). Immature flaviviral particles are produced when the C protein associates with the viral genome and membranes derived from the ER that are decorated with prM and E proteins. Such assembled particles are passed to the *trans*-Golgi network to undergo maturation; here, conformational changes and glycosylation of the E protein and cleavage of prM by furin take place. In fact, prM molecules function at this step to protect the fusion peptide on E from undergoing premature fusion before release. At this step, the morphology of non-infectious and immature viral particles is spiky. Then, before the release, immature particles are converted into smooth infectious virions as they pass through the secretory pathway into the neutral pH of the extracellular environment (Fischl and Bartenschlager 2011).

Flaviviral Proteins and Their Functions

E Protein

E protein is a glycoprotein with more than one potential N-glycosylation motifs. These motifs are frequently observed in domain I at asparagine residue 153 (N153) or N154. (Johnson et al. 1994; Kuhn et al. 2002). E protein forms the shell of all flaviviruses and is the most exposed structural viral glycoprotein. The molecular mass of this protein is 53 kDa, consisting of 495 amino acids; E proteins of all flaviviruses share approximately 40% amino acid identity. This protein is in charge of important functions, such as receptor binding, which allow internalization of viral particles through endocytosis (Ishak et al. 1988; van der Schaar et al. 2007). This glycoprotein interacts through its N-glycosylation sites with carbohydrate receptors. Furthermore, complex structural changes in the E protein due to the acidic environment induce the dissociation of the E dimer, thereby allowing exposure of the fusion loop, which is inserted into the membrane of endosomes, and enabling the release of viral RNA into the cytoplasm.

In all flaviviruses, the E protein is organized in dimers at the virus surface, and depending on its atomic structure (Rey et al. 1995; Zhang et al. 2003; Zhang et al. 2004; Modis et al. 2005; Nybakken et al. 2006), it folds into three structural and functional β-barrel-shaped domains: domain I (DI), domain II (DII), and domain III (DIII). DI is located in the N-terminus, and the hinge region is connected with

DII. This domain is conformational and possesses a discontinuous sequence involved in dimerization. In addition, DII contains a sequence that is highly conserved in the *Flavivirus* genus and contains 13 hydrophobic amino acids, which constitute the fusion loop (Allison et al. 2001; Stiasny et al. 2007). DIII is an immunoglobulin-like domain that forms small protuberances on the surface of virions and is located in the N-terminus (Kuhn et al. 2002; Zhang et al. 2004). E proteins are anchored at the viral membrane and organized in mature virions into 90 antiparallel dimers with rigid icosahedral surfaces (Modis et al. 2004).

E protein is the principal target of the immune response. The majority of neutralizing antibodies are directed against this protein; monoclonal antibodies directed against DII can neutralize the virus, affecting virus-cell membrane fusion by inducing structural changes in E protein. Moreover, cumulative evidence has indicated that natural infection with dengue virus (DENV) or other flaviviruses induces the production of cross-reactive antibodies directed against the fusion peptide in DII of the E protein. These antibodies exert weak neutralizing activity at minimum (Gollins and Porterfield 1986; Goncalvez et al. 2004; Beltramello et al. 2010). DIII contains multiple serotype-specific, conformation-dependent neutralizing epitopes, and an antibody against this domain inhibits the binding of virus to the host cell (Crill and Roehrig 2001). Furthermore, protein E is considered the primary candidate target of subunit vaccines.

C Protein

Compared with other flaviviral proteins, capsid C is small (approximately 100 amino acid residues), with a molecular weight of approximately 12 kDa. This protein possesses affinity for viral RNA and lipid membranes. The main role of this protein is to form part of the nucleocapsid as it is incorporated into the new virions (Dokland et al. 2004; Ivanyi-Nagy et al. 2008). According to crystallography studies (Dokland et al. 2004; Ma et al. 2004), the C protein is organized into four structurally conserved α-helices and an internal hydrophobic domain. The C protein forms dimers in solution; furthermore, studies of different members of the *Flaviviridae* family have described the translocation of C protein from the cytoplasm to the ER membrane or accumulation in lipid droplets in the nucleus and nucleolus of infected cells. Despite its small size, C protein possesses three functional nuclear localization sequences (NLSs), which are required for the interaction with importin-α to regulate its nuclear translocation. The identified NLS is quite variable, except for a GP motif conserved among flaviviruses. By substituting the G residue for an A, the nuclear localization of C is completely abolished (Sangiambut et al. 2008).

PrM Protein

In most flaviviruses, prM is a small protein with a molecular weight of approximately 19–21 kDa; this protein is glycosylated at asparagine N69 and is organized in 7 beta-stranded structures (Lorenz et al. 2002; Lobigs et al. 2010; Setoh et al. 2012). This protein is very important in the biogenesis of the viral particle since prM protein is required for prevention of premature fusion of the E protein within acidic compartments (Mukhopadhyay et al. 2005). Thus, when the viral particle matures, the prM protein is cleaved by a cellular furin in the *trans*-Golgi network. The pr remains associated with the E protein until the mature virion is released; following furin cleavage, liberating the pr peptide and the M protein is approximately 10 kDa (Stadler et al. 1997; Yu et al. 2008). Thus, prM protein serves as a chaperone for E protein, facilitating translocation into the ER lumen during polyprotein translation. The protein is then directed to the ER lumen by an N-terminal signal sequence immediately downstream from NS1 protein (Mackenzie and Westaway 2001).

NS1

Flaviviral NS1 is a multifunctional 48-kDa glycoprotein that is cleaved from NS2A by an unknown ER-resident host proteinase (Falgout et al. 1989). Following cleavage, NS1 is translocated to the ER lumen and cotranslationally processed by a host signal peptidase (Falgout and Markoff 1995). This protein is expressed in different cellular compartments, e.g. within intracellular membranes, or at the cell surface (Muller and Young 2013; Akey et al. 2014). Furthermore, NS1 may be a component of the viral replicase since NS1 colocalizes with double-stranded RNA (dsRNA) and interacts with NS4A and NS4B (Mackenzie et al. 1996, Lindenbach and Rice 1999; Youn et al. 2012); *NS1* gene deletion completely abrogates replication, but ectopic expression of NS1 in trans efficiently rescues ΔNS1 viruses (Lindenbach and Rice 1997; Khromykh et al. 1999).

The flaviviral NS1 protein is a critical protein involved in the induction of protection in mouse models, controlling viral spread by complement-mediated lysis of the infected cells (Gould et al. 1986; Henchal et al. 1988; Falgout et al. 1990; Despres et al. 1991; Chung et al. 2006a, b; Chung et al. 2007). In contrast, NS1 protein may also contribute to viral pathogenesis by increasing the permeability of capillaries (Beatty et al. 2015; Modhiran et al. 2015). Notably, NS1 elicits the production of autoantibodies that react with platelets and extracellular matrix proteins (Falconar 1997; Chang et al. 2002; Sun et al. 2007) with high possibilities to damage endothelial cells via antibody-dependent complement-mediated cytolysis (Lindenbach and Rice 2003). Recently, the presence of flaviviral NS1 protein in the infected host sera was shown to enhance viral acquisition by mosquitoes, by enabling the virus to overcome the immune barrier in the mosquito midgut. This is an example of an evolutionary mechanism whereby flaviviruses adapt to multiple host environments (Liu et al. 2016).

NS2A

Flaviviral NS2A is a 22–25-kDa hydrophobic transmembrane protein (Chambers et al. 1989; Xie et al. 2013) that plays a critical role in the viral life cycle (Mackenzie et al. 1998; Kummerer and Rice 2002; Leung et al. 2008). The NS2A N-terminus is processed in the ER lumen by a membrane-bound host protease; the C-terminus is processed in the cytoplasm by the NS2B/NS3 viral protease (Falgout et al. 1991; Falgout and Markoff 1995). Recently, a topological study of DENV NS2A has revealed that NS2A possesses eight predicted transmembrane segments (pTMS1–8) and five integral transmembrane segments (pTMS3, pTMS4, pTMS6–8) that span the lipid bilayer of the ER membrane (Xie et al. 2013).

The DENV2 NS2A protein has been shown to be involved in DENV RNA synthesis and virion assembly/maturation. The NS2A protein is also a member of the viral replication complex; in Kunjin virus (KUNV), NS2A was shown to interact with the 3′UTR of viral RNA and other components of the replication complex, such as NS4A, NS3, and NS5 (Mackenzie et al. 1998). This protein also participates in virion assembly and secretion (Kummerer and Rice 2002; Liu et al. 2003; Leung et al. 2008; Xie et al. 2013; Xie et al. 2015) and was found to associate with dsRNA in vesicle packets (Mackenzie et al. 1998). Published studies strongly suggest a model whereby NS2A is directly involved in transporting RNA from sites of RNA replication, across virus-induced membranes, to adjacent sites of virus assembly, with NS2A acting as a viroporin (Liu et al. 2003; Leung et al. 2008). The concept that NS2A may act as a viroporin is supported by the observation that both JEV and DENV NS2A proteins enhance membrane permeability (Chang et al. 1999; Shrivastava et al. 2017).

NS2B

In all flaviviruses, NS2B protein is a small integral membrane protein, with a molecular mass of 14 kDa (130 amino acids). This protein contains a conserved central hydrophilic region flanked by three hydrophobic domains located at the N- and C-termini (Li et al. 2015). The central hydrophilic region of NS2B is necessary and sufficient for the activation of NS3 protease; the hydrophobic regions may be involved in the oligomerization of the viral protease (Choksupmanee et al. 2012). In addition, some studies suggest that NS2B affects membrane permeability on account of its viroporin-like activity (Chang et al. 1999; Leon-Juarez et al. 2016). A recent study showed that NS2B plays an important role in viral replication and assembly by interacting with the NS2A protein; however, the mechanism is still unclear (Li et al. 2016).

NS3

Flaviviral NS3 protein is the second largest viral protein of the *Flavivirus* genus, with a molecular weight of 69 kDa. This molecule plays essential roles in the viral life cycle. The protein has two functional domains: an N-terminal protease with a chymotrypsin-like domain (amino acids 1–69), which is able to perform both *cis* and *trans* cleavage of the viral polyprotein precursor to generate nonstructural proteins (Li et al. 2014), and the C-terminal NTPase/RNA helicase (amino acids 180–618).

NS3 requires NS2B as a cofactor for protease activity. Thus, the association between the two proteins is mediated by the C-terminus of NS2B through a conserved β-turn hairpin interaction with a hydrophobic motif in NS3. The correct folding of the protease domain of NS3 depends on the interaction with NS2B. The protease catalytic triad (His51, Asp75, and Ser135) is localized in the central cleft of NS3.

Structural studies of the NS3 C-terminal domain identified three subdomains, of which the first and second adopt a RecA-like fold, with eight conserved motifs that are essential for RNA binding and ATP hydrolysis activities. The third subdomain is indispensable for synthesis of the ssRNA binding groove. Furthermore, evidence suggests that the third subdomain mediates the interaction between NS3 and NS5. This interaction is very important during viral replication. Moreover, NS3 may be able to resolve secondary structures of genomic RNA and separate dsRNA intermediates that are transiently formed during the polymerization reaction catalysed by NS5 (Brecher et al. 2013; Luo et al. 2015).

NS4A

All flaviviral NS4A proteins are small, multifunctional, and hydrophobic, with a molecular mass of approximately 16 kDa, although they may form larger structures upon self-oligomerization (Mackenzie et al. 1998). NS4A membrane topology suggests the presence of an N-terminal region, three transmembrane domains (TMD1–3), and a C-terminal tail; the N-terminal region resides in the cytoplasm, while the C-terminal region localizes in the ER lumen. NS4A associates with the ER membranes via four internal hydrophobic regions and colocalizes with double-stranded RNA (dsRNA), promoting membrane rearrangements that support viral replication complexes (Miller et al. 2007). Furthermore, the protein possesses a 2k fragment that plays a crucial role in ER membrane rearrangements and, consequently, in replication (Roosendaal et al. 2006; Miller et al. 2007). Besides its role in membrane proliferation, NS4A forms part of the WNV-KUN replication complex and interacts with the replicative intermediates dsRNA, NS1, NS2A, and NS5 (Chu and Westaway 1992).

NS4A has been shown to form an amphipathic helix within the first 20 amino acid residues of the cytoplasmic domain, which is essential for oligomerization and intracellular replication (Stern et al. 2013). Additionally, this cytoplasmic region of NS4A interacts with the cellular scaffolding protein vimentin, which is required for efficient replication (Teo and Chu 2014). NS4A is crucial during flaviviral replication, since it protects DENV-infected cells by activating phosphatidylinositol 3-kinase-dependent autophagy (McLean et al. 2011). Furthermore, in human foetal neural stem cells infected with ZIKV, NS4A inhibits the Akt/mTOR (mammalian target of rapamycin) signalling pathway, disrupting neurogenesis and inducing autophagy (Liang et al. 2016).

NS4B

The small hydrophobic protein NS4B, with a molecular mass of 30 kDa, is post-translationally modified by the host cell signalase to generate the mature form of the protein (molecular mass of 27 kDa). The cleaved peptide 2k is important for correct membrane insertion and folding of the NS4B protein (Miller et al. 2006). Topology analysis of this protein predicts the presence of five integral transmembrane domains; the N-terminal transmembrane regions reside in the ER lumen (TM1 and TM2), and TM2 may be N-glycosylated. Other transmembrane domains (TM2, TM4, and TM5) are localized in the C-terminus of NS4B and are part of a transmembrane α-helix inserted in lipid membranes (Roosendaal et al. 2006; Zmurko et al. 2015). The domains have different functions, e.g. TM1 and TM2 participate in antagonizing the activity of IFN-α/β (Evans and Seeger 2007) and TM4 and TM5 are important for dimerization of NS4B, which is crucial for assembly of the viral replication complex. Finally, different regions of NS4B are important for interactions with other viral proteins, such as NS3, NS1, and NS4A, which are essential components of viral replication machinery (Youn et al. 2012; Zou et al. 2014).

NS5

NS5 is the largest and most conserved protein synthesized by flaviviruses, with a 60% similarity among all flaviviruses. This protein is composed of approximately 900 amino acids and has a molecular weight of 105 kDa (Perera and Kuhn 2008). Moreover, NS5 is organized into two functionally and structurally different domains. At the N-terminus, NS5 possesses a methyltransferase (MTase) domain involved in viral cap formation (Egloff et al. 2002); the C-terminus contains an RNA-dependent RNA polymerase (RdRp) domain, which is involved in genome replication (Tan et al. 1996). NS5 has been shown to form a complex with NS3, and this interaction is required for viral replication (Kapoor et al. 1995; Cui et al. 1998).

The MTase domain consists of three subdomains: an N-terminal subdomain, the core subdomain, and a C-terminal subdomain. Additionally, the MTase domain possesses three main ligand-binding sites: a GTP binding site in the N-terminal subdomain and RNA and S-adenosyl-methionine (SAM) binding sites in the core subdomain. The MTase domain is involved in forming and attaching the cap at the 5′ end of the viral genome and has three activities, namely, guanylyltransferase (GTase), N7 MTase, and 2-O MTase activities (Egloff et al. 2002; Kanai et al. 2006; Issur et al. 2009).

The RdRp domain is more highly conserved among flaviviruses than other viral proteins because of its critical role in viral replication. This domain possesses a conventional RdRp conformation, with the shape of a cupped right hand, and three subdomains that form the fingers, thumb, and palm. The RdRp activity of NS5 is specific for flaviviral RNA and has been shown to initiate negative-strand RNA synthesis de novo, i.e. without the use of a primer (Ackermann and Padmanabhan 2001). NS5 possesses two nuclear localization sequences that allow NS5 to interact with importins and to enter the nucleus of infected cells. These sequences are not conserved among flaviviruses. NS3 and importin-β both bind to the region termed the bNLS; this suggests that in the case of some flaviviruses, NS5 is bound to NS3 for replication purposes and then disassociates from NS3 to bind to importin-β and enter the nucleus (Kapoor et al. 1995; Brooks et al. 2002). These events are regulated by phosphorylation. Casein kinase I (CKI) has been shown to be responsible for phosphorylating NS5 (Quintavalle et al. 2007; Bhattacharya et al. 2009). NS5 is also involved in the inhibition of the host immune response, as will be described later in the chapter. Furthermore, NS5 is a multifunctional protein and is expected to have additional functions that have not yet been identified.

The Innate Immune Response to Flaviviruses

The innate immune system constitutes the first line of host defence against pathogens. Once a virus finds the target cell, active crosstalk takes place between the pathogen and the cell. The crosstalk is initiated when the cell senses the viral pathogen through specialized molecules present in both the cytoplasm and endosomes. Thus, the optimum detection of flavivirus and the induction of the innate antiviral response define the consequences of infection (Suthar et al. 2013; Green et al. 2014; Castillo Ramirez and Urcuqui-Inchima 2015).

Cells possess many molecules that act as cellular pattern recognition receptors (PRRs) and are in charge of sensing a broad array of molecular components of pathogens, defined as pathogen-associated molecular patterns (PAMPs), such as proteins, carbohydrates, nucleic acids, and lipids. Thus far, three different PRR families have been recognized, including cytoplasmic or endosomal membrane-bound Toll-like receptors (TLRs), cytosolic RIG-I and MDA5, and cytosolic DNA sensors, e.g. NOD-like receptors (NLRs) (Brennan and Bowie 2010, Takeuchi and Akira 2010).

TLRs and TLR Signalling in Flaviviral Infections

TLR molecules contain leucine-rich repeats (LRRs) responsible for PAMP recognition, located in the ectodomain of the molecule. TLRs play distinct roles in triggering a response to an invading pathogen, since TLR expression is restricted to specific subtypes of immune cells (Uematsu and Akira 2007). TLRs can detect either extracellular or vesicle-bound PAMPs.

Once the extracellular ligands bind to the TLRs, the signal transduction cascade is initiated through a Toll/interleukin (IL)-1 receptor (TIR) homologous domain located in the cytoplasmic region of the protein (Kawai and Akira 2010). TLRs may then trigger signalling by two different pathways; the first is dependent on the recruitment of the adapter protein myeloid differentiation primary response gene 88 (MyD88), and the second proceeds via the adapter TIR-domain-containing adapter-inducing IFN-β (TRIF), which is independent of MyD88. TLR3 signalling triggers the recruitment of the adaptor molecule TRIF, which further interacts with tumour necrosis factor receptor-associated factors 3 and 6 (TRAF3 and TRAF6), activating the serine/threonine-protein kinase (TBK1) and inhibitor of κB kinase (IKK) and stimulating nuclear factor κ enhancer light chains of activated B cells (NF-κB), IFN regulatory factor (IRF) 3, and cytokine secretion. In addition, TRAF6 also interacts with RIP, which stimulates the transforming growth factor beta-activated kinase 1 (TAK-1) complex and promotes the activation of NF-κB and mitogen-activated protein kinase (MAPK) (Hemmi et al. 2002; Kawasaki and Kawai 2014). During TLR7 engagement, MyD88 is recruited and forms a complex with IL-1 receptor-associated kinases 1 and 4 (IRAK1 and IRAK4), TRAF3, and TRAF6. These complexes stimulate MAPK, IKK, and TBK, which activate and promote the nuclear translocation of activator protein 1, NF-κB, IRF3, and IRF7 transcription factors, resulting in induction of IFN-I, pro-inflammatory cytokine, and chemokine expression (Fig. 17.1.3) (Hemmi et al. 2002; Lin et al. 2010; von Bernuth et al. 2012; Zhou et al. 2013; Kawasaki and Kawai 2014). Therefore, activation of TLR3 or TLR7 promotes the secretion of IFNs and pro-inflammatory cytokines. Activation of TLR3 and TLR7 is involved in the immune response against flaviviruses; these proteins recognize dsRNA- and ssRNA-containing uridine-rich motifs, respectively. However, highly ordered structures containing viral RNA enhance TLR7 responses to ssRNA (Diebold et al. 2004; Lund et al. 2004; Wang et al. 2006). Along with RLS stimulation, the TLR pathway leads to a multivalent signalling cascade culminating in the production of IFN-α/β and inflammatory cytokines, which in turn triggers the maturation of DCs and the establishment of an antiviral response (Severa and Fitzgerald 2007). Therefore, the engagement of different TLRs is dependent on the specific viral infection. (Fig. 17.3).

Both TLR3 and TLR7 have been shown to be involved in sensing DENV. When human monocytes are infected with DENV and TLR3 is silenced, the response to DENV infection decreases (Tsai et al. 2009). Another study demonstrated that when TLR3 is overexpressed, production of copious amounts of cytokines is triggered, inhibiting DENV replication (16). Furthermore, TLR7-specific inhibitors reduce

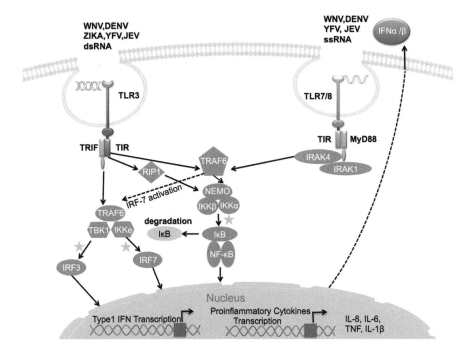

Fig. 17.3 Schematic showing the TLRs involved in flavivirus detection. TLR3 and TLR7/8 located in endosomal compartment detect viral dsRNA and ssRNA. TLR3 in turn phosphorylates and activates IRF3/7 through TRIF-TRAF3/TBK1/IKK kinase complex. Activated IRFs then drive the transcription of type I interferon genes. TLR7/8 through MyD88 activates NF-κB through the IRAKs-TRAF6-IKK pathway, which subsequently drives the expression of pro-inflammatory cytokines

IFN-α/β production in plasmacytoid DCs (pDCs) in response to DENV (Wang et al. 2006). Additionally, the detection of DENV by TLR7 is thought to be coupled to viral fusion and uncoating, since TLR7 signalling and viral entry into the cells require acidification of the endocytic vesicles (Tsai et al. 2009).

TLR3 and TLR7 have also been shown to be associated with WNV recognition. Experimental evidence has shown that TLR3 is essential for WNV penetration into the brain and for the induced inflammatory response. TLR7 and TLR3 contribute to antiviral responses to WNV, although the specific roles of these pathways in WNV-mediated pathogenesis remain elusive (Wang et al. 2004; Daffis et al. 2008; Welte et al. 2009).

Reduced levels of cytokine production in response to YFV-17D were observed in DCs recovered from single deletions of MyD88, TLR2, TLR7, or TLR9 in mice. In addition, TLR8 has been shown to provide a robust response to YFV-17D in human fibroblasts, suggesting that this TLR is also capable of detecting YFV (Querec et al. 2006).

During ZIKV infection, upregulation of TLR3, but not TLR7, is observed in human fibroblasts (Hamel et al. 2015). Another study demonstrated that ZIKV reduces the growth of cerebral organoids from human embryonic stem cells, by targeting neural progenitors. In addition, ZIKV was shown to activate the TLR3-mediated innate immune response, leading to deregulation of a network of 41 genes involved in neurogenesis, axon guidance, apoptosis, and differentiation (Dang et al. 2016).

Studies of JEV have demonstrated induced neuroinflammation and lethality via triggering of TLR3 and TLR4 signalling pathways (Han et al. 2014). In contrast, in another study, enhancement of JEV replication was observed in *TLR3*-gene-silenced, JEV-infected Neuro2a cells, suggesting that TLR3 serves a protective role against JEV (Fadnis et al. 2013). The role of TLR7 in the mouse brain during JEV infection was also demonstrated. Significant decreases in IFN-α and antiviral protein levels were also observed in TLR7-knockout mice and were accompanied by increased viral loads in the brain. TLR7 promotes IFN-I production and generation of the antiviral cellular state, contributing to the protective effects of the systemic response to infection following JEV infection (Nazmi et al. 2014a, b).

In summary, multiple PRRs are involved in initiation of the antiviral response to most flaviviruses; however, the pathways engaged during the infection are virus-dependent pathways.

Cytosolic Receptor RLR Family and Flaviviral Infections

The helicase family of enzymes plays many roles in cellular RNA metabolism and is therefore a ubiquitous protein expressed in many cells. These molecules bind to or remodel RNA or ribonucleoprotein (RNPs) complexes in an ATP-dependent manner. Among these, RLRs are soluble cytoplasmic receptors involved in viral recognition. Currently, three members of this family, i.e. retinoic acid-inducible gene I (RIG-I), melanoma differentiation-associated gene 5 (MDA5), and laboratory of genetics and physiology 2 (LGP2), have been characterized (Yoneyama et al. 2005).

Multiple lines of evidence have suggested that both RIG-I and MDA5 mediate the host's intracellular response to viral infection and are the two crucial innate immune receptors that bind specifically to dsRNA in the cytosol (Rawling and Pyle 2014). RIG-I, MDA5, and LGP2 share molecular and structural architecture, including two caspase-activation recruitment domains (CARDs). CARDs are located in tandem in the protein N-terminus and are critical for the interaction of the protein with the adaptor molecule mitochondrial antiviral signalling protein (which is known with the three names, MAVS, VISA, IPS-1). Furthermore, RIG-I and MDA5 contain a centrally located DExD/H motif helicase domain, which contains two RecA-like domains (Hel1 and Hel2). This helicase domain binds to dsRNA and contributes to ATP hydrolysis and RNA binding (Jiang et al. 2011) (Fig. 17.4). Additionally, both RIG-I and LGP2 contain a repressor domain that blocks signalling in the absence of ligand binding (5). Interestingly, RLR receptors can distinguish dsRNA bearing an uncapped 5′-triphosphate end (5′-ppp) that is at least 20-nt

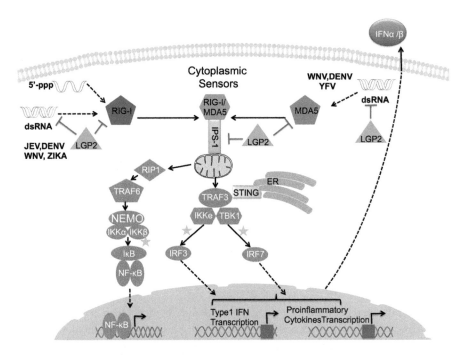

Fig. 17.4   RIG-I and MDA5 serve as cytosolic sensors and detect flavivirus dsRNA. RIG-I or MDA5 further activates TRAF3/TBK1/IKK kinase complex through RIG-I or MDA5/IPS-1/STING complex, which subsequently promotes the activation of NF-κB and phosphorylation of IRF3/7. Activated NF-κB and IRF3/7 translocate to the nucleus and induce the production of interferons and pro-inflammatory cytokines

long, via their RNA-binding domain. In contrast, MDA5 recognizes long dsRNA that lacks the uncapped triphosphate terminus (Hornung et al. 2006). Additionally, recent studies have suggested that LGP2 binds to blunt-ended dsRNA of different lengths (Li et al. 2009, Bruns and Horvath 2015). (Fig. 17.4).

Although the cytoplasmic molecules RIG-I and MDA5 are involved in sensing flaviviral genomic RNA, RIG-I was more frequently reported to be responsible for the recognition of the *Flavivirus* genus. Interestingly, flaviviruses, such as DENV and WNV, are sensed by both MDA5 and RIG-I. Initial studies showing the role of RIG helicases in the recognition of RNA viruses were published by Kato et al. (2006).

In hepatocytes infected with DENV, TLR3 synergises with RIG-I and MDA5 to induce IFN expressions. Moreover, in RIG-I- and MDA5-knock down Huh7 cells, the intracellular RNA virus sensors RIG-I, MDA5, and TLR3 are activated upon DENV infection, modulating the host antiviral defence system (Nasirudeen et al. 2011). Furthermore, both cytoplasmic (RIG-I and MDA5) and vesicular (TLR3) PRRs are upregulated during ZIKV infection and may play a synergistic role in enhancing the antiviral response to ZIKV infection (Hamel et al. 2015). RIG-I has been described to be essential for the production of IFNs in response to different

RNA viruses, including JEV. Many studies have suggested that RIG-I and MDA5 recognize different RNA viruses and are critical for host antiviral responses. Additionally, both RIG-I and MDA5 have been shown to be responsible for the induction of IFN, operating in synergy to establish an antiviral cellular state and mediate an IFN amplification loop that supports expression of immune effector genes during WNV infection (Fredericksen et al. 2008). The RIG-I molecule signalling pathway is involved in initiating the antiviral state against JEV, triggering the production of IFN-I and pro-inflamatory response (Chang et al. 2006; Kato et al. 2006; Nazmi et al. 2014a, b).

Interestingly, both RIG-I and MDA5 have been implicated in sensing and responding to flavivirus infection. An issue that must be elucidated is whether each RLR molecule is redundant or working at different moments and with different antiviral programme during flavivirus infection (Suthar et al. 2013).

NLR Molecules

NLR family members identify intracellular danger-associated molecular patterns (DAMPs) and PAMPs (Meunier and Broz 2017). Structurally, the members of the NLR family are multidomain proteins composed of a common central NOD domain, C-terminal LRRs, and N-terminal CARD or pyrin (PYD) domains. LRRs play a crucial role in ligand sensing and autoregulation processes that initiate NLR signalling, whereas the CARD and PYD domains mediate homotypic protein-protein interactions for downstream signalling.

NLRs recognize microbial products or intracellular signals, triggering the recruitment of a molecular platform as a signalling complex, e.g. inflammasomes and NOD signalosomes. These complexes promote the secretion of pro-inflammatory cytokines, including IL-1β, IL-18, and IL-33 (Shrivastava et al. 2016).

NLR family members can be classified into three subfamilies (NOD, NLRP, and IPAF) based on the sequences of their NACHT domains (Lupfer and Kanneganti 2013). However, few studies have investigated the involvement of NLRs in flaviviral infection (Suthar et al. 2013; Nazmi et al. 2014a, b). Studies performed in humans infected with WNV and DENV have shown elevated systemic IL-1β levels, suggesting that the viral infection activates inflammasome signalling (Ramos et al. 2012). Furthermore, caspase-1 activity and subsequent IL-1β production were identified in different cellular models of DENV, ZIKV, WNV, and JEV infections (Kumar et al. 2016; Wu et al. 2017). However, the exact activation signal that promotes the production of IL-1β during NLRP3 activation is not well understood. Future studies are needed to determine the role, if any, of viral proteins or virus-induced cell stress signals in triggering the activation of the NLRP3 inflammasome during flavivirus infection (Fig. 17.5).

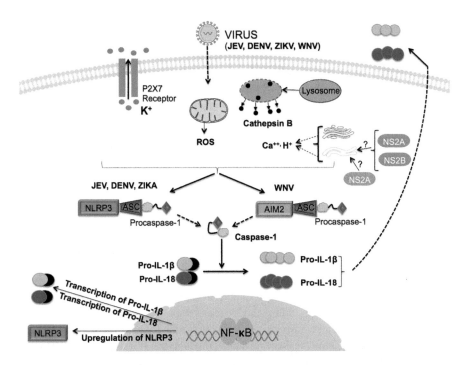

Fig. 17.5 NLR activation by *Flavivirus* families. Flavivirus RNA sensing together with other stress signals (K+ efflux, ROS, increased Ca++ and/or H+, released cathepsin B) activates inflammasome complex, inducing the activation of caspase-1 activation, which further cleaves zymogen pro-IL-1β and pro-IL-18 secretion into mature IL-1β and IL-18, respectively

Cyclic GMP-AMP Synthetase (cGAS) Is Involved in Flavivirus Sensing

Canonical molecules, such as TLR, MDA, RIG-I, and NLR, are responsible for sensing the presence of RNA viruses. However, DNA sensors, such as cGAS, have also been identified. cGAS is a member of the nucleotidyl transferase family that, once activated, produces the secondary messenger cyclic di-GMP-AMP (cGMP-AMP), which in turn activates the adaptor protein stimulator of IFN gene protein (STING) (Caiet al. 2014). This protein activates TBK1 (Kato et al. 2006), which phosphorylates IRF3 and activates NF-κB to transcriptionally induce IFN-I (Yie et al. 1997). IRF3 binds to IFN-stimulated response elements (ISREs) to induce additional antiviral genes. The two main kinases responsible for IRF3 activation, IKKε and TBK1, which are involved in both RLR and cGAS-STING pathways, are also targeted by *Flavivirus* (Yie et al. 1997). STING serves as both a PRR for cyclic dinucleotides and an adaptor for PRRs of cytoplasmic dsDNA (Cai et al. 2014).

cGAS is involve in flavivirus recognition; the DNA sensor mediates an antiviral response, inhibiting viral replication (van Gent and Gack 2017). The early work of (Schoggins et al. 2014) provided insights into the role of cGAS in sensing positive ssRNA. WNV-infected cGAS-deficient mice showed higher infection and mortality rates than wild-type mice. Moreover, (Aguirre and Fernandez-Sesma 2017) showed that DENV2 induces cGAS signalling, leading to increased production of IFN-I and consequently controlling the propagation of DENV in the cell monolayer. However, the mechanisms involved in ssRNA virus recognition may be indirect, e.g. through mitochondrial DNA that is released in the cytosol to activate cGAS during DENV infection (Sun et al. 2017).

IFN-I and Induction of the Antiviral State

Activation of IFN Signalling by Flaviviruses

Once the viruses are detected by different PRRs, an immunological alert is mounted to initiate the antiviral state. This occurs via the secretion of cytokines, such as IFN, secreted by the infected cells or through paracrine effects, to protect other cells. IFNs stimulate uninfected cells to produce proteins that inhibit the replication of various types of viruses, including flaviviruses. Rapid detection of viral infection and induction of the innate antiviral response are crucial for the resolution of infection. IFNs can be divided into three groups, namely, types I, II, and III, according to their amino acid sequence and receptors (Goodbourn et al. 2000).

IFN-I ($\alpha\beta$) is a key element of the innate immune response and is induced directly in response to viral infections. IFN-III (λ) comprises IFN-λ1, IFN-λ2, and IFN-λ3, which are also referred to as IL-29, IL-28A, and IL-28B, respectively. The IFN-I response could limit viral replication through different mechanisms, such as restricting the spread of progeny viruses to neighbouring cells, reducing the overall viral replication or both (Shresta et al. 2006).

The timing of the IFN response is crucial in the control of flaviviruses. Pretreatment of DENV-infected cells with IFN-I exerts protective effects, and the resulting viral titre is low; the same treatment after the infection did not confer efficient protection against the virus (Diamond and Harris 2001). Furthermore, IFNA$_R$-deficient (AG-129) or STAT-deficient mouse models revealed the essential role of IFN-I against infection, e.g. with WNV, DENV, YFV, and ZIKV (Shresta et al. 2006; Ashour et al. 2010; Tripathi et al. 2017) (Fig. 17.5).

IFN-α/β and IFN-λ constitute the first line of defence against flaviviruses. The signalling cascade is initiated when IFN binds to an IFN receptor (IFNAR). The IFN-α/β receptor is composed of two major subunits, IFNAR1 and IFNAR2, which are produced in all nucleated cells (de Weerd and Nguyen 2012). Each of these receptor subunits interacts with a member of the JAK family. Prior to stimulation, the cytoplasmic domains of IFNAR1 and IFNAR2 are associated with the Janus tyrosine kinase Tyk2 (Velazquez et al. 1992; Colamonici et al. 1994) and JAK1 (Silvennoinen et al. 1993; Novick et al. 1994), respectively. IFNAR2 is also associated with STAT2 (Wen et al. 1995; Goodbourn et al. 2000; Murray 2007). Upon IFN-α/β binding, IFNAR1 and IFNAR2 associate, facilitating the transphosphorylation and activation of Tyk2 and JAK1 (Ihle 1995; Platanias 2003). Tyk2 then phosphorylates tyrosine 466 of IFNAR1 (Velazquez et al. 1992), creating a new docking site for STAT2 through the latter's SH2 domain (Darnell et al. 1994, 1997). STAT2 is next phosphorylated by Tyk2 at tyrosine 690. The phosphorylated STAT1/STAT2 heterodimers dissociate from the receptor and are translocated to the nucleus, where they associate with the DNA-binding protein p48 or IRF9 (Darnell Jr 1997; Platanias 2003) to form a ternary complex called IFN-stimulated gene (ISG) factor (ISGF) 3, which binds ISREs of the IFN-α-/IFN-β-responsive genes (Levy et al. 1989; Fu et al. 1990).

Optimal induction of the *IFN-β* gene requires binding of a c-Jun/ATF-2 heterodimer to its promoter, as well as the enhancer region located immediately upstream of the core promoter (Panne et al. 2004). The IRF3, NF-κB, and c-Jun/ATF-2 complexes assemble at the promoter in a cooperative manner to form the so-called enhanceosome; formation of this complex is aided by the high-mobility group (HMG) chromatin-associated protein HMGI (Y), also known as HMGA (Yie et al. 1997).

In general, ssRNA viruses are highly sensitive to the antiviral effects of IFN-I, which is part of the overall innate immune response, evoked to effectively block early virus replication; hence, IFN induces a generalized antiviral state.

After the flaviviral RNA is recognized by RIG-I and MDA5 PRRs, these molecules change conformation to associate with a promoter stimulator 1 molecule and initiate signal transduction to activate the transcription factors IRF3 and NF-κB, consequently inducing the synthesis of IFNs. The production of IFN during viral infection leads to the induction of at least three transcription factors (IRF1, IRF7, and IRF9) (Wathelet et al. 1998). Binding of secreted IFN to its receptor stimulates the JAK/STAT signalling pathway in target cells, leading to transcriptional activation of numerous ISGs. Most ISG products, including PKR, 2′5′-oligoadenylate synthetase (OAS) (Tanaka et al. 2004), virus inhibitory protein, and ER-associated, IFN-inducible (viperin) protein (Helbig et al. 2013), are able to interfere with viral infections.

PKR is one of the most extensively studied ISGs. Analysis of PKR amino acid sequences from different species has revealed a common regulatory domain in the N-terminal region and a dsRNA-binding motif (dsRBM) (Weber et al. 2006). PKR induces translational arrest to prevent production of proteins required for the assembly of new viral particles, thus limiting viral spread. PKR is initially activated by the interaction of dsRNA (and other specific types of RNA) with the binding motif on PKR (Sadler and Williams 2007). Activated PKR phosphorylates the only substrate known thus far: eukaryotic initiation factor α (eIF2α) (Sadler and Williams 2007). Phosphorylated eIF2α prevents the recycling of GDP to GTP by the guanine nucleotide exchange factor eIF2B, resulting in global inhibition of protein synthesis (Samuel 1993). Furthermore, PKR promotes mRNP aggregation induced by G3BP protein (Li et al. 2013; Tsai and Lloyd 2014). Li et al. showed that silencing of PKR in A549 cells infected with DENV provokes a decrease in IFN-β at the translational level (Li et al. 2013). Additionally, PKR and RNase L are important effector molecules that protect mice against WNV (Samuel et al. 2006).

IFN-Induced OAS

The OAS family of proteins includes OAS1, OAS2, OAS3, and OAS-like (OASL) protein. This family is regulated by both IFN-I and IFN-II and includes proteins that are critical for the immune response to viral infection. The antiviral mechanism of OAS is modulated in two ways. First, OAS3 protein binds dsRNA and is activated to produce 2′,5′-oligoadenylate, leading to the activation of a latent form of RNase L, which degrades cellular and viral RNAs (Tanaka et al. 2004, Silverman 2007, Zhu et al. 2014). The second antiviral mechanism is stimulated by these cleavage products, which activate RIG-I and MDA5 (Zhu et al. 2014). Interestingly, replication of JEV in vitro is inhibited by porcine OASL (Zheng et al. 2016). Furthermore, the isoforms OAS1 p42/p46 and OAS3 p100 are likely to contribute to the host defence against DENV infection and play a role in determining the outcomes of DENV infection severity.

Viperin

Viperin has a molecular weight of 42 kDa. The protein possesses three domains: an N-terminal amphipathic α-helix, which enables its localization in the ER and in lipid droplets (Hinson and Cresswell 2009), a central radical SAM domain, and a C-terminal domain with unknown function. Viperin possesses multiple activities, e.g. immuno-regulatory effects in the production of type I IFN. Moreover, viperin is involved in intracellular signalling and has broad antiviral activity in pDCs (Seo et al. 2011).

Viperin plays an antiviral role against some flaviviruses, e.g. WNV and DENV. Additionally, viperin restricts WNV infection in a tissue- and cell-type-specific manner. Although the antiviral mechanism of viperin is not clear (Duschene and Broderick 2010), available experimental evidence suggests that this protein binds to the enzyme farnesyl diphosphate synthase, which is involved in cholesterol biosynthesis; cholesterol, in turn, is a key component of lipid droplets and lipid rafts, where replication and assembly occur (Zhang et al. 2000, Ono and Freed 2005). Furthermore, JEV-infected cells show upregulation of viperin mRNA (Chan et al. 2008). Interestingly, viperin possesses a central radical SAM, which is associated with strong antiviral effects against TBEV (Upadhyay et al. 2014); the C-terminal viperin domain has the ability to interact with nonstructural proteins, such as NS3, of DENV (Fukasawa 2010; Wang et al. 2012; Helbig et al. 2013).

IFN-Inducible Transmembrane (IFITMs) Proteins

IFITMs are involved in the inhibition of early flaviviral replication. At least three human IFITM proteins have been identified (IFITM1, IFITM2, and IFITM3). The first two are expressed ubiquitously in humans; their expression is induced by both

IFN-I and IFN-II (REF). The hallmark of this protein group is the prevention of infection before a virus traverses the lipid bilayer of the cell. IFITMs inhibit the replication of all flaviviruses tested to date, including WNV, DENV, and ZIKV (Jiang et al. 2010; Savidis et al. 2016). Recently, (Savidis et al. 2016) showed that IFITMs strongly block ZIKV replication. Additionally, IFITM3 can prevent ZIKV-induced cell death.

ISG15 is a pleiotropic molecule that acts as both a cytokine and protein modifier. This molecule can be covalently conjugated to different proteins in a manner similar to that of ubiquitin, causing protein degradation (Hishiki et al. 2014). In the case of DENV, the NS3 and NS5 proteins are subjected to ISGylation, which suppresses DENV particle release. Additionally, ISGylation also affects WNV (Hermann and Bogunovic 2017).

Inhibition of IFN Signalling and Evasion of Host Signalling by Flavivirus Proteins

IFN-I proteins are the most effective molecules involved in the antiviral response of the innate immune system; however, viruses have developed specific strategies to evade this response in order to establish an infection in the host.

Members of the *Flaviviridae* family are prominent IFN-I inhibitors. The inhibition takes place at the level of IFN production and secretion, inhibition of IFN binding to receptors by competition, inhibition or proteolysis of activators of the JAK/STAT signalling pathway, or inhibition of proteins with antiviral activity and products of the transcription of ISGs (Haller et al. 2006; Taylor and Mossman 2013; Ma and Suthar 2015). Other mechanisms, dependent on the accumulation of subgenomic flaviviral RNA, have also been reported (Schuessler et al. 2012; Bidet et al. 2014; Manokaran et al. 2015); however, these mechanisms are beyond the scope of this chapter.

Interestingly, most of the known strategies are dependent on nonstructural proteins, suggesting that the viral genome has to be transcribed and translated, with the polyprotein processed into individual proteins to counteract IFN-I activity.

Multiple flaviviral nonstructural proteins, including NS2A, NS4A, NS4B, and NS5 (Ma and Suthar 2015), have been shown to act as IFN-I inhibitors. However, NS5 is the most prominent effector involved in inhibition of IFN-I activity. This protein may block IFN-I production and effectors via different pathways, depending on the flavivirus. This variability in NS5 evasion strategies is counterintuitive considering the high homology of the MTase and RdRp domains in the different members of the *Flavivirus* genus.

In the context of DENV infection, the JAK/STAT pathway is inhibited by phosphorylation of Tyk2, STAT1, and STAT2 (Ho et al. 2005; Jones et al. 2005). Thus, DENV NS5 facilitates the ubiquitination of STAT2, directing it to undergo proteasome degradation. The interaction between dengue NS5 and STAT2 was proven to be functional since the ISG transcription activity was also blocked in IFN reporter model under the control of IFN-sensitive response elements (ISRE) in vitro (Ashour et al. 2009). Furthermore, the effect of DENV NS5 on STAT2 is dependent on the complete proteolytic maturation of the N-terminus (NS5 region) of the polyprotein accomplished by the viral protease NS2B/NS3 (Ashour et al. 2009; Mazzon et al. 2009). This strategy is supported by the interaction of the first amino acids of DENV NS5 with N-recognin 4 (UBR4), an E3 ubiquitin ligase, acting as a bridge between STAT2 and the ubiquitination machinery inducing the degradation of STAT2/NS5 complex (Morrison et al. 2013). It is well known that flaviviral NS5 is a multifunctional protein of the viral replication cycle, suggesting a fine balance between the amount of NS5 designated for the replication complex, for the nuclear localization, and serving as an IFN-I antagonist. The plasticity of DENV NS5 implies the temporal nature of NS5 allocation in the infected cell, participating in innate immune evasion depending on the balance between viral resources and modulation of the immune response and manipulating the cell machinery to achieve productive infection.

ZIKV NS5 also targets STAT2 for proteasomal degradation; however, the mechanism of STAT2 degradation is not dependent on the proteolytic maturation of the N-terminus of NS5, and UBR4 does not seem to be involved (Grant et al. 2016; Kumar et al. 2016). Nevertheless, it is highly probable that STAT2 is ubiquitinated by E3 ubiquitinase and directed for proteasomal degradation. In contrast, the NS5 protein of Spondweni virus, which is closely related to ZIKV, does not inhibit the activation of STAT1 or STAT2 after IFN-I treatment; instead, this protein seems to inhibit IFN-I activity at the transcription level of the ISRE-dependent genes (Grant et al. 2016). The molecular mechanisms through which transcription is inhibited remain to be elucidated; however, the overlapping of the two functions of NS5, i.e. nuclear localization and IFN-I antagonism, is important to the function of flavivirus throughout the replication cycle.

YFV NS5 also targets the JAK/STAT signalling pathway, albeit in an inducible manner. The N-terminus of YFV NS5 is ubiquitinated by TRIM23, inducing its binding to STAT2 after IFN-I stimulation. This association does not induce degradation by the proteasome; instead, the NS5-STAT2 complex inhibits the binding of ISGF3 to the promoters of ISGs (Laurent-Rolle et al. 2014). The mechanism of inhibition by YFV NS5 is interesting since TRIM23 contributes to Sendai virus infection-triggered NF-κB and IRF3 activation by targeting NF-κB essential modulator (NEMO) (Arimoto et al. 2010). In contrast, proteins that contribute to the antiviral innate immune response are hijacked by YFV NS5, resulting in antagonism.

Other *Aedes*-borne flaviviruses, such as Yokose (YOKV), Iguape (IGUV), and Usutu (USUV) viruses, also utilize their NS5 proteins to block the JAK/STAT signalling pathway. Current evidence has suggested that these NS5 proteins inhibit the phosphorylation of STAT1 and STAT2 (Grant et al. 2016). USUV and YOKV NS5 proteins prevent STAT1/2 nuclear translocation after IFN treatment; in contrast, the expression of IGUV NS5 protein does not prevent IFN-mediated translocation of STAT1 or STAT2 (Grant et al. 2016). The mechanism of this inhibition of phosphorylation and nuclear translocation is still unknown; more experiments in infection models are needed to identify the interactions between specific host proteins and viral NS5 (Fig. 17.6).

Tick-borne flaviviruses include 12 recognized species that are divided into two groups, the mammalian tick-borne virus group and the seabird tick-borne virus group (Grard et al. 2007). The first evidence for the role of flaviviral NS5 as an IFN-I antagonist was reported for Langat virus (LGTV) (Best et al. 2005). Infection with LGTV inhibited the transcription from IFN-α- and IFN-γ-responsive promoters in reporter assays. Furthermore, in infected human cells, inhibition of JAK1 and Tyk2 phosphorylation during IFN-α stimulation and inhibition of JAK1 phosphorylation following IFN-γ stimulation were observed without direct interactions between these kinases and LGTV NS5 (Best et al. 2005). Additionally, the loss of IFNAR1 but not IFNAR2 expression in LGTV monocyte-derived DCs was observed (Best et al. 2005). Inhibition of the expression of IFNAR1/IFNAR2 in LGTV-infected cells seems to be associated with the interaction with hScrib, a membrane protein that belongs to the LAP, LRR, and PSD/Dlg/ZO-1 family (Werme et al. 2008), and with the interaction with the host protein prolidase (PEPD), a metabolic enzyme belonging to the metalloproteinase family responsible for the hydrolysis of imidodipeptides containing C-terminal proline or hydroxyproline residues required for the glycosylation of IFNAR1. This inhibition of IFNAR1 expression depends on the interaction of NS5 with PEPD during WNV infection (Lubick et al. 2015). WNV is transmitted by the *Culex* mosquito and shares a vector with JEV. For both viruses, evasion of the IFN-I signalling pathway increases their fitness and virulence (Keller et al. 2006; Liang et al. 2009). Although NS5 proteins from both viruses act as IFN-I antagonists, the mechanisms differ. For WNV, as mentioned before, the proposed mechanism involves the inhibition of the glycosylation of IFNAR1 through interaction with PEPD. In contrast, in JEV-infected cells, IFNAR1/IFNAR 2 expression is not affected; instead, the JEV NS5 IFN-antagonizing activity involves protein tyrosine phosphatases, as the IFN-antagonizing activity can be pharmacologically inhibited by sodium orthovanadate (Lin et al. 2006). Inhibition of the translocation of STAT1 by JEV NS5 also correlates with the downregulation of calreticulin in neuroblastoma cells (Yang et al. 2013).

All available evidence suggests a dynamic host-pathogen interaction between NS5 and the infected cell. The regulation and temporality of IFN-I signalling antagonism, which modulate the multiple functions of NS5 during the viral cycle, are unknown; however, the plasticity of this nonstructural protein renders the virus adaptable to different microenvironments. The mechanisms that regulate the activation of the diverse functions of flaviviral NS5 in different lineages of immune cells or haematopoietic cells remain unclear. Despite this, it is possible that the IFN-antagonizing effect of NS5 may be inducible depending on the abundance of the protein, which is dependent on the permissiveness and susceptibility of the cells (Fig. 17.6).

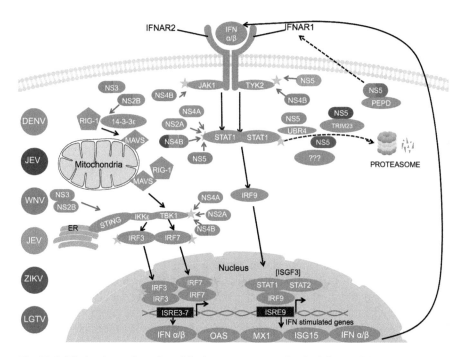

Fig. 17.6 Mechanisms of evasion of the immune response by flaviviruses. Schematic representation of the mechanisms utilized by different flaviviruses to inhibit the immune response. Upon binding of type I interferon to its receptor, both subunits, IFNAR1 and IFNAR2, recruit JAK1 and Tyk2 and phosphorylate them. NS4B of WNV inhibits both phosphorylations, whereas NS5 of JEV inhibits Tyk2 phosphorylation. Phosphorylation of JAK1 and Tyk2 induces binding of STAT1 and STAT2, and these are phosphorylated by JAK1 and Tyk2. STAT1 phosphorylation is inhibited by DENV NS2A, NS4A, and NS4B, as well as YFV NS4B and WNV NS4B and NS5. STAT2 is a target for proteasomal degradation by NS5 of DENV, YFV, and ZKV, but this effect is dependent on different host factors, DENV NS5 targets STAT2 in the presence of UBR4, YFV targets STAT2 in the presence of TRIM23, and ZIKV host factor has not been yet determined. RIG-I helicase is translocated to the mitochondria in the presence of 14–3-3ε chaperone. This translocation is inhibited by the binding of the NS3-NS2B complex of DENV and WNV. RIG-I binds to IPS-1, and IPS-1 activates IKKε, which phosphorylates TBK1 and, together, phosphorylates IRF3 and IRF7. The NS3-NS2B complex inhibits IKKε and prevents TBK1 phosphorylation. DENV NS2A, NS4A, and NS4B and WNV NS4B inhibit TBK1 phosphorylation. The NS3-NS2B complex of DENV cleaves STING and inhibits the cGAS pathway. NS5 of WNV and LGTV binds to PEPD and inhibits IFNAR1 maturation and expression. Once phosphorylated, STAT1 and STAT2 recruit IRF9 and together translocate to the nucleus to bind to ISRE9. Host cellular proteins are represented in grey, whereas viral proteins are represented in different colours depending on the flavivirus: DENV in turquoise, YFV in red, WNV in orange, JEV in green, ZKV in purple, and LGTV in blue. Viral proteins from different flaviviruses that share the same function are depicted in split colours. Continuous black lines represent the normal pathway, and dashed lines represent the effect of flaviviral proteins, whereas short red lines represent inhibition. Phosphorylation is represented as a star

Other Flaviviral Proteins with IFN-I Antagonist Activity

Other flaviviral proteins have also been described as IFN-I antagonists, either in the context of infection or exogenously expressed in the absence of virus. Flavivirus proteins other than NS5 and with IFN-I antagonistic activity may play a redundant role to ensure the inhibition of IFN-I antiviral activity. The mechanisms through which the remaining NS proteins block the IFN signalling pathway are also diverse.

Multiple DENV NS proteins act as IFN antagonists and seem to work synergistically. For example, NS4B blocks the IFN-induced signal transduction cascade by interfering with STAT1 phosphorylation; this mechanism is also used by KUNV NS2A, NS2B, NS3, NS4A, and NS4B (Liu et al. 2005). NS4A and NS2A also block IFN signalling; the cumulative effect of DENV NS4B, NS4A, and NS2A completely inhibits IFN signalling. The polyprotein of DENV needs to be cleaved between NS4A and NS4B in order for these proteins to exert IFN-antagonizing effects, suggesting the need for polyprotein translation (Munoz-Jordan et al. 2003; Munoz-Jordan et al. 2005). In addition to this antagonistic activity, DENV NS2A and NS4B inhibit TBK1 and IRF3 phosphorylation, suggesting that regulation occurs at the level of TBK1 complex activation. In this study, serotype-specific antagonistic activity was also described; NS4A protein from DENV1, but not DENV2 or DENV4, uniquely inhibits TBK1-directed IFN-β transcription (Dalrymple et al. 2015). Additionally, DENV NS2B/3 serine protease plays an important role, blocking the protein kinase domain of the IKKε kinase and inhibiting nuclear translocation and phosphorylation of IRF3 (Anglero-Rodriguez et al. 2014). DENV NS2B/3 also cleaves the human mediator of IRF3 activation, blocking its association with TBK1 and preventing the phosphorylation of IRF3 (Yu et al. 2012).

References

Ackermann M, Padmanabhan R (2001) De novo synthesis of RNA by the dengue virus RNA-dependent RNA polymerase exhibits temperature dependence at the initiation but not elongation phase. J Biol Chem 276(43):39926–39937

Aguirre S, Fernandez-Sesma A (2017) Collateral damage during dengue virus infection: making sense of DNA by cGAS. J Virol 91(14)

Akey DL, Brown WC, Dutta S, Konwerski J, Jose J, Jurkiw TJ, DelProposto J, Ogata CM, Skiniotis G, Kuhn RJ, Smith JL (2014) Flavivirus NS1 structures reveal surfaces for associations with membranes and the immune system. Science 343(6173):881–885

Allison SL, Schalich J, Stiasny K, Mandl CW, Heinz FX (2001) Mutational evidence for an internal fusion peptide in flavivirus envelope protein E. J Virol 75(9):4268–4275

Anglero-Rodriguez YI, Pantoja P, Sariol CA (2014) Dengue virus subverts the interferon induction pathway via NS2B/3 protease-IkappaB kinase epsilon interaction. Clin Vaccine Immunol 21(1):29–38

Arimoto K, Funami K, Saeki Y, Tanaka K, Okawa K, Takeuchi O, Akira S, Murakami Y, Shimotohno K (2010) Polyubiquitin conjugation to NEMO by triparite motif protein 23 (TRIM23) is critical in antiviral defense. Proc Natl Acad Sci U S A 107(36):15856–15861

Ashour J, Laurent-Rolle M, Shi PY, Garcia-Sastre A (2009) NS5 of dengue virus mediates STAT2 binding and degradation. J Virol 83(11):5408–5418

Ashour J, Morrison J, Laurent-Rolle M, Belicha-Villanueva A, Plumlee CR, Bernal-Rubio D, Williams KL, Harris E, Fernandez-Sesma A, Schindler C, Garcia-Sastre A (2010) Mouse STAT2 restricts early dengue virus replication. Cell Host Microbe 8(5):410–421

Bartholomeusz AI, Wright PJ (1993) Synthesis of dengue virus RNA in vitro: initiation and the involvement of proteins NS3 and NS5. Arch Virol 128(1–2):111–121

Beatty PR, Puerta-Guardo H, Killingbeck SS, Glasner DR, Hopkins K, Harris E (2015) Dengue virus NS1 triggers endothelial permeability and vascular leak that is prevented by NS1 vaccination. Sci Transl Med 7(304):304ra141

Beltramello M, Williams KL, Simmons CP, Macagno A, Simonelli L, Quyen NT, Sukupolvi-Petty S, Navarro-Sanchez E, Young PR, de Silva AM, Rey FA, Varani L, Whitehead SS, Diamond MS, Harris E, Lanzavecchia A, Sallusto F (2010) The human immune response to dengue virus is dominated by highly cross-reactive antibodies endowed with neutralizing and enhancing activity. Cell Host Microbe 8(3):271–283

von Bernuth H, Picard C, Puel A, Casanova JL (2012) Experimental and natural infections in MyD88- and IRAK-4-deficient mice and humans. Eur J Immunol 42(12):3126–3135

Best SM, Morris KL, Shannon JG, Robertson SJ, Mitzel DN, Park GS, Boer E, Wolfinbarger JB, Bloom ME (2005) Inhibition of interferon-stimulated JAK-STAT signaling by a tick-borne flavivirus and identification of NS5 as an interferon antagonist. J Virol 79(20):12828–12839

Bhattacharya D, Ansari IH, Striker R (2009) The flaviviral methyltransferase is a substrate of casein kinase 1. Virus Res 141(1):101–104

Bidet K, Dadlani D, Garcia-Blanco MA (2014) G3BP1, G3BP2 and CAPRIN1 are required for translation of interferon stimulated mRNAs and are targeted by a dengue virus non-coding RNA. PLoS Pathog 10(7):e1004242

Blitvich BJ, Firth AE (2015) Insect-specific flaviviruses: a systematic review of their discovery, host range, mode of transmission, superinfection exclusion potential and genomic organization. Virus 7(4):1927–1959

Brecher M, Zhang J, Li H (2013) The flavivirus protease as a target for drug discovery. Virol Sin 28(6):326–336

Brennan K, Bowie AG (2010) Activation of host pattern recognition receptors by viruses. Curr Opin Microbiol 13(4):503–507

Brooks AJ, Johansson M, John AV, Xu Y, Jans DA, Vasudevan SG (2002) The interdomain region of dengue NS5 protein that binds to the viral helicase NS3 contains independently functional importin beta 1 and importin alpha/beta-recognized nuclear localization signals. J Biol Chem 277(39):36399–36407

Bruns AM, Horvath CM (2015) LGP2 synergy with MDA5 in RLR-mediated RNA recognition and antiviral signaling. Cytokine 74(2):198–206

Cai X, Chiu YH, Chen ZJ (2014) The cGAS-cGAMP-STING pathway of cytosolic DNA sensing and signaling. Mol Cell 54(2):289–296

Castillo Ramirez JA, Urcuqui-Inchima S (2015) Dengue virus control of type I IFN responses: a history of manipulation and control. J Interf Cytokine Res 35(6):421–430

Chambers TJ, McCourt DW, Rice CM (1989) Yellow fever virus proteins NS2A, NS2B, and NS4B: identification and partial N-terminal amino acid sequence analysis. Virology 169(1):100–109

Chan YL, Chang TH, Liao CL, Lin YL (2008) The cellular antiviral protein viperin is attenuated by proteasome-mediated protein degradation in Japanese encephalitis virus-infected cells. J Virol 82(21):10455–10464

Chang YS, Liao CL, Tsao CH, Chen MC, Liu CI, Chen LK, Lin YL (1999) Membrane permeabilization by small hydrophobic nonstructural proteins of Japanese encephalitis virus. J Virol 73(8):6257–6264

Chang HH, Shyu HF, Wang YM, Sun DS, Shyu RH, Tang SS, Huang YS (2002) Facilitation of cell adhesion by immobilized dengue viral nonstructural protein 1 (NS1): arginine-glycine-aspartic acid structural mimicry within the dengue viral NS1 antigen. J Infect Dis 186(6):743–751

Chang TH, Liao CL, Lin YL (2006) Flavivirus induces interferon-beta gene expression through a pathway involving RIG-I-dependent IRF-3 and PI3K-dependent NF-kappaB activation. Microbes Infect 8(1):157–171

Choksupmanee O, Hodge K, Katzenmeier G, Chimnaronk S (2012) Structural platform for the autolytic activity of an intact NS2B-NS3 protease complex from dengue virus. Biochemistry 51(13):2840–2851

Chu PW, Westaway EG (1992) Molecular and ultrastructural analysis of heavy membrane fractions associated with the replication of Kunjin virus RNA. Arch Virol 125(1–4):177–191

Chung KM, Liszewski MK, Nybakken G, Davis AE, Townsend RR, Fremont DH, Atkinson JP, Diamond MS (2006a) West Nile virus nonstructural protein NS1 inhibits complement activation by binding the regulatory protein factor H. Proc Natl Acad Sci U S A 103(50):19111–19116

Chung KM, Nybakken GE, Thompson BS, Engle MJ, Marri A, Fremont DH, Diamond MS (2006b) Antibodies against West Nile virus nonstructural protein NS1 prevent lethal infection through fc gamma receptor-dependent and -independent mechanisms. J Virol 80(3):1340–1351

Chung KM, Thompson BS, Fremont DH, Diamond MS (2007) Antibody recognition of cell surface-associated NS1 triggers fc-gamma receptor-mediated phagocytosis and clearance of West Nile virus-infected cells. J Virol 81(17):9551–9555

Colamonici O, Yan H, Domanski P, Handa R, Smalley D, Mullersman J, Witte M, Krishnan K, Krolewski J (1994) Direct binding to and tyrosine phosphorylation of the alpha subunit of the type I interferon receptor by p135tyk2 tyrosine kinase. Mol Cell Biol 14(12):8133–8142

Crill WD, Roehrig JT (2001) Monoclonal antibodies that bind to domain III of dengue virus E glycoprotein are the most efficient blockers of virus adsorption to Vero cells. J Virol 75(16):7769–7773

Cui T, Sugrue RJ, Xu Q, Lee AK, Chan YC, Fu J (1998) Recombinant dengue virus type 1 NS3 protein exhibits specific viral RNA binding and NTPase activity regulated by the NS5 protein. Virology 246(2):409–417

Daffis S, Samuel MA, Suthar MS, Gale M Jr, Diamond MS (2008) Toll-like receptor 3 has a protective role against West Nile virus infection. J Virol 82(21):10349–10358

Dalrymple NA, Cimica V, Mackow ER (2015) Dengue virus NS proteins inhibit RIG-I/MAVS Signaling by blocking TBK1/IRF3 phosphorylation: dengue virus serotype 1 NS4A is a unique interferon-regulating virulence determinant. MBio 6(3):e00553-00515

Dang J, Tiwari SK, Lichinchi G, Qin Y, Patil VS, Eroshkin AM, Rana TM (2016) Zika virus depletes neural progenitors in human cerebral organoids through activation of the innate immune receptor TLR3. Cell Stem Cell 19(2):258–265

Darnell JE Jr (1997) STATs and gene regulation. Science 277(5332):1630–1635

Darnell JE Jr, Kerr IM, Stark GR (1994) Jak-STAT pathways and transcriptional activation in response to IFNs and other extracellular signaling proteins. Science 264(5164):1415–1421

Despres P, Dietrich J, Girard M, Bouloy M (1991) Recombinant baculoviruses expressing yellow fever virus E and NS1 proteins elicit protective immunity in mice. J Gen Virol 72(Pt 11):2811–2816

Diamond MS, Harris E (2001) Interferon inhibits dengue virus infection by preventing translation of viral RNA through a PKR-independent mechanism. Virology 289(2):297–311

Diamond MS, Pierson TC (2015) Molecular insight into dengue virus pathogenesis and its implications for disease control. Cell 162(3):488–492

Diebold SS, Kaisho T, Hemmi H, Akira S, Reis e Sousa C (2004) Innate antiviral responses by means of TLR7-mediated recognition of single-stranded RNA. Science 303(5663):1529–1531

Dokland T, Walsh M, Mackenzie JM, Khromykh AA, Ee KH, Wang S (2004) West Nile virus core protein; tetramer structure and ribbon formation. Structure 12(7):1157–1163

Duschene KS, Broderick JB (2010) The antiviral protein viperin is a radical SAM enzyme. FEBS Lett 584(6):1263–1267

Egloff MP, Benarroch D, Selisko B, Romette JL, Canard B (2002) An RNA cap (nucleoside-2'-O-)-methyltransferase in the flavivirus RNA polymerase NS5: crystal structure and functional characterization. EMBO J 21(11):2757–2768

Evans JD, Seeger C (2007) Differential effects of mutations in NS4B on West Nile virus replication and inhibition of interferon signaling. J Virol 81(21):11809–11816

Fadnis PR, Ravi V, Desai A, Turtle L, Solomon T (2013) Innate immune mechanisms in Japanese encephalitis virus infection: effect on transcription of pattern recognition receptors in mouse neuronal cells and brain tissue. Viral Immunol 26(6):366–377

Falconar AK (1997) The dengue virus nonstructural-1 protein (NS1) generates antibodies to common epitopes on human blood clotting, integrin/adhesin proteins and binds to human endothelial cells: potential implications in haemorrhagic fever pathogenesis. Arch Virol 142(5):897–916

Falgout B, Markoff L (1995) Evidence that flavivirus NS1-NS2A cleavage is mediated by a membrane-bound host protease in the endoplasmic reticulum. J Virol 69(11):7232–7243

Falgout B, Chanock R, Lai CJ (1989) Proper processing of dengue virus nonstructural glycoprotein NS1 requires the N-terminal hydrophobic signal sequence and the downstream nonstructural protein NS2a. J Virol 63(5):1852–1860

Falgout B, Bray M, Schlesinger JJ, Lai CJ (1990) Immunization of mice with recombinant vaccinia virus expressing authentic dengue virus nonstructural protein NS1 protects against lethal dengue virus encephalitis. J Virol 64(9):4356–4363

Falgout B, Pethel M, Zhang YM, Lai CJ (1991) Both nonstructural proteins NS2B and NS3 are required for the proteolytic processing of dengue virus nonstructural proteins. J Virol 65(5):2467–2475

Fischl W, Bartenschlager R (2011) Exploitation of cellular pathways by dengue virus. Curr Opin Microbiol 14(4):470–475

Fredericksen BL, Keller BC, Fornek J, Katze MG, Gale M Jr (2008) Establishment and maintenance of the innate antiviral response to West Nile virus involves both RIG-I and MDA5 signaling through IPS-1. J Virol 82(2):609–616

Fu XY, Kessler DS, Veals SA, Levy DE, Darnell JE Jr (1990) ISGF3, the transcriptional activator induced by interferon alpha, consists of multiple interacting polypeptide chains. Proc Natl Acad Sci U S A 87(21):8555–8559

Fukasawa M (2010) Cellular lipid droplets and hepatitis C virus life cycle. Biol Pharm Bull 33(3):355–359

Gebhard LG, Filomatori CV, Gamarnik AV (2011) Functional RNA elements in the dengue virus genome. Virus 3(9):1739–1756

van Gent M, Gack MU (2017) Viral pathogenesis: dengue virus takes on cGAS. Nat Microbiol 2:17050

Gerold G, Bruening J, Weigel B, Pietschmann T (2017) Protein interactions during the Flavivirus and Hepacivirus life cycle. Mol Cell Proteomics 16(4 suppl 1):S75–S91

Gollins SW, Porterfield JS (1986) A new mechanism for the neutralization of enveloped viruses by antiviral antibody. Nature 321(6067):244–246

Goncalvez AP, Purcell RH, Lai CJ (2004) Epitope determinants of a chimpanzee fab antibody that efficiently cross-neutralizes dengue type 1 and type 2 viruses map to inside and in close proximity to fusion loop of the dengue type 2 virus envelope glycoprotein. J Virol 78(23):12919–12928

Goodbourn S, Didcock L, Randall RE (2000) Interferons: cell signalling, immune modulation, antiviral response and virus countermeasures. J Gen Virol 81(Pt 10):2341–2364

Gould EA, Buckley A, Barrett AD, Cammack N (1986) Neutralizing (54K) and non-neutralizing (54K and 48K) monoclonal antibodies against structural and non-structural yellow fever virus proteins confer immunity in mice. J Gen Virol 67(Pt 3):591–595

Grant A, Ponia SS, Tripathi S, Balasubramaniam V, Miorin L, Sourisseau M, Schwarz MC, Sanchez-Seco MP, Evans MJ, Best SM, Garcia-Sastre A (2016) Zika virus targets human STAT2 to inhibit type I interferon Signaling. Cell Host Microbe 19(6):882–890

Grard G, Moureau G, Charrel RN, Lemasson JJ, Gonzalez JP, Gallian P, Gritsun TS, Holmes EC, Gould EA, de Lamballerie X (2007) Genetic characterization of tick-borne flaviviruses: new insights into evolution, pathogenetic determinants and taxonomy. Virology 361(1):80–92

Green AM, Beatty PR, Hadjilaou A, Harris E (2014) Innate immunity to dengue virus infection and subversion of antiviral responses. J Mol Biol 426(6):1148–1160

Haller O, Kochs G, Weber F (2006) The interferon response circuit: induction and suppression by pathogenic viruses. Virology 344(1):119–130

Hamel R, Dejarnac O, Wichit S, Ekchariyawat P, Neyret A, Luplertlop N, Perera-Lecoin M, Surasombatpattana P, Talignani L, Thomas F, Cao-Lormeau VM, Choumet V, Briant L, Despres P, Amara A, Yssel H, Misse D (2015) Biology of Zika virus infection in human skin cells. J Virol 89(17):8880–8896

Han YW, Choi JY, Uyangaa E, Kim SB, Kim JH, Kim BS, Kim K, Eo SK (2014) Distinct dictation of Japanese encephalitis virus-induced neuroinflammation and lethality via triggering TLR3 and TLR4 signal pathways. PLoS Pathog 10(9):e1004319

Helbig KJ, Carr JM, Calvert JK, Wati S, Clarke JN, Eyre NS, Narayana SK, Fiches GN, McCartney EM, Beard MR (2013) Viperin is induced following dengue virus type-2 (DENV-2) infection and has anti-viral actions requiring the C-terminal end of viperin. PLoS Negl Trop Dis 7(4):e2178

Hemmi H, Kaisho T, Takeuchi O, Sato S, Sanjo H, Hoshino K, Horiuchi T, Tomizawa H, Takeda K, Akira S (2002) Small anti-viral compounds activate immune cells via the TLR7 MyD88-dependent signaling pathway. Nat Immunol 3(2):196–200

Henchal EA, Henchal LS, Schlesinger JJ (1988) Synergistic interactions of anti-NS1 monoclonal antibodies protect passively immunized mice from lethal challenge with dengue 2 virus. J Gen Virol 69(Pt 8):2101–2107

Hermann M, Bogunovic D (2017) ISG15: in sickness and in health. Trends Immunol 38(2):79–93

Hinson ER, Cresswell P (2009) The N-terminal amphipathic alpha-helix of viperin mediates localization to the cytosolic face of the endoplasmic reticulum and inhibits protein secretion. J Biol Chem 284(7):4705–4712

Hishiki T, Han Q, Arimoto K, Shimotohno K, Igarashi T, Vasudevan SG, Suzuki Y, Yamamoto N (2014) Interferon-mediated ISG15 conjugation restricts dengue virus 2 replication. Biochem Biophys Res Commun 448(1):95–100

Ho LJ, Hung LF, Weng CY, Wu WL, Chou P, Lin YL, Chang DM, Tai TY, Lai JH (2005) Dengue virus type 2 antagonizes IFN-alpha but not IFN-gamma antiviral effect via down-regulating Tyk2-STAT signaling in the human dendritic cell. J Immunol 174(12):8163–8172

Holbrook MR (2017) Historical perspectives on Flavivirus research. Virus 9(5)

Hollidge BS, Weiss SR, Soldan SS (2011) The role of interferon antagonist, non-structural proteins in the pathogenesis and emergence of arboviruses. Virus 3(6):629–658

Hornung V, Ellegast J, Kim S, Brzozka K, Jung A, Kato H, Poeck H, Akira S, Conzelmann KK, Schlee M, Endres S, Hartmann G (2006) 5′-triphosphate RNA is the ligand for RIG-I. Science 314(5801):994–997

Iglesias NG, Gamarnik AV (2011) Dynamic RNA structures in the dengue virus genome. RNA Biol 8(2):249–257

Ihle JN (1995) The Janus protein tyrosine kinase family and its role in cytokine signaling. Adv Immunol 60:1–35

Ishak R, Tovey DG, Howard CR (1988) Morphogenesis of yellow fever virus 17D in infected cell cultures. J Gen Virol 69(Pt 2):325–335

Issur M, Geiss BJ, Bougie I, Picard-Jean F, Despins S, Mayette J, Hobdey SE, Bisaillon M (2009) The flavivirus NS5 protein is a true RNA guanylyltransferase that catalyzes a two-step reaction to form the RNA cap structure. RNA 15(12):2340–2350

Ivanyi-Nagy R, Lavergne JP, Gabus C, Ficheux D, Darlix JL (2008) RNA chaperoning and intrinsic disorder in the core proteins of Flaviviridae. Nucleic Acids Res 36(3):712–725

Jiang D, Weidner JM, Qing M, Pan XB, Guo H, Xu C, Zhang X, Birk A, Chang J, Shi PY, Block TM, Guo JT (2010) Identification of five interferon-induced cellular proteins that inhibit west nile virus and dengue virus infections. J Virol 84(16):8332–8341

Jiang F, Ramanathan A, Miller MT, Tang GQ, Gale M Jr, Patel SS, Marcotrigiano J (2011) Structural basis of RNA recognition and activation by innate immune receptor RIG-I. Nature 479(7373):423–427

Johnson AJ, Guirakhoo F, Roehrig JT (1994) The envelope glycoproteins of dengue 1 and dengue 2 viruses grown in mosquito cells differ in their utilization of potential glycosylation sites. Virology 203(2):241–249

Jones M, Davidson A, Hibbert L, Gruenwald P, Schlaak J, Ball S, Foster GR, Jacobs M (2005) Dengue virus inhibits alpha interferon signaling by reducing STAT2 expression. J Virol 79(9):5414–5420

Kanai R, Kar K, Anthony K, Gould LH, Ledizet M, Fikrig E, Marasco WA, Koski RA, Modis Y (2006) Crystal structure of west nile virus envelope glycoprotein reveals viral surface epitopes. J Virol 80(22):11000–11008

Kapoor M, Zhang L, Ramachandra M, Kusukawa J, Ebner KE, Padmanabhan R (1995) Association between NS3 and NS5 proteins of dengue virus type 2 in the putative RNA replicase is linked to differential phosphorylation of NS5. J Biol Chem 270(32):19100–19106

Kato, H., O. Takeuchi, S. Sato, M. Yoneyama, M. Yamamoto, K. Matsui, S. Uematsu, A. Jung, T. Kawai, K. J. Ishii, O. Yamaguchi, K. Otsu, T. Tsujimura, C. S. Koh, C. Reis e Sousa, Y. Matsuura, T. Fujita and S. Akira (2006). "Differential roles of MDA5 and RIG-I helicases in the recognition of RNA viruses." Nature 441(7089): 101–105

Kawai T, Akira S (2010) The role of pattern-recognition receptors in innate immunity: update on toll-like receptors. Nat Immunol 11(5):373–384

Kawasaki T, Kawai T (2014) Toll-like receptor signaling pathways. Front Immunol 5:461

Keller BC, Fredericksen BL, Samuel MA, Mock RE, Mason PW, Diamond MS, Gale M Jr (2006) Resistance to alpha/beta interferon is a determinant of West Nile virus replication fitness and virulence. J Virol 80(19):9424–9434

Khromykh AA, Sedlak PL, Guyatt KJ, Hall RA, Westaway EG (1999) Efficient trans-complementation of the flavivirus kunjin NS5 protein but not of the NS1 protein requires its coexpression with other components of the viral replicase. J Virol 73(12):10272–10280

Kuhn RJ, Zhang W, Rossmann MG, Pletnev SV, Corver J, Lenches E, Jones CT, Mukhopadhyay S, Chipman PR, Strauss EG, Baker TS, Strauss JH (2002) Structure of dengue virus: implications for flavivirus organization, maturation, and fusion. Cell 108(5):717–725

Kumar A, Hou S, Airo AM, Limonta D, Mancinelli V, Branton W, Power C, Hobman TC (2016) Zika virus inhibits type-I interferon production and downstream signaling. EMBO Rep 17(12):1766–1775

Kummerer BM, Rice CM (2002) Mutations in the yellow fever virus nonstructural protein NS2A selectively block production of infectious particles. J Virol 76(10):4773–4784

Laurent-Rolle M, Morrison J, Rajsbaum R, Macleod JM, Pisanelli G, Pham A, Ayllon J, Miorin L, Martinez-Romero C, tenOever BR, Garcia-Sastre A (2014) The interferon signaling antagonist function of yellow fever virus NS5 protein is activated by type I interferon. Cell Host Microbe 16(3):314–327

Leon-Juarez M, Martinez-Castillo M, Shrivastava G, Garcia-Cordero J, Villegas-Sepulveda N, Mondragon-Castelan M, Mondragon-Flores R, Cedillo-Barron L (2016) Recombinant dengue virus protein NS2B alters membrane permeability in different membrane models. Virol J 13(1)

Leung JY, Pijlman GP, Kondratieva N, Hyde J, Mackenzie JM, Khromykh AA (2008) Role of nonstructural protein NS2A in flavivirus assembly. J Virol 82(10):4731–4741

Levy DE, Kessler DS, Pine R, Darnell JE Jr (1989) Cytoplasmic activation of ISGF3, the positive regulator of interferon-alpha-stimulated transcription, reconstituted in vitro. Genes Dev 3(9):1362–1371

Li X, Ranjith-Kumar CT, Brooks MT, Dharmaiah S, Herr AB, Kao C, Li P (2009) The RIG-I-like receptor LGP2 recognizes the termini of double-stranded RNA. J Biol Chem 284(20):13881–13891

Li Y, Xie J, Wu S, Xia J, Zhang P, Liu C, Zhang P, Huang X (2013) Protein kinase regulated by dsRNA downregulates the interferon production in dengue virus- and dsRNA-stimulated human lung epithelial cells. PLoS One 8(1):e55108

Li K, Phoo WW, Luo D (2014) Functional interplay among the flavivirus NS3 protease, helicase, and cofactors. Virol Sin 29(2):74–85

Li Y, Li Q, Wong YL, Liew LS, Kang C (2015) Membrane topology of NS2B of dengue virus revealed by NMR spectroscopy. Biochim Biophys Acta 1848(10 Pt A):2244–2252

Li XD, Deng CL, Ye HQ, Zhang HL, Zhang QY, Chen DD, Zhang PT, Shi PY, Yuan ZM, Zhang B (2016) Transmembrane domains of NS2B contribute to both viral RNA replication and particle formation in Japanese encephalitis virus. J Virol 90(12):5735–5749

Liang JJ, Liao CL, Liao JT, Lee YL, Lin YL (2009) A Japanese encephalitis virus vaccine candidate strain is attenuated by decreasing its interferon antagonistic ability. Vaccine 27(21):2746–2754

Liang Q, Luo Z, Zeng J, Chen W, Foo SS, Lee SA, Ge J, Wang S, Goldman SA, Zlokovic BV, Zhao Z, Jung JU (2016) Zika virus NS4A and NS4B proteins deregulate Akt-mTOR Signaling in human Fetal neural stem cells to inhibit neurogenesis and induce autophagy. Cell Stem Cell 19(5):663–671

Lin RJ, Chang BL, Yu HP, Liao CL, Lin YL (2006) Blocking of interferon-induced Jak-Stat signaling by Japanese encephalitis virus NS5 through a protein tyrosine phosphatase-mediated mechanism. J Virol 80(12):5908–5918

Lin SC, Lo YC, Wu H (2010) Helical assembly in the MyD88-IRAK4-IRAK2 complex in TLR/IL-1R signalling. Nature 465(7300):885–890

Lindenbach BD, Rice CM (1997) Trans-complementation of yellow fever virus NS1 reveals a role in early RNA replication. J Virol 71(12):9608–9617

Lindenbach BD, Rice CM (1999) Genetic interaction of flavivirus nonstructural proteins NS1 and NS4A as a determinant of replicase function. J Virol 73(6):4611–4621

Lindenbach BD, Rice CM (2003) Molecular biology of flaviviruses. Adv Virus Res 59:23–61

Liu WJ, Chen HB, Khromykh AA (2003) Molecular and functional analyses of Kunjin virus infectious cDNA clones demonstrate the essential roles for NS2A in virus assembly and for a non-conservative residue in NS3 in RNA replication. J Virol 77(14):7804–7813

Liu WJ, Wang XJ, Mokhonov VV, Shi PY, Randall R, Khromykh AA (2005) Inhibition of interferon signaling by the New York 99 strain and Kunjin subtype of West Nile virus involves blockage of STAT1 and STAT2 activation by nonstructural proteins. J Virol 79(3):1934–1942

Liu J, Liu Y, Nie K, Du S, Qiu J, Pang X, Wang P, Cheng G (2016) Flavivirus NS1 protein in infected host sera enhances viral acquisition by mosquitoes. Nat Microbiol 1(9):16087

Lobigs M, Lee E, Ng ML, Pavy M, Lobigs P (2010) A flavivirus signal peptide balances the catalytic activity of two proteases and thereby facilitates virus morphogenesis. Virology 401(1):80–89

Lorenz IC, Allison SL, Heinz FX, Helenius A (2002) Folding and dimerization of tick-borne encephalitis virus envelope proteins prM and E in the endoplasmic reticulum. J Virol 76(11):5480–5491

Lubick KJ, Robertson SJ, McNally KL, Freedman BA, Rasmussen AL, Taylor RT, Walts AD, Tsuruda S, Sakai M, Ishizuka M, Boer EF, Foster EC, Chiramel AI, Addison CB, Green R, Kastner DL, Katze MG, Holland SM, Forlino A, Freeman AF, Boehm M, Yoshii K, Best SM (2015) Flavivirus antagonism of type I interferon Signaling reveals Prolidase as a regulator of IFNAR1 surface expression. Cell Host Microbe 18(1):61–74

Lund JM, Alexopoulou L, Sato A, Karow M, Adams NC, Gale NW, Iwasaki A, Flavell RA (2004) Recognition of single-stranded RNA viruses by toll-like receptor 7. Proc Natl Acad Sci U S A 101(15):5598–5603

Luo D, Vasudevan SG, Lescar J (2015) The flavivirus NS2B-NS3 protease-helicase as a target for antiviral drug development. Antivir Res 118:148–158

Lupfer C, Kanneganti TD (2013) The expanding role of NLRs in antiviral immunity. Immunol Rev 255(1):13–24

Ma DY, Suthar MS (2015) Mechanisms of innate immune evasion in re-emerging RNA viruses. Curr Opin Virol 12:26–37

Ma L, Jones CT, Groesch TD, Kuhn RJ, Post CB (2004) Solution structure of dengue virus capsid protein reveals another fold. Proc Natl Acad Sci U S A 101(10):3414–3419

Mackenzie JM, Westaway EG (2001) Assembly and maturation of the flavivirus Kunjin virus appear to occur in the rough endoplasmic reticulum and along the secretory pathway, respectively. J Virol 75(22):10787–10799

Mackenzie JM, Jones MK, Young PR (1996) Immunolocalization of the dengue virus nonstructural glycoprotein NS1 suggests a role in viral RNA replication. Virology 220(1):232–240

Mackenzie JM, Khromykh AA, Jones MK, Westaway EG (1998) Subcellular localization and some biochemical properties of the flavivirus Kunjin nonstructural proteins NS2A and NS4A. Virology 245(2):203–215

Maier CC, Delagrave S, Zhang ZX, Brown N, Monath TP, Pugachev KV, Guirakhoo F (2007) A single M protein mutation affects the acid inactivation threshold and growth kinetics of a chimeric flavivirus. Virology 362(2):468–474

Manokaran G, Finol E, Wang C, Gunaratne J, Bahl J, Ong EZ, Tan HC, Sessions OM, Ward AM, Gubler DJ, Harris E, Garcia-Blanco MA, Ooi EE (2015) Dengue subgenomic RNA binds TRIM25 to inhibit interferon expression for epidemiological fitness. Science 350(6257):217–221

Mazzon M, Jones M, Davidson A, Chain B, Jacobs M (2009) Dengue virus NS5 inhibits interferon-alpha signaling by blocking signal transducer and activator of transcription 2 phosphorylation. J Infect Dis 200(8):1261–1270

McLean JE, Wudzinska A, Datan E, Quaglino D, Zakeri Z (2011) Flavivirus NS4A-induced autophagy protects cells against death and enhances virus replication. J Biol Chem 286(25):22147–22159

Meunier E, Broz P (2017) Evolutionary convergence and divergence in NLR function and structure. Immunol, Trends

Miller S, Sparacio S, Bartenschlager R (2006) Subcellular localization and membrane topology of the dengue virus type 2 non-structural protein 4B. J Biol Chem 281(13):8854–8863

Miller S, Kastner S, Krijnse-Locker J, Buhler S, Bartenschlager R (2007) The non-structural protein 4A of dengue virus is an integral membrane protein inducing membrane alterations in a 2K-regulated manner. J Biol Chem 282(12):8873–8882

Modhiran N, Watterson D, Muller DA, Panetta AK, Sester DP, Liu L, Hume DA, Stacey KJ, Young PR (2015) Dengue virus NS1 protein activates cells via toll-like receptor 4 and disrupts endothelial cell monolayer integrity. Sci Transl Med 7(304):304ra142

Modis Y, Ogata S, Clements D, Harrison SC (2004) Structure of the dengue virus envelope protein after membrane fusion. Nature 427(6972):313–319

Modis Y, Ogata S, Clements D, Harrison SC (2005) Variable surface epitopes in the crystal structure of dengue virus type 3 envelope glycoprotein. J Virol 79(2):1223–1231

Morrison J, Laurent-Rolle M, Maestre AM, Rajsbaum R, Pisanelli G, Simon V, Mulder LC, Fernandez-Sesma A, Garcia-Sastre A (2013) Dengue virus co-opts UBR4 to degrade STAT2 and antagonize type I interferon signaling. PLoS Pathog 9(3):e1003265

Mukhopadhyay S, Kuhn RJ, Rossmann MG (2005) A structural perspective of the flavivirus life cycle. Nat Rev Microbiol 3(1):13–22

Muller DA, Young PR (2013) The flavivirus NS1 protein: molecular and structural biology, immunology, role in pathogenesis and application as a diagnostic biomarker. Antivir Res 98(2):192–208

Munoz-Jordan JL (2010) Subversion of interferon by dengue virus. Curr Top Microbiol Immunol 338:35–44

Munoz-Jordan JL, Sanchez-Burgos GG, Laurent-Rolle M, Garcia-Sastre A (2003) Inhibition of interferon signaling by dengue virus. Proc Natl Acad Sci U S A 100(24):14333–14338

Munoz-Jordan JL, Laurent-Rolle M, Ashour J, Martinez-Sobrido L, Ashok M, Lipkin WI, Garcia-Sastre A (2005) Inhibition of alpha/beta interferon signaling by the NS4B protein of flaviviruses. J Virol 79(13):8004–8013

Murray PJ (2007) The JAK-STAT signaling pathway: input and output integration. J Immunol 178(5):2623–2629

Nasirudeen AM, Wong HH, Thien P, Xu S, Lam KP, Liu DX (2011) RIG-I, MDA5 and TLR3 synergistically play an important role in restriction of dengue virus infection. PLoS Negl Trop Dis 5(1):e926

Nazmi A, Dutta K, Hazra B, Basu A (2014a) Role of pattern recognition receptors in flavivirus infections. Virus Res 185:32–40

Nazmi A, Mukherjee S, Kundu K, Dutta K, Mahadevan A, Shankar SK, Basu A (2014b) TLR7 is a key regulator of innate immunity against Japanese encephalitis virus infection. Neurobiol Dis 69:235–247

Ng WC, Soto-Acosta R, Bradrick SS, Garcia-Blanco MA, Ooi EE (2017) The 5′ and 3′ untranslated regions of the Flaviviral genome. Virus 9(6)

Novick D, Cohen B, Rubinstein M (1994) The human interferon alpha/beta receptor: characterization and molecular cloning. Cell 77(3):391–400

Nybakken GE, Nelson CA, Chen BR, Diamond MS, Fremont DH (2006) Crystal structure of the West Nile virus envelope glycoprotein. J Virol 80(23):11467–11474

Ono A, Freed EO (2005) Role of lipid rafts in virus replication. Adv Virus Res 64:311–358

Panne D, Maniatis T, Harrison SC (2004) Crystal structure of ATF-2/c-Jun and IRF-3 bound to the interferon-beta enhancer. EMBO J 23(22):4384–4393

Pardigon N (2017) Pathophysiological mechanisms of Flavivirus infection of the central nervous system. Transfus Clin Biol

Pastorino B, Nougairede A, Wurtz N, Gould E, de Lamballerie X (2010) Role of host cell factors in flavivirus infection: implications for pathogenesis and development of antiviral drugs. Antivir Res 87(3):281–294

Perera R, Kuhn RJ (2008) Structural proteomics of dengue virus. Curr Opin Microbiol 11(4):369–377

Pierson TC, Kielian M (2013) Flaviviruses: braking the entering. Curr Opin Virol 3(1):3–12

Platanias LC (2003) The p38 mitogen-activated protein kinase pathway and its role in interferon signaling. Pharmacol Ther 98(2):129–142

Querec T, Bennouna S, Alkan S, Laouar Y, Gorden K, Flavell R, Akira S, Ahmed R, Pulendran B (2006) Yellow fever vaccine YF-17D activates multiple dendritic cell subsets via TLR2, 7, 8, and 9 to stimulate polyvalent immunity. J Exp Med 203(2):413–424

Quintavalle M, Sambucini S, Summa V, Orsatti L, Talamo F, De Francesco R, Neddermann P (2007) Hepatitis C virus NS5A is a direct substrate of casein kinase I-alpha, a cellular kinase identified by inhibitor affinity chromatography using specific NS5A hyperphosphorylation inhibitors. J Biol Chem 282(8):5536–5544

Ramos HJ, Lanteri MC, Blahnik G, Negash A, Suthar MS, Brassil MM, Sodhi K, Treuting PM, Busch MP, Norris PJ, Gale M Jr (2012) IL-1beta signaling promotes CNS-intrinsic immune control of West Nile virus infection. PLoS Pathog 8(11):e1003039

Rawling DC, Pyle AM (2014) Parts, assembly and operation of the RIG-I family of motors. Curr Opin Struct Biol 25:25–33

Rey FA, Heinz FX, Mandl C, Kunz C, Harrison SC (1995) The envelope glycoprotein from tick-borne encephalitis virus at 2 a resolution. Nature 375(6529):291–298

Roby JA, Setoh YX, Hall RA, Khromykh AA (2015) Post-translational regulation and modifications of flavivirus structural proteins. J Gen Virol 96(Pt 7):1551–1569

Roosendaal J, Westaway EG, Khromykh A, Mackenzie JM (2006) Regulated cleavages at the West Nile virus NS4A-2K-NS4B junctions play a major role in rearranging cytoplasmic membranes and Golgi trafficking of the NS4A protein. J Virol 80(9):4623–4632

Sadler AJ, Williams BR (2007) Structure and function of the protein kinase R. Curr Top Microbiol Immunol 316:253–292

Samsa MM, Mondotte JA, Iglesias NG, Assuncao-Miranda I, Barbosa-Lima G, Da Poian AT, Bozza PT, Gamarnik AV (2009) Dengue virus capsid protein usurps lipid droplets for viral particle formation. PLoS Pathog 5(10):e1000632

Samuel CE (1993) The eIF-2 alpha protein kinases, regulators of translation in eukaryotes from yeasts to humans. J Biol Chem 268(11):7603–7606

Samuel MA, Whitby K, Keller BC, Marri A, Barchet W, Williams BR, Silverman RH, Gale M Jr, Diamond MS (2006) PKR and RNase L contribute to protection against lethal West Nile virus infection by controlling early viral spread in the periphery and replication in neurons. J Virol 80(14):7009–7019

Sangiambut S, Keelapang P, Aaskov J, Puttikhunt C, Kasinrerk W, Malasit P, Sittisombut N (2008) Multiple regions in dengue virus capsid protein contribute to nuclear localization during virus infection. J Gen Virol 89(Pt 5):1254–1264

Savidis G, Perreira JM, Portmann JM, Meraner P, Guo Z, Green S, Brass AL (2016) The IFITMs inhibit Zika virus replication. Cell Rep 15(11):2323–2330

van der Schaar HM, Rust MJ, Waarts BL, van der Ende H, Metselaar RJK, Wilschut J, Zhuang X, Smit JM (2007) Characterization of the early events in dengue virus cell entry by biochemical assays and single-virus tracking. J Virol 81(21):12019–12028

Scherwitzl I, Mongkolsapaja J, Screaton G (2017) Recent advances in human flavivirus vaccines. Curr Opin Virol 23:95–101

Schoggins JW, MacDuff DA, Imanaka N, Gainey MD, Shrestha B, Eitson JL, Mar KB, Richardson RB, Ratushny AV, Litvak V, Dabelic R, Manicassamy B, Aitchison JD, Aderem A, Elliott RM, Garcia-Sastre A, Racaniello V, Snijder EJ, Yokoyama WM, Diamond MS, Virgin HW, Rice CM (2014) Pan-viral specificity of IFN-induced genes reveals new roles for cGAS in innate immunity. Nature 505(7485):691–695

Schuessler A, Funk A, Lazear HM, Cooper DA, Torres S, Daffis S, Jha BK, Kumagai Y, Takeuchi O, Hertzog P, Silverman R, Akira S, Barton DJ, Diamond MS, Khromykh AA (2012) West Nile virus noncoding subgenomic RNA contributes to viral evasion of the type I interferon-mediated antiviral response. J Virol 86(10):5708–5718

Seo JY, Yaneva R, Cresswell P (2011) Viperin: a multifunctional, interferon-inducible protein that regulates virus replication. Cell Host Microbe 10(6):534–539

Setoh YX, Prow NA, Hobson-Peters J, Lobigs M, Young PR, Khromykh AA, Hall RA (2012) Identification of residues in West Nile virus pre-membrane protein that influence viral particle secretion and virulence. J Gen Virol 93(Pt 9):1965–1975

Severa M, Fitzgerald KA (2007) TLR-mediated activation of type I IFN during antiviral immune responses: fighting the battle to win the war. Curr Top Microbiol Immunol 316:167–192

Shresta S, Sharar KL, Prigozhin DM, Beatty PR, Harris E (2006) Murine model for dengue virus-induced lethal disease with increased vascular permeability. J Virol 80(20):10208–10217

Shrivastava G, Leon-Juarez M, Garcia-Cordero J, Meza-Sanchez DE, Cedillo-Barron L (2016) Inflammasomes and its importance in viral infections. Immunol Res 64(5–6):1101–1117

Shrivastava G, García-Cordero J, León-Juárez M, Oza G, Tapia-Ramírez J, Villegas-Sepulveda N, Cedillo-Barrón L (2017, October 3) NS2A comprises a putative viroporin of Dengue virus 2. Virulence 8(7):1450–1456. https://doi.org/10.1080/21505594.2017.1356540. Epub 2017 Jul 19

Silvennoinen O, Ihle JN, Schlessinger J, Levy DE (1993) Interferon-induced nuclear signalling by Jak protein tyrosine kinases. Nature 366(6455):583–585

Silverman RH (2007) Viral encounters with 2′,5′-oligoadenylate synthetase and RNase L during the interferon antiviral response. J Virol 81(23):12720–12729

Stadler K, Allison SL, Schalich J, Heinz FX (1997) Proteolytic activation of tick-borne encephalitis virus by furin. J Virol 71(11):8475–8481

Stern O, Hung YF, Valdau O, Yaffe Y, Harris E, Hoffmann S, Willbold D, Sklan EH (2013) An N-terminal amphipathic helix in dengue virus nonstructural protein 4A mediates oligomerization and is essential for replication. J Virol 87(7):4080–4085

Stiasny K, Kossl C, Lepault J, Rey FA, Heinz FX (2007) Characterization of a structural intermediate of flavivirus membrane fusion. PLoS Pathog 3(2):e20

Sun DS, King CC, Huang HS, Shih YL, Lee CC, Tsai WJ, Yu CC, Chang HH (2007) Antiplatelet autoantibodies elicited by dengue virus non-structural protein 1 cause thrombocytopenia and mortality in mice. J Thromb Haemost 5(11):2291–2299

Sun B, Sundstrom KB, Chew JJ, Bist P, Gan ES, Tan HC, Goh KC, Chawla T, Tang CK, Ooi EE (2017) Dengue virus activates cGAS through the release of mitochondrial DNA. Sci Rep 7(1):3594

Suthar MS, Aguirre S, Fernandez-Sesma A (2013) Innate immune sensing of flaviviruses. PLoS Pathog 9(9):e1003541

Takeuchi O, Akira S (2010) Pattern recognition receptors and inflammation. Cell 140(6):805–820

Tan BH, Fu J, Sugrue RJ, Yap EH, Chan YC, Tan YH (1996) Recombinant dengue type 1 virus NS5 protein expressed in Escherichia Coli exhibits RNA-dependent RNA polymerase activity. Virology 216(2):317–325

Tanaka N, Nakanishi M, Kusakabe Y, Goto Y, Kitade Y, Nakamura KT (2004) Structural basis for recognition of 2′,5′-linked oligoadenylates by human ribonuclease L. EMBO J 23(20):3929–3938

Taylor KE, Mossman KL (2013) Recent advances in understanding viral evasion of type I interferon. Immunology 138(3):190–197

Teo CS, Chu JJ (2014) Cellular vimentin regulates construction of dengue virus replication complexes through interaction with NS4A protein. J Virol 88(4):1897–1913

Tripathi S, Balasubramaniam VR, Brown JA, Mena I, Grant A, Bardina SV, Maringer K, Schwarz MC, Maestre AM, Sourisseau M, Albrecht RA, Krammer F, Evans MJ, Fernandez-Sesma A, Lim JK, Garcia-Sastre A (2017) A novel Zika virus mouse model reveals strain specific differences in virus pathogenesis and host inflammatory immune responses. PLoS Pathog 13(3):e1006258

Tsai WC, Lloyd RE (2014) Cytoplasmic RNA granules and viral infection. Annu Rev Virol 1(1):147–170

Tsai YT, Chang SY, Lee CN, Kao CL (2009) Human TLR3 recognizes dengue virus and modulates viral replication in vitro. Cell Microbiol 11(4):604–615

Uematsu S, Akira S (2007) Toll-like receptors and type I interferons. J Biol Chem 282(21):15319–15323

Upadhyay AS, Vonderstein K, Pichlmair A, Stehling O, Bennett KL, Dobler G, Guo JT, Superti-Furga G, Lill R, Overby AK, Weber F (2014) Viperin is an iron-sulfur protein that inhibits genome synthesis of tick-borne encephalitis virus via radical SAM domain activity. Cell Microbiol 16(6):834–848

Velazquez L, Fellous M, Stark GR, Pellegrini S (1992) A protein tyrosine kinase in the interferon alpha/beta signaling pathway. Cell 70(2):313–322

Wang QY, Shi PY (2015) Flavivirus entry inhibitors. ACS Infect Dis 1(9):428–434

Wang T, Town T, Alexopoulou L, Anderson JF, Fikrig E, Flavell RA (2004) Toll-like receptor 3 mediates West Nile virus entry into the brain causing lethal encephalitis. Nat Med 10(12):1366–1373

Wang JP, Liu P, Latz E, Golenbock DT, Finberg RW, Libraty DH (2006) Flavivirus activation of plasmacytoid dendritic cells delineates key elements of TLR7 signaling beyond endosomal recognition. J Immunol 177(10):7114–7121

Wang S, Wu X, Pan T, Song W, Wang Y, Zhang F, Yuan Z (2012) Viperin inhibits hepatitis C virus replication by interfering with binding of NS5A to host protein hVAP-33. J Gen Virol 93(Pt 1):83–92

Wathelet MG, Lin CH, Parekh BS, Ronco LV, Howley PM, Maniatis T (1998) Virus infection induces the assembly of coordinately activated transcription factors on the IFN-beta enhancer in vivo. Mol Cell 1(4):507–518

Weber F, Wagner V, Kessler N, Haller O (2006) Induction of interferon synthesis by the PKR-inhibitory VA RNAs of adenoviruses. J Interf Cytokine Res 26(1):1–7

de Weerd NA, Nguyen T (2012) The interferons and their receptors--distribution and regulation. Immunol Cell Biol 90(5):483–491

Welsch S, Miller S, Romero-Brey I, Merz A, Bleck CK, Walther P, Fuller SD, Antony C, Krijnse-Locker J, Bartenschlager R (2009) Composition and three-dimensional architecture of the dengue virus replication and assembly sites. Cell Host Microbe 5(4):365–375

Welte T, Reagan K, Fang H, Machain-Williams C, Zheng X, Mendell N, Chang GJ, Wu P, Blair CD, Wang T (2009) Toll-like receptor 7-induced immune response to cutaneous West Nile virus infection. J Gen Virol 90(Pt 11):2660–2668

Wen Z, Zhong Z, Darnell JE Jr (1995) Maximal activation of transcription by Stat1 and Stat3 requires both tyrosine and serine phosphorylation. Cell 82(2):241–250

Werme K, Wigerius M, Johansson M (2008) Tick-borne encephalitis virus NS5 associates with membrane protein scribble and impairs interferon-stimulated JAK-STAT signalling. Cell Microbiol 10(3):696–712

Wu Y, Liu Q, Zhou J, Xie W, Chen C, Wang Z, Yang H, Cui J (2017) Zika virus evades interferon-mediated antiviral response through the co-operation of multiple nonstructural proteins in vitro. Cell Discov 3:17006

Xie X, Gayen S, Kang C, Yuan Z, Shi PY (2013) Membrane topology and function of dengue virus NS2A protein. J Virol 87(8):4609–4622

Xie X, Zou J, Puttikhunt C, Yuan Z, Shi PY (2015) Two distinct sets of NS2A molecules are responsible for dengue virus RNA synthesis and virion assembly. J Virol 89(2):1298–1313

Yang TC, Li SW, Lai CC, Lu KZ, Chiu MT, Hsieh TH, Wan L, Lin CW (2013) Proteomic analysis for type I interferon antagonism of Japanese encephalitis virus NS5 protein. Proteomics 13(23–24):3442–3456

Yie J, Liang S, Merika M, Thanos D (1997) Intra- and intermolecular cooperative binding of high-mobility-group protein I(Y) to the beta-interferon promoter. Mol Cell Biol 17(7):3649–3662

Yoneyama M, Kikuchi M, Matsumoto K, Imaizumi T, Miyagishi M, Taira K, Foy E, Loo YM, Gale M Jr, Akira S, Yonehara S, Kato A, Fujita T (2005) Shared and unique functions of the DExD/H-box helicases RIG-I, MDA5, and LGP2 in antiviral innate immunity. J Immunol 175(5):2851–2858

Youn S, Li T, McCune BT, Edeling MA, Fremont DH, Cristea IM, Diamond MS (2012) Evidence for a genetic and physical interaction between nonstructural proteins NS1 and NS4B that modulates replication of West Nile virus. J Virol 86(13):7360–7371

Yu IM, Zhang W, Holdaway HA, Li L, Kostyuchenko VA, Chipman PR, Kuhn RJ, Rossmann MG, Chen J (2008) Structure of the immature dengue virus at low pH primes proteolytic maturation. Science 319(5871):1834–1837

Yu CY, Chang TH, Liang JJ, Chiang RL, Lee YL, Liao CL, Lin YL (2012) Dengue virus targets the adaptor protein MITA to subvert host innate immunity. PLoS Pathog 8(6):e1002780

Zhang J, Pekosz A, Lamb RA (2000) Influenza virus assembly and lipid raft microdomains: a role for the cytoplasmic tails of the spike glycoproteins. J Virol 74(10):4634–4644

Zhang Y, Corver J, Chipman PR, Zhang W, Pletnev SV, Sedlak D, Baker TS, Strauss JH, Kuhn RJ, Rossmann MG (2003) Structures of immature flavivirus particles. EMBO J 22(11):2604–2613

Zhang Y, Zhang W, Ogata S, Clements D, Strauss JH, Baker TS, Kuhn RJ, Rossmann MG (2004) Conformational changes of the flavivirus E glycoprotein. Structure 12(9):1607–1618

Zhang X, Ge P, Yu X, Brannan JM, Bi G, Zhang Q, Schein S, Zhou ZH (2013) Cryo-EM structure of the mature dengue virus at 3.5-a resolution. Nat Struct Mol Biol 20(1):105–110

Zheng S, Zhu D, Lian X, Liu W, Cao R, Chen P (2016) Porcine 2′, 5′-oligoadenylate synthetases inhibit Japanese encephalitis virus replication in vitro. J Med Virol 88(5):760–768

Zhou H, Yu M, Fukuda K, Im J, Yao P, Cui W, Bulek K, Zepp J, Wan Y, Kim TW, Yin W, Ma V, Thomas J, Gu J, Wang JA, DiCorleto PE, Fox PL, Qin J, Li X (2013) IRAK-M mediates toll-like receptor/IL-1R-induced NFkappaB activation and cytokine production. EMBO J 32(4):583–596

Zhu J, Zhang Y, Ghosh A, Cuevas RA, Forero A, Dhar J, Ibsen MS, Schmid-Burgk JL, Schmidt T, Ganapathiraju MK, Fujita T, Hartmann R, Barik S, Hornung V, Coyne CB, Sarkar SN (2014) Antiviral activity of human OASL protein is mediated by enhancing signaling of the RIG-I RNA sensor. Immunity 40(6):936–948

Zmurko J, Neyts J, Dallmeier K (2015) Flaviviral NS4b, chameleon and jack-in-the-box roles in viral replication and pathogenesis, and a molecular target for antiviral intervention. Rev Med Virol 25(4):205–223

Zou J, Xie X, Lee le T, Chandrasekaran R, Reynaud A, Yap L, Wang QY, Dong H, Kang C, Yuan Z, Lescar J, Shi PY (2014) Dimerization of flavivirus NS4B protein. J Virol 88(6):3379–3391

Printed by Printforce, the Netherlands